Disaster Resilience and Green Growth

Over the years, the relationship between environment and disasters has received significant attention. This is largely due to the emerging recognition that environmental changes - climate change, land-use and natural resource degradation make communities more vulnerable to disaster impacts. There is a need to break this nexus through environment based and sustainability inclusive interventions. Science – technology and economic measures for disaster risk management, hence, need to adapt more integrated approaches for infrastructure and social resilience. Environmental and anthropogenic factors are key contributors to hazard, risk, and vulnerability and, therefore, should be an important part of determining risk-management solutions.

Green growth approaches have been developed by emphasizing sustainability inclusion and utilizing the benefits of science-technology interventions along policy-practice linkages with circular economy and resource efficiency. Such approaches recognize the perils of traditional material-oriented economy growth models that tend to exploit natural resources, contribute to climate change, and exacerbate disaster vulnerabilities, Green growth integrated approaches are rapidly becoming as preferred investment avenue for mitigating climate change and disaster risks and for enhancing resilience. This includes ecosystem-based and nature-based solutions with potential to contribute to the resilience of infrastructure, urban, rural and peri-urban systems, livelihoods, water, and health. They can lead to food security and can further promote people-centric approaches.

Some of the synergistic outcomes of green growth approaches include disaster risk reduction, climate change mitigation and adaptation, resilient livelihoods, cities, businesses and industry. The disaster risk reduction and resilience outcome of green growth approaches deserve special attention, both for the academic and policy communities. Scholars and professionals across the domains of DRR, CCA, and green growth are in need of publications that fulfill their knowledge needs concerning the disaster resilience outcomes of green growth approaches. Keeping the above background in view, the book series offers comprehensive coverage combining the domains of environment, natural resources, engineering, management and policy studies for addressing disaster risk and resilience in the green growth context in an integrated and holistic manner. The book series covers a range of themes that highlight the synergistic outcomes of green growth approaches.

The book series aims to bring out the latest research, approaches, and perspectives for disaster risk reduction along with highlighting the outcomes of green growth approaches and including Science-technology-research-policy-practice interface, from both developed and developing parts of the world under one umbrella. The series aims to involve renowned experts and academicians as volume-editors and authors from all the regions of the world. It is curated and developed by authoritative institutions and experts to serve global readership on this theme.

Y.D. Imran Khan • Mrinalini Goswami
Sunil Nautiyal
Editors

Ecosystem-based Approaches for Resilience Building in Himalayan Landscapes

Insights for Climate Change and Disaster Risk Reduction

Springer

Editors
Y.D. Imran Khan
WOTR- Centre for Resilient Studies (W-CReS)
Watershed Organisation Trust (WOTR)
Pune, Maharashtra, India

Mrinalini Goswami
Researcher in Sustainability Science
Bengaluru, Karnataka, India

Sunil Nautiyal
Centre for Ecological Economics and Natural Resources (CEENR)
Institute for Social and Economic Change (ISEC)
Bengaluru, Karnataka, India

G. B. Pant National Institute of Himalayan Environment, Kosi - Katarmal, Almora, Uttarakhand, India

ISSN 2662-4885 ISSN 2662-4893 (electronic)
Disaster Resilience and Green Growth
ISBN 978-981-95-2006-0 ISBN 978-981-95-2007-7 (eBook)
https://doi.org/10.1007/978-981-95-2007-7

This Springer imprint is published by the registered company Springer Nature Singapore Pte Ltd.
The registered company address is: 152 Beach Road, #21-01/04 Gateway East, Singapore 189721, Singapore

Foreword

The Himalayan region is globally recognized for its ecological significance, constituting an unparalleled repository of natural capital. It serves as a vital lifeline for millions people across Asia, encompassing vast glacial systems, biodiversity hotspots, and critical hydrological networks. Nevertheless, this region represents one of Earth's most delicate ecosystems, increasingly susceptible to the deleterious effects of climate change and anthropogenic pressures. Various phenomena, such as melting of glaciers, erratic weather patterns, loss of biodiversity, deforestation, and an increase in the frequency of disasters such as landslides and floods, have exacerbated the ecological and socioeconomic vulnerabilities of the region. In this context, ecosystem-based approaches (EbA) offer a comprehensive and sustainable framework for enhancing resilience, concurrently addressing climate change adaptation and disaster risk reduction.

The book, *Ecosystem-based Approaches for Resilience Building in Himalayan Landscapes: Insights for Climate Change and Disaster Risk Reduction*, provides an exhaustive analysis of how ecosystem based approaches can mitigate the multifaceted challenges confronting the Himalaya. By integrating ecological restoration, sustainable land management practices, and participatory community engagement, EbA emerges as a pragmatic and inclusive strategy for enhancing the resilience of both ecosystems and dependent human populations. This work is illustrated by its robust synthesis of scientific data, indigenous knowledge systems, and empirical case studies, yielding a holistic understanding of Himalayan ecosystems and their perceived threats. The text highlights as how ecosystem restoration and management can be seamlessly incorporated into disaster risk reduction frameworks. A salient feature is its emphasis on harmonizing traditional ecological knowledge with cutting-edge technological innovations, such as remote sensing and climate modeling, thereby enhancing the efficacy and cultural relevance of adaptation measures.

The significance of this work is its commitment to participatory and inclusive methodologies. Resilience-building in the Himalaya necessitates the active involvement of local communities, who are often the primary responders to environmental perturbations and disasters. The book underscores the pivotal role of indigenous and local knowledge systems, which are intricately intertwined with the biodiversity and ecological dynamics of the region. By empowering these communities and promoting a sense of stewardship, EbA can deliver sustainable and enduring outcomes. Furthermore, the book highlights the critical role of policy and governance frameworks for scaling up EbA initiatives. Given the transboundary nature of Himalayan challenges, the book advocates for regional cooperation and harmonized policy frameworks to ensure cohesive and impactful resilience-building endeavors.

The book also provides actionable recommendations for policymakers, researchers, and practitioners, delineating a strategic blueprint for the implementation of ecosystem-based approaches within the Himalayan context. As global community increasingly embraces ecosystem-based approaches for addressing climate change and disaster risks, this book represents a timely contribution. It calls upon a diverse array of stakeholders, ranging from governmental bodies and international organizations to grassroots communities to prioritize the conservation and restoration of Himalayan ecosystems. The insights and frameworks articulated herein are not only pertinent to the Himalaya but also hold broad applicability to other mountains and vulnerable ecosystems worldwide. The urgency of these challenges is obvious, and this work impressively reflects that urgency. While the obstacles are formidable, the opportunities for transformative change are equally significant. Ecosystem-based approaches hold the potential to safeguard the unique natural heritage of Himalaya and secure a sustainable future for generations to come. I commend the authors and contributors for their dedication and vision in producing this invaluable resource, which will serve as an essential guide and reference for anyone working toward a resilient and sustainable Himalayan region.

Director, G.B. Pant National Institute
of Himalayan Environment (NIHE)
Almora, Uttarakhand, India
July 2025

Indra Dutt Bhatt

Preface

The Himalayan region, the "Water Tower of Asia" due to its vast glacial reserves, is an ecological treasure trove that sustains millions of people and harbors unique biodiversity. However, it is also one of the most vulnerable regions to climate change, with rising temperatures, glacial retreat, erratic weather patterns, and increasing frequency of natural disasters posing significant threats to both human and ecological systems. As the impacts of climate change intensify, it becomes imperative to explore sustainable solutions that not only mitigate these risks but also enhance the resilience of the communities and ecosystems that depend on this fragile landscape.

This book, *Ecosystem-Based Approaches for Resilience Building in Himalayas: Insights on Climate Change and Disaster Risk Reduction*, emerges as a critical contribution to the ongoing discourse on climate adaptation and disaster risk management in the Himalayas. It delves into the concept of ecosystem-based approach (EbA) as a holistic strategy that integrates natural systems into climate change mitigation and disaster preparedness. Through evidence-based research, case studies, and expert insights, this volume highlights how EbA can serve as a viable and sustainable mechanism to safeguard the Himalayan environment and its inhabitants against the escalating climate crisis.

The rationale for compiling this book stems from the urgent need to bridge the gap between policy, research, and on-the-ground implementation. While there exists a growing body of literature on climate change and disaster risk reduction, the application of ecosystem-based approaches in the Himalayan context remains relatively underexplored. This book seeks to fill that void by providing a comprehensive understanding of the interconnections between ecosystems, climate resilience, and human well-being in this unique geographical setting. It offers not only theoretical perspectives but also practical insights, making it a valuable resource for policymakers, researchers, environmentalists, and practitioners working in the field of climate adaptation and disaster risk reduction.

Designed to be both informative and solution-oriented, this book presents a diverse range of topics, including the role of forests, water systems, and agroecosystems in enhancing resilience, the impact of glacial melt on downstream communities, and the significance of traditional ecological knowledge in adaptive strategies. It also showcases successful case studies of community-led conservation efforts, nature-based solutions, and policy interventions that have demonstrated tangible benefits in strengthening ecosystem resilience in the Himalayas.

Our primary audience includes researchers and practitioners who are committed to addressing the multifaceted challenges posed by climate change in the region. However, this book is not limited to academic circles alone. It is equally relevant for government agencies, nongovernmental organizations, and local communities who seek actionable strategies to promote resilience against climate-induced disasters. Furthermore, it serves as an invaluable guide for students and educators looking to deepen their understanding of ecosystem-based approaches in the context of mountain landscapes.

As the world grapples with the growing uncertainties of climate change, the need for adaptive, nature-based solutions has never been more urgent. We hope this book serves as a catalyst for further research, dialogue, and collaboration among stakeholders working toward the common goal of safeguarding the Himalayan ecosystem. By implementing ecosystem-based approaches, we can collectively work toward a more resilient future—one that ensures the sustainability of the region's natural resources while enhancing the well-being of the communities that call the Himalayas home.

It is with great optimism and a shared sense of responsibility that we present this book to you. May it inspire new perspectives, strengthen meaningful discussions, and contribute to the global effort of building resilience in one of the most ecologically significant and climate-sensitive regions of the world.

Pune, Maharashtra, India Y.D. Imran Khan
Bengaluru, Karnataka, India Mrinalini Goswami
Bengaluru, Karnataka, India Sunil Nautiyal

Acknowledgments

The completion of this book, *Ecosystem-Based Approaches for Resilience Building in Himalayas: Insights on Climate Change and Disaster Risk Reduction*, would not have been possible without the invaluable contributions and unwavering support of Watershed Organisation Trust (WOTR), Pune, Maharashtra; Institute for Social and Economic Change (ISEC), Bengaluru, Karnataka and Govind Ballabh Pant "National Institute of Himalayan Environment" (NIHE), Almora, Uttarakhand.

First and foremost, we sincerely thank our contributors—researchers, practitioners, and experts—who have generously shared their knowledge, insights, and experiences. Their dedication to addressing the pressing issues of climate change, ecosystem conservation, and disaster risk reduction in the Himalayas has enriched this book immensely. Their rigorous research and case studies provide a solid foundation for advancing ecosystem-based approach strategies in the Himalayas.

We are immensely grateful to Crispino Lobo, Co-Founder of WOTR, and Dr. Marcella D'Souza, Director of W-CReS, WOTR, for supporting this initiative. Their commitment to environmental sustainability and resilience-building has facilitated the research and discussions underpinning this book.

Special thanks go to the Series Editors Prof. Anil Kumar Gupta, Prof. Manish K. Goyal, Dr. André Lindner and their team, whose meticulous efforts in reviewing, refining, and structuring the content have been invaluable. Their guidance and hard work in ensuring the coherence and clarity of the chapters has significantly enhanced the quality of this publication. The Production team's dedicated efforts and efficient support during the creation of this volume have not gone unnoticed and are deeply appreciated.

We extend our heartfelt appreciation to our families and colleagues for their patience, encouragement, and unwavering belief in the importance of this work. Their support has been a constant source of motivation throughout this journey. This book stands as a testament to the collaborative spirit, dedication, and shared vision that have guided this endeavour from its inception to its culmination. With profund gratitude, we acknowledge the contributions of all who have made this initiative a resounding success.

Finally, we acknowledge the resilience and knowledge of the communities living in the Himalayas. Their traditional knowledge, adaptation practices, and deep connection with nature continue to inspire efforts toward sustainable and ecosystem-based solutions. This book is dedicated to them, hoping to contribute to building a more resilient and sustainable future for the region.

Y.D. Imran Khan
Mrinalini Goswami
Sunil Nautiyal

Prologue

Introduction and Contextual Background

Mountain regions provide critical resources and ecosystem services—from water and timber to biodiversity conservation, mineral extraction, flood protection, electricity generation, and recreational amenities—directly supporting 40% of people worldwide (Schild 2008). The Himalayan region, spanning multiple nations and encompassing diverse ecosystems, is one of Earth's most critical ecological frontiers. This vast mountain range plays a pivotal role in regulating regional climate patterns and supporting the livelihoods of millions. Recent studies have highlighted the increasing vulnerability of this region to climate change impacts (Thompson et al. 2020; Vaidya et al. 2019). The steep terrain, tectonically active geological formations, high altitudes, heavy rainfall, and melting glaciers make the region more prone to the risk of disasters. The Himalayan region is also subjected to anthropogenic degradation like any other ecosystem, which undergoes deforestation, unsustainable land use change, fragmentation, pollution, overexploitation of resources, etc. These impacts, coupled with the impacts of climate change, weaken the resilience of the Himalayan environment. These changes directly affect water security, agriculture, livelihoods, and disaster risks. The growing pressures have created an urgent need for innovative, sustainable approaches to building resilience in these mountain ecosystems.

The Hindu Kush Himalayan (HKH) region faces severe climate change impacts that vary across its geographical extent. In Nepal, accelerated glacier retreat has led to the formation of numerous glacial lakes, increasing glacial lake outburst floods (GLOFs) risks (Bolch et al. 2022). In India's Himalayan states, changing precipitation patterns have disrupted traditional agricultural systems, with rainfall variability increasing by 50% in some regions (Singh et al. 2011). The Indian Himalayan region has experienced temperature increases of more than 0.2°C per decade, higher than global averages (Krishnan et al. 2019). Himalayan urban agglomerates are reported to have an increasing frequency of warmer days and a reduced number of cold days and nights (Diksha et al. 2022). Bhutan faces unique challenges due to its heavy dependence on hydropower, with glacier retreat threatening long-term energy security. Studies in HKH project that glacier volumes will decline by up to 90%, resulting in a reduction in glacier-fed river flows in the twenty-first century, and snowline changes are expected to be as high as 900 m (Bolch et al. 2019). In Pakistan's Hindu Kush region, the increased frequency of extreme weather events has led to devastating floods and landslides. Climate change has also intensified water scarcity issues in Pakistan's mountain communities. The Tibet Autonomous Region of China has experienced significant permafrost degradation, affecting infrastructure stability and natural ecosystems. This has led to increased landslide risks and changes in alpine meadow ecosystems.

Given the fragile nature of the Himalayan ecosystem and its role in regional climate regulation, there is an urgent need to integrate sustainable, nature-based solutions into climate adaptation and disaster risk reduction strategies. Governments, researchers, and conservationists are now exploring ecosystem-based approaches as an essential framework for addressing these challenges. This book, therefore, attempts to provide a timely and crucial exploration of the status of resilience and resilience-building efforts in the Himalayan region in the context of climate change and disaster risk reduction. In this book, we delve into the critical topics of ecosystem-based adaptation (EbA) and its importance in addressing these climate-related challenges. This book provides innovative strategies, case studies, and solutions that empower the region's policymakers, researchers, environmentalists, and practitioners. It also strives to delineate a roadmap for sustainable, ecosystem-focused approaches to tackle the increasing risks and vulnerabilities faced by the Himalayan communities.

This book is expected to provide

- Scientific knowledge from multidisciplinary and trans-disciplinary studies on climate change impacts, adaptation, and resilience building in Himalayan ecosystems
- Exchange of experiences and know-hows and documentation of pathways for climate change adaptation strategies, barriers, and possible solutions to adopt different adaptation options
- Discussion and development of understanding among a wider audience from science, policy, and practice on the vulnerability of different socioeconomic sections and varied ecological settings to climate-related risks with special reference to the Himalayan region
- Collation of initiatives for disaster risk reduction and their effectiveness to identify strategic requirements for policy recommendation

Ecosystem-Based Approaches

EbA has gained immense importance in the last decade, although the discussion started long before. These approaches address the interplay among climate change, biodiversity, and sustainable resource management to enable society and enhance ecosystem services for climate change mitigation and adaptation in better ways. Successful implementation of EbA cannot be achieved in Silo; it needs a policy mix of both development and environmental policies, which will lead to an adaptive transition (Scarano 2017). Several scientists use multiple definitions and structures to explain EbA. Among those, the following two can be quoted as notable:

> The use of biodiversity and ecosystem services (BES) as part of an overall adaptation strategy to help people to adapt to the adverse effects of climate change (CBD 2009)

> EbA is people's use of natural capital to adapt to climate change impacts, which can also have multiple co-benefits for mitigation, protection of livelihoods and poverty alleviation. It is an approach that is applicable across both developed and developing countries (Munang et al. 2013)

Ecosystem-based approaches for disaster risk reduction (DRR) and climate change adaptation (CCA) have potential solutions for building resilience in diverse environments, including coastal, riverine, and mountain areas (Renaud et al. 2016). These approaches align with international DRR, CCA, and sustainable development agreements, facilitating their uptake and implementation. In the Himalayas, community-based adaptation strategies have emerged as effective means to build local resilience against climate change impacts (Chaudhary et al. 2021). The integration of ecosystem-based approaches and community-based adaptation can provide a comprehensive framework for addressing the complex challenges faced by Himalayan landscapes, enhancing their resilience to natural hazards and climate change while promoting sustainable development (Renaud et al. 2016; Vaidya et al. 2019; Chaudhary et al. 2021).

These solutions can mitigate hazards, increase community resilience, and reduce exposure and vulnerabilities of the ecosystem to extreme events, enhancing sustainability in diverse sectors, including agriculture, forestry, energy, and water. In Nepal, EbA has shown potential to enhance adaptive capacity, provide ecosystem services, and offer economic viability (Poudel et al. 2021). However, mainstreaming these approaches faces challenges due to policy, institutional, and political obstacles. The Hindu Kush Himalayan region is prone to climate-induced disasters like flash floods; the study emphasized the need for harmonization between built and natural environments for effective risk reduction (Nibanupudi et al. 2015). Recent international agreements have facilitated the uptake of ecosystem-based approaches, which can contribute to achieving multiple goals related to DRR, CCA, and sustainable development (Renaud et al. 2016). Despite their potential, the implementation of ecosystem-based solutions remains slow, necessitating improved dialogue between scientists, practitioners, and policymakers.

This approach encompasses sustainable management, conservation, and restoration of ecosystems to provide services that enable people to adapt to climate change impacts while maintaining ecological resilience. Indigenous knowledge systems deserve special consideration in resilience-building frameworks. Traditional ecological practices, particularly those governing sacred landscapes, provide substantive insights for conservation strategies. Integrating these time-tested local governance mechanisms can significantly enhance institutional approaches to resource stewardship for better management. EbA can operate through multiple pathways, including:

1. Maintenance and restoration of natural buffer systems
2. Sustainable management of watersheds and forest resources
3. Conservation of agricultural biodiversity
4. Enhancement of ecosystem connectivity
5. Integration of traditional ecological knowledge with modern conservation practices

Relevance of EbA to the Himalayan Region

The Hindu Kush Himalayan region presents a compelling case for EbA implementation due to its unique ecological vulnerability and socioeconomic challenges. Research indicates that EbA strategies are particularly effective in mountain ecosystems where traditional practices and natural resource management are deeply intertwined (Chaudhary et al. 2020). In the Himalayan context, EbA demonstrates significant relevance through multiple dimensions, which are discussed below.

Ecological Resilience

EbA contributes to maintaining ecological resilience by preserving critical ecosystem services. Studies in Nepal's mid-hills have shown that EbA initiatives in community forests have enhanced biodiversity conservation and carbon sequestration capacity and delivered livelihood benefits (Pandey et al. 2018). These approaches have increased forest cover while simultaneously improving ecosystem services such as water regulation and soil stability.

Traditional Ecological Knowledge

A study reveals that mountain communities in the Himalayan region have a rich traditional knowledge base that can support sustainable development through environmentally sustainable and economically viable farming systems (Kala and Nautiyal 2023). Ecosystem-based adaptation strategies, when combined with local knowledge, can show promising results in building community resilience.

Socioeconomic Benefits

The socioeconomic relevance of EbA in the Himalayas is particularly noteworthy. Research conducted across multiple Himalayan states demonstrates that communities implementing EbA strategies show enhanced adaptive capacity and reduced vulnerability to climate change impacts (Klein et al. 2019). For instance, in Uttarakhand, India, EbA interventions in agricultural systems have led to a 40% increase in household income through diversified farming practices and improved ecosystem services.

Cost-Effectiveness

EbA presents a cost-effective alternative to conventional infrastructure-based adaptation approaches. Analysis of adaptation strategies in Nepal indicates that EbA measures require less financial investment while providing multiple co-benefits compared to traditional engineering solutions.

Cultural Integration

The Himalayan region's rich cultural heritage and traditional ecological knowledge align well with EbA principles. Studies show that communities practicing traditional resource management systems integrated with EbA demonstrate greater resilience to climate change impacts. This cultural resonance enhances community acceptance and long-term sustainability of adaptation measures.

Policy Integration

Recent policy frameworks in Himalayan nations increasingly recognize the importance of EbA. For example, Nepal's National Adaptation Plan explicitly incorporates EbA principles, while India's National Mission for Sustaining Himalayan Ecosystems emphasizes ecosystem-based solutions.

References

Bolch T et al (2019) Status and change of the cryosphere in the extended Hindu Kush Himalaya Region. In: Wester P, Mishra A, Mukherji A, Shrestha A (eds) The Hindu Kush Himalaya assessment. Springer, Cham. https://doi.org/10.1007/978-3-319-92288-1_7

Bolch T, Yao T, Bhattacharya A, Hu Y, King O, Liu L, Pronk JB, Rastner P, Zhang G (2022) Earth observation to investigate occurrence, characteristics and changes of glaciers, glacial lakes and rock glaciers in the Poiqu River Basin (Central Himalaya). Remote Sens 14(8):1927. https://doi.org/10.3390/rs14081927

Chaudhary S, Adhikari BR, Chaudhary P, Dorji T, Poudel R (2021) Ecosystem-based adaptation (EbA) in the Hindu Kush Himalaya: status, progress and challenges. In: Mukherjee M, Shaw R (eds) Ecosystem-based disaster and climate resilience, Disaster and risk research: GADRI book series. Springer, Singapore. https://doi.org/10.1007/978-981-16-4815-1_2

Diksha, Kumar A, Lal P (2022) Analysing climatic variability and extremes events in the Himalayan regions focusing on mountainous urban agglomerations. Geocarto Intl 37(26):14148–14170. https://doi.org/10.1080/10106049.2022.2086635

Kala CP, Nautiyal S (2023) Traditional food knowledge of local people and its sustainability in mountains of Uttarakhand State of India. J Soc Econom Develop 25(1):32–51

Klein JA, Tucker CM, Steger CE, Nolin A, Reid R, Hopping KA et al (2019) An integrated community and ecosystem-based approach to disaster risk reduction in mountain systems. Environ Sci Pol 94:143–152

Mishra A, Appadurai AN, Choudhury D, Regmi BR, Kelkar U, Alam M et al (2019) Adaptation to climate change in the Hindu Kush Himalaya: stronger action urgently needed. The Hindu Kush Himalaya assessment: mountains, climate change, sustainability and people, pp 457–490

Munang R, Thiaw I, Alverson K, Mumba M, Liu J, Rivington M (2013) Climate change and ecosystem-based adaptation: a new pragmatic approach to buffering climate change impacts. Curr Opin Environ Sustain 5(1):67–71

Pandey SS, Cockfield G, Maraseni TN (2018) Assessing the roles of community forestry in climate change mitigation and adaptation: a case study from Nepal. For Ecol Manage 415–416:60–71

Scarano FR (2017) Ecosystem-based adaptation to climate change: concept, scalability and a role for conservation science. Perspect Ecol Conserv 15(2):65–73

Schild A (2008) ICIMOD's position on climate change and mountain systems. Mount Res Develop 28(3):328–331

Secretariat of the Convention on Biological Diversity (2009) Connecting biodiversity and climate change mitigation and adaptation: Report of the second ad hoc technical expert group on biodiversity and climate change. CBD technical series

Singh SP, Bassignana-Khadka I, Karky BS, Sharma E (2011) Climate change in the Hindu Kush-Himalayas: the state of current knowledge. ICIMOD, Kathmandu

Thompson I, Shrestha M, Chhetri N, Agusdinata DB (2020) An institutional analysis of glacial floods and disaster risk management in the Nepal Himalaya. Intl J Disas Risk Reduct 47, Article 101567. https://doi.org/10.1016/j.ijdrr.2020.101567

Vaidya RA et al (2019) Disaster risk reduction and building resilience in the Hindu Kush Himalaya. In: Wester P, Mishra A, Mukherji A, Shrestha A (eds) The Hindu Kush Himalaya assessment. Springer, Cham. https://doi.org/10.1007/978-3-319-92288-1_11

Zhang G, Yao T, Bolch T, Yang W, Chen W (2020) Glacial lake inventory of High Mountain Asia based on Landsat images from 2015 to 2017 and comparison with ICIMOD inventory. Sci Total Environ 718:137212

Pune, Maharashtra, India — Y.D. Imran Khan
Bengaluru, Karnataka, India — Mrinalini Goswami
Bengaluru, Karnataka, India — Sunil Nautiyal

Contents

About the Editors and Contributors

About the Editors

Y.D. Imran Khan is a Programme Manager and Thematic Lead (Ecology) at the WOTR Centre for Resilience Studies (W-CReS), Watershed Organisation Trust (WOTR) in Pune, India. He holds a Ph.D. in Zoology (specializing in Ecology) from Bangalore University, Bengaluru, and an M.Sc. in Zoology from Sri Krishnadevaraya University, Anantapur. He also earned a Bachelor of Education from Sri Krishnadevaraya University, Anantapur, Andhra Pradesh. Prior to joining WOTR, he worked at the Institute for Social and Economic Change (ISEC), Bengaluru; the National Institute of Rural Development and Panchayati Raj (NIRD&PR), Hyderabad; and the Tata Institute of Social Sciences (TISS), Mumbai. He has held various positions, including Research Assistant, Junior Research Fellow, Senior Research Fellow, Research Associate, and Training Manager in the field of Natural Resource Management and Rural Development. His areas of interest include conservation biology, socio-ecology, rural livelihoods, and watershed Ecology. He is dedicated to developing an integrated interdisciplinary approach that supports sustainable landscape development. This involves creating an operational framework that combines ecological and social science perspectives to evaluate the impact of policy and behavioral changes on natural resource conservation, management, and community livelihoods. Dr. Khan collaborates with interdisciplinary and transdisciplinary teams of scholars from various national and international organizations to integrate ecological and social science approaches in assessing the impact of policy and behavioral changes.

Mrinalini Goswami is a researcher in ecological science, associated with the GB Pant National Institute of Himalayan Environment and the Centre for Ecological Economics and Natural Resources at the Institute for Social and Economic Change, Bengaluru. She received her Ph.D. in Environmental Science from the University of Mysore, Mysuru, Karnataka, for her research on the sustainability of ecosystem-based livelihoods in peri-urban landscapes. Her research spans a range of topics, including environmental sustainability, climate change, natural resource management, water management, and urban development. Dr. Mrinalini has published scientific papers and books in the fields of socio-ecology, environment, and ecology. Her substantial experience in the development sector has inspired her interdisciplinary approach to environmental research.

Sunil Nautiyal is a Professor at the Centre for Ecological Economics and Natural Resources (CEENR), Institute for Social and Economic Change (ISEC), Bengaluru, and an Honorary Professor at the University of Ladakh. He completed his PhD research at the G.B. Pant National Institute of Himalayan Environment (NIHE) and received his doctorate from HNB Garhwal University. He has also served as Director of the GB Pant National Institute of Himalayan Environment (NIHE), Kosi-Katarmal, Almora, Uttarakhand, India. His areas of specialization include natural resource management and biodiversity conservation, production system analysis, protected area management and human-wildlife conflict and coexistence, landscape dynamics, ecological modeling, GIS and remote sensing for landscape research, the urban-rural interface and sustainability, climate change, sustainable livelihoods, and socio-ecological development. Dr. Nautiyal develops integrated, interdisciplinary approaches for sustainable landscape development, creating operational frameworks that combine ecological and social science methods. His work evaluates how policy and behavioral changes related to natural resource conservation and management impact community livelihoods across different agro-ecological regions of India. This research has significant policy implications for socio-ecological development and biodiversity conservation. He has published over 250 scientific papers and 16 books. He has been honored with several prestigious national and international fellowships and awards (AvH, JSPS, NIE). Dr. Nautiyal serves on the editorial and reviewer boards of multiple international and national journals and collaborates closely with scientists from India, Germany, the UK, Australia, Japan, Bangladesh, and the USA.

Contributors

Fatima Amin Integrated Centre for Adaptation, Disaster Risk Resilience & Sustainability, Indian Institute of Technology, Roorkee, Greater Noida Extension Centre, Greater Noida, India

Hemant K. Badola G.B. Pant National Institute of Himalayan Environment, Sikkim Regional Centre, Gangtok, Sikkim, India

D 408, Aditya Doonshire Apartments, Sailok Phase IIGMS Road, Dehradun, Uttarakhand, India

Surendra S. Bargali Department of Botany, DSB campus, Kumaun University, Nainital, Uttarakhand, India

Joram Beda Meghalaya Climate Change Centre, Government of Meghalaya, Shillong, India

Meghalaya Basin Development Authority, Government of Meghalaya, Shillong, India

Planning, Investment Promotion and Sustainable Development Department, Government of Meghalaya, Shillong, India

Dattatray Bhandalkar Amity Institute of Liberal Arts (AILA) Amity University Mumbai (AUM), Mumbai, Maharashtra, India

Sami Ullah Bhat Department of Environmental Science, University of Kashmir, Srinagar, India

Mahendra Pratap Singh Bisht Uttarakhand Space Application Centre, Dehradun, India

Debojyoti Chakraborty Department of Forest Growth, Silviculture and Genetics, Austrian Research Centre for Forests, Vienna, Austria

Nakul Chettri International Centre for Integrated Mountain Development, Kathmandu, Nepal

Bishal Chhetri Southfield College, Darjeeling, West Bengal, India

Albert Chiang Meghalaya Climate Change Centre, Government of Meghalaya, Shillong, India

Dada. R. Dadas School of Social Work, Tata Institute of Social Sciences, Mumbai, Maharashtra, India

W-CReS, Watershed Organisation Trust, Pune, Maharashtra, India

Anandan Das Department of Zoology, Pandit Deendayal Upadhyaya Adarsha Mahavidyalaya, Eraligool, India

Hemen Deka Ecology and Environmental Remediation Laboratory, Department of Botany, Gauhati University, Guwahati, India

Anjumoni Devee Department of Entomology, Assam Agricultural University, Jorhat, India

Jyotsna Dubey Department of Advanced Study in Geology, Lucknow University, Lucknow, Uttar Pradesh, India

Surajit Dutta Department of Geography, Dr. Shyama Prasad Mukherjee University, Ranchi, Jharkhand, India

Kailash S. Gaira G.B. Pant National Institute of Himalayan Environment, Sikkim Regional Centre, Gangtok, Sikkim, India

Junmoni Gayon Department of Entomology, Assam Agricultural University, Jorhat, India

Sanjay Ghildiyal Department of History D.S.B. Campus, Kumaun University, Nainital, Uttarakhand, India

Mrinalini Goswami Researcher in Sustainability Science, Bengaluru, Karnataka, India

V. Guhan Meteorological Centre, Hyderabad, Telangana, India

Anil K. Gupta Integrated Centre for Adaptation, Disaster Risk Resilience & Sustainability, Indian Institute of Technology, Roorkee, Greater Noida Extension Centre, Greater Noida, India

Reneema Hazarika Institute of Silviculture, University of Natural Resources and Life Sciences (BOKU), Vienna, Austria

Atif Hussain Department of Forestry and Natural Resources, HNB Garhwal University, Srinagar, Uttarakhand, India

Department of Wildlife Protection, Wildlife Division, Shopian, J&K, India

Kaiser Iqbal Department of Forestry and Natural Resources, HNB Garhwal University, Srinagar, Uttarakhand, India

Department of Wildlife Protection, Wildlife Division, Shopian, J&K, India

Soha ul Islam Department of Environmental Science, University of Kashmir, Srinagar, India

Rajesh Joshi G.B. Pant National Institute of Himalayan Environment, Sikkim Regional Centre, Gangtok, Sikkim, India

Harshit Pant Centre for Socioeconomic Development (CSED), G.B. Pant National Institute of Himalayan Environment (NIHE), Almora, Uttarakhand, India

Dharmveer Kaintura Department of Forestry and Natural Resources, HNB Garhwal University, Srinagar, Uttarakhand, India

Nabajit Kalita Ecology and Environmental Remediation Laboratory, Department of Botany, Gauhati University, Guwahati, India

C. Karpagam ICAR-National Research Centre for Banana, Ariyavoor, Tamil Nadu, India

Deeksha Katyal University School of Environment Management, Guru Gobind Singh Indraprastha University, New Delhi, India

S. Kaviraju Department of Agricultural Economics, Hyderabad, Telangana, India

Y. D. Imran Khan The WOTR-Centre for Resilience Studies, Watershed Organisation Trust, Pune, India

Vijay Kumar ICAR-Directorate of Poultry Research, Hyderabad, Telangana, India

Jhony Lepcha G.B. Pant National Institute of Himalayan Environment, Sikkim Regional Centre, Gangtok, Sikkim, India

Kabita Lepcha University of Gour Banga, Malda, West Bengal, India

Vivek Lyngdoh Meghalaya Climate Change Centre, Government of Meghalaya, Shillong, India

Meghalaya Basin Development Authority, Government of Meghalaya, Shillong, India

Marbakor Mary Lynrah Meghalaya Climate Change Centre, Government of Meghalaya, Shillong, India

Meghalaya Basin Development Authority, Government of Meghalaya, Shillong, India

Subhasis Mahato Department of Forestry and Natural Resources, HNB Garhwal University, Srinagar, Uttarakhand, India

Department of Forestry, P.R.M.S. Mahavidyalaya, Bankura, West Bengal, India

Rakesh K. Maikhuri Department of Environment Science, H.N.B. Garhwal University, Srinagar Garhwal, Uttarakhand, India

Ajay Maletha Amity Institute of Forestry and Wildlife, Amity University, Noida, Uttar Pradesh, India

Tabina Manzoor Department of Environmental Science, University of Kashmir, Srinagar, India

Shaik N. Meera ICAR-Agricultural Technology Application Research Institute Zone-X, Hyderabad, Telangana, India

P. Mooventhan ICAR-National Institute of Biotic Stress Management, Baronda, Chhattisgarh, India

P. P. Murugan Tamil Nadu Agricultural University TNAU, Coimbatore, India

Sunil Nautiyal Centre for Ecological Economics and Natural Resources (CEENR), Institute for Social and Economic Change (ISEC), Bengaluru, Karnataka, India

G. B. Pant National Institute of Himalayan EnvironmentKosi - Katarmal, Almora, Uttarakhand, India

Govind Singh Negi Uttarakhand Space Application Centre, Dehradun, India

Vikram S. Negi Department of Botany and Microbiology, H.N.B. Garhwal University, Pauri Garhwal, Uttarakhand, India

Nikhil Nikam W-CReS, Watershed Organisation Trust, Pune, Maharashtra, India

Amica L. Nongrang Meghalaya Climate Change Centre, Government of Meghalaya, Shillong, India

Meghalaya Basin Development Authority, Government of Meghalaya, Shillong, India

Vikas Painuly Uttarakhand Space Application Centre, Dehradun, India

Mehul Pant National Law Institute University, Bhopal, Madhya Pradesh, India

Shiva Pokhrel Central Department of Botany, Tribhuvan University, Kirtipur, Nepal

Research Centre for Applied Science and Technology (RECAST), Tribhuvan University, Kirtipur, Nepal

Center for Environmental and Sustainable Agriculture Research, Pokhara, Nepal

P. Prashanth Agricultural Extension, Jayashankar Telangana Agricultural University (PJTAU), Hyderabad, Telangana, India

Saurabh Purohit WOTR-Centre for Resilience Studies (W-CReS), Watershed Organization Trust (WOTR), Pune, Maharashtra, India

Tanjil Rahman Faculty of Agricultural Sciences and Technology, Assam Downtown University, Guwahati, India

Rishav Rawal Department of History D.S.B. Campus, Kumaun University, Nainital, Uttarakhand, India

Neelam Rawat Uttarakhand Space Application Centre, Dehradun, India

M. Jaganmohan Reddy Extension Education Institute (EEI), Hyderabad, Telangana, India

M. R. Reddy ICAR-Directorate of Poultry Research, Hyderabad, Telangana, India

Niranjan Roy Department of Economics, Assam University, Silchar, India

Shubhadeep Roychoudhury Department of Life Science and Bioinformatics, Assam University, Silchar, India

Faraz Rupani W-CReS, Watershed Organisation Trust, Pune, Maharashtra, India

D. R. K. Saikanth Agricultural Extension, Jayashankar Telangana Agricultural University (PJTAU), Hyderabad, Telangana, India

Momi Sarma Department of Obstetrics and Gynaecology, College of Agriculture, Assam Agricultural University, Guwahati, India

Rupak Kr. Sarma Department of Botany, Nalbari College, Nalbari, Assam, India

Nomi Sarmah Department of Entomology, Assam Agricultural University, Jorhat, India

Sarswati Prakash Sati Department of Basic Sciences, College of Forestry, Ranichauri, Tehri Garhwal (Veer Chandra Singh Garhwali Uttarakhand University of Horticulture and Forestry, Bharsar, Pauri Garhwal), Bharsar, Uttarakhand, India

Chungla Sherpa Forest Research and Training Center Gandaki Province, Pokhara, Nepal

Dhruv Sen Singh Department of Advanced Study in Geology, Lucknow University, Lucknow, Uttar Pradesh, India

Rahul Kumar Singh Department of Forestry and Natural Resources, HNB Garhwal University, Srinagar, Uttarakhand, India

Department of Forestry, P.R.M.S. Mahavidyalaya, Bankura, West Bengal, India

Sangeeta Singh Department of Education, University of Allahabad, Allahabad, Uttar Pradesh, India

Mohd Sharjeel Sofi Department of Environmental Science, University of Kashmir, Srinagar, India

Shikha Subba Southfield College, Darjeeling, West Bengal, India

Independent Researcher, Darjeeling, West Bengal, India

Khan Tahama Indian Institute of Geomagnetism, Navi Mumbai, Maharashtra, India

Pavan Tikhile Nawegaon Nagzira Tiger Reserve, Gondia, Maharashtra, India

D. Uttej ICAR-Agricultural Technology Application Research Institute Zone-X, Hyderabad, Telangana, India

Evenstone Wahlang Meghalaya Climate Change Centre, Government of Meghalaya, Shillong, India

Meghalaya Basin Development Authority, Government of Meghalaya, Shillong, India

Nivanaliza Wahlang Meghalaya Climate Change Centre, Government of Meghalaya, Shillong, India

Meghalaya Basin Development Authority, Government of Meghalaya, Shillong, India

Taufique Warsi Raintree Foundation, Mumbai, Maharashtra, India [illegible] Department of Entomology, Assam Agricultural University, [illegible], India

[illegible] Prakash [illegible] Department of [illegible] College of Forestry, [illegible] Garhwal, [illegible] University [illegible]

[illegible]

[illegible]

[illegible]

[illegible]

[illegible] Department of [illegible] University [illegible], India

[illegible] Department of Environmental Sciences, University of [illegible], India

[illegible]

[illegible]

[illegible]

[illegible]

[illegible] Shillong, India

[illegible]

[illegible]

[illegible]

[illegible]

Part I
Climate Change Impacts and Ecosystem Dynamics

Chapter 1
Cryosphere Response to Climate Change and Disaster Risk Reduction in the Arid Northwestern Ladakh Himalaya

Jyotsna Dubey, Dhruv Sen Singh, and Sarswati Prakash Sati

Abstract The Ladakh Himalaya, which is characterised by an average elevation of about 3000 meters, is considered as one of the climate hotspots in the Hindu Kush Himalayan (HKH) region. Primarily located in the rain shadow zone (Trans-Himalaya), around 90% of the population residing in rural areas are reliant on agriculture which is sustained by glacial and snowmelt. It is increasingly being realised that climate change is affecting the Himalayan cryosphere, thus threatening the local livelihoods. For example, the receding glaciers due to global warming are expanding the glacial lakes and thus posing threat to the glacial lake outburst floods (GLOFs). Due to the receding glaciers particularly after the Little Ice Age (LIA), large quantities of unconsolidated paraglacial sediments are sequestered in the relict glaciated valleys. These sediments are readily mobilised during extreme weather events and inflict severe damage to the various infrastructures besides modifying the terrain morphology and making it unsafe for habitation. Therefore, to ensure safety and security of the terrain and the people, amid climate change, it is important that environmentally friendly methodologies must be put in place. This paper aims to provide a broad overview of the existing challenges posed to the cryosphere in the HKH region, with special reference to the Ladakh Himalaya including glaciers, lakes, permafrost, and paraglacial and their associated processes. Additionally, the paper explores strategies for mitigating the potential impacts on ecosystem sustainability and livelihoods.

Keywords Ladakh Himalaya · Glaciers · Permafrost · Glacial lakes · Paraglacial sediments

---

J. Dubey · D. S. Singh
Department of Advanced Study in Geology, Lucknow University, Lucknow, Uttar Pradesh, India

S. P. Sati (✉)
Department of Basic Sciences, College of Forestry, Ranichauri, Tehri Garhwal (Veer Chandra Singh Garhwali Uttarakhand University of Horticulture and Forestry, Bharsar, Pauri Garhwal), Uttarakhand, India

Y. D. I. Khan et al. (eds.), *Ecosystem-based Approaches for Resilience Building in Himalayan Landscapes*, Disaster Resilience and Green Growth,
https://doi.org/10.1007/978-981-95-2007-7_1

1.1 Introduction

The collision of Indian and Eurasian plates led to the creation of spectacular topography ranging in elevation from <1000 m to >7000 m which is popularly known as the Hindu Kush Himalaya (HKH). Spread over more than eight countries (60.85–105.04° E longitude and 15.95–39.31° N latitude), the HKH encompasses more than 4,192,000 km^2 of area (Bajracharya and Shrestha 2011). HKH known as the Third Pole is one of the most dynamic mountain systems, known for its fragility and immense complexity consisting of highest agglomeration of glaciers outside the polar region (Dyhrenfurth 1955). The HKH ranges play a crucial role in the hydrological cycle, global atmospheric circulation, and water resource availability and provide a wide range of ecosystem services (Xu et al. 2019). Glaciers originating from the HKH feed ten largest rivers which together support nearly 20% of the world's population (e.g. Bookhagen and Burbank 2006).

The dramatic change in altitude across the HKH dictates the earth surface processes, climate, and human activities. Across the south to north, the climate transitions from the sub-tropical humid conditions (in the southern sub-Himalaya) to the alpine tundra-like climate of the snow-covered ridges in the northern hinterland. The growing population has increased the consumption of fossil fuel demand, causing an increase in the greenhouse gas emissions which in turn is enhancing the global warming. Particularly, the black carbon deposition on the HKH glaciers has become a matter of serious concern. Their detrimental impact on the Himalayan cryosphere is profoundly causing the global warming and reduced albedo (Ramanathan and Carmichael 2008; Negi et al. 2019). In addition, at higher altitudes the rate of warming is rapidly increasing due to higher absorption of aerosols that further indirectly amplifies the melting of snowpacks and glaciers (Barnett 2005: Ramanathan and Carmichael 2008; Das et al. 2010; Sarangi et al. 2020). The study from the Gangotri glacier has shown the presence of black carbon (BC) aerosol in the surrounding air. According to Negi et al. (2019), the presence of BC in the Higher Himalayan environment is significantly going to affect the radiation budget, air quality, hydrological cycles, monsoon circulation, and natural resources. An increasing number of studies from the Indian Himalaya Region (IHR) indicate that the temperature rise in the Himalayas is at a faster rate than the nearby Indian land mass (Sabin et al. 2020; Banerjee et al. 2021; Bhutiyani et al. 2007; Tewari et al. 2017). There are already signals of increased air temperature over the higher elevations. During the first week of August 2024, Leh town in the Ladakh region witnessed exceptionally high temperatures, which soared up to 33.5 °C. This led to the cancellation of numerous flights because higher air temperature makes the high-altitude air thin—a condition unfavourable for the safe operation of the aircraft (The Indian Express, 31 July 2024). The recent warming rates are heterogeneous over the HKH region which is attributed to the elevation-dependent warming (EDW) (e.g. Shrestha et al. 1999; Thompson et al. 2003; Liu et al. 2009; Ren et al. 2017). Although the mechanism of EDW is still being debated due to the lack of high-elevation meteorological stations and climate models, we cannot adequately capture the climate complexities involved EDW due to the multiple physical processes involved (Bolch et al. 2019). Let us have a look at the cryosphere response in recent times and the likely trend in times to come with emphasis on the nature of threat perceptions in the Ladakh Himalaya.

1.2 Study Area

The present study focuses on the Higher Himalaya in the south and the Karakoram Himalaya in the north. The latter is collectively recognised as the Ladakh Himalaya, which covers the Zanskar ranges in the south, the Ladakh ranges in the middle, and the Karakoram ranges in the north (Fig. 1.1).

1.2.1 Geology and Geomorphology

The Ladakh Himalayas comprise of the (i) Zanskar ranges (ii) the Ladakh Batholith, and (iii) the Karakoram ranges (Thakur 1981; Fig. 1.2). The rocks of Zanskar constitute the northern margin of the Indian plate. This range is dominated by the Phanerozoic sediments including the Upper Palaeozoic volcanic rocks (equivalent of Panjal volcanics). These sediments overlie the Precambrian basement comprising Higher Himalayan crystalline rocks (Thakur 1981). The Zanskar ranges in the north

Fig. 1.1 Satellite picture showing the bird's-eye view of the NW Himalayas. From south to north are the Zanskar ranges, the Ladakh ranges, and the Karakoram ranges. Also marked are a few important locations and geomorphic features. (Source: Christoph Hormann, http://earth.imagico.de)

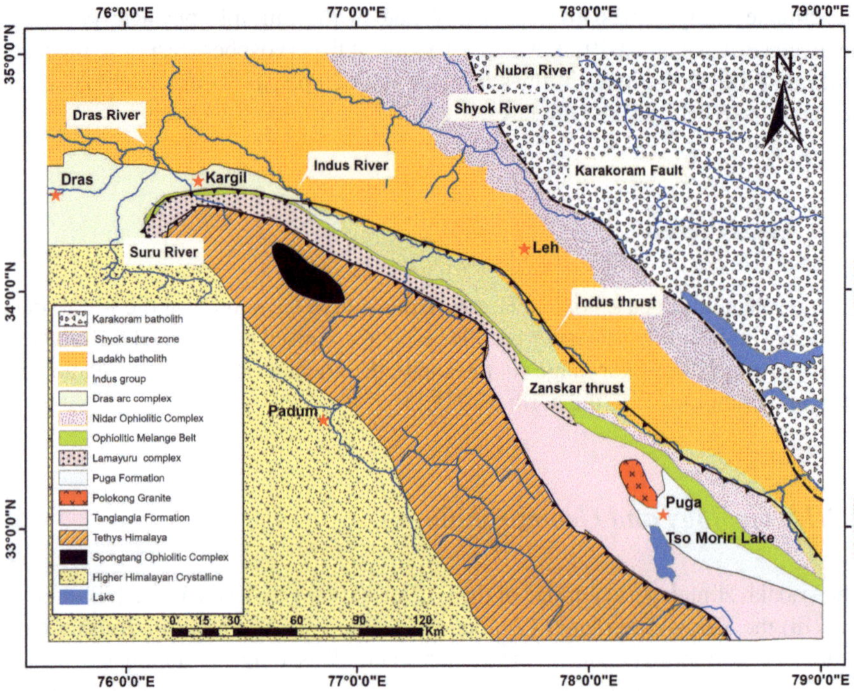

Fig. 1.2 Geology and structure of the Ladakh Himalaya, including major lithologies and structures. (Modified after Thakur 1981)

are differentiated from the Ladakh and Karakoram Batholith by the Indus suture zone (Searle et al. 1990). The piles of Tertiary molasses occur parallel to the Indus Suture Zone and represent the post-collision, continental deposits (Searle et al. 1990).

Geomorphologically, glacial processes dominated the terrain and are responsible for sculpturing the present-day landscape. Wide U-shaped valleys carry the evidences of past glacial advances, which in many places are plugged with the thick pile of the paraglacial sediments. The fluvial processes dominate around 3000 m altitude where seasonal temperature variability governs both the fluvial discharge and the sediment supply. In summers, rivers become turbid, and flash floods are common in the region. This is discernible by the presence of assorted sediments found along the major river's course (Pant et al. 2005). In addition, rivers are one of the major geomorphic agents after glaciers that cause significant denudation. These denuded sediments are deposited under different geo-environmental settings including glacial, glacio-fluvial, alluvial fan (debris flow), lacustrine-aeolian, and mass movements (Searle 1991).

1.2.2 Climate

Due to the presence of Higher Himalayan Range in the south, the Ladakh Himalaya is obstructed from the northward penetration of the Indian summer Monsoon (ISM). Overall, the climate is marked by cold winters and moderately warm summers. The mean annual precipitation of 100–300 mm dominated by the western disturbances (Government of Jammu and Kashmir 2007). Thus, the low rainfall and high-altitude topography make it the cold and arid to hyper-arid terrain in the HKH region. Interestingly, the observational data along with the inverse glacier mass balance modelling indicates that precipitation above 5000 m can be significantly high as compared to the downstream valleys (Hewitt 2005; Immerzeel et al. 2015).

Despite being in the cold and arid desert, infrequent cloud burst-induced torrential rainstorms occasionally cross over the southern Higher Himalayan ranges. Because the Ladakh Himalaya is influenced by the low-pressure systems associated with the westerly disturbances which occasionally penetrate the HKH ranges during the winter and spring seasons (Lang and Barros 2004). The contribution of the moistures associated with the westerly disturbances decreases significantly to the east and southeast of the Ladakh Himalaya (Hartmann 1999). Since the precipitation is low, irrigation is largely dependent upon the meltwater from ice packs, permafrost thaw, and small glaciers (Nüsser et al. 2012). In fact the distribution of habitational sites is governed by the presence of snow/glacial fed streams which flow through the stabilised alluvial fans (between 3000 m and 4500 m). The alluvial fans are important geomorphic features that formed after the recession of the tributary glaciers and are currently fed by the streams that have a source in the hanging (cirque) glaciers. These streams are the lifeline for the people of Ladakh, who are currently threatened by an increasing warming trend (Fig. 1.3). The region is sparsely populated, with the registered population growing at an annual rate by 1.5% between 2001 and 2011 (Census of India 2011).

Cryosphere in Ladakh Himalayas plays a very important role in the sustainability of the ecosystem and the nature and distribution of the habitation sites. The cryosphere is well-defined by the presence of frozen water in various forms, such as snow, permafrost, glaciers, ice sheets, ice caps, river, and lake ice. Freshwater is not only crucial in local and regional hydrology, agriculture, and ecology but also for generation of hydroelectric power. Therefore, variations in the cryosphere status will lead to serious challenges to the society. Therefore, this paper aims to provide a broad summary of the major cryosphere elements (both active and relict) and associated processes in the Hindu Kush Himalaya (HKH) in general and Ladakh Himalaya in particular. In addition to this, an attempt has also been made to discuss the complex relationship between four major elements of the cryosphere (glaciers, lakes, permafrost, and paraglacial sediments) with emphasis on their geomorphic response to the threat posed by climate change (global warming). Finally, strategies for combating the anticipated impacts on the ecosystem sustainability and livelihoods in the Ladakh Himalaya are discussed.

Fig. 1.3 Typical vegetated alluvial fan surfaces in Shyok Valley. Most of the rural and urban settlements are located on these fans. These fans are fed by perennial meltwater streams, which are fed by the cirque glaciers and have remained largely confined to their course for centuries. However, these streams have become unstable in the last few years due to accelerated melting and large-scale reworking of the paraglacial sediments

1.3 Result

1.3.1 Glacier

Glaciers are one of the most visible components of the cryosphere and the most important proxy for studying climate change. Studies indicate that Himalayan glaciers began to retreat since the Little Ice Age (LIA) terminated around 1850 AD (Oerlemans 2005). The receding glaciers left behind a large volume of unconsolidated sediment called moraines, which are visible in several glacially vacated Himalayan valleys, located hundreds of meters below the current snout position of the glaciers (Fig. 1.4; Chand et al. 2017; Singh et al. 2022). It has been suggested that the LIA glacial advance (retreat) provides a benchmark for understanding natural climate variability, providing a basis to assess anthropogenic climate change since the beginning of the twentieth century. A compilation of studies on glacier mass and area changes indicates that glaciers in most regions are shrinking and losing mass. However, in regions like eastern Pamir, Karakoram, and western Kunlun Shan, there seems to be relatively less glacier mass loss. It is feared that if the current greenhouse gas emission scenario persists, there are likely chances that Himalayan glacier are going to lose significant mass in near future. This will cause upward migration of regional equilibrium line altitudes (ELAs). As a consequence, at lower altitudes there would be complete disappearance of debris-free

Fig. 1.4 An example of the area vacated by the glacier since the Little Ice Age (LIA). **(a)** Holocene moraine east of the Siachen glacier snout. The yellow circle shows snout of Siachen glacier. **(b)** The closeup of the Siachen glacier snout. This glacier feeds the Nubra River in which large chunks of the detached glacier ice can be seen these days probably due to the rise in air temperature

lower-elevation glaciers and increased volume losses from glaciers with high-elevation accumulation areas (Singh et al. 2017; Bolch et al. 2019).

It is essential to monitor the pattern of glacier responses to climate change in the Himalayas. It is extremely important to assess the threat perception posed by the receding glaciers to the valleys located downstream, particularly the glacial lake outburst floods (GLOFs). Also, the vacating glaciers leave behind a large volume of unconsolidated sediments (moraines and associated alluvium), collectively known as paraglacial sediments, and are subjected to frequent mobilisation during extreme weather events (Singh and Mishra 2002; Singh et al. 2019). Therefore, studies should provide a scientifically tenable assessment of the glacier loss (both length and thickness). Although the current methods provide a broad indication that the glaciers in HKH (barring a few) are showing signs of recession, scientifically rigorous methods need to be employed to infer the role of climate. For instance, the most common method is recording the annual position of the glacier terminus (snout). However, such an observation may represent/indicate transient response to short-term climate variability and is location–/site-specific. The rate of retreat is influenced by various factors, involving elevation, topographic slope and aspect, ice thickness, and debris cover (Singh and Mishra 2002; Armstrong 2010). It has been found that researchers involved in understanding the health of Himalayan glaciers rarely consider the cumulative impact of the above variables when they report retreat rates. There are some studies which suggest that the higher retreat rate is associated with those glaciers which are located at the lowest altitudes on gentle slopes, with thin ice near the terminus, and a debris-cover thickness accelerates rather than retarding melt. There is general agreement that smaller glaciers (<5 km) situated at the lower elevations with a southerly aspect along with hanging glaciers 'cut off from a substantial accumulation area are likely to first disappear' (Armstrong 2010).

Compared to the glacier length, a more comprehensive and scientifically rigorous method involves measurements of glacier mass balance (Kaser et al. 2006). A glacier can be divided into two distinct zones: the upper zone of accumulation, where the glacier gains net mass annually, and the lower zone of ablation, where the glacier loses net mass. The boundary between these two zones is defined as the equilibrium line altitude (ELA), where the glacier's annual net mass balance is zero and is highly sensitive to climate variations (Kaser et al. 2006). Since changes in glacier length are delayed response to climate change and are also influenced by debris cover, glacier mass balance is considered a more immediate and direct indicator of glacier's response to climate variability (Bhattacharya et al. 2016a, b).

In central Ladakh, study pertaining to multi-temporal remote sensing data (1969–2016) by Schmidt and Nüsser (2017) has shown that over 1800 glaciers cover an area of 997 km^2. Nearly all glaciers are devoid of debris cover; among these, 79% glaciers are relatively small (<0.75 km^2). Majority of the glaciers (almost 90%) in Ladakh are characterised by their high altitudinal termini, situated above 5200 m, with the exception Siachen glacier in Nubra valley where the snout descended down to an elevation of around 3570 m (Fig. 1.4b). Nevertheless, in the Central Zanskar and Central Ladakh Ranges, about 20% and around 24%, respectively, of the snout reach down to elevations below 5200 m. In all other regions, the snout terminates above 5200 m with a maximum glacierised area above 5800 m. The average equilibrium line altitude (ELA) is approximately 5690 m, increasing from west to northeast. The Central Zanskar Range and Central Ladakh have a mean ELA of ~5500 m. In contrast, the averaged ELA in the Korzog and Pangong Range is located at an altitude of about 5950 and 5970 m, respectively (Braithwaite and Raper 2009).

1.3.2 Glacial Lakes

Glacier lakes are associated with the glacier snout which are supported by glacier mass loss (Fig. 1.5).

Studies indicate that the glacial lakes in the HKH are growing rapidly. According to Shugar et al. (2020 and references therein), between 1990–1999 and 2015–2018, the volume of the lake in the HKH amplified by ~45%, whereas the number of glacial lakes nearly doubled in northern India, Nepal, and Bhutan together with parts of southwestern China. According to them, there is not a significant increase in absolute terms. Still, the rise in storage capacity is likely to heighten the possibilities

Fig. 1.5 The development of the proglacial lake is evidence of sub-recent glacier recession. This is an example from the Drang Drung glacier in the southern Zanskar Ranges

of GLOF hazards in the Himalayan region. The eastern and central Himalayas exhibit the largest total glacial lake area and the highest absolute growth rate of glacial lakes (Bajracharya and Shrestha 2011). In a recent study by Harrison et al. (2018) a delayed response of GLOF triggered from the moraine-dammed lake has been postulated. According to them, the risk of GLOF involves (i) glacier recession, (ii) lake development and expansion, and (iii) finally, the GLOF triggers. The above sequence of events is based on a global GLOF peak flood study generated from moraine-dammed lakes in the 1960s to 1980s, which, according to them, was a delayed response to the glacier recession following the Little Ice Age. The study implies that a future GLOF peak will probably lag by several decades in the responses of glacier lake systems to current warming. According to Shugar et al. (2020), most of the earlier studies were focused on the first step (glacier recession/thinning), whereas their work represents the second stage (glacial lake expansion). Hence, they predict a third stage of increased GLOFs as glacier systems would now respond to contemporary climate warming. Many glacial lakes are growing and likely to accelerate due to the increasing global warming trend; particularly, the smaller lakes may grow more rapidly. The bursting of glacial lakes (Singh 2014) and landslide lake outburst floods (LLOFs) in the near future are likely to modify the HKH landforms significantly (Singh and Mishra 2002; Singh 2014), which would impact the ambitious river valley projects as observed during the June 2013 Kedarnath disaster (Sundriyal et al. 2015). Further, model prediction indicates that lakes that are decoupled from the glaciers will grow rather slowly, whereas others will drain or gradually fill with sediment.

According to a study, there were around 266 glacier lakes mapped in 2014 in the Ladakh Range (Narama et al. 2015), and the history of GLOFs from the Ladakh region is reasonably well documented. A report published in *Nature* (1932) mentions that the first recorded damming of Shyok River by the glacier ice dates to 1779 AD. Subsequently, great ice dams formed across the Upper Shyok River in 1924–1925, which sequentially busted in 1926, 1929, and 1932, sending flood surges from the release of the large volumes of water pent up behind the ice dam downstream in the Indus valley (Grant and Mason 1940). The above-mentioned ice dam was created in a constricted Shyok gorge section (~60 km wide) by the surging of the Kichik Kumdan glacier. Again during 1929 and 1936, the Kichik Kumdan glacier surged around 2.5 km, obstructing the Shyok River (Grant and Mason 1940). Although proglacial lakes in the Ladakh Himalaya are relatively small compared to those in the eastern Himalaya (Mool et al. 2001; Ukita et al. 2011), GLOFs have still caused significant damage in the valleys of the Ladakh Range. To give an example, in 1971, a GLOF triggered by a lake located around 18 km upstream from Nymmo Village generated a flash flood, damaging the residential areas and causing few deaths. Similarly, in year 2003, a GLOF affected Domkhar village, located around 80 km west of Leh, and inflicted damage upon bridges, farmland, and water mills (Narama et al. 2010). The most recent GLOF event is the 2014 Gya flood which was reported in the northern Zanskar Ranges. According to Schmidt et al. (2020), a proglacial Gya lake developed in front of a cirque glacier at an altitude of 5400 m by a 400-m-wide and 100-m-high moraine having a surface area of around 100 m^2. A scrutiny of the area indicated that the lake was created on the thermokarst hollow, implying that the permafrost thawed recently, thus suggesting terrain response to the recent rise in mean annual air temperature (Johannessen et al. 2004; Séjourné et al. 2015). The lake was not drained by a spillover but rather a subsurface (tunnelling) drainage facilitated by thawing ice cores in the moraine. This is important because the moraine-dammed lakes are usually breached by the spillovers triggered by snow avalanches or increased hydrostatic pressure from increased lake volume. However, the occurrence of the GLOF indicates that the rise in air temperature at higher altitudes (probably owing to EDW) was the factor responsible for the eventual breach by melting of the dead ice within the moraine (subsurface tunnelling).

1.3.3 Paraglacial Sediment

These sediments are generated from paraglacial environment, which according to Church and Ryder (1972) refers to the non-glacial processes that occur in response to the exposure of extensive glaciogenic drift resulting from glacier retreat. Once the glaciers began to recede, hillslopes can be in a metastable state because of the de-buttressing effect. This makes the glaciogenic sediment vulnerable to erosion, transportation, and sedimentation largely by debris flow processes. The paraglacial processes are so strong that they modify the landforms/landscape and create confusion in identifying diamictons (Singh 2004; Singh et al. 2017). Church and Ryder (1972) associated the process to a paraglacial period, characterised by relatively rapid sedimentation during or immediately following deglaciation when the environment is transitioning from predominantly glacial to dominantly fluvial conditions. However, this was contested by Ballantyne and Benn (1996), who acknowledged the delayed or renewed reworking of paraglacial sediments (glaciogenic) regardless of the environmental context in which such sediment reworking occurs (Singh et al. 2022). The paraglacial sediments are still being reworked in the HKH during extreme weather events. The long-term inheritance of glacial landforms and sediments continues to undergo mobilisation during extreme weather events. That is what is being recently observed in the HKH implying that, regardless of the time elapsed since deglaciation, the presence of glaciogenic sediment should be regarded as the defining characteristic of paraglacial activity (Curry 1998).

The paraglacial sediments are transported as debris flows or sediment gravity flows (Johnson 1970). These flows are not confined to talus or drift slopes nor are they limited to alpine or periglacial environments. Instead, they are regarded as being responsible for much of the reworking of glaciogenic deposits in recently deglaciated valleys and their re-sedimentation as coalescing debris cones and fans (Derbyshire and Owen 1990; Ballantyne and Benn 1994). The flows are initiated by a combination of rainfall and temperature supported by meltwater. In fact a combination of the processes is involved such as unusual rainfall (Schneuwly-Bollschweiler and Stoffel 2012; Singh et al. 2022), glacial meltwater flow or ice particle ablation (Arenson and Springman 2005), or GLOF (Chiarle et al. 2007; Singh 2014). In

HKH studies suggest that the extreme rainfall events which enter into the paraglacial valleys are the major factors responsible for triggering the paraglacial debris flow (Decaulne and Saemundsson 2007; Takahashi 2007; Kean et al. 2013). The majority of the paraglacial debris flows occur during the summer when the insolation summer monsoon is stronger, and the thawing of glaciers and permafrost facilitate percolation of meltwater into the unconsolidated glacial tills. This process makes the sediments susceptible to flowage like that of the antecedent rainfall (Rahardjo et al. 2008). Studies have shown that when the meltwater volume is high in early summer, debris flows can be triggered even by low-intensity rainfall. However, in late summer and early autumn, when meltwater volume is lower, more intense rainstorms are required to initiate debris flows (Stoffel et al. 2011; Schneuwly-Bollschweiler and Stoffel 2012). Attempts have been made to assess the rainfall threshold (tipping point for debris flow initiation), which according to Chen et al. (2011) varies from 0.6 to 6.3 mm h^{-1} or 3.0–19.4 mm/24 h, which is assisted by the increase in air temperature as at times found to be associated with an increase in the glacial melt (Liu et al. 2014). Given the fact that climate warming on high alpine environments is steadily increasing, it is anticipated that paraglacial debris flows are likely to become more prevalent in time to come (Deng MingFeng et al. 2017).

1.3.4 Permafrost

Permafrost is typically defined as soil or rock that remains below 0 °C for at least two consecutive years occurs. It is commonly found at high latitudes but also occurs in cold, high-elevation environments at lower latitudes (Haeberli et al. 2010). The subsurface ice can be present within the pores of rock or soil as well as in discrete layers. When this ice melts, significant changes occur in ground's characteristics, such as hydraulic permeability and mechanical strength, in response to both environmental changes and anthropogenic disturbances. With seasonal fluctuations, a surface layer of ground experiences cycle of freezing and thawing. The section of the ground that thaws during summer is referred to as the active layer. Compared to the understanding of glaciers and snow and their geographical extent (mapped based mainly on remote sensing), permafrost is poorly documented because, in remote sensing products, they cannot be readily visible (Haeberli et al. 2010).

In the HKH, with increasing elevation there is a corresponding decrease in the air temperature; hence, given the conducive geomorphological setting, it is likely that the permafrost would occur where the surface air temperature is below 1 °C. However, the situation is not entirely straightforward, as a thick winter snow cover could insulate the ground from a cool atmosphere, thus resulting in higher ground temperatures. Field observations in higher elevations indicate that low surface snow cover allows the cool atmosphere to interact with and favours the development of permafrost conditions and vice versa. This is one of the reasons why prevalence of permafrost beneath debris and soil rises with aridity. Therefore, in cold and dry landscapes like the NW Ladakh Himalaya, the area covered by permafrost is higher than glaciers compared to their humid central Himalayan counterparts (Ran et al. 2012). Under the climate change scenarios, the potential threat due to accelerated thawing of the permafrost in the Ladakh Himalaya would be a significant contributor towards the flash floods, debris flow, and slope instability.

1.4 Discussion

In comparison to the monsoon-dominated central and eastern Himalaya, some glaciers in the northwestern Karakoram Himalaya are either found to be advancing since the 1990s (Hewitt 2005; Rankl et al. 2014) or are standstill since the 1970s (Bolch et al. 2017). In the moisture-deficit arid and semi-arid Ladakh Himalaya, glaciers are generally small (<0.75 km^2) and are primarily confined to high-altitude regions. Most of the glacier snouts (~90%) terminate above 5200 m, barring the debris-covered Siachen Glacier in Karakoram, which terminates around 3570 m (Schmidt and Nüsser 2017). Regardless of their small size, these glaciers are the lifeline of the people inhabiting various valleys in this region. The meltwater ensuing from these glaciers provides the water for agriculture, which is fundamental to food security in the region (Nüsser et al. 2012). The small glaciers in Ladakh are going to respond too sensitively to climate change and, hence, the availability of water storage and runoff (Barnett 2005), particularly during lean periods. The time series temperature data for the period between 1901 and 2014 indicate that the annual mean temperature across the HKH region was more than 0.3 °C per decade; in addition, in the northwest Himalaya (Ren et al. 2017) and Karakoram, it was at least 0.7 °C higher (You et al. 2017) for which anthropogenically contributed greenhouse gases are implicated (Krishnan et al. 2020). Additionally, the EDW is further likely to exacerbate the higher altitude (Higher Himalayan) temperature compared to the lower plains. Although the causes of EDW are still being debated, it is believed that factors like the melting of snow/ice along with upslope tree line shift are likely to decrease surface albedo in a certain elevation band. Since the warmer atmosphere can hold more moisture, it promotes greater latent heat release, particularly at high elevations in the Himalayas, further accelerating the melting processes (Liu et al. 2009; Ren et al. 2017).

Ladakh Himalaya has a mean annual precipitation of around 100 mm; agriculture relies upon streams fed by the glacial or snow melt during the agricultural season, which is spring and summer (Nüsser et al. 2012). In the last few decades, the stream flow has become quite unpredictable due to the growing incidences of rain-on-snow-induced floods during spring and autumn. Studies pertaining to the consequence of climate change on glaciers in addition to associated streamflow in Ladakh Himalaya are still in its embryonic stage (Nüsser and Schmidt 2018). Nevertheless, water crises in the region are not new; since time immemorial, a tradition of creating artificial ice reservoirs existed. These reservoirs stored the water in the autumn and winter, until spring, when it melts and provides the much-needed irrigation water to the local inhabitants (Nüsser and Schmidt 2018). Time has come that the traditional water harvesting practices although initiated at an individual level must be strengthened and propagated in the region. The water crises under the warm earth scenario are going to be extremely acute, so much so that it might force the human exodus to seek climate refuge. The pulse is already being felt in the terrain between Kargil and Lamayuru, which presents a harrowing picture caused by the steady decline in meltwater from snow and ice (Fig. 1.6).

GLOFs have emerged as a serious hazard in the mountain region in recent decades. They are threatening human life and infrastructures in the Himalayan region, and the Ladakh Himalaya is no exception. Notable instances include GLOFs and LLOFs in the Gangotri glacier area (Singh and Mishra 2002) and Kedarnath (Dobhal et al. 2013; Singh 2013). While mountain communities have dealt with GLOF risks since centuries, climate change is aggravating physical and social vulnerability (IPCC, Summary for Policymakers 2019). The processes leading to GLOFs are reasonably well understood; what is lacking is creating a socio-economic, cultural, and institutional interference towards creating effective strategies to mitigate the GLOF risks (Carey et al. 2015). Considering the geological, geomorphological, and climate complexities of the Ladakh Himalaya, studies pertaining to the GLOF risk anticipation, reduction, and scientific investigations (geological, geophysical, and metrological aspects) need to be strengthened. At the same time, the administrative machinery should be sensitised and the local population educated on threat perception associated with GLOFs under the climate change scenarios. GLOF events can traverse vast distances, through various districts, state, and even national boundaries. Therefore, a unified methodological approach is essential to identify the risks and foresee emerging challenges for Himalayan region before it actually happens. Mitigating current GLOF risks using early warning systems, emphasising community awareness and preparedness, and devising sustainable land use practices would be noteworthy steps towards adapting to future challenges in many cases. We must learn from the Bhutan Himalayas, where local people, along with the experts, became Bhutan's local inhabitants, calling them the 'army' against the impacts of climate change. Their mission is to enlarge and deepen the outlet channel from lakes formed by the rapidly melting Thorthormi glacier (pictured above, to the right of the lake). By accelerating the drainage of water, Bhutanese officials aim to avert potentially catastrophic flood.

Ladakh Himalaya was extensively glaciated in the recent geological past (Ali and Juyal 2013). The receding glaciers have left behind a large volume of unconsolidated glaciogenic sediments, particularly the moraines, which can be seen plugging many amphitheatre valleys (Fig. 1.7). Together glacial and fluvial processes play crucial roles in modifying landscapes in high mountain environments. As we are aware glaciers are extremely effective erosional agents and producers of large volumes of sediment, though these are spatially restricted, while rivers can transmit such sediment through landscapes and flush them out of mountain belts and at times into the sedimentary basins (Hobley et al. 2010). Considering that the arid to semi-arid regions are geomorphic threshold areas, the sparsely vegetated, steep hillslopes and channel networks in these regions are highly sensitive to climatic changes.

Fig. 1.6 Being a cold and arid desert, the meltwater ensuing from glaciers/snowpack is the main source of irrigation in the Ladakh region; this provides regional food security. Unfortunately, in the last few decades, the glaciers have shown a decrease in their volume, whereas the snowfall has also reduced due to the rise in winter temperature. Consequently, people are at places forced to desert their ancestral habitation sites in search of better climate refugia. (**a**) A village is on the verge of destruction just below Fotu La Pass between Kargil and Lamayuru. (**b**) The streams, once fed by the snowpacks, are virtually dry, giving an impression of the Martian landscape

Fig. 1.7 Ladakh Himalaya was extensively glaciated in the recent geological past; hence, majority of the glacially vacated valleys are plugged with large volume of unconsolidated paraglacial sediments in the amphitheatre valleys. (**a**) Anthropogenic encroachments in these valleys were responsible for amplifying the impact of August 2010 Leh cloud burst. (**b**) Not learning from the past disaster, people have again clogged these valleys and reoccupied these valleys inviting another disaster. (Picture taken during October 2023)

Ladakh Himalaya therefore can be considered as one of the hotspots when it comes to the geomorphic response to climate change. It has been observed that during extreme weather events, the flow strength of threshold is exceeded along sparsely vegetated, steep hillslopes in the arid and semi-arid regions of the Himalayas, leading to the formation of new channels in the headwaters through debris flows (Bookhagen et al. 2005; Singh 2014). As a case study, we would like to discuss the August 2010 Leh disaster, which is a testimony to demonstrate the vulnerability of the arid and semi-arid Trans-Himalaya to extreme weather events. This unfortunate disaster must be taken as an eye opener for understanding the arid and semi-arid Trans-Himalayan terrain response to climate change.

In the early morning hours of 6 August 2010, the Leh valley was devastated by a cloud burst that probably lasted less than 3 h (Juyal 2010). Leh town is located on the southern slope of the Ladakh Batholith which is 2–4 km above the valley floor and part of the western end of the Trans-Himalayan Batholith system (Weinberg and Dunlap 2000). The ridgeline of the Batholith runs ~350 km broadly in southeast-northwest direction and is lying between the Shyok suture zone to the north and the Indus-Tsangpo suture zone to the south (Searle et al. 1990). The northwest-trending Indus River broadly follows the suture and, at places, cuts across the Indus molasses (Hobley et al. 2010). Glacial and periglacial processes dominated the south-trending valleys in the geological past, which have carved wide 'U'-shaped valleys plugged with moraines/debris-laden south-facing slopes around Leh. The retreated glaciers have also given rise to innumerable dried channels that are trending from NW to SE around Leh. At present, little ice patches are present in the distal parts of the relict glaciated valleys, particularly along the Khardungla ridge in the north. These ice patches feed some of the perennial streams that eventually drain into the Indus River. During the cloud burst rainfall-induced runoff caused failure of debris-laden slopes around Leh. The granular slope sediments, along with the fluid, lifted large boulders and transported them downslope due to buoyancy in the early morning hours of 6 August 2010, causing large-scale destruction around the Leh Valley. According to Juyal (2010), from a process point of view, it was a simple debris flow triggered by torrential rain, whereas the destruction was focused in parts situated proximal to the mountain slopes and dried relict glacial melt-out channels. A rapid assessment of the cause of cloud burst-induced destruction around Leh by Juyal (2010) made the following observations: (i) The natural flow path of the dried channels was obstructed due to the man-made constructions that resulted in temporary obstruction and deviation of the flow from its natural courses. (ii) Maximum destruction was observed along the new Leh road. This road winds alongside the course of a dried channel that emanates from the southwest trending amphitheatre valleys, which inflicted maximum damage to the buildup structures (new bus stand, hospital, government school, roads, and residential houses). (iii) Compared to this, the old Leh City (belonging to the sixteenth century) remained largely unaffected. This part of the town is located on a rather elevated ground protected by a rocky ledge in the north, on which the majestic Leh fort is situated. This suggests that in the past our ancestors could appreciate the sensitivity of the terrain towards unusual weather events; hence, the habitation sites were judiciously selected, particularly keeping in mind the safety and security of people from natural calamities (Fig. 1.8). This consideration needs to be rejuvenated considering the accelerated pace of human growth and the associated infrastructure requirement in the ecologically sensitive Trans-Himalayan valleys of Ladakh region.

Even though detailed data are few, existing measurements, particularly from the Tibetan Plateau, point towards permafrost warming, accompanied by an increase in the depth of the active layer. It is anticipated that if the present trend of warming

Fig. 1.8 Picture of the old Leh town taken in 1873 by Edward Francis Chapman and is a part of the Dunlop Smith Collection. As can be seen, the town is located on a rather elevated ground protected by a rocky ledge in the north (like an eyebrow), on which the majestic Leh fort is situated. This suggests that in the past, our ancestors could appreciate the sensitivity of the terrain towards unusual weather events; hence, the habitation sites were judiciously selected, particularly keeping in mind the safety and security of people from natural calamities

continues, the permafrost areas will have an adverse impact. This would lead to mountain slope destabilisation (Huggel et al. 2013; Krautblatter et al. 2013), increasing the possibility of rock falls, debris flows, and/or landslides (Fort 2015; Gruber et al. 2017). Slope failures can affect infrastructure and settlements both directly and indirectly, for instance when they impact glacial lakes, which are typically situated below or near steep slopes (Haeberli et al. 2016). Additionally, permafrost thaw can cause ground movement through thaw settlement and change geotechnical characteristics via ground-ice melt. This is particularly imperative as it can lead to issues with infrastructure constructed in permafrost regions (Bommer et al. 2010; Doré et al. 2016). Some signs of the impact of permafrost degradation began to show up in various valleys where the deepening of the permafrost active layer is not only exposing the locked organic carbon to the atmosphere, but the accelerated thawing is creating the Thermokarst lakes in the region (Fig. 1.9). These lakes together with the ever-increasing number of proglacial lakes likely to exacerbate the impact of GLOFs in the region.

Fig. 1.9 Climate change is impacting the permafrost areas in the Ladakh Himalayas. This picture shows the development of thermokarst in the Suru valley. (Southern Zanskar Ranges)

1.5 Conclusions

The Himalayan cryosphere will be severely impacted if the course corrections are not made. There are growing numbers of studies from the Indian Himalayan Region (IHR) that indicate that the Himalayas are warming at a rate faster than that of the adjacent Indian land mass (Sabin et al. 2020). The observations also show that recent warming rates are heterogeneous over the HKH region, which is altitude-dependent and is attributed to the EDW (e.g. Liu et al. 2009; Ren et al. 2017; Shrestha et al. 1999; Thompson et al. 2003). A recent study by Pandey et al. (2024) from monsoon-dominated central Himalaya provided conjectural support for the suggested EDW based on the temporal changes in the snow glacier ice, which would have resulted in the reduction in albedo at higher elevations. The study postulates that this might lead to an increasing risk of cryosphere-related hazards (e.g. rock, ice, snow avalanches, and glacial lake outbursts).

We tend to emphasise that for disaster risk reduction in the Ladakh Himalaya, it is important to understand the complex relationship between four major elements of the cryosphere (glaciers, lakes, permafrost, and paraglacial sediment mobilisation mechanisms). One cannot view them in isolation because they are interdependent. For example, the GLOF cannot happen if the glaciers are not receding, leaving behind the paraglacial sediments (particularly the moraine), thus obstructing temporarily the meltwater discharge. Similarly, the debris flow is triggered only when either the ice core moraines thaw due to a rise in air temperature, destabilisation of the valley slopes due to the de-buttressing of the valley wall due to the shrinking and recession of the valley glaciers, or the increase in the moisture availability due to the extreme weather events (including rain on snow). Therefore, to reduce the disaster risk reduction in the Ladakh Himalaya, the following suggestions emerge from the present study:

1. The glaciers should be continuously monitored, particularly the temporal changes in the meltwater discharge as a function of seasonality. In addition, it is high time to have continuous automated sampling of the black carbon and mineral aerosols so that their impact on the glaciers and snow can be evaluated with less uncertainty (radiative forcing).
2. The mapping, distribution, and temporal changes in the glacial lakes (area, volume, and the nature of damming ice, moraine, or landslide/alluvial fan) need to be documented to assess the sustainability of the lakes. Based on geological, geomorphological, and hydrological criteria, these lakes should be categorised under different threat perception categories and monitored using multi-temporal satellite data with selected few field validations. The anticipated breaching-induced down valley inundation must be marked, and the information must be shared among the stakeholders, which include the administration, village panchayats, or any other community-based organisation.
3. It is imperative to recognise that in the Himalayas, during flood events, it is not solely the water but rather the associated sediment that inflicts severe damage to the infrastructure, as observed in multiple flash flood-induced disasters, including the August 2010 Leh cloud burst. Therefore, the glacially vacated valleys (the amphitheatre valleys) must be mapped, and the sediment sequestered in these valleys should be quantified. This would help in assessing the severity of the damage during an eventuality like a cloud burst in the sediment surplus and transport-limited Ladakh Himalaya.
4. Most importantly, the glacially vacated valley outlets should be kept free from anthropogenic encroachments so that during extreme weather events, the water and sediments find a free-flow path towards the regional slope. Therefore, all the habitational sites must be mapped for the nature of deposits and the regional slopes. The planning for the infrastructure must consider all these factors.
5. Elevation-dependent warming is a new buzzword, although logically, it seems to be acting as a catalyst for enhancing the temperature at higher altitudes. However, for a more realistic assessment of the EDM contribution to the vital cryosphere, there is a need for stalling the high-altitude metrological stations so that real-time data can be gathered for more authentic model simulation for a terrain like Ladakh.

Acknowledgement We are thankful to the Head, Centre of Advanced Study in Geology, University of Lucknow, Uttar Pradesh, and Department of Basic Sciences, College of Forestry, Ranichauri, Tehri Garhwal (Veer Chandra Singh Garhwali Uttarakhand University of Horticulture and Forestry, Bharsar, Pauri Garhwal), Uttarakhand, for providing the working facilities. JD acknowledges the Anusandhan National Research Foundation (ANRF N-PDF), Government of India, for financial assistance (PDF/2023/003328). Dr. Navin Juyal is thanked for providing the field photographs.

References

Ali SN, Juyal N (2013) Chronology of late quaternary glaciations in Indian Himalaya: a critical review. J Geol Soc India 82:628–638

Arenson LU, Springman SM (2005) Mathematical descriptions for the behaviour of ice-rich frozen soils at temperatures close to 0 C. Can Geotech J 42(2):431–442

Armstrong RL (2010) The glaciers of the Hindu Kush-Himalayan region: a summary of the science regarding glacier melt/retreat in the Himalayan, Hindu Kush, Karakoram, Pamir, and Tien Shan mountain ranges, p 16

Bajracharya SR, Shrestha BR (2011) In: Bajracharya SR, Shrestha B (eds) The status of glaciers in the Hindu Kush-Himalayan region. International Centre for Integrated Mountain Development (ICIMOD), Kathmandu, p 128

Ballantyne CK, Benn DI (1994) Paraglacial slope adjustment and resedlmenfation following recent glacier retreat, Fåbergstølsdalen, Norway. Arct Alp Res 26(3):255–269

Ballantyne CK, Benn DI (1996) Paraglacial slope adjustment during recent deglaciation and its implications for slope evolution in formerly glaciated environments. In: Advances in hillslope processes. Wiley, Chichester, pp 1173–1195

Banerjee A, Dimri AP, Kumar K (2021) Temperature over the Himalayan foothill state of Uttarakhand: present and future. J Earth Syst Sci 130:33. https://doi.org/10.1007/s12040-020-01527-5

Barnett TP (2005) Potential impacts of a warming climate on water availability in snow-dominated regions. Nature 438:7066. https://doi.org/10.1038/nature04141

Bhattacharya A, Bolch T, Mukherjee K et al (2016a) Overall recession and mass budget of Gangotri Glacier, Garhwal Himalayas, from 1965 to 2015 using remote sensing data. J Glaciol 62(236):1115–1133

Bhattacharya A, Bolch T, Mukherjee K et al (2016b) Overall recession and mass budget of Gangotri Glacier, Garhwal Himalayas, from 1965 to 2015 using remote sensing data. J Glaciol 62(236):1115–1133. https://doi.org/10.1017/jog.2016.96

Bhutiyani MR, Kale VS, Pawar NJ (2007) Long-term trends in maximum, minimum and mean annual air temperatures across the Northwestern Himalaya during the twentieth century. Clim Chang 85(1–2):159–177

Bolch T, Pieczonka T, Mukherjee K et al (2017) Brief communication: Glaciers in the Hunza catchment (Karakoram) have been nearly in balance since the 1970s. Cryosphere 11(1):531–539

Bolch T, Shea JM, Liu S et al (2019) Status and change of the cryosphere in the extended Hindu Kush Himalaya region. In: Wester P, Mishra A, Mukherji A, Shrestha AB (eds) The Hindu Kush Himalaya assessment: mountains, climate change, sustainability and people. Springer, Cham, pp 209–255. https://doi.org/10.1007/978-3-319-92288-1

Bommer C, Phillips M, Arenson LU (2010) Practical recommendations for planning, constructing and maintaining infrastructure in mountain permafrost. Permafr Periglac Process 21(1):97–104

Bookhagen B, Burbank DW (2006) Topography, relief, and TRMM-derived rainfall variations along the Himalaya. Geophys Res Lett 33(8):L08405

Bookhagen B, Thiede RC, Strecker MR (2005) Abnormal monsoon years and their control on erosion and sediment flux in the high, arid northwest Himalaya. Earth Planet Sci Lett 231(1–2):131–146

Braithwaite RJ, Raper SCB (2009) Estimating equilibrium-line altitude (ELA) from glacier inventory data. Ann Glaciol 50(53):127–132

Carey M, McDowell G, Huggel C et al (2015) Integrated approaches to adaptation and disaster risk reduction in dynamic socio-cryospheric systems. In: Haeberli W, Whiteman C (eds) Snow and ice-related hazards, risks, and disasters, Elsevier. Academic Press, pp 219–261. https://doi.org/10.1016/B978-0-12-394849-6.00008-1

Census of India (2011) Census of India 2011. Retrieved January 20, 2020 from https://www.census2011.co.in/census/district/621-leh.html

Chand P, Sharma MC, Bhambri R et al (2017) Reconstructing the pattern of the Bara Shigri Glacier fluctuation since the end of the Little Ice Age, Chandra valley, north-western Himalaya. Prog Phys Geogr 41(5):643–675. https://doi.org/10.1177/0309133317728017
Chen NS, Zhou HB, Hu GS (2011) Development rules of debris flow under the influence of climate change in Nyingchi. Adv Clim Chang Res 7(6):412
Chiarle M, Iannotti S, Mortara G et al (2007) Recent debris flow occurrences associated with glaciers in the Alps. Glob Planet Chang 56(1–2):123–136
Church M, Ryder JM (1972) Paraglacial sedimentation: a consideration of fluvial processes conditioned by glaciation. Geol Soc Am Bull 83(10):3059–3072
Curry AM (1998) Paraglacial modification of drift-mantled hillslopes. Doctoral dissertation, University of St Andrews
Das SK, Dobhal DP, Juyal N (2010) Variability of aerosol optical depth and recent recessional trend in Dokriani Glacier, Bhagirathi Valley, Garhwal Himalaya. Curr Sci 99(12):1816–1821
Decaulne A, Sæmundsson Þ (2007) Spatial and temporal diversity for debris-flow meteorological control in subarctic oceanic periglacial environments in Iceland. Earth Surf Process Landf 32(13):1971–1983
Deng MingFeng DM, Chen NingSheng CN, Mei L (2017) Meteorological factors driving glacial till variation and the associated periglacial debris flows in Tianmo Valley, south-eastern Tibetan Plateau. Nat Hazards Earth Syst Sci 17:345–356
Derbyshire E, Owen LA (1990) Quaternary alluvial fans in the Karakoram Mountains. In: Rachocki AH, Church M (eds) Alluvial fans: a field approach. Wiley, Chichester, pp 27–53
Dobhal DP, Gupta AK, Mehta M et al (2013) Kedarnath disaster: facts and plausible cause. Curr Sci 105:171–174
Doré G, Niu F, Brooks H (2016) Adaptation methods for transportation infrastructure built on degrading permafrost. Permafr Periglac Process 27(4):352–364
Dyhrenfurth GO (1955) To the third pole: the history of the High Himalaya, 1st edn. Ex Libris, Werner Laurie, London
Fort M (2015) Natural hazards versus climate change and their potential impacts in the dry, northern Himalayas: focus on the upper Kali Gandaki (Mustang District, Nepal). Environ Earth Sci 73:801–814
Government of Jammu and Kashmir (2007) Environment and social management framework for Participatory Watershed Management Project (PWMP) J&K. IWDPHills II, 130 pp., Annex
Grant IHL, Mason K (1940) The upper Shyok glaciers 1939. Himal J 12:1–23
Gruber S, Fleiner R, Guegan E et al (2017) Inferring permafrost and permafrost thaw in the mountains of the Hindu Kush Himalaya region. Cryosphere 11(1):81–99
Haeberli W, Noetzli J, Arenson L et al (2010) Mountain permafrost: development and challenges of a young research field. J Glaciol 56(200):1043–1058
Haeberli W, Schaub Y, Huggel C (2016) Increasing risks related to landslides from degrading permafrost into new lakes in de-glaciating mountain ranges. Geomorphology 293:405–417. https://doi.org/10.1016/j.geomorph.2016.02.009
Harrison S, Kargel JS, Huggel C et al (2018) Climate change and the global pattern of moraine-dammed glacial lake outburst floods. Cryosphere 12(4):1195–1209
Hartmann H (1999) Studies on the flora and vegetation in eastern Trans-Himalaya of Ladakh (India). Candollea 54:171–230
Hewitt K (2005) The Karakoram anomaly? Glacier expansion and the elevation effect, Karakoram Himalaya. Mt Res Dev 25(4):332–340
Hobley DE, Sinclair HD, Cowie PA (2010) Processes, rates, and time scales of fluvial response in an ancient postglacial landscape of the northwest Indian Himalaya. Bulletin 122(9–10):1569–1584
Huggel C, Stone D, Auffhammer M (2013) Loss and damage attribution. Nat Clim Chang 3(8):694–696
Immerzeel WW, Wanders N, Lutz AF et al (2015) Reconciling high-altitude precipitation in the upper Indus basin with glacier mass balances and runoff. Hydrol Earth Syst Sci 19(11):4673–4687

IPCC, Summary for Policymakers (2019) In: IPCC special report on the ocean and cryosphere in a changing climate. https://doi.org/https://www.ipcc.ch/report/srocc/
Johannessen O, Bengtsson L, Miles M et al (2004) Arctic climate change: observed and modelled temperature and sea-ice variability. Tellus A Dyn Meteorol Oceanogr 56(4):328–341
Johnson AM (1970) Physical processes in geology. Freeman, Cooper, San Francisco, p 577
Juyal N (2010) Cloud burst-triggered debris flows around Leh. Curr Sci 99(9):1166–1167
Kaser G, Cogley JG, Dyurgerov MB et al (2006) Mass balance of glaciers and ice caps: consensus estimates for 1961–2004. Geophys Res Lett 33:L19501. https://doi.org/10.1029/2006GL027511
Kean JW, McCoy SW, Tucker GE et al (2013) Runoff-generated debris flows: observations and modeling of surge initiation, magnitude, and frequency. J Geophys Res Earth 118(4):2190–2207
Krautblatter M, Funk D, Günzel FK (2013) Why permafrost rocks become unstable: a rock–ice-mechanical model in time and space. Earth Surf Process Landf 38(8):876–887. https://doi.org/10.1002/esp.3374
Krishnan R, Gnanaseelan C, Sanjay J et al (2020) Introduction to climate change over the Indian region. In: Assessment of climate change over the Indian Region: a report of the Ministry of Earth Sciences (MoES). Government of India, pp 1–20
Lang TJ, Barros AP (2004) Winter storms in the central Himalayas. J Meteorol Soc Jpn Ser II 82(3):829–844
Liu X, Cheng Z, Yan L et al (2009) Elevation dependency of recent and future minimum surface air temperature trends in the Tibetan Plateau and its surroundings. Glob Planet Chang 68(3):164–174
Liu JJ, Cheng ZL, Su PC (2014) The relationship between air temperature fluctuation and Glacial Lake outburst floods in Tibet, China. Quatern Int 321:78–87
Mool PK, Wangda D, Bajracharya SR et al (2001) Inventory of glaciers, glacial lakes and glacial lake outburst floods: monitoring and early warning systems in the Hindu Kush-Himalayan Region, Bhutan. International Centre for Integrated Mountain Development, Kathmandu
Narama C, Duishonakunov M, Kääb A et al (2010) The 24 July 2008 outburst flood at the western Zyndan glacier lake and recent regional changes in glacier lakes of the Teskey Ala-Too range, Tien Shan, Kyrgyzstan. Nat Hazards Earth Syst Sci 10(4):647–659
Narama C, Daiyrov M, Kazehare S et al (2015) Glacier lake inventory of the northern Tien Shan–Kyrgyz. Kungoy, and Teskey Ala-Too Ranges–Report of Mountain Research Group of Niigata University, Niigata, Japan, Niigata Printing
Negi PS, Pandey CP, Singh N (2019) Black carbon aerosols in the ambient air of Gangotri Glacier valley of north-western Himalaya in India. Atmos Environ 214:116879
Nüsser M, Schmidt S (2018) Nanga Parbat revisited: Evolution and dynamics of sociohydrological interactions in the Northwestern Himalaya. In: Mountains: physical, human-environmental, and sociocultural dynamics. Routledge, pp 166–178. https://doi.org/10.4324/9781315158914-15
Nüsser M, Schmidt S, Dame J (2012) Irrigation and development in the Upper Indus Basin: characteristics and recent changes of a socio-hydrological system in Central Ladakh, India. Mt Res Dev 32(1):51–61
Oerlemans J (2005) Extracting a climate signal from 169 glacier records. Science 308(5722):675–677
Pandey P, Ali SN, Bhardwaj A et al (2024) Pattern and imprints of elevation-dependent warming on central Himalayan cryosphere as revealed by Earth observation datasets. Results Earth Sci 2:100021. https://doi.org/10.1016/j.rines.2024.100021
Pant RK, Phadtare NR, Chamyal LS et al (2005) Quaternary deposits in Ladakh and Karakoram Himalaya: a treasure trove of the palaeoclimate records. Curr Sci 88(11):1789–1798
Rahardjo H, Leong EC, Rezaur RB (2008) Effect of antecedent rainfall on pore-water pressure distribution characteristics in residual soil slopes under tropical rainfall. Hydrol Process Int J 22(4):506–523
Ramanathan V, Carmichael G (2008) Global and regional climate changes due to black carbon. Nat Geosci 1(4):221–227

Ran Y, Li X, Cheng G et al (2012) Distribution of permafrost in China: an overview of existing permafrost maps. Permafr Periglac Process 23(4):322–333
Rankl M, Kienholz C, Braun M (2014) Glacier changes in the Karakoram region mapped by multimission satellite imagery. Cryosphere 8(3):977–989
Ren YY, Ren GY, Sun XB et al (2017) Observed changes in surface air temperature and precipitation in the Hindu Kush Himalayan region over the last 100-plus years. Adv Clim Chang Res 8(3):148–156. https://doi.org/10.1016/j.accre.2017.08.001
Sabin TP, Krishnan R, Vellore R et al (2020) Climate change over the Himalayas. In: Assessment of climate change over the Indian region: a report of the Ministry of Earth Sciences (MoES). Government of India, pp 207–222
Sarangi C, Qian Y, Rittger K et al (2020) Dust dominates high-altitude snow darkening and melt over high-mountain Asia. Nat Clim Chang 10(11):1045–1051
Schmidt S, Nüsser M (2017) Changes of high altitude glaciers in the Trans-Himalaya of Ladakh over the past five decades (1969–2016). Geosciences 7(2):27
Schmidt S, Nüsser M, Baghel R et al (2020) Cryosphere hazards in Ladakh: the 2014 Gya glacial lake outburst flood and its implications for risk assessment. Nat Hazards 104:2071–2095
Schneuwly-Bollschweiler M, Stoffel M (2012) Hydrometeorological triggers of periglacial debris flows in the Zermatt valley (Switzerland) since 1864. J Geophys Res Earth 117(F2)
Searle MP (1991) Geology and tectonics of the Karakoram Mountains. Wiley, New York, p 358
Searle MP, Pickering KT, Cooper DJW (1990) Restoration and evolution of the intermontane Indus Molasse basin, Ladakh Himalaya, India, Tectonophysics 174(3–4): 301–314
Séjourné A, Costard F, Fedorov A et al (2015) Evolution of the banks of thermokarst lakes in Central Yakutia (Central Siberia) due to retrogressive thaw slump activity controlled by insolation. Geomorphology 241:31–40
Shrestha AB, Wake CP, Mayewski PA et al (1999) Maximum temperature trends in the Himalaya and its vicinity: an analysis based on temperature records from Nepal for the period 1971–94. J Clim 12(9):2775–2786
Shugar DH, Burr A, Haritashya UK et al (2020) Rapid worldwide growth of glacial lakes since 1990. Nat Clim Chang 10(10):939–945
Singh DS (2004) Late Quaternary morpho-sedimentary processes in the Gangotri Glacier area, Garhwal Himalaya, India. Geol Surv India Spec Publ 80:97–103
Singh DS (2013) Causes of Kedarnath tragedy and human responsibilities. J Geol Soc India 82(3):303–304
Singh DS (2014) Surface Processes during flash floods in the glaciated terrain of Kedarnath, Garhwal Himalaya and their role in the modification of landforms. Curr Sci 106(4):594–597
Singh DS, Mishra A (2002) Role of Tributary glaciers on landscape modification in the Gangotri Glacier area, Garhwal Himalaya, India. Curr Sci 82(5):101–105
Singh DS, Tangri AK, Kumar D et al (2017) Pattern of retreat and related morphological zones of Gangotri Glacier, Garhwal Himalaya, India. Quat Int 444:172–181. https://doi.org/10.1016/j.quaint.2016.07.025
Singh DS, Dubey CA, Kumar D et al (2019) Monsoon variability and major climatic events between 25 and 0.05 ka BP using sedimentary parameters in the Gangotri Glacier region, Garhwal Himalaya, India. Quat Int 507:148–155. https://doi.org/10.1016/j.quaint.2019.02.018
Singh DS, Singh AK, Dubey CA et al (2022) Multi-Proxy analysis in the Gangotri Glacier region, Garhwal Himalaya: Glacial Stratigraphy and the overview of snout retreat, geomorphic evolution, and palaeoclimate signatures. J Palaeontol Soc India 67(1):158–182
Stoffel M, Bollschweiler M, Beniston M (2011) Rainfall characteristics for periglacial debris flows in the Swiss Alps: past incidences–potential future evolutions. Clim Chang 105(1):263–280
Sundriyal YP, Shukla AD, Rana N et al (2015) Terrain response to the extreme rainfall event of June 2013: evidence from the Alaknanda and Mandakini River Valleys, Garhwal Himalaya, India. Episodes J Int Geosci 38(3):179–188
Takahashi T (2007) Debris flow: mechanics, prediction and countermeasures, 1st edn. Taylor & Francis. https://doi.org/10.1201/9780203946282

Tewari VP, Verma RK, Gadow KV (2017) Climate change effects in the Western Himalayan ecosystems of India: evidence and strategies. For Ecosyst 4:13. https://doi.org/10.1186/s40663-017-0100-4

Thakur VC (1981) An overview of thrusts and nappes of western Himalaya. Geol Soc Lond Spec Publ 9(1):381–392

Thompson LG, Mosley-Thompson E, Davis ME et al (2003) Tropical glacier and ice core evidence of climate change on annual to millennial time scales. In: Diaz HF (ed) Climate variability and change in high elevation regions: past, present & future. advances in global change research, vol 15. Springer, Dordrecht, pp 135–155. https://doi.org/10.1007/978-94-015-1252-7_8

Ukita J, Narama C, Tadono T et al (2011) Glacial lake inventory of Bhutan using ALOS data: methods and preliminary results. Ann Glaciol 52(58):65–71

Weinberg RF, Dunlap WJ (2000) Growth and deformation of the Ladakh Batholith, Northwest Himalayas: implications for timing of continental collision and origin of calc-alkaline batholiths. J Geol 108(3):303–320

Xu J, Badola R, Chettri N et al (2019) Sustaining biodiversity and ecosystem services in the Hindu Kush Himalaya. In: The Hindu Kush Himalaya assessment: Mountains, climate change, sustainability and people, pp 127–165. https://doi.org/10.1007/978-3-319-92288-1_5

You QL, Ren GY, Zhang YQ et al (2017) An overview of studies of observed climate change in the Hindu Kush Himalayan (HKH) region. Adv Clim Chang Res 8(3):141–147. https://doi.org/10.1016/j.accre.2017.04.001

Chapter 2
Climate Change Impacts on Himalayan Agriculture

Nomi Sarmah, Anjumoni Devee, Junmoni Gayon, Momi Sarma, and Tanjil Rahman

Abstract The alarming climatic changes calls for the dire attention towards the possible consequences and their solution. Such changes have brought unpredicted crop management problems. In Himalayan region, the resilient crops are encouraged to grow. The crop development along with the occurrence of crop pests and diseases have also shown huge alteration putting forward the need of dynamic crop development system in field level. The Himalayan region specifically the temperature, humidity and irrigation requirement increasingly disrupted by climate change. Fluctuations in rainfall patterns, irregular snowfall, and rising temperatures are placing additional stress on the agroecosystem. Therefore, the need to stabilize crop production under these changing conditions becomes paramount. This may involve integrating climate-smart agricultural practices, improving irrigation efficiency, adopting stress-tolerant crop varieties, and strengthening pest and disease monitoring systems. Ultimately, addressing these challenges requires a holistic approach that combines scientific innovation, traditional knowledge, and sustainable resource management to secure the agricultural future of the Himalayan region. The chapter gives an insight of current prospectives of Himalayan Agriculutural ecosystem and pathways executed to establish the stable production.

Keywords Climate change · Global warming · Resilient crop · Crop suitablity · Food security · United Nations Framework Convention on Climate Chnage (UNFCCC)

N. Sarmah (✉) · A. Devee · J. Gayon
Department of Entomology, Assam Agricultural University, Jorhat, India
e-mail: nomi.sarmah@aau.ac.in

M. Sarma
Department of Obstetrics and Gynaecology, College of Agriculture, Assam Agricultural University, Guwahati, India

T. Rahman
Faculty of Agricultural Sciences and Technology, Assam Downtown University, Guwahati, India

Y. D. I. Khan et al. (eds.), *Ecosystem-based Approaches for Resilience Building in Himalayan Landscapes*, Disaster Resilience and Green Growth,
https://doi.org/10.1007/978-981-95-2007-7_2

2.1 Introduction

Climate change, driven by human activities such as fossil fuel combustion, has caused significant and irreversible shifts in global weather patterns and temperatures. These changes threaten global food security (Homer-Dixon 1991; Phalkey et al. 2015). Over the past few decades, from 1990 to 2020, rising temperatures and sea levels have been attributed to greenhouse gas emissions (Shivanna 2022). International efforts have been undertaken to mitigate climate change in response to this crisis. The United Nations Framework Convention on Climate Change (UNFCCC) was established in 1990, followed by the Kyoto Protocol 1997, which committed industrialised nations to reducing greenhouse gas emissions. The Paris Agreement of 2015 further strengthened global commitment by aiming to limit global warming to below 2 °C above preindustrial levels (Afionis 2017; Grubb 2019). These initiatives underscore the urgent need to address climate change, particularly in the agricultural sector, which is highly vulnerable to environmental disruptions.

2.2 Impact of Climate Change on Agriculture

Climate change significantly affects agriculture and livestock farming. Rising temperatures cause heat stress in staple crops like wheat, maize, and rice, reducing yields and disrupting planting seasons (Gomez-Zavaglia 2020). The prolonged warm season alters the crop life cycle and intensifies pest infestations, further decreasing productivity. Water availability is another critical concern; prolonged droughts and erratic rainfall patterns interfere with irrigation schedules, negatively impacting crop growth and yields (Du et al. 2018). Frequent floods and extreme weather events also threaten agricultural output.

Beyond direct crop damage, climate change promotes the spread of pests and diseases, making plants more vulnerable to infestations (Skendžić et al. 2021). Additionally, soil degradation due to higher temperatures and drought conditions leads to erosion and nutrient loss, further harming productivity (Gomiero 2016). Changing climate conditions may render traditional crops unsuitable, forcing farmers to adopt new agricultural practices and switch to more resilient crops. However, this transition requires significant investment and adaptation strategies.

Livestock farming is equally affected, with high temperatures leading to reduced reproduction rates and milk production (Das et al. 2016). Adverse climatic conditions can also disrupt livestock feed supply, increasing costs and reducing production efficiency. These agricultural challenges contribute to food price volatility, threatening food security at local and global levels (Naylor and Falcon 2010). In extreme cases, persistent climate impacts may force farmers to abandon their lands, leading to rural displacement and migration.

Given agriculture's critical role in food security, adaptation strategies such as crop diversification, improved water management, and sustainable farming techniques are essential. Developing countries, particularly those with limited infrastructure and resources, are especially vulnerable and require targeted interventions to mitigate climate change's impact on agricultural livelihoods (Maharjan and Joshi 2013).

2.3 Agriculture in the Himalayan Region

The Himalayan region is home to millions who depend on agriculture for their livelihoods, including Indian, Nepali, Bhutia, and Tibetan-Chinese communities (Micevska and Rahut 2008). This region's agricultural system is shaped by its unique geography and traditional farming practices. However, climate change poses serious environmental, economic, and social challenges, threatening agricultural sustainability.

Despite these challenges, there is significant potential for revitalising Himalayan agriculture through strategic investments in climate-resilient crops, soil conservation, and advanced farming technologies (Kumar et al. 2021). Strengthening local farmers' adaptive capacity through training and infrastructure development can help sustain agriculture in this fragile and biodiverse region.

2.3.1 Characteristics of Himalayan Agriculture

Heterogeneous Agro-ecological Zones The Himalayas support diverse climatic conditions, from tropical lowlands to alpine highlands, allowing the cultivation of a wide variety of crops, including rice, maize, barley, millet, potatoes, and vegetables.

Terrace Farming Due to the region's steep slopes, terracing has been practised for centuries to prevent soil erosion and make land suitable for cultivation.

Livestock Farming Livestock, including cattle, goats, sheep, and yaks, play a crucial role in the region's agriculture, providing milk, meat, and wool. The yak, in particular, is vital in high-altitude areas.

Traditional Farming Practices Most Himalayan farmers engage in subsistence farming, focusing on food production for local consumption rather than large-scale commercial agriculture.

Crop Diversity Farmers cultivate a variety of crops, including cereals (wheat, maize, barley), pulses (lentils, beans), vegetables (cabbage, potatoes, cauliflower), fruits (apples, apricots, pears), and spices (ginger, turmeric).

Integrated Farming Systems Many farming systems integrate crops and livestock to maintain ecological balance and reduce risks.

However, Himalayan agriculture faces significant challenges due to climate change. Erratic weather patterns, unseasonal rainfall, droughts, and shifting monsoon cycles disrupt crop yields and livestock health (Srinivasarao et al. 2020). The retreat of Himalayan glaciers, a key source of irrigation water, threatens water availability, exacerbating agricultural difficulties. Soil erosion, often driven by deforestation and overgrazing, reduces land fertility, while biodiversity loss further undermines agricultural sustainability.

Additionally, rural out-migration is reshaping the demographic landscape, leaving behind ageing populations unable to sustain agricultural productivity. Limited access to modern farming technologies, irrigation systems, and financial resources further hampers agricultural development in remote areas. The reliance on traditional farming methods, though culturally significant, may not be viable in the face of climate change and environmental degradation.

2.4 Opportunities for Sustainable Agriculture

Despite the challenges in the region, there are numerous opportunities to enhance and integrate sustainable agricultural practices (Kumar et al. 2021). Farmers can greatly benefit from agroforestry and sustainable land use. Integrating trees with crops not only prevents soil erosion but also yields timber and enhances biodiversity. Additionally, it contributes to increased carbon sequestration, thereby mitigating climate change.

To further support sustainable agriculture, practices such as contour ploughing, terracing, and organic farming can help conserve soil, maintain its quality, and prevent erosion. Climate-smart agriculture (CSA) plays a crucial role in this transformation (Gavali et al. 2023). By adopting climate-smart principles such as planting drought-resistant crop varieties, employing rainwater harvesting techniques, using drip irrigation, and switching to more resilient crops (Subedi et al. 2019), farmers can improve their resilience to climate variability.

Agroecological practices, including crop rotation, organic manure application, and biological pest management, enhance soil health and make farming systems more resilient to climate stress. In addition, emerging trends such as the cultivation of medicinal herbs, organic tea, and exotic fruits offer profitable opportunities for organic farming. Farmers catering to niche markets can increase their income while promoting environmentally sustainable agriculture.

Another promising avenue is agri-tourism, which has gained popularity in the Himalayan region. The picturesque landscapes provide a unique opportunity for visitors to experience farming life while directly supporting local economies. Improved transportation networks and communication systems can further bolster farmers' access to larger markets, reducing post-harvest losses and ensuring better price realisation. Additionally, agricultural extension services, modern farming techniques, and climate change-related information can equip farmers with the knowledge needed to adapt to emerging challenges.

2.5 Climate-Resilient Agriculture (CRA)

Building upon these opportunities, climate-resilient agriculture (CRA) is essential for ensuring food security and sustainability amid changing climatic conditions. CRA encompasses a range of strategies that enhance the adaptability of farming systems, minimise negative climate impacts, and strengthen resilience to extreme weather events (Rai et al. 2018).

A fundamental aspect of CRA is the diversification of crops and livestock to reduce dependence on a single species, which may be vulnerable to climate risks (Chand 1996; Kumar et al. 2019). Practices such as crop rotation, intercropping, and agroforestry contribute to improved soil health, increased biodiversity, and reduced risks associated with climate extremes.

Water management is another key pillar of CRA. Efficient irrigation systems, rainwater harvesting, and drought-resistant crop varieties enable farmers to cope with water scarcity while maintaining soil fertility and minimising water wastage. Integrated pest management (IPM) is also crucial, as it combines biological, cultural, and chemical methods to control pests and diseases without excessive reliance on chemical inputs, which may become ineffective with changing weather patterns (Goswami et al. 2023).

Soil health is central to climate resilience. Techniques such as no-till farming, cover cropping, and organic fertilisation improve soil structure, enhance water retention, and reduce erosion. These approaches are particularly beneficial in regions prone to heavy rainfall or drought. Furthermore, precision agriculture tools and climate-resistant technologies can enhance productivity while minimising environmental degradation.

Ultimately, transitioning to climate-resilient agriculture involves adopting sustainable farming practices that address climate change challenges and contribute to food security and ecological well-being.

2.6 Ecosystem-Based Adaptation (EbA)

Complementing CRA, ecosystem-based adaptation (EbA) leverages natural ecosystems and biodiversity to build resilience against climate change (Nalau and Becken 2018). In agriculture, EbA involves working with nature to enhance resilience, reduce vulnerabilities, and restore disrupted ecological functions.

One of the most effective EbA strategies is agroforestry, which integrates trees, shrubs, and diverse vegetation into farming systems. This practice enhances biodiversity, improves soil and water retention, mitigates soil erosion, and provides habitat for beneficial wildlife. Additionally, agroforestry plays a significant role in carbon sequestration, thereby contributing to climate change mitigation. Wetlands also serve as critical components of EbA. Marshes and mangroves act as natural buffers against flooding and storm surges while offering ecosystem services such as water filtration, wildlife habitat provision, and carbon storage. Integrating wetland conservation into agricultural landscapes can protect farmlands from extreme weather events like floods and droughts. Wetland plants help stabilise soil, improve groundwater recharge, and promote overall agricultural resilience.

Sustainable soil management is another essential aspect of EbA. Healthy soils function as carbon sinks, retain water efficiently, and resist erosion. Restoring ecosystems within agricultural landscapes can lead to increased yields while reducing water usage and enhancing long-term productivity. Integrated pest management (IPM) through ecological processes is gaining traction in climate adaptation strategies. Instead of relying heavily on chemical pesticides, EbA encourages natural pest control by promoting plant diversity and maintaining habitats for natural predators such as beneficial insects and birds. This approach reduces pesticide dependency, minimises pollution, and enhances farm resilience against climate-induced pest outbreaks.

Watershed management is another key element of EbA. Maintaining riparian zones and restoring catchment forests help regulate water flow, reduce flooding, and ensure a steady year-round water supply for agriculture. Sustainable land-use practices that integrate agricultural land within broader ecological systems contribute to improved water quality and availability. Sustainable livestock grazing also plays a critical role in EbA. By implementing rotational grazing and protecting natural grasslands, farmers can prevent overgrazing, restore pasturelands, and maintain soil health. This approach not only combats desertification but also enhances the resilience of pasture ecosystems to climate-induced stresses.

By recognising the interdependence between agriculture and ecosystems, EbA promotes a holistic approach that benefits both farmers and the environment. This strategy not only supports climate adaptation but also preserves ecological integrity for future generations, ensuring a more sustainable and resilient agricultural system.

2.7 Government of India's Initiatives to Promote Agriculture and Allied Sectors

The Government of India has recognised climate-resilient agriculture and has undertaken many activities, especially to mitigate the effects of climate change on the agriculture sector (Goswami et al. 2023). Such measures essentially build adaptive capacity among farmers and promote sustainable farming with water management and agriculture innovations in the fold. They include the following major initiatives (Prasad and Kochher 2009; Khan et al. 2022; Thawani 2023).

National Mission for Sustainable Agriculture (NMSA)
Under the National Action Plan on Climate Change, NMSA has been launched to promote sustainable agriculture practices by ensuring soil health, water conservation, and reducing agriculture's carbon footprint. The primary goals are promoting organic farming, agroforestry, and conservation tillage; water-use efficiency and rainwater harvesting; soil health improvement with integrated nutrient management; and climate-resilient crop and varieties.

Pradhan Mantri Krishi Sinchayee Yojana (PMKSY)
PMKSY aims to improve the irrigation potential of any agricultural land. Improvement of irrigation potential will ensure that agricultural productivity is maintained regardless of rainfall patterns. PMKSY focuses on the Accelerated Irrigation Benefit Programme (AIBP) by expanding irrigation coverage through the modernisation of irrigation systems, ensuring that every field is individually self-reliant with adequate water supply so that there would be no undue dependence on erratic rainfall, rainwater harvesting, and water-use efficiency through promotion of micro-irrigation systems, drip, and sprinkler irrigation to enhance water-use efficiency.

National Adaptation Fund for Climate Change (NAFCC)
The NAFCC is a fund that provides funding for projects that build climate resilience in the vulnerable sector, particularly the agricultural sector. This fund also provides support to adaptation projects that enhance farmers' ability to deal with the impacts of climate changes, such as promoting climate-resilient agricultural practices in drought-prone regions, developing early warning systems for weather-related disasters, and encouraging drought-resistant crop varieties.

Soil Health Management and Fertiliser Use
Launched by the government, the Soil Health Management (SHM) program aims at rehabilitating and enhancing soil health and improving agricultural production. The distribution of soil health cards to farmers promotes balanced fertiliser and organic input use. Key components include integrated Nutrient Management (INM) for reduction in the application of chemical fertilisers with the promotion of organic and biofertilisers and promotion of vermicomposting and green manure to improve marred soil fertility sustainably.

Climate-Resilient Agriculture for Vulnerable Regions
This would continue the focus of climate-resilient practice on semi-arid and drought-prone regions most affected by climate change. This initiative encourages crop diversification to reduce risks and ensure food needs; it also trains farmers in climate-smart agriculture through conservation tillage, agroforestry, and water-efficient farming. The government established the Kisan Vikas Kendras, which provided farmers with information, tools, and training on climate-resilient practices. These centres aim to disseminate knowledge about new agricultural technologies that could withstand extreme situations, promote climate-resilient seed varieties, and inform farmers about disaster management techniques and weather forecasts.

Pradhan Mantri Fasal Bima Yojana (PMFBY)
PMFBY is a crop insurance scheme that covers farmers against yield loss due to extreme weather events like floods, droughts, and unseasonal rains. Key features include an affordable premium made available to small and marginal farmers, prompt and timely compensation of financial losses in case of any crop loss, assessment of crop damage through satellite images, and quick payments to farmers.

Climate-Smart Village (CSV) Programme
The CSV Programme is an innovative project on climate-resilient agriculture for selected villages. Under this programme, pilot projects are undertaken to test and demonstrate the viability of different climate-smart agricultural practices: drought-resistant crops, integrated pest management, and water conservation; to create awareness among farmers about climate change as well as adaptation measures; and to set up weather-based agro-advisories on time for farmers in sowing and harvesting.

National Action Plan on Climate Change (NAPCC)
NAPCC, which came in 2008, provides a comprehensive plan for climate change in India. The main component is the National Mission for Sustainable Agriculture. Other related missions include the National Solar Mission, which promotes the use of solar energy in farming operations for a smaller carbon footprint. The National Water Mission promotes water conservation and efficient irrigation practices.

Farmer's Welfare and Financial Assistance
The government has initiated various welfare programmes to help farmers adapt to climate-resilient agriculture practices, such as PM-KISAN (Pradhan Mantri Kisan Samman Nidhi), direct cash transfer to farmers for financial stability and additional support during periods of economic stress due to natural calamities or during climate-induced stresses, and Kisan Credit Cards (KCC), easy access to credit for farmers, owing to which they can invest in climate-resilient agricultural technologies and agricultural practices.

Promotion of Organic Farming
The government is promoting organic farming to reduce reliance on chemical fertilisers and pesticides to contribute to a more sustainable and climate-resilient farming system. Some initiatives include the National Programme for Organic Production (NPOP), which provides organic certificates to farmers supporting organic clusters, and farmer-producer organisations (FPOs) that help farmers adopt organic practices.

Use of Technology and Data
The government employs modern technological applications to enhance weather forecasts and provide timely warnings of extreme meteorological events such as floods, droughts, and cyclones. The India Meteorological Department, in partnership with the Agri-Tech Management Agency, furnishes timely weather reports and advisories.

Application in Mobile Phones Some mobile applications, such as M-Kisan, AgriCloud, and Digital Green, give real-time information on the weather, which practices can be undertaken, pest management, and market prices.

All these schemes launched by the Government of India to encourage climate-resilient agriculture will alleviate climatic stress on vulnerable farmers, raise agricultural productivity, and ensure food security entailed by climate variability (Khan et al. 2022). These are working towards water management, soil health improvement, promotion of sustainable agricultural practices, and financial assistance throughout their implementation; their success will depend on continued support, practical implementation, and a strong focus on local-level adaptation strategies for the long haul.

The Sustainable Development Goals are a UN set of goals for the global community to eliminate hunger and secure food through sustainable agriculture (Viana et al. 2022). Apart from ensuring food security, SDG 2: Zero Hunger also aims to build agriculture resilience in the face of climate change. By promoting sustainable farming practices, reducing food waste, providing better access to productive resources for smallholder farmers, and increasing agricultural productivity, countries will significantly advance towards achieving the goal of zero hunger by the year 2030.

The United Nations' Sustainable Development Goals (SDGs) set the global framework to end hunger and achieve food security through sustainable agriculture. SDG 2: Zero Hunger ensures everyone has access to sufficient, nutritious food while promoting sustainable agricultural practices that are resilient to climate change. By prioritising sustainable farming methods, reducing food waste, improving access to resources for smallholder farmers, and enhancing agricultural productivity, countries can make significant progress toward achieving Zero Hunger by 2030.

2.8 Conclusion

Sustainable and climate-resilient agriculture is essential in addressing environmental challenges while ensuring food security and economic stability. Farmers can enhance soil health, increase biodiversity, and mitigate climate impacts by integrating agroforestry, climate-smart practices, and ecosystem-based adaptation. Efficient water management, integrated pest control, and precision agriculture technologies further strengthen resilience. Additionally, emerging opportunities such as agritourism and niche organic markets benefit farming communities economically. Government initiatives are crucial in supporting these efforts through policy interventions, infrastructure development, and knowledge dissemination. A collaborative approach between farmers, policymakers, and researchers is necessary to create a productive and sustainable agricultural system. Adopting innovative and ecologically sound farming practices will be critical in ensuring long-term resilience, fostering environmental conservation, and enhancing rural livelihoods.

References

Afionis S (2017) The European Union in international climate change negotiations. Routledge

Chand R (1996) Diversification through high-value crops in Western Himalayan Region: evidence from Himachal Pradesh. Indian J Agric Econ 51:652–663

Das R, Sailo L, Verma N et al (2016) Impact of heat stress on health and performance of dairy animals: a review. Vet World 9:260

Du L, Mikle N, Zou Z et al (2018) Global patterns of extreme drought-induced loss in land primary production: identifying ecological extremes from rain-use efficiency. Sci Total Environ 628:611–620

Homer-Dixon TF (1991) On the threshold: environmental changes as causes of acute conflict. Int Secur 16(2):76–116

Gavali RS, Babu VS, Kakumanu KR, Mukate SV, Imran Khan YD, Patil B et al (2023) Scalable adaptation model for sustainable agriculture livelihoods under changing climate: a case study from Bihar and Madhya Pradesh. In: The Palgrave handbook of socio-ecological resilience in the face of climate change: contexts from a developing country. Springer, Singapore, pp 499–526. https://doi.org/10.1007/978-981-99-2206-2_29

Gomez-Zavaglia A, Mejuto JC, Simal-Gandara J (2020) Mitigation of emerging implications of climate change on food production systems. Food Res Int 134:109256

Gomiero T (2016) Soil degradation, land scarcity and food security: reviewing a complex challenge. Sustainability 8:281

Goswami M, Gupta AK, Kishan R, Baidya S, Khan YDI, Prakash S, Premkumar A, Nautiyal S (2023) An evaluation of climate resilient agricultural practices in India: a narrative synthesis of literature. Environ Sustain 6(1):7–23. https://doi.org/10.1007/s42398-022-00255-4

Grubb M (2019) United Nations framework convention on climate change. In: The Earth summit agreements: a guide and assessment, pp 61–74

Khan YDI, Goswami M, Nautiyal S, Gupta AK, Premkumar A, Baidya S, Prakash S (2022) Promoting climate smart agriculture through policies and schemes in India: temporal and strategic overview. Int J Ecol Environ Sci 48(5):547–569

Kumar MA, Kumar RA, Rangnamei KL et al (2019) Crop diversification for enhancing the productivity for food and nutritional security under the Eastern Himalayas. Indian J Agric Sci 89:1157–1161

Kumar P, Sharma PK, Kumar P et al (2021) Agricultural sustainability in Indian Himalayan region: constraints and potentials. Indian J Ecol 48:649–661

Maharjan KL, Joshi NP (2013) Climate change, agriculture and rural livelihoods in developing countries. Japan

Micevska M, Rahut DB (2008) Rural nonfarm employment and incomes in the Himalayas. Econ Dev Cult Change 57:163–193

Nalau J, Becken S (2018) Ecosystem-based adaptation to climate change: review of concepts. Griffith Institute for Tourism, Research Griffith University, Queensland

Naylor RL, Falcon WP (2010) Food security in an era of economic volatility. Popul Dev Rev 36:693–723

Phalkey RK, Aranda-Jan C, Marx S et al (2015) Systematic review of current efforts to quantify the impacts of climate change on under nutrition. Proc Natl Acad Sci 112(33):E4522–E4529

Prasad HA, Kochher JS (2009) Climate change and India – some major issues and policy implications. Department of Economic Affairs, Ministry of Finance, Government of India

Rai RK, Bhatta LD, Acharya U, Bhatta AP (2018) Assessing climate-resilient agriculture for smallholders. Environ Dev 27:26–33

Shivanna KR (2022) Climate change and its impact on biodiversity and human welfare. Proc Indian Natl Sci Acad 88:160–171

Skendžić S, Zovko M, Živković IP, Lešić V, Lemić D (2021) The impact of climate change on agricultural insect pests. Insects 12:440

Srinivasarao CH, Rao KV, Gopinath KA et al (2020) Agriculture contingency plans for managing weather aberrations and extreme climatic events: development, implementation and impacts in India. Adv Agron 159:35–91

Subedi R, Bhatta LD, Udas E et al (2019) Climate-smart practices for improvement of crop yields in mid-hills of Nepal. Cogent Food Agric 5:1631026

Thawani MM (2023) An evaluation of efforts made by the Government of India to combat the adverse effects of climate change. Chelonian Res Found 18:99–112

Viana CM, Freire D, Abrantes P et al (2022) Agricultural land systems importance for supporting food security and sustainable development goals: a systematic review. Sci Total Environ 806:150718

Chapter 3
Ecology and Economy of Himalayan Pastoralism: Evidence from the Changpa Community from Changthang Region, Ladakh

Dada. R. Dadas, Faraz Rupani, and Nikhil Nikam

Abstract Pastoralism is a complex socio-ecological system practised by the Changpa community in fragile cold desert ecosystems. Pastoral livelihood practice maintains a balance between ecology and economy; however, there is limited rational understanding of the pastoral livelihood practice. The study was conducted to explore the economy and ecology of the Changpa community by understanding pashmina-based livelihood practices that depend on pasture and water resources. The study used an ethnographic approach. The Changpa community has been adapting to harsh climatic conditions, geopolitical challenges and scarce natural resources to sustain its livelihood; however, findings show that many profound changes have taken place in and around Korzok, including infrastructural development in the village and the border area, increasing pashmina population, expansion of tourism, restrictions on movement due to protected areas and security issues at the border affecting access to traditional pasture and water resources. Climatic changes have also badly resulted in increasing temperatures, drying up of pasture and reduced water, thus creating many livelihood vulnerabilities. More markets for pashmina are intensifying pashmina goats, changing livestock composition and creating pressure on limited pastureland. These changes also disrupt the balance between ecology and economy and threaten traditional livelihoods. In the case of the economy, pastoral livelihood is diversifying in tourism, labour and small economic activities. Evidence from ecology and economy suggests a need for integrated policies focusing on conserving ecology and protecting the livelihood of the Changpa community. Therefore,

D. R. Dadas (✉)
WOTR Centre for Resilience Studies (W-CReS), Watershed Organisation Trust, Pune, Maharashtra, India

School of Social Work, Tata Institute of Social Sciences (TISS), Mumbai, Maharashtra, India

F. Rupani · N. Nikam
WOTR Centre for Resilience Studies (W-CReS), Watershed Organisation Trust, Pune, Maharashtra, India

Y. D. I. Khan et al. (eds.), *Ecosystem-based Approaches for Resilience Building in Himalayan Landscapes*, Disaster Resilience and Green Growth,
https://doi.org/10.1007/978-981-95-2007-7_3

the paper attempts to focus on understanding the economy and ecology of the Changpa community for the sustainability of pastoralism.

Keywords Pastoralism · Ecology · Economy · Pasture · Water · Intensification of Pashmina · Pastoralism and Ladakh

3.1 Introduction

Nomadic communities are defined as "social groups with no permanent habitation or residence" and have insisted on an ecological perspective in the study of nomadism (Spooner 1971). These communities move from one place to another rather than being settled permanently at one single location. At the same time, its cultures are predominantly categorised into three economic specialisations, i.e. hunter-gatherers, pastoral nomads and peripatetic nomads. Nomadic hunting, gathering and pastoralism are followed seasonally and are the oldest human subsistence methods (Dadas et al. 2025). Peripatetic nomads offer trade or craft skills to those who travel among them, which are most common in industrialised nations. Here, the discourse focuses on pastoralism, which is largely neglected in research and academia.

Pastoralism is an old socio-ecological system. The origins of pastoralism can be traced back to 6000 BP in the Andes of South America and even as early as 9000 BP in Northeast Africa (MacDonald and MacDonald 2000), the exact year of origin being a topic of debate among scholars till now. Some documented literature mentions Amorites were the people who first appeared as pastoralists and started herding cattle, sheep, goats and donkeys in the Near East during the first half of the second millennium BC (Cribb 1991). On the other hand, some scholars think that nomadic pastoralism originated from multiple centres worldwide. The definition of pastoralism differs significantly on the lines of purposes and focuses; the two standard definitions used for pastoralism have been derived from the production, which focuses on animal husbandry or the livelihood perspective, focusing on tending herds of large animals (Blench 2001).

Pastoralists predominantly live in arid and semi-arid rangelands and continue to have global importance (Scanes and Toukhsati 2018), with 25% of the earth's land area engaged in extensive pastoralism (Dong 2016) and a recent estimation suggests that around 500 million people across 100 different countries in the world depend on pastoralism for their livelihoods (Manzano et al. 2021). A significant population of 30 to 40 million pastoralists practise nomadic pastoralists (Blench 2001), which is a common practice in regions with limited arable land, typically in the drylands and highlands of the world. As per the global pastoralism map developed by the World Initiative for Sustainable Pastoralism, the pastoral communities are currently found in sub-Saharan Africa, southern Africa, northern Europe (including the Russian Arctic), Central Asia (including the Himalayas), central South America, western North America and Australia. South Asian countries are home to a vibrant diversity of pastoral communities (Rao and Casimir 2003; Agrawal and Saberwal 2007; Dong et al. 2016; Sharma 2013).

Pastoralism is a form of natural resource management that directly links natural resources and their users (Pratt et al. 1997), deriving most of their income or sustenance from keeping domestic livestock. Herds and households move from one place to another in search of grazing resources in patterns that usually avoid depleting pastures beyond their ability to recover and provide various valued services and products such as biodiversity, tourism and raw materials. These natural resource management systems are practised across remotest, resource-poor and mountainous regions (Nori and Davies 2007) and are called by names such as grazers or herders (Dong 2016; Johnsen et al. 2019). As pastoralists operate on vast ecological terrains covering a quarter of the planet's land surface, they are known as the "stewards of rangelands" (Bensada 2017). Nomadic pastoralism is believed to have originated in the Near East during the Neolithic period, as communities transitioned from mixed farming to specialised animal husbandry practices.

Moreover, mobile pastoralism practised mainly in the drylands and highlands of the world, from South America to the Central Asian steppes and from Europe to sub-Saharan Africa, has shown considerable economic value and untapped potential in rangeland environments and is central to the livelihoods and well-being of millions of the world's poor. The policies emerging due to a lack of understanding lead to the devaluation of mobile pastoralism, which often has a significant cost to national economies and the natural environment. Policymakers are trying to encourage investment in pastoralism, mainly in Europe and Asia (Davies and Hatfield 2007), considering its importance for sustainable environmental management in the rangelands (Hatfield and Davies 2006).

Many cultures have traditionally been nomadic, but due to industrialisation, traditional nomadic culture seems to be becoming increasingly rare, with many countries converting open grasslands into cropland (Brogaard and Xueyong 2002). One-third of Mongolia's population lives as nomads, depending on livestock husbandry for livelihood (Dagvadorj et al. 2009).

Despite a growing body of evidence that highlights the economic and environmental importance of pastoralism, the discourse and knowledge generation has been weak (Hatfield and Davies 2006), with only a few governments ready to tolerate mobile livestock production, while many are pursuing explicit or inadvertent policies of settlement (Davies and Hatfield 2007). The policy of sedentarisation, particularly in the drylands, has shown time and again to result in increased environmental degradation, reduced economic potential and eroded social and cultural systems (quoted in Davies and Hatfield 2007).

3.1.1 Pastoral Nomads in the Indian Context

India has a rich cultural diversity coupled with unique traditions, languages and food varying among social groups (Zehol 2008; Ford and Kelly 2005). Pastoralists are defined in the Indian context as "members of a caste or ethnic groups with a strong traditional association with livestock-keeping, where a substantial proportion of the group derive over 50% of household consumption from livestock products or their sale, and where over 90% of animal consumption is from natural pasture or browse, and where households are responsible for the full cycle of livestock breeding (Sharma et al. 2003)". According to Blench (2001), Indian pastoralism is the worst documented by far, with confused descriptions of pastoral systems and confused terminology for pastoral ethnic groups (Sharma 2013).

India has the most significant number of poor livestock keepers (Birthal 2022), and according to the Annual Report 2023–2024 by the Ministry of Fisheries, Animal Husbandry and Dairying (2023), the livestock sector contributes around 5.50% to the country's total gross value added (GVA). As per an estimate, 13 million people, or 1 per cent of India's population, are engaged in pastoralism, and more than 50 per cent of the milk and 70 per cent of the meat produced come from pastoral production.

India presents a unique array of nomadism connected with its ancient civilisation (Prasad 1994), which plays a vital role in Indian society. Pastoral nomads are communities that move from one place to another rather than settling permanently in one location. This has been practised in India since at least 4000 BC (Allchin 1963) and continues to be practised in almost all parts of the country. For thousands of years, the nomads in India have wandered in the forests and hills in search of livelihood. It is estimated that the Meat Atlas 2021 consists of more than 200 tribes (Sharma et al. 2003), and historical evidence shows that political, economic, cultural and geographical factors, such as topography, climate, vegetation and hydrology, have combined to give rise to nomadism as a way of life (Dadas and Akhup 2025).

The spread of pastoral nomads is found across India from the drylands of western India, where the Rabari and Bharwad communities from the state of Gujarat are engaged in rearing sheep, goats, cattle and small livestock and the Raika/Rabari and Gujjar communities from Rajasthan raise camels, sheep and goats. In the Deccan Plateau state of Andhra Pradesh, communities like the Gollas and Kurumas engage in cattle and sheep rearing. In Karnataka, the Kuruba and Dhangar communities usually rear sheep. In high-altitude beautiful regions of Himachal Pradesh and the Western Himalayas are the Gaddi, Gujjar and Bakarwal communities who rear sheep, goats and buffalo, while the Bhutias of Sikkim in the Eastern Himalayas and the Changpa community of Ladakh in the western side of the Himalayas raise yaks (Sharma 2013). With such a large population engaged in and a vast history of pastoralism, the Government of India (GOI) still does not officially categorise "pastoralism", resulting in a lack of data being collected (Meat Atlas 2021). This chapter focuses on the ecology and economy of the Changpa community in the Indian Himalayas.

3.1.2 Pastoralism in the Indian Himalayas

The Indian Himalayan region spans 2500 km, covering 13 states and union territories. The majority of the Himalayan ranges are found in Ladakh, Jammu and Kashmir whose ethnographic and archaeological findings state that pastoralism was developed as early as 4000 BP (Spate 2019; Yatoo 2012) and the Himalayan ranges are also found in Uttarakhand and Himachal Pradesh (Chatterjee and Bishop 2023). The alpine meadows found in Uttarakhand, also known as "Bugyals", have been used as a significant type of rangeland by livestock for centuries during the summer season (Farooquee 1994). Livestock rearing has been an integral part of livelihood for communities in the Himalayan region, and its reliance is known to increase with increased elevation (Sundriyal 1995).

After the domestication in the Neolithic period at 5000–3000 BP (Betts et al. 2019; Pokharia et al. 2018; Sharma 2013), pastoralism practice has since been followed involving cyclical movements from lowlands to highlands to take advantage of seasonally available pastures at different elevations (Bhasin 1988), leading to a type of management practice. These management practices rely on social and cultural factors such as communal herding based on inter- and intra-household cooperation, making it labour-intensive (Kreutzmann 2015; Schmidt 2000). These pastoral techniques may have developed in response to herd management and indirectly paved the way for establishing a network of cultures (Spate 2019) via the "Inner Asian Mountain Corridor" (IAMC) (Frachetti 2012).

The pastoral communities in the western Indian Himalayas include several tribes engaged in different varieties of livestock rearing. In the Ladakh region, the Changpa and Brokpa tribes rear sheep, cashmere goats, yaks and horses (Chaudhari 2000; Singh et al. 2013), while in Himachal Pradesh, Jammu and Kashmir and Uttarakhand, the tribal communities of the Gaddis rear large flocks of sheep and goats; the nomadic goat herder tribes of the Bakarwals and Van Gujjars are known for rearing buffalo (Rao and Casimir 2003) prevalently seen in these states during seasonal movements (Bhasin 2011a, b; Kreutzmann 2012; Namgay et al. 2013; Rao and Casimir 2003; Yamaguchi 2011).

The agro-pastoral communities in the Trans-Himalayan region cover Lahaul and Spiti in Himachal Pradesh, where sheep, goats, horses, cattle and yaks are reared (Mishra 2001; Singh et al. 2020). Moreover, the Bhutia tribe of North of Sikkim and the Brokpa tribe of Arunachal Pradesh are known to raise herds of yaks (Verma and Khadka 2016), while the Gurung and Chettri tribes are sheep and cattle herders, respectively, who are found in the state of Sikkim (Bhasin 2011a). Even with such a wide presence of pastoralists across the Indian Himalayas, there has been poor documentation of the exact number of populations engaged in it (Derville and Bonnemaire 2010).

3.2 Materials and Methods

3.2.1 Study Area

Ladakh, situated in the Himalayas on India's far northern frontiers, is one of the coldest regions in the world, with winter temperatures dropping below −40 degrees Celsius (Dame and Nüsser 2011). The region experiences low levels of oxygen, high-intensity winds, higher solar radiation (Namgail et al. 2004) and excessive dryness due to limited rainfall, ranging between 20 mm and 145 mm (Soheb et al. 2022) with inhibition of fewer than 5 people per square kilometre (Dame and Nüsser 2011). The region of Ladakh is divided into two districts, namely, Leh and Kargil, and is governed by the Autonomous Hill Development Councils, which are responsible for working with village panchayats to make decisions on education, economic development, land use, healthcare, local governance and taxation. This present study was conducted in the Korzok village which is part of the Changthang region of the Ladakh (Fig. 3.1).

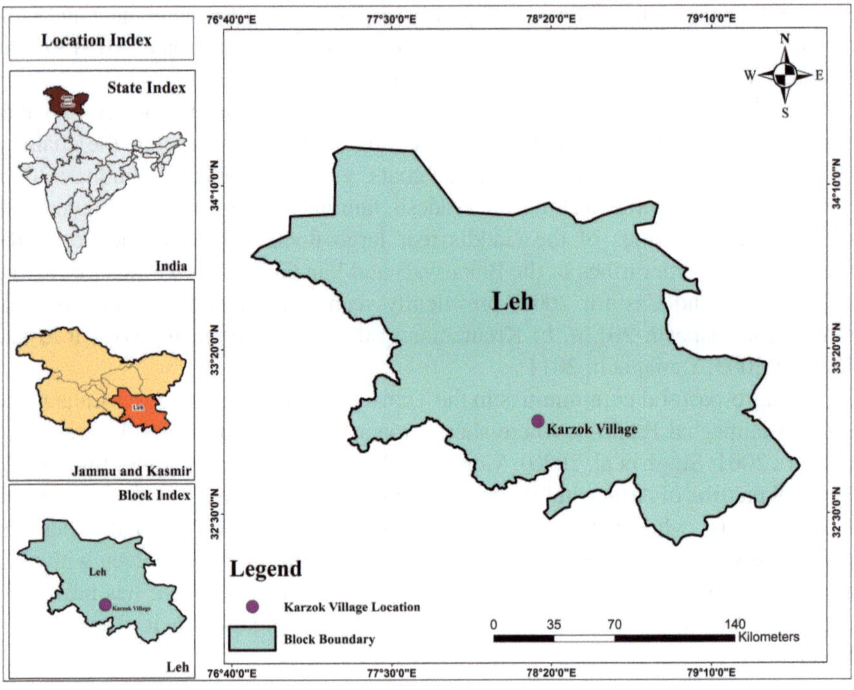

Fig. 3.1 Geographical location of Korzok

3.2.2 About the Changpa Community

The Changpas, meaning "northerners" in Tibetan (Rizvi 1999), are found in the Changthang plateau of Ladakh, located east of Leh on the India-China border, a western extension of the Tibetan plateau (Goldstein and Beall 1990; Mann 2002) and is home to wild animals and rare birds. The Changpas migrated initially from Tibet in the eighth century AD (Jina 1995) and are divided into two subgroups: the Fangpa (sedentary) and the Phalpa (nomads) classified as a Scheduled Tribe by the Indian government.

In 1989, half a million nomads lived in Changthang (Goldstein and Beall 1991) who initially practised settled agriculture and livestock rearing but gradually gave up agriculture and completely shifted to transhuman lives. The transhumant life is characterised by the seasonal migration from one migratory side to another migratory side with their livestock. Winter pastures are focused in lower-elevation valleys, while in summer, the pastoralists move in the highland or alpine meadows (Chundawat and Qureshi 1999). Unlike many other nomadic pastoralists, the Changpas do not move from one climatic region to another; this allows them to move shorter distances in many cases, in the range of about 10–40 miles (Goldstein and Beall 1990).

3.2.3 Methodology

This chapter provides a critical discussion that attempts to understand ecology and economy and the associated challenges and opportunities. To get the context, a comprehensive review was conducted to understand the global, national and Himalayan overview of pastoralism. The process begins with intensive literature reviews based on published articles, policy papers and case studies on pastoralism, key issues and challenges associated with the ecology and economy of the Changpas and state responses to the livelihood challenges of the Changpa community.

3.3 Ecological Association of the Changpa Community

The Himalayan glaciers have almost shrunk 11% of the total surface area from 1990 to 2015 in 25 years (Ji et al. 2020). The Changpa tribes have intimate ethnobotanical knowledge of the landscape's biodiversity and ecology, depending upon pasturelands for livestock grazing. The changing climate is leading to shifts in the weather patterns, increasing temperatures, intense storms and precipitation patterns affecting the transhumance cycles (Bhattacharya et al. 2004; Herrero et al. 2016; Dimri et al. 2021).

3.3.1 Ecological Challenges

3.3.1.1 Shortage of Grazing Pasture

Though the Changthang area is vast, there is a noticeable increase in the pressures on the pastoralist lands stemming out of intense competition for its usage and access (Robinson and Flintan 2022) not only limiting the land for grazing pastures but also increasing it's degradation at an alarming rate..

3.3.1.2 Locust Attacks

Locust attacks are leading to gradual desertification of the area, stressing the livelihood of the local population. The locust swarm attack in 2007 led to 124,530 sheep and goats and 10,391 large animals, primarily yaks, directly affected by pasture scarcity which ranged between 15 and 60 per cent of the total grazing land in different regions of Ladakh (Morup 2008).

3.3.1.3 Highland Wildlife Sanctuary and Perceived Grazing Control

The decision by the then Jammu and Kashmir's government to declare a large tract of land in the Changthang region as a high altitude wildlife sanctuary had severe repercussions for the Changpa nomads. For centuries, the Changpas have lived in harmony with Changthang's fragile ecosystem, and have been using the land to graze their livestock, including their famous pashmina goats. However, the creation of the sanctuary imposed strict restriction on areas that the nomads had traditionally used for grazing.

A major flaw in the establishment of the protected area was the absence of consultation with the local communities. At that time, sanctuaries could be notified without settling the land rights of the people living inside protected areas. As a result, the Changpas were deeply concerned that they would lose their traditional livelihood options and be evicted from their ancestral lands. They reiterated that

> *"Many tents are near the lake, and they fear that the government will not allow livestock grazing there."*

3.4 Economy of the Changpa Community

Swift's definition (1988) on the economics of pastoralism states, it is the system in which around 50% of the gross incomes of households come from pastoralism or its related activities, or more than 15% of households' food energy consumption involves the milk or dairy products they produce. The visible contributions of pastoralism to the economy are generally limited to measuring livestock sales and some by-products, such as dairy, hair and hides. While these are challenging to quantify, they do not capture the full value of pastoralism. Two important points need to be recognised: there are multiple and extensive sets of values associated with pastoralism of which some are tangible, but many are not; some can be measured, but many cannot, and those that can be measured are often underestimated. Assessing an economic activity's total contribution to the national economy is one tool to identify, quantify and aggregate all values associated with that activity (Hatfield and Davies 2006).

In the context of Ladakh, livestock production is central to the economy and livelihoods in Changthang being the primary income source for 90–95% of households (Chundawat and Qureshi 1999; Dame and Nüsser 2011; Akand et al. 2019). With 7–15 animals, on average, including goats, sheep, yaks, horses, donkeys and dzomos (crossbreeds), households depend significantly on pastoralism compared with national figs. (2–3 animals per household), reflecting the remote geography where animal husbandry focused on vast rangelands is the sole reliable production system (Ahmad et al. 2016).

3.4.1 *Trade Economy of the Changpas*

A Changpa's contact and interaction with non-Changpas was at the level of barter trade, specifically raw wool as their chief commercial product. The men travel long distances, seeking favourable prices for their wares, which consist of salt, dry fruits, cultured pearls and semi-precious stones. In return, they get tea, tobacco, grain, sugar and other essential goods. The region was well connected with an intricate trade link based on the barter system prevalent among the Changpas of the Changthang region and the Zanskar Valley and villages situated along the Indus Valley. The people of Ladakh were also engaged in cross-border trade with Tibet and Xinjiang (formerly Sin Kiang). Members of these communities also collect salt from the impure deposits on the northern shore of the Tso Kar Lake in Rupshu which is sold all over Ladakh and, in the past, was even sent to the Kashmir Valley.

3.4.2 Animal Husbandry

As far as the economy of Ladakh is concerned, traditionally, the economy of Ladakh has been subsistence agriculture in the lowland villages situated along the Indus and animal husbandry in the Changthang region. The livestock population is on the rise all over the Changthang region. It is estimated at c. 200,000 in 1999, which is twice that in the late 1970s (Khan 2003; Bhatnagar et al. 2006). The increase is most dramatic for goats, which the Changpas prefer due to the increasing demand for cashmere wool (Ahmed 2004). In any case, this trend is consistent with that of other pastoral nomads of Ladakh. The increase in livestock population in Changthang began with the arrival of the Tibetan refugees (TRs).

3.4.3 Pashmina Cashmere Wool

Fine cashmere is produced in high and cold deserts, and the major fibre-producing countries are China, Mongolia, Iran, Afghanistan, Pakistan, Nepal and India. China produces around 70% of total cashmere production. In contrast, India produces around 1%, mainly in the Changthang region, with the finest quality of pashmina (12–14 microns) globally and has received the geographical indication (GI) tag due to its unique significance and quality. The pashmina goat is the Changpas' primary economic source and is reared for meat or cashmere wool. A significant percentage of pashmina goats are reared in the Changthang region of Ladakh. The herders received a market value of Rs. 3500 per kg in 2020–2021, a significant increase from 2004 to 2005 due to the exponential demand for pashmina shawls globally. The herders sell the raw wool to a production facility owned by the Pashmina Growers' Cooperative Society and Looms of Ladakh—one of the key producers that make the pashmina shawls.

3.4.4 Role of Government in Promoting Pashmina Production

The government is making great efforts to improve the welfare of nomadic communities. This includes coaching, health insurance, livelihood initiatives at the community level and financial assistance for building houses.

The Government of India has been facilitating pashmina production by providing incentives to the Changpas for several decades in the form of supplementary cattle feed during severe winters, subsidised food provisions, etc. The herds of Hanle Valley pastoralists consist predominantly of sheep and goats. They also keep a variety of livestock: yak is used for meat, milk and transportation, while horses are used for riding, carrying loads and, most recently, ploughing on agricultural land.

Since the early 1970s, many families have also kept cows for milk. All the livestock except cows are grazed in riverine pastures near the human settlements.

The Government of India has implemented a reconstruction plan for the UT of Jammu and Kashmir and UT of Ladakh during the financial year 2017–2018 to 2021–2022. The Ministry of Textiles has approved the Pashmina Wool Development Scheme under the Integrated Wool Development Programme (IWDP) with a budget allocation of Rs. 29.25 Cr. for implementation during FY 2021–2022 to 2025–2026. Provision has been made to provide financial assistance in the form of a revolving fund for procurement/marketing of pashmina wool in UT of Ladakh and UT of J and K. A project to set up a dehairing plant along with ancillary machines at Leh, UT of Ladakh, with a project cost of Rs. 19.75 Cr. has been sanctioned and out of it Rs. 12.92 Cr. has been released. A provision has been made under the HRD component of IWDP regarding awareness and training to wool growers in the UT of Jammu and Kashmir and pashmina nomads of the UT of Ladakh region (PIB 2023).

3.4.5 *Economic Challenges*

Pastoralism is the main economic contributor for the local Changpas which now faces severe economic challenges.

3.4.5.1 Impact of Tourism

The biodiversity of the Changthang rangeland is increasingly threatened by tourism. The changpas claim that these activities are threating the fragile ecology, mainly due to the growing construction of tourist accommodation and rise in plastic litter, both of which are harming the environment. Yet a lot of nomads are thinking of shifting toward tourism business hoping for better livelihoods.

> *This year (2024) the tourist flow is less and most of tourist based businesses witnessed Non-Performing Assets and were forced into debt. They could not even pay the salaries of workers. Most of nomads from Korzok are attracting towards tourism and looking as an alternative economic activity. Tough tourism has potential, it is uncertain economic activity and somebody should convince them that tourism is uncertain so pastoralism should be carried along with tourism. ~ Government official (name withheld on request).*

3.4.5.2 Outbreak of Livestock Disease

With the advancement of time, the opening of this isolated area to the external world (development in tourism and supply of sheep and goat breeds of southern India by the military department) has created a threat of new diseases, an emerging problem for livestock. Recently, there have been several outbreaks of contagious diseases like Peste des petits ruminants (PPR), Contagious caprine pleuropneumonia (CCPP), Foot and Mouth Disease (FMD), which were uncommon in this place.

3.4.5.3 Fluctuating Prices of Pashmina Cashmere

The village elders have informed us that due to the smuggling of pashmina from China through Nepal, the prices have dwindled, which is very concerning for their income. But nowadays, a cooperative society has been founded for pashmina breeders, and they have great hope for good returns.

3.4.5.4 Leaving of Herding and Migration to Leh

Frequent loss of livestock due to diseases, lack of healthcare, border conflict and shrinking grazing land worsened by extreme climatic changes has forced hundreds to migrate from sparsely populated villages to mainly urban clusters in the region known for its sublime mountain landscape and the expensive wool (Hussain 2022). As per the 2011 census of India, the total livestock population is more than 2.6 million, outnumbering the human population (Hamadani et al. 2022). However, there is a trend of declining livestock pastoral population in 8.9% of pastoral communities (Directorate of Census Operations in Jammu and Kashmir n.d.); this decline is attributed to increased inclination of urban migration influenced by factors like loss of grazing lands, market integration and livelihood transitions (Pandit et al. 2024).

> *The cost of living and everything is become expensive and we don't find pastoralism as a sustainable livelihood option. Earlier there was no road, there was no car. Today everyone has a car, so number of draft animals like Yak, Donkey are reducing. Nowadays everyone is running after tourism and pashmina*

3.5 Recommendation

There are many challenges and difficulties in changing the lifestyle of pastoral nomadic communities. There is vast scope for bringing change in their socio-economic lifestyle. So, while formulating the policies for the overall development of the pastoral nomadic communities from an Indian perspective, the various areas needed to be covered. The answer to the massive problem of the pastoral nomads lies in an appropriate and integrated national land-use plan.

Regarding policy issues concerning the pastoral nomads, the Government of India has ultimately taken cognisance of the problems faced by the pastoralists regarding grazing and livestock management. With the review of evolving a national grazing and livestock management policy, the Government of India, the National Mission on Wasteland Development and the Ministry of Environment and Forests constituted an interdisciplinary advisory group on policy issues relating to grazing and livestock management. The group was advised to consider and suggest policy prescriptions on grazing, fodder production and livestock management with specific reference to the following:

- Management of common pasture grazing.
- Supplementary fodder production strategy from agricultural lands.
- Organisational linkages between the concerned central ministries/departments and the state department regarding research, extension, infrastructural and other support.
- Reduction of higher cattle-land ratio to increase the productivity of the pasture land through substitution with improved lots.
- Pasture development and fodder production corporations should be established.
- Fodder development bank.
- Cooperatives to be capacitated to support value addition services for pashmina cashmere and wool.
- The pastoral economy should be strengthened through cooperative dairy farming.
- These communities should be provided with alternative livelihood sources to cope with livelihood shocks.
- Community-driven climate vulnerability assessment.
- Participatory grassland and water assessment.

3.6 Conclusion

Human and climatic factors together have accelerated land degradation, leading to reduced pasture production (Daba and Mammo 2024). Intensifying land use changes (Bhasin et al. 2022) have severely affected pastoral communities, forcing them to alter traditional migration routes as pasturelands continue to shrink (Tundup et al. 2016) Driven by economic incentives, many Changpa households are increasingly shifting towards rearing Pashmina goats. While this shifts enhances short term income, it has dramatically increased the demand for year-round grazing (Bhatnagar et al. 2006). Changing weather patterns further exacerbate the problem, limiting available grazing areas for other livestock and heightening competition for resources. To ensure long term resilience, an approach which balance economic aspiration with ecological preservation needs to be followed. Integrating traditional knowledge with improved technological and modern conservation access can help the communities to improve conservation practices and create an ecological balance. Without such balance and inclusive strategies, the fragile ecosystem of Changthang risks facing severe ecological imbalance in the near future.

References

Agrawal A, Saberwal VK (2007) Whither South Asian pastoralism? An introduction. Nomadic Peoples 8(2):36–53. https://doi.org/10.3167/082279404780446113

Ahmed M (2004) The politics of pashmina: the Changpas of eastern Ladakh. Nomadic Peoples 8(2):89–106

Ahmad S, Gangoo S, Sultan SM, Dar M (2016) Ethnoveterinary practices and use of herbal medicine by pastoralists of Himalaya: a review. Ann Biol 32(2):260–268

Akand AH, Singh B, Chander M, Ahmed N (2019) Livestock produce and their disposal pattern in Ladakh region of Jammu and Kashmir. Indian J Anim Sci 89(9):1025–1029

Allchin FR (1963) Neolithic cattle-keepers of South India: a study of the Deccan ashmounds. University of Cambridge Oriental Publications, No. 9. xvi, 189 pp., Cambridge: University Press

Bensada A (2017) Putting sustainable pastoralism on the global agenda. UNEP, Nature Action. http://www.unep.org/news-and-stories/story/putting-sustainablepastoralism-global-agenda

Betts A, Yatoo M, Spate M, Fraser J, Kaloo Z, Rashid Y, Pokharia A, Zhang G (2019) The northern neolithic of the Western Himalayas: new research in the Kashmir valley. Archaeol Res Asia 18:17–39

Bhasin V (2011a) Pastoralists of Himalayas. J Hum Ecol 33(3):147–177

Bhasin V (2011b) Status of women in transhumant societies. Sociol J Br Sociol Assoc 2(1):1–22

Bhasin V (1988) Himalayan ecology, transhumance, and social organisation. India, Kamla-Raj Enterprises

Bhasin A, Dolker P, Raina P, Ghosal S (2022) Land use and land cover change detection using remote sensing in the Trans Himalayan Region of Ladakh, India. ECS Trans 107:2985–2997. https://doi.org/10.1149/10701.2985ecst

Bhattacharya T, Misra S, Sheikh F, Kumar P, Sharma A (2004) Changthangi goats: A rich source of pashmina production in Ladakh. Anim Genetic Resour 35:75–85

Bhatnagar YV, Wangchuk R, Prins HH, Van Wieren SE, Mishra C (2006) Perceived conflicts between pastoralism and conservation of the kiang equus kiang in the Ladakh Trans-Himalaya, India. Environ Manag 38(6):934–941

Birthal PS (2022) Livestock, agricultural growth and poverty alleviation. NABARD Research and Policy Series (7)

Blench R (2001) You can't go home again: Pastoralism in the new millennium. FAO report

Brogaard S, Xueyong Z (2002) Rural reforms and changes in land management and attitudes: a case study from Inner Mongolia, China. AMBIO J Hum Environ 31(3):219–225

Chatterjee SP, Bishop BC (2023) Himalayas. Encyclopedia Britannica. https://www.britannica.com/place/Himalayas

Chaudhari A (2000) Change in changthang: to stay or to leave? Econ Polit Wkly 35(1):52–58

Chundawat R, Qureshi Q (1999) Planning Wildlife Conservation in Leh and Kargil Districts of Ladakh, Jammu and Kashmir. Unpublished report submitted to the Wildlife Institute of India, Dehradun

Cribb R (1991) Nomads in archaeology. Cambridge University Press, Cambridge

Daba B, Mammo S (2024) Rangeland degradation and management practice in Ethiopia: a systematic review paper. Environ Sustain Indic 23:100413. https://doi.org/10.1016/j.indic.2024.100413

Dadas DR, Khan YDI, Bhandalkar D, Indurkar C, Gokhale K (2025) Ecological disruption and its impact on Indian pastoralism: ecosystem-based adaptation is way forward. In Handbook of nature-based solutions to mitigation and adaptation to climate change. Cham: Springer Nature Switzerland, pp 1–22

Dadas D, Akhup A (2025) Pastoral livelihood vulnerabilities in the himalayan region: negotiating for community resilient practice. In Social work research: methodologies in fields of practice. Singapore: Springer Nature Singapore, pp 81–99

Dagvadorj D, Natsagdorj L, Dorjpurev J, Namkhainyam B (2009) Mongolia: assessment report on climate change. Ministry of Nature, Environment and Tourism, Ulaanbaatar

Dame J, Nüsser M (2011) Food security in high mountain regions: agricultural production and the impact of food subsidies in Ladakh, Northern India. Food Secur 3:179–194
Davies J, Hatfield R (2007) The economics of mobile pastoralism: a global summary. Nomadic Peoples 11(1):91–116
Derville M, Bonnemaire J (2010) Marginalization of yak herders in Bhutan: can public policy generate new stabilities that support the transformation of their skills and organization? Innovation and Sustainable Development in Agriculture and Food. ISDA, Montpellier, pp 1–10
Dimri A, Allen S, Huggel C, Mal S, Ballesteros-Canovas J, Rohrer M, Shukla A, Tiwari P, Maharana P, Bolch T (2021) Climate change, cryosphere and impacts in the Indian Himalayan region. Curr Sci 120(5):774–790
Directorate of Census Operations in Jammu and Kashmir (n.d.) Leh district – population 2011–2024. Census Organization of India. https://www.census2011.co.in/census/district/621-leh.html
Dong S (2016) Overview: pastoralism in the world. https://doi.org/10.1007/978-3-319-30732-9_1
Dong S, Kassam KA, Tourrand JF, Boone RB (eds) (2016) Building resilience of human-natural systems of pastoralism in the developing world. Springer International Publishing. https://doi.org/10.1007/978-3-319-30732-9
Farooquee NA (1994) Transhumance in the central Himalaya: a study of its impact on environment. Ph.D. Thesis, HNB Garhwal University, Srinagar, India
Ford ME, Kelly PA (2005) Conceptualising and categorising race and ethnicity in health services research. Health Serv Res 40(5 Pt 2):1658–1675. https://doi.org/10.1111/j.1475-6773.2005.00449.x
Frachetti MD (2012) Multiregional emergence of mobile pastoralism and nonuniform institutional complexity across Eurasia. Curr Anthropol 53(1):2–38
Goldstein MC, Beall CM (1990) Nomads of Western Tibet: the survival of a way of life. University of California Press
Goldstein M, Beall C (1991) Nomads of Western Tibet. University of California Press, Berkeley, p 60. ISBN 0-520-07211-1
Hamadani AG, Rather NA, Shanaz MA, Ayaz S, Mansoor A, Nazir S (2022) Livestock and poultry breeds of Jammu and Kashmir and Ladakh. Indian J Anim Sci 92(4):409–416
Herrero MA, Bedelian J, Carabine C, Havlık E, Henderson P, Van De Steeg B, Thornton J (2016) Climate change and pastoralism: impacts, consequences, and adaptation. Rev Sci Tech 35(2):417–433
Hussain A (2022) Climate migration: nomads move to towns in Warming Ladakh. The Diplomat
Hatfield R, Davies J (2006) Global review of the economics of pastoralism. https://agritrop.cirad.fr/549237/1/document_549237.pdf. Accessed on 25 March 2025
Ji Q, Dong J, Li HR, Qin Y, Liu R, Yang T (2020) Glacier variations in the Himalaya from 1990 to 2015 based on remote sensing. The Cryosphere Discussions, 2020, 1–28
Johnsen K Niamir-Fuller I, Bensada M, Waters-Bayer A (eds) (2019) A case of benign neglect: knowledge gaps about sustainability in pastoralism and rangelands. United Nations Environment Programme and GRID-Arendal, www.grida.no.
Jina PS (1995) High pasturelands of Ladakh Himalaya. Indus Publishing
Khan AG (2003) NASSD background paper: rangelands and livestock. IUCN Pakistan, Northern Areas Progamme, Gilgit. viii+47 pp
Kreutzmann H (2012) Pastoral practices in transition: animal husbandry in high Asian contexts. Pastoral practices in High Asia: agency of 'development' effected by modernisation, resettlement and transformation, pp 1–29
Kreutzmann H (2015) Boundaries and space in Gilgit-Baltistan. Contemporary South Asia 23(3):276–291
MacDonald KC, MacDonald RH (2000) The origins and development of domesticated animals in arid West Africa. In: Blench RM, MacDonald KC (eds) The origin and development of African livestock. University College Press, London, pp 127–162
Mann RS (2002) Ladakhi then and now: cultural, ecological and political. Mittal Publication, New Delhi

Manzano P, Burgas D, Cadahía L, Eronen JT, Fernández-Llamazares Á, Bencherif S, Holand Ø, Seitsonen O, Byambaa B, Fortelius M, Fernández-Giménez ME, Galvin KA, Cabeza M, Stenseth NC (2021) Toward a holistic understanding of pastoralism. One Earth 4(5):651–665. https://doi.org/10.1016/j.oneear.2021.04.012

Mishra C (2001) High altitude survival: conflicts between pastoralism and wildlife in the Trans-Himalaya. PhD Thesis, Wageningen University, The Netherlands

Morup T (2008) Pastures gone. Down To Earth. https://www.downtoearth.org.in/environment/pastures-gone-5131. Accessed on 12 March 2025

Namgail T, Fox JL, Bhatnagar YV (2004) Habitat segregation between sympatric Tibetan argali Ovis ammon hodgsoni and blue sheep Pseudois nayaur in the Indian trans-Himalaya. J Zool 262(1):57–63

Namgay K, Millar J, Black R, Samdup T (2013) Transhumant agro-pastoralism in Bhutan: exploring contemporary practices and socio-cultural traditions. Pastoralism Res Policy Practice 3(1):13

Nori M, Davies J (2007) Change of wind or wind of change? Climate change, adaptation and pastoralism. The World Initiative for Sustainable Pastoralism, IUCN

Pandit A, Mir M, Mir M, Wani Y, Bisati I, Nisa S, Khan HM, Shah RA (2024) Pastoralism in Changthang, Ladakh: adaptations, challenges, and pathways for sustainability. Mt Res Dev 44(1):A1–A7. https://doi.org/10.1659/mrd.2023.00028

Pokharia AK, Mani BR, Spate MB, A and Srivastava, A. (2018) Early Neolithic agriculture (2700–2000 BC) and Kushan period developments (AD 100–300): macrobotanical evidence from Kanispur in Kashmir, India. Veg Hist Archaeobotany 27:477–491

Pratt DJ, Le Gall F, De Haan C (1997) Investing in pastoralism: sustainable natural resource use in arid Africa and the Middle East. World Bank Technical Paper 365, World Bank, Washington, DC, USA

Press Information Bureau (PIB) (2023) Ministry of Textiles approved Pashmina Wool Development Scheme under Integrated Wool Development Programme (IWDP) with a budget allocation of Rs. 29.25 Cr for 2021–22 to 2025–26. Press Release, Press Information Bureau

Rao A, Casimir J (eds) (2003) Nomadism in South Asia. Oxford University Press

Rizvi J (1999) Trans-Himalayan caravans: merchant princes and peasant traders in Ladakh. Oxford University Press

Robinson L, Flintan F (2022) Can formalisation of pastoral land tenure overcome its paradoxes? Reflections from East Africa. Pastoralism 12:34. https://doi.org/10.1186/s13570-022-00250-8

Scanes CG, Toukhsati SR (2018) The Neolithic revolution, animal domestication, and early forms of animal agriculture. In: Animals and human society. Academic Press, pp 103–131. https://doi.org/10.1016/B978-0-12-805247-1.00006-X

Schmidt M (2000) Pastoral systems in Shigar/Baltistan: communal herding management and pasturage rights. In: Ehlers E, Kreutzmann H (eds) High mountain pastoralism in northern Pakistan. Steiner, Stuttgart

Sharma A (2013) Excavation at Gufkral (Jammu and Kashmir). BR Publishing Corporation

Sharma VP, Rollefson IK, Morton J (2003) Pastoralism in India: a scoping study. Centre for Management on Agriculture, IIM, Ahmedabad

Singh NJ, Bhatnagar YV, Lecomte N, Fox JL, Yoccoz NG (2013) No longer tracking greenery in high altitudes: pastoral practices of Rupshu nomads and their implications for biodiversity conservation. Pastoralism: Res Policy Practice 3(1):1. https://doi.org/10.1186/2041-7136-3-16

Singh R, Sharma RK, Babu S, Bhatnagar YV (2020) Traditional ecological knowledge and contemporary changes in upper Spiti landscape agro-pastoral system, Indian trans-Himalayas. Pastoralism Res Policy Practice 10(15):1–14

Soheb M, Ramanathan A, Bhardwaj A, Coleman M, Rea B, Spagnolo M, Sam L (2022) Multitemporal glacier inventory revealing four decades of glacier changes in the Ladakh region. Earth System Science Data Discussions 2022:1–26

Spate M (2019) Mountain and meadow: a reconstruction of long-term pastoralist ecology in the Kashmir Valley. Doctoral dissertation, Department of Archaeology, Faculty of Arts and Social Sciences, University of Sydney

Sundriyal RC (1995) Grassland forage production and management in the Himalayas: a review. J Hill Res 8(2):135–150
Spooner B (1971) Towards a generative model of nomadism. Anthropol Q 198–210
Tundup P, Wani MA, Dawa S, Hussain S, Rai AP (2016) Influence of climate change on traditional agriculture of cold desert region-Ladakh. Global J Biosci Biotech 5(4):449–452
Verma R, Khadka M (eds) (2016) Gender and pastoralism in the rangelands of the Hindu Kush Himalayas: knowledge, culture, and livelihoods at the margins of the margins. Kathmandu, Nepal
Yamaguchi T (2011) Transition in mountain pastoralism: an agro-diversity analysis of the livestock population and herding strategies in Southwest Tibet, China. Hum Ecol 39(2):141–154
Yatoo M (2012) Characterising material culture to determine settlement patterns in North West Kashmir. Doctoral dissertation, School of Archaeological Studies, University of Leicester
Zehol L (2008) Ethnic tension and conflicts: north eastern experience. In: Fernandes W (ed) Search for peace with justice: issues around conflicts in Northeast India. North Eastern Social Research Centre, Guwahati, pp 44–65

Chapter 4
Landscape Dynamics and Expansion of Himalayan Birch (*Betula utilis* D. Don) in Timberline Ecotone of the Nanda Devi Biosphere Reserve, Western Himalayas

Ajay Maletha, Rakesh K. Maikhuri, Surendra S. Bargali, Sunil Nautiyal, and Vikram S. Negi

Abstract This study focuses on land use and land cover change and expansion of the Himalayan birch (*Betula utilis* D. Don) forest in the Nanda Devi Biosphere Reserve (NDBR) during a period of 23 years (1990–2013). It is based on two points of time, drawing on Landsat remote sensing data acquired from two different sensors: the Thematic Mapper (TM) in the year 1990 and the Operational Land Imager (OLI)/Thermal Infrared Sensor (TIRS) in 2013. The study revealed negative changes in vegetation cover, snow cover and river area. However, grasslands, open lands, agricultural lands and human settlements showed an increase in areas within the boundaries of NDBR. The results of the present study confirm a fair degree of variability in climate during the last few decades, leading to an increase in the density and expansion of pure and mixed Himalayan birch forest to the extent of 268 ha and 224 ha within the timberline ecotone (3200–4200 m Above sea level (Above sea level (ASL))). However, the timberline showed no advancement except for the

A. Maletha (✉)
Amity Institute of Forestry and Wildlife, Amity University, Noida, Uttar Pradesh, India

R. K. Maikhuri
Department of Environment Science, H.N.B. Garhwal University, Srinagar Garhwal, Uttarakhand, India

S. S. Bargali
Department of Botany, DSB campus, Kumaun University, Nainital, Uttarakhand, India

S. Nautiyal
Centre for Ecological Economics and Natural Resources (CEENR), Institute for Social and Economic Change (ISEC), Bengaluru, Karnataka, India

G.B. Pant National Institute of Himalayan Environment, Kosi-Katarmal, Almora, Uttarakhand, India

V. S. Negi
Department of Botany and Microbiology, H.N.B. Garhwal University, Pauri Garhwal, Uttarakhand, India

Y. D. I. Khan et al. (eds.), *Ecosystem-based Approaches for Resilience Building in Himalayan Landscapes*, Disaster Resilience and Green Growth,
https://doi.org/10.1007/978-981-95-2007-7_4

increase in the number of Himalayan birch individuals within the existing boundaries of the timberline ecotone. The availability and abundance of seedlings/saplings beyond the treeline suggest great possibilities for the upward movement of the treeline in future. There is a need to explore the potential of new generation sensors with higher spatial or spectral resolution to accurately map even small patches of vegetation dynamics and produce accurate information on the differentiation between various species and vegetation types.

Keywords Himalaya · Timberline ecotone · Himalayan birch · Landscape dynamics · Remote sensing and GIS

4.1 Introduction

The Himalayan mountain ranges are renowned for their spectacular biodiversity and unique ecological characteristics, supporting a diverse array of forests that exhibit varying species composition across altitudinal gradients and geographic regions. These forests are not only a reservoir of endemic flora and fauna but also play a pivotal role in maintaining the ecological balance of one of the most fragile environments on Earth. The Indian Himalayan region, rich in natural resources, faces significant challenges due to land use changes, population pressures and the pervasive impacts of climate change, which have drastically affected the livelihoods of local communities and heightened their economic and environmental vulnerabilities (Ramakrishnan et al. 2003; Sharma and Roy 2007).

A comprehensive analysis of global land cover suggests that nearly one-third of the region is forested, while approximately two-thirds accounts for farmland, with the remainder comprising rural and urban habitats, as well as areas less suited for agriculture (Dutta and Zade 2003). Given the intricate interplay of various drivers of change, continuous monitoring of land use and land cover transitions is crucial. The integration of empirical data with remote sensing technologies offers a powerful framework for understanding and assessing these dynamics. Landsat-TM images have been instrumental in providing a rich historical record of the Earth's surface, enabling researchers to evaluate land cover changes over the past three decades (USGS 2014).

In the Western Himalayas, species such as *Betula utilis* (Himalayan birch) dominate the timberline and serve as both foundation and keystone species, forming ecotones between coniferous forests and sub-alpine/alpine zones. This forest type is vital for supporting diverse ecological communities, which undergo significant changes in response to environmental factors. While numerous studies utilising remote sensing techniques have contributed to understanding the forest vegetation of the Indian Himalayas, many have focused primarily on broad vegetation patterns rather than providing detailed, species-level analyses (Rao and Pant 2001; Sen et al. 2002; Raturi and Bhatt 2004; Joshi and Gairola 2004).

The past century has witnessed a global increase in average surface temperatures, with projections estimating a rise between 1.4 °C and 5.8 °C by 2100, particularly affecting high-altitude and high-latitude regions (IPCC 2007; COP-21 2015). Such rapid climate changes induce significant biophysical impacts, leading to alterations in species composition, shifts in distribution ranges and changes in phenological patterns (Root et al. 2003; Parolo and Rossi 2008; Chen et al. 2011; Gottfried et al. 2012; Kirdyanov et al. 2012; Pauli et al. 2012; Webb et al. 2012; Singh et al. 2011; Bharti et al. 2011). The high-altitude timberline provides an ideal indicator for monitoring early signs of climate change, serving as a barometer for the structural and compositional shifts within these ecosystems (Kullman 1998).

The need for enhanced understanding and management of these ecosystems is urgent, particularly as recorded temperature increases have been observed across different regions of the Western Himalayas (Shekhar et al. 2010; Dimri and Dash 2011). Remote sensing and geographic information systems (GIS) are pivotal in mapping vegetation dynamics, assessing species distribution shifts and understanding the underlying climatic and topographical influences on forest change (Theurillat and Guisan 2001).

This investigation aims to provide insights into various land use classes and to evaluate the expansion of Himalayan birch within the context of ecosystem-based approaches for resilience building. It focuses on employing remote sensing data for Nanda Devi National Park and its buffer zone, contributing to the broader understanding of climate change impacts and disaster risk reduction in the Himalayan ecosystems. Through this research, we seek to highlight the importance of integrating ecological insights with adaptive management strategies, ensuring the conservation and sustainable use of these invaluable ecosystems in the face of ongoing environmental challenges.

4.2 Study Area

Nanda Devi Biosphere Reserve (NDBR) is northwest of the Himalayas and comprises three districts of Uttarakhand, viz. Chamoli, Bageshwar and Pithoragarh. In this manuscript, the words NDBR, Niti Valley, Bhyundar Valley and the reserve are used interchangeably based on the context of the sentences. The Biosphere Reserve, located between 30°05′ N to 31°02′ N latitudes and 79°12′ E to 80°19′ E longitudes, spreads over 5860.69 km^2 of land area with two core zones, viz. Nanda Devi National Park (624.62 km^2) and the world-famous “Valley of Flowers National Park” (87.50 km^2) (Fig. 4.1). It covers a wide altitude range from 1700 to 7817 m Above sea level (ASL). The buffer zone covers 47 villages, out of which 34 fall under Chamoli district. The NDBR is home to Indo-Mongoloid (Bhotias, Tolchhas) and Indo-Aryan groups (Garhwalis) and boasts rich biodiversity and a tremendous array of flora and fauna. The floristic and faunal excursions into the Nanda Devi Biosphere Reserve have been undertaken by several authors, viz. Hajra (1983), Hajra and Balodi (1995), Naithani (1984), Gaur et al. (1999), Maikhuri et al. (2000a,

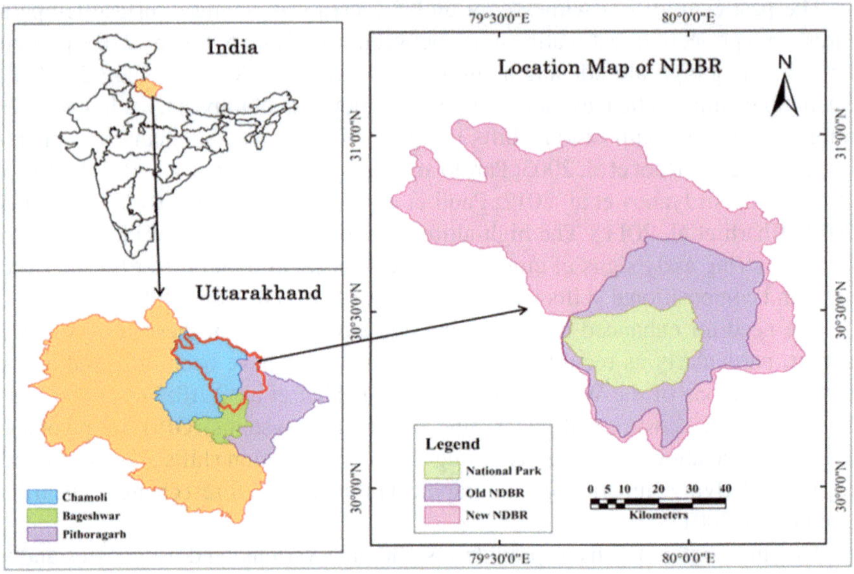

Fig. 4.1 Map of Nanda Devi Biosphere Reserve with Nanda Devi National Park (NDNP) and Valley of Flowers National Park (VoFNP)

b), Kala (2005), Rawal and Rawat (2012), Kumar et al. (2013), etc. The climatic year consists of three seasons: summer (April–June), rainy season (June–September) and winter (October–March) with an average annual rainfall of 928.81 mm. About 47.8% of the annual rainfall occurs over 2 months (July–August), featuring a strong monsoonic influence. The maximum temperature ranges from 11°C to 24°C, and the minimum from 3°C to 7.5°C. The altitudinal range of the biosphere reserve varies from 2100 m Above sea level (ASL) to 7817 m Above sea level (ASL).

4.3 Methodology

To study land use, land cover and vegetation change, geographic information system (GIS) and remote sensing data were used to verify the results obtained through empirical field studies and visual observation. The study's first goal was to delineate the boundary of the study area. To demarcate the boundary, 18 toposheets from the Survey of India (SOI) at a 1:50000 scale was procured to cover the entire NDBR region (Table 4.1).

The toposheets were scanned and converted into a digital copy for further processing. In all the scanned toposheets, geo-referencing was performed to assign the real-world coordinates with the geographic coordinate system (WGS 1984) before converting the same into a projected coordinate system (WGS 1984, UTM Zone 44 N). After geo-referencing the toposheets, the area of interest was clipped, and the

Table 4.1 Information about different datasets used in the study

Toposheet no.						Sources
53 N/5, 6, 9, 10, 11, 12, 13, 14, 15 and 16						Survey of India
62B/1, 2, 3, 4, 5, 6, 7 and 8						
Satellite data						
Satellite	Sensor	Path	Row	Spatial resolution (m)	Year	Sources
Landsat-5	TM	145	039	30 multispectral	November 1990	Glovis and EarthExplorer (USGS)
Landsat-8	OLI/ TIRS	145	039	30 multispectral 15 PAN	December 2013	
Digital elevation model (DEM)						Sources
SRTM 90 m DEM, version 4						CGIAR-CSI

mosaics of all the clipped images were made. The study area was divided into three zones: National Park as the core zone, old NDBR as the transitional zone and new NDBR as an outer buffer zone. A land use/land cover map was considered one of the important components of the study to understand vegetation dynamics in the study region. Landsat-5 satellite images were acquired from the Thematic Mapper (TM) sensor for thematic mapping in 1990. For 2013, Landsat-8 datasets were used and acquired from two sensors: an Operational Land Imager (OLI) and a Thermal Infrared Sensor (TIRS). The Landsat satellites have a 30-metre spatial resolution for the multispectral band, and from ETM+ onward, a 15-metre panchromatic band was available additionally. Various atmospheric correction methods were performed to acquire superior information from the images, and the conversion of digital number (DN) to reflectance was done for each band to arrive at an appropriate signature value for vegetation and other features. The formulas and methods mentioned in Chander and Markham's (2003) publication were applied for the Landsat-TM image analysis. Landsat-8 data were converted using the formula mentioned on the USGS website (http://landsat.usgs.gov/Landsat8_Using_Product.php). After converting DN to reflectance, a layer stacking process was used to generate a multispectral image for the datasets of both periods. For 2013, the panchromatic band of Landsat-8 was used to convert a multispectral 30-metre spatial resolution image into a 15-metre resolution image using merge techniques available in ERDAS 2013. The multispectral image of both periods was a subset, using a subset and clip tool and a digitised shape file of the study region boundary. For all the three zones, the area of interest was subset separately.

A classification technique was used to gather land use and land cover information from the satellite data. Land cover refers to the area under different forests, water bodies, grasslands, etc. In contrast, land use refers to how humans utilise land for purposes like settlement, agriculture and other developmental activities. The satellite images consist of pixels with a code based on reflectance. Reflectance is an important property for identifying features in digital images. The present study used the supervised classification technique, with training sites defined by the user. A training site is a set of instructions where the user defines the range of reflectance values or digital numbers to assign to a particular land cover category or land use. The software classifies the images or classes based on the user-defined signature set for different classes. The supervised classification technique used the maximum likelihood method as a parametric rule. The maximum likelihood method is based on Bayes theorem on decision-making. The maximum likelihood method, while assigning classes, considers the variance and covariance of the signature set in conveying each pixel value to one of the classes defined in the signature set. The generated thematic maps are cross-checked with the collected ground truth data for accuracy assessment. The land use and land cover maps of the study area for 1990 and 2013 were prepared using a hybrid classification technique, including supervised and vector classification. The thematic maps were converted the same way into a vector layer, while misplaced pixels were identified and merged into the relevant classes. The thematic map for both periods was overlaid on the DEM. At the same time, the expansion of vegetation cover and the selected field plots were demarked to evaluate the expansion of vegetation cover in the study region along with elevation. The Himalayan birch community was assessed to evaluate the timberline ecotone based on their natural habitat ranging between 3200 m and 4200 m Above sea level (ASL) and visual observations and findings from some earlier studies. The detailed approach, methodology and process followed are presented in the flowchart (Fig. 4.2). For the bioresource utilisation pattern of the villages located in the buffer zone of the reserve, personal interviews and group discussions were carried out in the Chamoli part of NDBR (Saxena et al. 2004; Maikhuri et al. 2009).

4.4 Results

In LULC analysis of NDBR, Landsat-5 and Landsat-8 satellite images were used for 1990 and 2013, respectively. The final LULC map was prepared using the ArcGIS 10.2 software. The total demarked area covered under the seven major land use classes (forest cover, grassland, open area, river, snow cover, agricultural crop and settlements) was analysed for change detection between 1990 and 2013 in NDBR. In NDNP, one of the core zones of the NDBR established in 1982, human activity inside the park area has now been banned, and restrictions imposed on the rights of local inhabitants for the collection of NTFPs and mountaineering/expeditions to Nanda Devi Peak. This has caused tensions between local people and forest authorities (Maikhuri et al. 2000a). Unlike the other protected areas in the Western Himalayan region, NDNP is free from disturbances due to the ban on human activity, its challenging terrain and poor accessibility. In the present study, the land use and land cover of NDBR was analysed across three zones, namely, NDNP, old NDBR and new NDBR (including VoFNP), between 1990 and 2013, while NDNP was selected to assess the expansion and shift in pure Himalayan birch forests (sites having pure patches of Himalayan birch stand) and mixed Himalayan birch forests (sites having associated tree species with dominant Himalayan birch species).

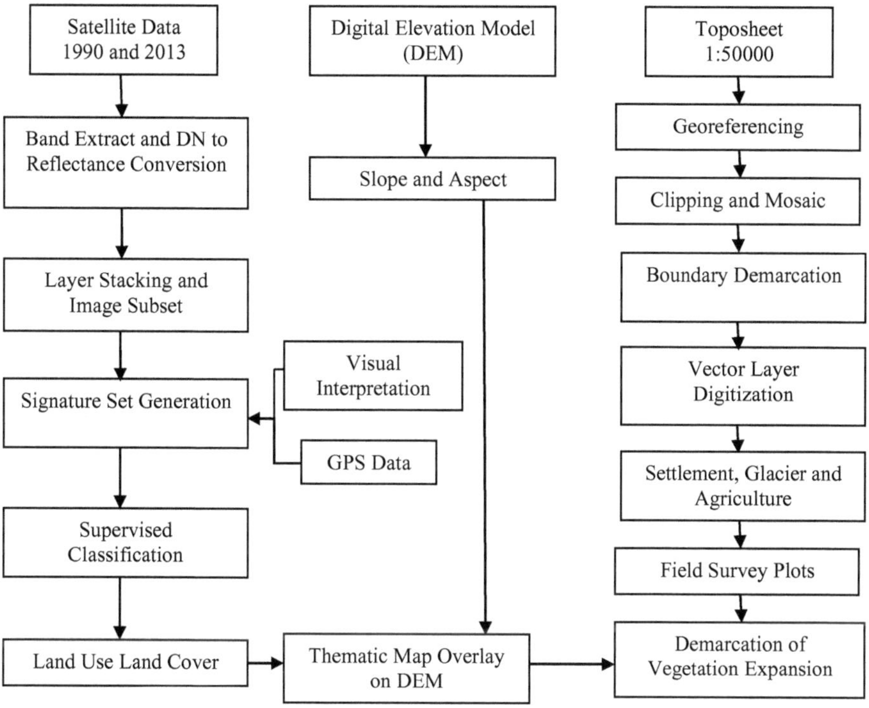

Fig. 4.2 Framework of the study

4.4.1 Population Dynamic of Himalayan Birch

Vegetation analysis of the Himalayan birch forest at the timberline of NDBR indicated that the forest contains mixed and pure evergreen-deciduous birch forests. However, *B. utilis*, *A. pindrow* and *R. campanulatum* form the major vegetation type in the pure Himalayan birch forest, and *B. utilis*, *C. deodara*, *A. pindrow*, *T. baccata*, *Picea smithiana*, *P. wallichiana*, *Populus ciliata*, *A. caesium*, *S. sikkimensis* and *R. campanulatum* form the mixed Himalayan birch forest in the NDBR. Tree density was observed between 1159 Ind. ha^{-1} and 1170.50 Ind. ha^{-1} in pure Himalayan birch forest, whereas in mixed forest it ranged between 575 Ind. ha^{-1} and 657 Ind. ha^{-1} at VoFNP and NDNP, respectively. Sapling layer exhibits higher number of individuals (1251 Ind. ha^{-1}) in pure Himalayan birch forest of VoFNP than NDNP (775.50 Ind. ha^{-1}). The seedling layer revealed the highest number of individuals ranging between 1280.50 Ind. ha^{-1} and 1544.50 Ind. ha^{-1} in pure Himalayan birch forest of VoFNP and NDNP (Fig. 4.3).

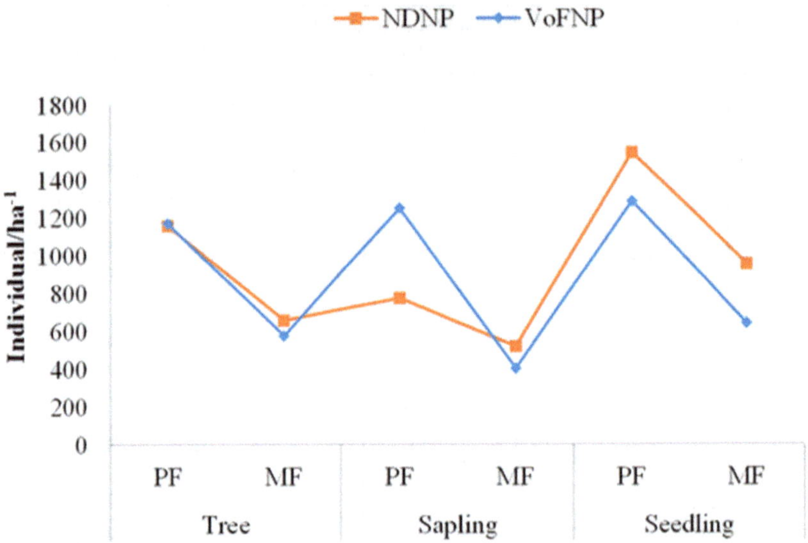

PF=Pure Himalayan birch forest; **MF**=Mixed Himalayan birch forest;
VoFNP= Valley of Flower National Park; **NDNP**= Nanda Devi National Park.

Fig. 4.3 Population dynamics of Himalayan birch in NDBR

4.4.2 LU and LC Area of Old and New NDBR

The old NDBR represents the area under NDNP and its buffer zone before VoFNP is established. Landsat data analysis shows a massive change occurred in the snow-covered region (−13,235 ha) between 1990 and 2013. Besides, the forest cover has also decreased by 29.6 ha. Consequently, the area under open land has increased from 43118.3 ha in 1990 to 55914.2 ha in 2013. In the buffer zone of the region, cropland and settlements have increased (Fig. 4.4). The decreased snow cover area is due to increased temperature in the region over the last 23 years (Dimri and Dash 2011). In 2005, the inclusion of VoFNP in the NDBR increased the region's importance with its unique biodiversity and cultural heritage. Therefore, NDBR now has two core zones (NDNP and VoFNP) and a vast buffer zone surrounding the core zones. Satellite imagery analysis again reveals a decline of 8977.17 ha in snow cover area, besides a decline in forest cover area (−713.9 ha). Similarly, due to snow cover loss, open land increased from 102,068 ha in 1990 to 111687.3 ha in 2013. The data also highlighted an increase of 178.02 ha for crop cultivation and 54.24 ha for human settlements (Fig. 4.5). This trend of increasing area under crops and settlements is a matter of concern as it increases anthropogenic pressure on the natural environment.

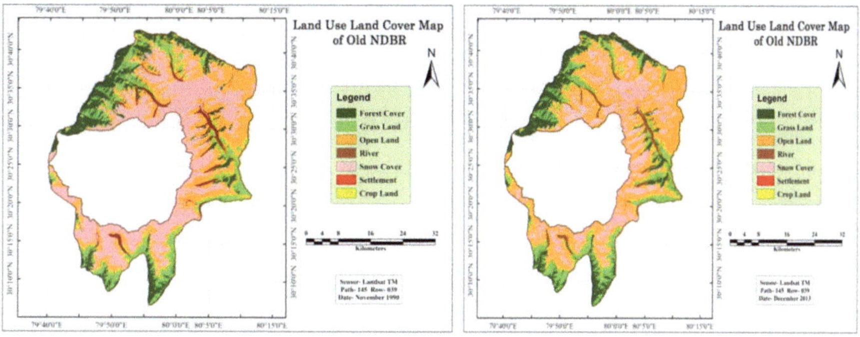

Fig. 4.4 Thematic map of old NDBR between 1990 and 2013

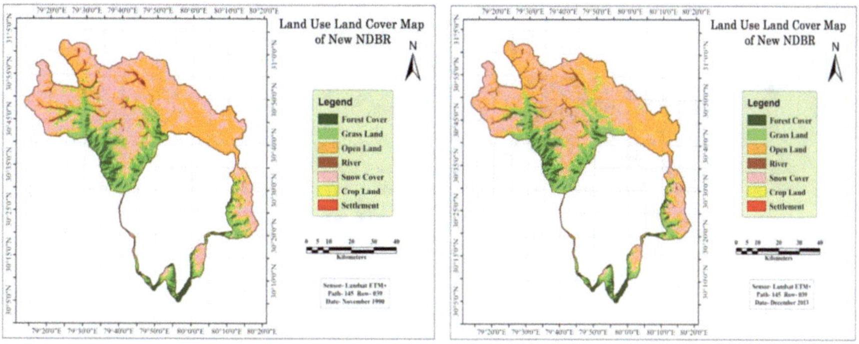

Fig. 4.5 Thematic map of new NDBR between 1990 and 2013

4.4.3 LULC and Expansion of Himalayan Birch Forests at the Timberline Zone of NDNP

The data reveals a huge change in the snow-covered area (−1063.1 ha) between 1990 and 2013. This can be attributed to the effects of warming in the higher Himalayan regions (Bhutiyani et al. 2007; Arora et al. 2008; Shekhar et al. 2010). The region's climate data also supports the fact that temperature increased by 0.6 °C and 1.3°C between 1975 and 2006 (Dimri and Dash 2011). Besides, due to a decrease in snow cover, the total forest cover of the NDNP, which was 11.19% of the total area in 1990, has also declined by 11.17%. Similarly, the open land area in the region has increased from 39.81% to 54.95%. This is because of a decline in the snow-covered area (Fig. 4.6).

The change in the distribution of pure and mixed Himalayan birch forest at the timberline was studied as a function of a shift in the altitude between 3200 m and 4200 m Above sea level (ASL) from 1990 to 2013. The shift and expansion of the

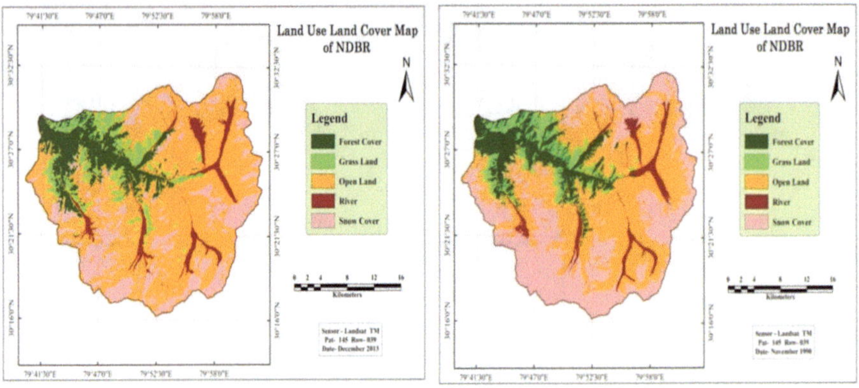

Fig. 4.6 Thematic map of Nanda Devi National Park, NDBR, between 1990 and 2013

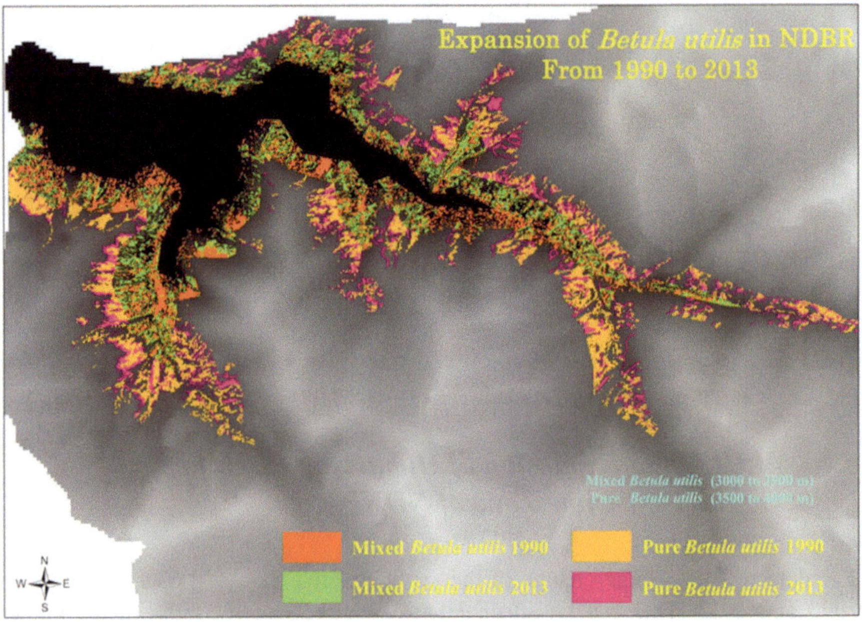

Fig. 4.7 Distribution and expansion of Himalayan birch (*Betula utilis*) at timberline zone of NDBR

area were computed with the help of GIS techniques. An equal number of station points were generated in the past, and now, mixed and pure Himalayan birch forests are overlaid to form the shift lines. The past and current nodes with elevation information extracted from DEM were subtracted to populate the shift value in the vector table. The surface length was also calculated using DEM as a base layer. The classified image shows mixed Himalayan birch forest in saffron and green colour and pure Himalayan birch forest in yellow and pink colour for 1990 and 2013, respectively (Fig. 4.7).

Table 4.2 Distribution and shift of *Betula utilis* forest in the timberline zone of NDNP

	1990		2013		
Forest types	Area (ha)	User accuracy	Areas (ha)	User accuracy	Area increased (ha)
Pure *Betula utilis* forests	761	0.73	1029	0.78	268
Mixed *Betula utilis* forests	883	0.74	1107	0.77	224

The green and pink patches depict an increased area of mixed and pure Himalayan birch forest during the last 23 years. In 1990, the proportion of pure Himalayan birch forest was estimated at 761 ha, while that of mixed forest was about 883 ha. Still, the digitisation of satellite imagery indicates an increase of 1029 and 1107 ha in 2013. Pure and mixed Himalayan birch forests increased by 268 ha and 224 ha between 1990 and 2013 at Nanda Devi National Park of NDBR (Table 4.2).

4.5 Discussion

The NDBR enjoys unique natural landscapes, endemic biodiversity, high inaccessibility and rich indigenous cultures (Maikhuri et al. 2001a). The aim of protected areas can be achieved when conservation actions and policies are coupled with the social development of local communities. However, a number of recent studies do not support the assumptions of protected area planners and policymakers (Maikhuri et al. 2000a, 2001b; Rao et al. 2000; Chandrasekhar et al. 2007). The NDBR belongs to India's Himalayan highlands bio-geographic province (2A) and offers much scope for promoting biodiversity conservation as it has diverse forest ecosystems with a mosaic of anthropogenic landscapes. The points of conflict include curtailment of peoples' usufruct rights in the area, loss of economic opportunities due to lack of regulated mountaineering/trekking, tourism in the core zone and livestock and crop depredation caused by wildlife. The alpine timberline of the Western Himalayas, considered a highly sensitive zone, may have suffered climate change, causing a significant impact on the distribution of the upper limit of vegetation cover. Collecting remotely sensed databases offers the opportunity for synoptic analysis of Earth-system functions, patterns and changes at the local, regional and global scales over time. Also, it provides a vital link between intensive, localised ecological research and regional, national and international conservation and management of biological diversity (Wilkie and Finn 1996). The present study compared whole datasets related to different spatial and temporal dynamics of the timberline zone. The land use and land cover change (LUCC) data of the National Park (NP), old Nanda Devi Biosphere Reserve (NDBR) and new NDBR of Chamoli district highlights that forest cover has decreased (713.9 ha) slightly between 1990 and 2013. This may be due to the over-exploitation of bioresources by villagers involved in *Ophiocordyceps sinensis* collection (a high-valued fungus), grazing, road construction, herding at higher ridges and degradation of forest due to natural calamities such as landslides, floods, avalanches and heavy snowfall that have led to significant loss of forest cover inside the buffer zone of the reserve area (Maletha et. al. 2020). The agricultural and human settlement areas have increased from 4.90 km^2 to 7.21 km^2 and 0.81 km^2 to 1.40 km^2 in old and new NDBR regions, respectively. The entire region's open land and river areas have increased due to increased temperatures, causing an excessive snow cover melting in the upper ridges. This has resulted in a decline in the area under snow cover (2.66%) between 1990 and 2013, thus showing a negative trend in snow accumulation in the region (Table 4.3). A comparison of different vegetation types based on altitude over the 23 years indicates an increase in pure and mixed Himalayan birch forest between 3000 m and 4200 m Above sea level (ASL). The user classification accuracies of Himalayan birch indicate 0.73 and 0.78 for pure forest and 0.74 and 0.77 for mixed forest between 1990 and 2013 derived from error matrices of NDNP. These changes were consistently observed in the field with Himalayan birch seedlings and saplings found above the timberline zone. However, the regeneration of Himalayan birch was found more in the forest edges than in the upper altitude gradients (Maletha et al. 2023; Adhikari 2004), indicating an increase in Himalayan birch's spread. Though the spread of seedlings and saplings has been gradually increasing, anthropogenic pressure exerted by *Ophiocordyceps sinensis* collectors poses a serious threat to the mature individual species of Himalayan birch at the timberline zone. Changes in the vegetation dynamics of the timberline have been observed during the last 23 years. Still, surprisingly, very low vegetation changes have been noticed in coniferous forests (*Abies-Cedrus* mixed forest) below the timberline (2400–3000 m Above sea level (ASL)) (Maletha et al. 2022). The rate of change in vegetation dynamics is very rapid, indicating a threat to diversity, and there is a disturbance to many important herbaceous species, which are being replaced by certain species like Himalayan birch locally and regionally in the timberline zone. Numerous studies confirm that high-altitude forest areas of the Indian Himalayas are the habitat for several rare, endemic and threatened species, which are very sensitive to climate change and may experience comparatively high rates of species loss (Samant 1993; Nautiyal et al. 2001; Rawat 2005; Kala 2005).

The vegetation map was developed during a study to assess the timberline zone before and after the extension of the reserve area (inclusion of VoFNP) in the region, which shows an adverse change in vegetation cover, snow cover and river area. A positive change has been observed in grassland, open land, agricultural land and settlements. In remote sensing studies, vegetation indices are normally used. Of the vegetation indices, the normalised difference vegetation index (NDVI) is widely used in detecting vegetation cover changes and vegetation status (Myneni et al. 1997). Change detection based on the differences in images relies on the phenological stage of a particular site at two different points in time. The result of the present study confirms a quite variability observed in micro-climate during the last few decades (Dimri and Dash 2011), leading to an increase in the density and expansion of pure and mixed Himalayan birch forest to the extent of 268 ha and 224 ha within

Table 4.3 Change in LULC area of old and new NDBR between 1990 and 2013

	Old NDBR					New NDBR				
	Area in 1990		Area in 2013			Area in 1990		Area in 2013		
LULC classes	ha	%	ha	%	Area change	ha	%	ha	%	Area change
Forest cover	20626.6	14.82	20,597	14.8	−29.6	30108.6	8.92	29394.7	8.71	−713.9
Grassland	17203.3	12.36	17,536	12.6	332.7	56143.3	16.63	56242.7	16.66	99.4
Open land	43118.3	30.98	55914.2	40.18	12795.9	102,068	30.24	111687.3	33.09	9619.3
River	5548.4	3.99	5626.44	4.04	78.04	10597.6	3.14	10337.21	3.06	−260.39
Snow cover	52,466	37.7	39,231	28.19	−13,235	138282.3	40.96	129305.2	38.3	−8977.17
Cropland	181.98	0.13	234.74	0.17	52.76	308.76	0.09	486.78	0.14	178.02
Settlement	17.05	0.01	21.95	0.02	4.9	63.99	0.02	118.23	0.04	54.24
Total area	139161.6		139161.3			337572.6		337572.1		

the timberline ecotone (Table 4.2). There was no geographical advancement of the mature trees except an augment in the density of the Himalayan birch, which occurred in the timberline ecotone (Hättenschwiler and Korner 1995; Paulsen et al. 2000; Klasner 2002; Zhang et al. 2009; Bharti et al. 2011). However, many seedlings (1544.50 Ind. ha^{-1}) and saplings (1251.50 Ind. ha^{-1}) of Himalayan birch have been observed above the timberline zone of NDNP. Besides, some studies reveal that the treeline of Himalayan birch will move upwards to 5914 m by 2080 compared to the present treeline at 5347 m Above sea level (ASL) in 2010. That is an average shift of 184 m per 7 decades (3 m/year) (Singh et al. 2011). Considerable uncertainties continue to surround the actual scenario of the timberline zone. Nevertheless, identifying its sensitive specialised habitats, distinct biological assemblages, native and endemic species, and the diversity of economically important flora and fauna is essential in the context of climate change. (Dhar et al. 2000), remote sensing satellite data has been found very effective in monitoring the timberline ecotone. The change in vegetation dynamics could be due to unplanned development and inadequate management interventions. Anthropogenic pressure, coupled with the invasion of local weed *Polygonum polystachyum* in the strictly protected VoFNP landscape, is responsible for reduced forest area and degraded forest quality, resulting in habitat loss in the NDBR (Nautiyal and Kaechele 2007).

The results presented here may be helpful for policy planners to redesign and reframe existing forest policies concerning timberline ecotone for conservation management in NDBR, in particular, and the Indian Himalaya, in general. There is a need to explore the potential of new generation sensors with higher spatial (e.g. Quickbird, IKONOS and WV-2) or spectral resolution (e.g. Hyperion) to map accurately even small patches of vegetation recovery and discriminate between different vegetation types and species (Shaw et al. 1998; Mumby and Harborne 1999; Greiwe and Ehlers 2005). Further, ecologically important keystone species such as Himalayan birch, often socioculturally important, are of value in rehabilitating disturbed forest ecosystems in and around the timberline ecotone. To achieve this, conservation-linked developmental strategies must be based on a value system in which people can understand, appreciate and participate. This should be the basis for promoting the conservation and management of the sensitive ecosystem in the Western Himalayas, a region highly endowed with traditional knowledge and wisdom. Such an approach would be a significant step in improving the knowledge of vegetation dynamics and future changes in the timberline ecotone of the Indian Himalayas.

4.6 Conclusion

Long-term research conducted in the Nanda Devi Biosphere Reserve (NDBR) underscores significant concerns related to the depletion of vegetation, snow cover and riverine ecosystems. These issues are primarily attributed to increased human activities focused on resource exploitation, compounded by substantial climatic variability. The observed positive changes in grasslands, open lands, agricultural zones and human settlements reflect a concerning trend that results in the decline of forested areas, ultimately leading to habitat loss for a diverse range of indigenous species. Despite these adverse impacts, our findings indicate an increase in the density and spatial expansion of both pure and mixed *Betula utilis* (Himalayan birch) forests within the timberline ecotone. Importantly, while no significant geographic shift of the ecotone has been noted, the emergence of Himalayan birch saplings and seedlings both above the timberline and within the krummholz zone suggests a potential for regeneration in these vital habitats. Moreover, to enhance our understanding of climate change effects on these sensitive high-altitude ecosystems, it is critical to establish regular monitoring protocols utilising high-resolution satellite imagery. Such technological innovations will facilitate more accurate assessments of ecological changes and inform the formulation of effective adaptive management strategies.

Given the distinct biophysical significance of Himalayan ecosystems and their pronounced vulnerability to climate change and anthropogenic disturbances, it is imperative for stakeholders to intensify their engagement and cooperative efforts. Adopting ecosystem-based approaches to resilience building is essential to mitigate current challenges and promote sustainable management practices. By integrating robust ecological insights with collaborative management strategies, we can bolster the resilience of Himalayan ecosystems, effectively support biodiversity conservation and navigate the complexities of ongoing environmental change.

Acknowledgement The authors are thankful to the Director, Govind Ballabh Pant National Institute of Himalayan Environment (GBPNIHE) Kosi-Katarmal, Almora, for providing facilities to undertake this work. We are also extremely thankful to Department of Science & Technology - Science and Engineering Research Board (DST-SERB) for financial support for conducting this study. We are also thankful to the Chief Wildlife Warden, Govt. of Uttarakhand, Conservator of Forest, Director, NDBR, and local people for their support and cooperation during the study in the VoFNP and NDNP (NDBR).

Conflicts of Interest The authors declare no conflicts of interest regarding the use of research contents and publication of this paper.

Data Availability The land use/land cover and remote sensing data used to support the findings of this study are available from the corresponding author upon request.

References

Adhikari BS (2004) Ecological attributes of vegetation in Nanda Devi National Park. Biodiversity monitoring expedition, Nanda Devi 2003. Wildlife Institute of India. Dehradun

Arora M, Singh P, Goel NK, Singh RD (2008) Climate variability influences on hydrological responses of a large Himalayan basin. Water Resour Manag 22:1461–1475. https://doi.org/10.1007/s11269-007-9237-1

Bharti RR, Adhikari BS, Rawat GS (2011) Assessing vegetation changes in timberline ecotone of Nanda Devi National Park, Uttarakhand. Int J Appl Earth Obs Geoinf 18:472–479. https://doi.org/10.1016/j.jag.2011.09.018

Bhutiyani MR, Kale VS, Pawar NJ (2007) Long-term trends in maximum, minimum and mean annual air temperatures across the northwestern Himalaya during the 20th century. Climate Change 85:159–177. https://doi.org/10.1007/s10584-006-9196-1

Chander G, Markham BL (2003) Revised Landsat-5 TM radiometric calibration procedures, and postcalibration dynamic ranges. IEEE Trans Geosci Remote Sens 41:2674–2677

Chandrasekhar K, Rao KS, Maikhuri RK, Saxena KG (2007) Ecological implications of traditional livestock husbandry and associated land-use practices: a case study from the trans-Himalaya, India. J Arid Environ 69:299–314. https://doi.org/10.1016/j.jaridenv.2006.09.002

Chen IC, Hill JK, Ohlemüller R, Roy DB, Thomas CD (2011) Rapid range shifts of species associated with high levels of climate warming. Science 333:1024–1026. https://doi.org/10.1126/science.1206432

CIP-21 (2015) Annual Conference of Parties (COP) http://www.cop21paris.org/about/cop21

Dhar U, Rawal RS, Upreti J (2000) Setting prioritization for conservation of medicinal plants: a case study in the Indian Himalaya. Biol Conserv 95:57–65. https://doi.org/10.1016/S0006-3207(00)00010-0

Dimri AP, Dash SK (2011) Wintertime climatic trends in the western Himalaya. Climate Change 111:775–800. https://doi.org/10.1007/s10584-011-0201-y

Dutta S, Zade M (2003) RISE – a distributed hydrologic model for rice agriculture: concept and evaluation. In: Singh VP, Yadava RN (eds) Watershed hydrology. Allied Publishers, pp 240–251

Gaur RD (1999) Flora of the District Garhwal: NW Himalaya (With Ethnobotanical Notes). *TransMedia*, Srinagar Garhwal

Gottfried M, Pauli H, Futschik A, Akhalkatsi M, Barancok P, Alonso JLB, Coldea G, Dick J, Erschbamer B, Calzado MRF, Kazakis G, Krajci J, Larsson P, Mallau M, Michelsen O, Moiseev D, Moiseev P, Molau U, Merzouki A, Nagy L, Nakhutsrishvili G, Pedersen B, Pelino G, Puscas M, Rossi G, Stanisci A, Theurillat JP, Tomaselli M, Villar L, Vittoz P, Vogiatzakis I, Grabherr G (2012) Continent-wide response of mountain vegetation to climate change. Nat Clim Chang 2:111–115. https://doi.org/10.1038/nclimate1329

Greiwe A, Ehlers M (2005) Combined analysis of hyperspectral and high resolution image data in an object oriented classification approach. Proc. ISPRS WG VII/1

Hajra PK (1983) A contribution to the botany of Nanda Devi National Park. BSI Calcutta

Hajra PK, Balodi B (1995) Plant wealth of Nanda Devi biosphere reserve. *Flora of India*. Series-4. BSI Dehra Dun

Hättenschwiler S, Korner C (1995) Responses to recent climate warming of *Pinus sylvestris* and *Pinus cembra* within their montane transition zone in the Swiss Alps. J Veg Sci 6:357–368. https://doi.org/10.2307/3236235

IPCC (2007) Climate change: the scientific basis. Regional climate projections. Contribution of working group I to the third assessment report of the intergovernmental panel on climate change. Cambridge University Press Cambridge

Joshi PK, Gairola S (2004) Land cover dynamics in Garhwal Himalayas – a case study of Balkhila subwatershed. J Indian Soc Remote Sens 32:199–208. https://doi.org/10.1007/BF03030876

Kala CP (2005) Indigenous uses, population density and conservation of threatened medicinal plants in protected areas of the Indian Himalayas. Conserv Biol 19:368–378. https://doi.org/10.1111/j.1523-1739.2005.00602.x

Kirdyanov AV, Hagedorn F, Knorre AA, Fedotova EV, Vaganov EA, Naurzbaev MM, Moiseev PA, Rigling A (2012) 20th century treeline advance and vegetation changes along an altitudinal transect in the Putorana Mountains, northern Siberia. Boreas 41:56–67

Klasner FL (2002) A half century of change in alpine patterns at glacier national park, Montana, USA. Arct Antarct Alp Res 34:49–56. https://doi.org/10.1080/15230430.2002.12003468

Kullman L (1998) Tree-limits and montane forests in the Swedish Scandes: sensitive biomonitors of climate change and variability. Ambio 27:312–321

Kumar A, Mitra M, Adhikari BS, Rawat GS (2013) Additions to plant wealth of Nanda Devi biosphere reserve, Western Himalaya. Indian Forester 139:959–961

Maikhuri RK, Nautiyal S, Rao KS, Chandrasekhar K, Gavali R, Saxena KG (2000a) Analysis and resolution of protected area-people conflicts in Nanda Devi biosphere reserve, India. Environ Conserv 27:43–53. https://doi.org/10.1017/S0376892900000060

Maikhuri RK, Nautiyal S, Rao KS (2000b) Indigenous knowledge of medicinal plants and wild edibles among three tribal sub-communities of central Himalaya, India. IKDM 8:7–13

Maikhuri RK, Nautiyal S, Rao KS (2001a) Medicinal plant cultivation practices of Bhotiyas in Nanda Devi biosphere reserve villages of Garhwal Himalaya. In: Samant SS, Dhar U, Palni LMS (eds) Himalayan medicinal plants: potential and prospects. Gyanodaya Prakashan, Nainital, pp 217–328

Maikhuri RK, Nautiyal S, Rao KS, Saxena KG (2001b) Conservation policy-people conflicts: a case study from Nanda Devi biosphere reserve (a world heritage site), India. Forest Policy Econ 2:335–365. https://doi.org/10.1016/S1389-9341(01)00037-5

Maikhuri RK, Rawat LS, Negi VS, Phondani PC, Bahuguna A, Chamoli KP, Farooquee N (2009) Impact of climate change and coping strategies in Nanda Devi biosphere, Central Himalayas, India. In: Proceeding of international mountain biodiversity conference. ICIMOD, Kathmandu, pp 138–146

Maletha A, Maikhuri RK, Bargali SS (2020) Criteria and Indicator for assessing threat on Himalayan birch (B. utilis) at timberline ecotone of Nanda Devi Biosphere Reserve: a world heritage site, Western Himalaya, India. Environ Sustain Indic 8(2020):100086

Maletha A, Maikhuri RK, Bargali SS, Sharma A, Negi VS, Rawat LS (2022) Vegetation dynamics and soil nutrient availability in a temperate forest along altitudinal gradient of Nanda Devi Biosphere Reserve, Western Himalaya, India. PLoS ONE 17(10):e0275051. https://doi.org/10.1371/journal.pone.0275051

Maletha A, Maikhuri RK, Bargali SS (2023) Population structure and regeneration pattern of Himalayan Birch (Betula utilis D.Don) at Timberline zone of Nanda Devi Biosphere Reserve, Western Himalaya, India. Geol Ecol Landsc 7(3):248–257

Mumby PJ, Harborne AR (1999) Development of a systematic classification scheme of marine habitats to facilitate regional management and mapping of Caribbean coral reefs. Biol Conserv 88:155–163. https://doi.org/10.1016/S0006-3207(98)00108-6

Myneni RB, Keeling CD, Tucker CJ, Asrar G, Nemani RR (1997) Increased plant growth in the northern high latitudes from 1981 to 1991. Nature 386:698–702. https://doi.org/10.1038/386698a0

Naithani BD (1984) Flora of Chamoli. Howrah: botanical survey of India, vol 1–2, p 789

Nautiyal S, Kaechele H (2007) Conserving the Himalayan forests: approaches and implications of different conservation regimes. Biodivers Conserv 16:3737–3754. https://doi.org/10.1007/s10531-007-9178-2

Nautiyal S, Maikhuri RK, Rao KS, Saxena KG (2001) Medicinal plant resources in Nanda Devi biosphere Reserve in the Central Himalaya, India. Herbs Spices Med Plants 8:47–64. https://doi.org/10.1300/J044v08n04_06

Parolo G, Rossi G (2008) Upward migration of vascular plants following a climate warming trend in the Alps. Basic Appl Ecol 9:100. https://doi.org/10.1016/j.baae.2007.01.005

Pauli H, Gottfried M, Dullinger S, Abdaladze O, Akhalkatsi M, Alonso JLB, Coldea G, Dick J, Erschbamer B, Calzado RF, Ghosn D, Holten JI, Kanka R, Kazakis G, Kollár J, Larsson P, Moiseev P, Moiseev D, Molau U, Mesa JM, Nagy L, Pelino G, Puscas M, Rossi G, Stanisci A, Syverhuset AO, Theurillat JP, Tomaselli M, Unterluggauer P, Villar L, Vittoz P, Grabherr G (2012) Recent plant diversity changes on Europe's mountain summits. Science 336:353–355. https://doi.org/10.1126/science.1219033

Paulsen J, Weber UM, Korner C (2000) Tree growth near treeline: abrupt or gradual reduction with altitude? Arct Antarct Alp Res 32:14–20. https://doi.org/10.1080/15230430.2000.12003334

Ramakrishnan PS, Saxena KG, Patnaik S (2003) Methodological issues in mountain research: socio-ecological system approach. New Delhi, UNESCO and Oxford & IBH, p 283

Rao KS, Pant R (2001) Land use dynamics and landscape change pattern in a typical micro watershed in the mid elevation zone of central Himalaya, India. Agric Ecosyst Environ 86:113–123. https://doi.org/10.1016/S0167-8809(00)00274-7

Rao KS, Nautiyal S, Maikhuri RK, Saxena KG (2000) Management conflicts in Nanda Devi biosphere reserve, India. Mt Res Dev 20:320–323

Raturi GP, Bhatt AB (2004) Vegetational pattern analysis in Rudraprayag district using remote sensing and GIS. J Indian Soc Remote Sens 32:217–224. https://doi.org/10.1007/BF03030878

Rawal RS, Rawat B (2012) Nanda Devi biosphere reserve. In: Palni LMS, Rawal RS, Rai RM, Reddy SV (eds) Compendium on biosphere reserve, progression during two decade of conservation. G.B. Pant Institute of Himalayan Environment and Development, Almora

Rawat GS (2005) Alpine meadows of Uttarakhand: ecology, land use and status of medicinal and aromatic plants. Dehradun, Bishen Singh Mahendra Pal Singh, p 219

Root TL, Price JT, Hall KR, Schneider SH, Rosenzweig C, Pounds JA (2003) Fingerprints of global warming on wild animals and plants. Nature 421:57–60. https://doi.org/10.1038/nature01333

Samant SS (1993) Scientific and ecological expedition to Nanda Devi: diversity and status of plants in Nanda Devi biosphere reserve. Report, Army Head Quarters, New Delhi, pp 54–85

Saxena KG, Ramakrishnan PS, Maikhuri RK, Rao KS, Patnaik S (2004) Assessment of vulnerability of Forest, Medows and mountain ecosystem due to climate change.Com -4E:/ Winrock/ agriculture, forestry Ch-14 pm6 2 nd proof 20-01-04. Agric For Nat Ecosys:163–168

Sen KK, Semwal RL, Rana U, Nautiyal S, Maikhuri RK, Rao KS, Saxena KG (2002) Patterns and implications of land use/cover change: a case study in Pranmati watershed (Garhwal Himalaya, India). Mt Res Dev 22:56–62

Sharma S, Roy PS (2007) Forest fragmentation in the Himalaya: a central Himalayan case study. Int J Sust Dev World 14:201–210. https://doi.org/10.1080/13504500709469720

Shaw DT, Malthus TJ, Kupiec JA (1998) High-spectral resolution data for monitoring scots pine (*Pinus sylvestris* L.) regeneration. Int J Remote Sens 19:2601–2608. https://doi.org/10.1080/014311698214668

Shekhar MS, Chand H, Kumar S, Srinivasan K, Ganju A (2010) Climate-change studies in the western Himalaya. Ann Glaciol 51:105–113. https://doi.org/10.3189/172756410791386508

Singh CP, Panigrahy S, Thapliyal A, Kimothi MM, Soni P, Parihar JS (2011) Monitoring the alpine treeline shift in parts of the Indian Himalaya using remote sensing. Curr Sci 102:559–562

Theurillat JP, Guisan A (2001) Potential impact of climate change on vegetation in the European Alps: a review. Climate Change 50:77–109. https://doi.org/10.1023/A:1010632015572

USGS (2014) Earth resource observation centre. http://glovis.usgs.gov

Webb LB, Whetton PH, Bhend J, Darbyshire R, Briggs PR, Barlow EWR (2012) Earlier winegrape ripening driven by climatic warming and drying and management practices. Nat Clim Chang 2:259–264. https://doi.org/10.1038/NCLIMATE1417

Wilkie DS, Finn JT (1996) Remote sensing imagery for natural resources monitoring. Columbia University Press, New York, p 295

Zhang YJ, Xu M, Adams J, Wang X (2009) Can Landsat imagery detect treeline dynamics? Int J Remote Sens 30:1327–1340. https://doi.org/10.1080/01431160802509009

Chapter 5
Landslide Susceptibility Analysis of the Chhota Rangit River Basin of Darjeeling Hill Area by Using the AHP Model

Kabita Lepcha, Bishal Chhetri, and Shikha Subba

Abstract Landslides are the downslope fall of dislodged soil and rock debris under the force of gravity. It is one of the most frequently occurring and significantly damaging geo-environmental disasters in the world's mountainous regions. It occurs due to the intricate relationships among climatic, geological, geomorphological, hydrological and human factors like LULC dynamics. Among these different factors, the impact of land use dynamics has been recognised as the most significant causes for the precipitation-triggered landslides throughout the globe. The occurrence of landslides is a primary geo-environmental distress in the parts of Darjeeling Himalayas, causing significant damage of lives and properties. The slope instability factors, along with the human interventions like clearing of forest cover for cultivation land and plantation, construction of roads in the steep slopes, expansion of settlements and construction of high-rise buildings, quarrying of stones from the steep slopes, etc., are some of the important reasons for recurrent landslide hazards in the region. This study attempts to generate the spatial landslide susceptibility zones of the river basin by analysing 18 landslide-causing factors in a GIS-based AHP (analytic hierarchy process) model. The production of the landslide susceptibility map would help to identify the vulnerable zones and minimise the effect through proper controlling methods. One of the mitigation measures can be an ecosystem-based approach. The ecosystem-based approach incorporates various measures like planting protective vegetative grasses, effective soil and water conservation methods, reforestation of barren lands and sustainable land use planning.

K. Lepcha
University of Gour Banga, Malda, West Bengal, India

B. Chhetri (✉) · S. Subba
Southfield College, Darjeeling, West Bengal, India

Y. D. I. Khan et al. (eds.), *Ecosystem-based Approaches for Resilience Building in Himalayan Landscapes*, Disaster Resilience and Green Growth,
https://doi.org/10.1007/978-981-95-2007-7_5

Keywords Analytic hierarchy process (AHP) · Darjeeling · Chhota Rangit · Himalayas · Landslides · Ecosystem-based approach

5.1 Introduction

Landslides are the downslope fall of dislodged soil and rock debris under the force of gravity (Mandal and Maiti 2015). It is one of the frequently occurring and significantly damaging geo-environmental disasters, especially in the hilly parts of the Earth (Ayalew and Yamagishi 2005; Bhattacharya 2016). In most cases, they are due to the intricate relationships among climatic, geological, geomorphic, hydrological and anthropogenic factors like LULC dynamics (Reichenbach et al. 2014). Among them, the impact of land use dynamics has been recognised as the most significant causes for the precipitation-triggered landslides throughout the globe in recent times (Glade 2003). In the studies of slope instability, the influence of physical characteristics such as geological, geomorphological and climatic factors can be considered constant for relatively more extended periods, but the impact of land use change varies within a very short period (Reichenbach et al. 2014; Pisano et al. 2017; Gunarathna et al. 2018). The impact of land use change-induced landslides, particularly in human habitats and densely populated regions, proves to be most dangerous and highly damaging (Vanacker et al. 2003; Reichenbach et al. 2014). Numerous scholars have studied the influence of LULC changes on the occurrence of landslides (Glade 2003). Thomas Glade (2003) studied the landslide as a result of LULC change in New Zealand. Mugagga et al. (2012) studied implications of land use modifications on the occurrences of landslides in Mount Elgon. Reichenbach et al. (2014) studied the impact of LULC modifications on landslide susceptibility zones in Italy. Gunarathna et al. (2018) focused their studies on the role of land use on the occurrences of landslides in Sabaragamuwa, Sri Lanka.

The occurrence of landslides has also become a primary geo-environmental concern in the regions of Darjeeling Himalayas. It is, in fact, the most destructive frequently occurring hazard the region has encountered over the years. The death and damages due to landslide over the years in the region has attracted numerous scholars around the world to study the causes of landslides in the area. Researchers like Dutta (1966), Basu and Sarkar (1985), Basu and Sarkar (1987), Basu and Ghatowar (1988), Froehlich and Starkel (1987), Mandal and Maiti (2015), Bhattacharya (2016), and others have put an effort for systematic studies of landslides in different parts of Darjeeling hill areas. Although most of the landslides in the region have been influenced by natural factors like geology, topography, soil, climate, etc., the impact of the increased human interventions like LULC changes seems to be the prime cause for the recent landslide episodes in the region[1] (Bhattacharya 2016). The different slope instability factors along with the human interventions like

[1] www.darjeeling.gov.in/geography.html

clearing of forest cover for cultivation land and plantation, construction of roads in the steep slopes, expansion of settlements and construction of high-rise buildings, quarrying of stones from the steep slopes as well as coal mining, etc. are some of the important causes for frequent landslide hazards in the region. Such interventions have increased the frequency and enormity of the landslides in the area and the spatial division of the landslide zones in the area, which has augmented the risk of such hazards to larger areas and greater populations. Considering the present scenario of increased spatial distribution and risk of landslide hazard in this part, it is of utmost necessity to produce landslide susceptibility zone maps with different magnitudes, which would help manage and minimise the damages. Landslide susceptibility means the possibility of a landslide taking place in the area based on the local topographic characteristics and the extent to which the area can be affected (Guzzetti et al. 1999; Reichenbach et al. 2014). There are various methodologies available for landslide susceptibility studies, such as different statistical analysis, AHP, frequency ratio model (FRM), artificial neural network (ANN), etc. (Basu and Pal 2017).

In recent times, this study analyses the effect of LULC dynamicity on landslides in the Chhota Rangit river basin. An effort has been made to classify and delineate the spatial distribution of landslide susceptibility areas in the Chhota Rangit river basin by considering eighteen (18) landslide-causing factors, including physical and human-induced factors. The Chhota Rangit river basin includes critical landslide-prone areas like Bijanbari and Relling in the Darjeeling district (Starkel and Basu 2000). The landslide susceptibility study of the Chhota Rangit river basin has been done using the analytic hierarchy process (AHP) by Saaty (1980) in the ArcGIS 10.3.1 environment.

5.2 Study Area

The area chosen for this research paper is the Chhota Rangit river basin, an integral part of the Darjeeling hill area in India (Fig. 5.1). Chhota Rangit river basin is well drained by the little Rangit and its left and right bank tributaries like Kahel Khola, Hospital Jhora, Relling Khola, Serjan and Neora Jhora, covering a distance of 370.59 km; the total basin area is 140.98 sq. km. The river basin extends between 26°57′ N to 27°20′ N latitude and 88°6′ E to 88°18′ E longitude and covers 22 mouzas[2] of Darjeeling Pulbazar Block and 8 mouzas of Jorebunglow Sukhiapokhri Block. Geo-environmentally, the river basin is highly vulnerable to various problems like landslides, soil erosion, ecosystem services loss, erratic and intense rainfall, drastic LULC changes and unscientific infrastructural developmental activities.

[2] A mouza refers to a specific geographical unit or administrative area within a revenue district. A mouza typically represents a defined area of land that may include one or more villages. It serves as a basic unit for land revenue assessment and administration.

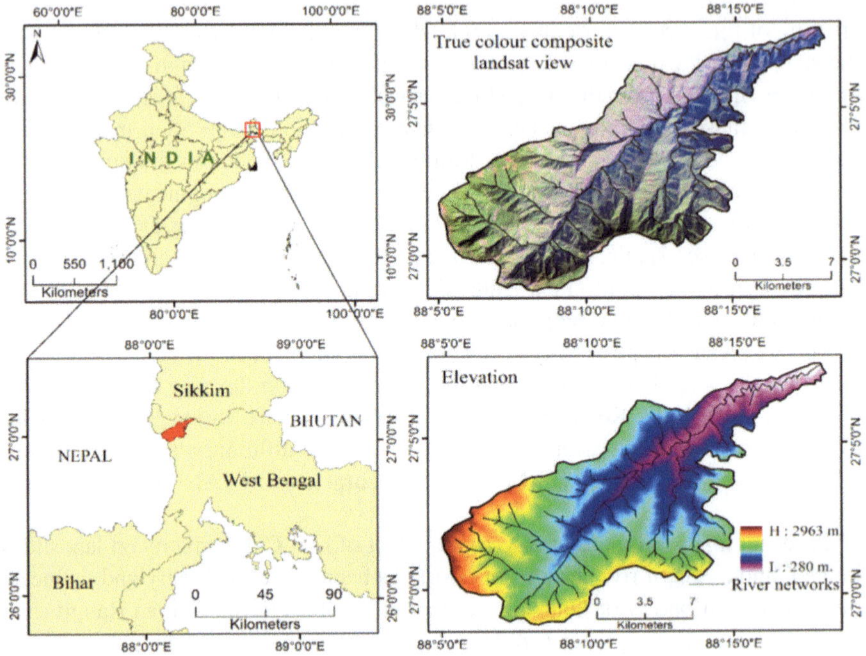

Fig. 5.1 Location of the study area

5.3 Materials and Methods

This research work utilised both primary and secondary data. The NDVI and the temporal LULC classification were conducted utilising satellite images. Landsat-5 TM data for 1991 and 2011, Landsat-7 for 2001 and Landsat-8 OLI for 2022 extracted from the USGS global visualisation viewer have been used (Table 5.1). The acquired Landsat data was registered in the UTM projection northern zone 45 datum WGS 84. Advanced Spaceborne Thermal Emission and Reflection Radiometer Digital Elevation Model (ASTER-DEM) data acquired from the USGS (Table 5.1) has been used to generate elevation maps, slope maps, slope curvatures, SPI (stream power index), TWI (topographic wetness index), slope aspect, etc. Precipitation data for river basin has been collected from Tropical Rainfall Measuring Mission (TRMM) data (mirador.gsfc.nasa.gov). Altogether, 30 soil samples were collected from the field; the sieving method was applied for the textural segregation of the gathered soil sample, and a texture map was prepared by inverse distance weighting techniques in ArcGIS 10.3.1. A lineament map of the study area was collected from bhuvan.nrsc.gov.in (ISRO) to make lineament density and distance from the lineament map. Road density and distance from the road maps were prepared using topographic maps 78 A/4, 78 A/8 and 78 B/1 and Google Earth.

Table 5.1 Data sources

Sl. no	Source	Types of data	Path/row	Date
1.	United States geological survey earth explorer	Landsat-5 TM	139/41	1991/12/29
2.		Landsat-7 ETM+	139/41	2001/12/20
3.		Landsat-5 TM	139/41	2011/12/08
		Landsat-8 OLI/TIRS	139/41	2022/12/04
4.		ASTER-DEM 30 m		2022/12/04

5.3.1 Analytic Hierarchy Process

The detection and delineation of landslide susceptibility with different magnitudes in the Chhota Rangit river basin has been done using the AHP model. This model has been extensively utilised for selecting suitable sites, regional planning and landslide susceptibility (Ayalew and Yamagishi 2005). The AHP is the semi-qualitative method that allows the direct involvement of the decision-makers in identifying the result (Basu and Pal 2017). 'This is a multi-objective, multi-criteria decision-making approach that includes building a hierarchy of decision elements; a comparison matrix is used to calculate the weight of the selected factors by calculating the matrix's eigen vectors' (Saaty 1980; Saaty 1994; Saaty and Vargas 2001). In this process, two chosen parameters are assessed together, and importance is given to one parameter above another based on skilled observation. Such observations are determined utilising a preference scale of 1–9, wherein a smaller number suggests less significant, and vice versa (Rehman et al. 2022). The preference score values for pairwise comparison given and used by Saaty (1990) have been used for this work.

The CR (consistency ratio) is computed to see the consistency of the pairwise comparisons (Kolat et al. 2012; Bera et al. 2019). A remarkable advantage of the AHP technique is its potentiality to hold inconsistent relationships while offering a CR as a measure of the level of consistency or inconsistency (Chen et al. 2010; Feizizadeh and Blaschke 2013; Bera et al. 2019). The CR is given by Eq. 5.1.

$$\mathrm{CR} = \frac{\mathrm{CI}}{\mathrm{RI}} \tag{5.1}$$

Where CR is the consistency ratio, the CI (consistency index) which can be measured using Eq. 5.2, RI (random index) is affected by the number of factors (n):

$$\mathrm{RI} = \frac{\lambda_{\max - n}}{n - 1} \tag{5.2}$$

where λmax is the largest eigen number of the matrix and 'n' is the order of the square matrix. The random consistency index (RI) was derived from a sample of arbitrarily produced reciprocal matrices as given and used by Saaty (1980) (Table 5.2).

Table 5.2 The priority weights of selected landslide-causing factors by analytic hierarchy process (AHP)

Parameters	RD	LU	SL	ND	EL	RA	LI	DFL	GO	DFR	ST	SC	DFS	SPI	PGA	TWI	SA	GA
RD	1	1	2	3	4	2	6	7	8	9	4	7	3	4	5	6	7	8
LU	1	1	1	1	2	3	4	5	6	7	8	9	1	7	9	4	5	6
SL	0.5	1	1	1	2	2	1	2	3	4	5	6	2	8	9	7	8	3
ND	0.33	1	1	1	4	2	6	7	8	9	4	7	3	4	5	6	7	8
EL	0.25	0.5	0.5	0.25	1	1	1	2	3	4	5	6	2	8	9	4	7	3
RA	0.5	0.33	0.5	0.5	1	1	4	5	6	7	8	9	1	6	3	4	5	6
LD	0.16	0.25	1	0.16	1	0.25	1	2	2	2	1	2	1	4	5	6	7	8
DFL	0.14	0.2	0.5	0.14	0.5	0.2	0.5	1	1	1	2	3	1	5	6	7	8	9
GO	0.12	0.16	0.33	0.12	0.33	0.16	0.5	1	1	1	2	3	1	5	6	7	8	9
DFR	0.11	0.14	0.25	0.11	0.25	0.14	0.5	1	1	1	1	2	1	4	5	6	7	8
ST	0.25	0.12	0.2	0.25	0.2	0.12	1	0.5	0.5	1	1	1	1	2	3	4	5	6
SC	0.14	0.11	0.16	0.14	0.16	0.11	0.5	0.33	0.33	0.5	1	1	1	2	2	1	2	3
DFS	0.33	1	0.5	0.33	0.5	1	1	1	1	1	1	1	1	4	5	6	7	8
SPI	0.25	0.14	0.12	0.25	0.12	0.16	0.25	0.2	0.2	0.25	0.5	0.5	0.25	1	1	1	2	3
PGA	0.2	0.11	0.11	0.2	0.11	0.33	0.2	0.16	0.16	0.2	0.33	0.5	0.2	1	1	1	2	2
TWI	0.16	0.25	0.14	0.16	0.25	0.25	0.16	0.14	0.14	0.16	0.25	1	0.16	1	1	1	2	2
SA	0.14	0.2	0.12	0.14	0.14	0.2	0.14	0.12	0.12	0.14	0.2	0.5	0.14	0.5	0.5	0.5	1	2
GA	0.12	0.16	0.33	0.12	0.33	0.16	0.12	0.11	0.11	0.12	0.16	0.33	0.12	0.33	0.5	0.5	0.5	1
Weights	0.10	0.17	0.13	0.09	0.07	0.09	0.04	0.04	0.04	0.03	0.02	0.01	0.04	0.01	0.01	0.01	0.009	0.009
CR	0.09																	

Note: *RD*, road density, *LU* land use/cover, *SL* slope, *ND* NDVI, *EL* elevation, *RA* annual rainfall, *LI* lineament density, *DFL* distance from the lineament, *GE* geology, *DFR* distance from the road, *ST* soil texture, *SC* slope curvature, *DFS* distance from the stream, SPI stream power index; PGA peak ground acceleration; TWI topographic wetness index, *SA* slope aspect and *GA* gravity anomaly
Source: Compiled by the authors

Table 5.3 Significant weights and consistency ratio of the selected parameters using pairwise comparison matrix

Factors	C.R	Weights
Geology	0.00	0.5
Elevation (in metres)	0.02	0.19
Slope (in degrees)	0.09	0.11
Aspect	0.08	0.10
Curvature	0.01	0.33
NDVI	0.01	0.19
Land use/land cover	0.06	0.14
Annual rainfall (in mm)	0.06	0.93
Lineament density (km/sq. km)	0.03	0.19
Distance from the lineament (in metres)	0.02	0.19
Road density (km/sq. km)	0.09	0.20
Distance from the road (in metres)	0.02	0.19
Distance from the stream (in metres)	0.02	0.19
Stream power index	0.02	0.19
Peak ground acceleration (g)	0.02	0.19
Topographic wetness index	0.09	0.19
Gravity anomaly (mgals)	0.02	0.19

Source: Compiled by the authors

The CR < 0.10 signifies an excellent level of consistency for recognising class weightages (Saaty 1996; Malczewski 1999; Bera et al. 2019). The calculated consistency ratio of each selected parameter in the Chhota Rangit river basin ranges from 0.01 to 0.09 (Table 5.3), which shows the good consistency among the compared factors and weight (Table 5.2) and are suitable for landslide susceptibility zones.

5.3.2 Validation of the Landslide's Susceptibility Map

Validation is performed using the landslide inventory maps to ascertain degree of accuracy of the final product or the produced landslide susceptibility maps. The validation process is always linked with direct incidences (Samy et al. 2014). For the validation, the landslide patches were recognised and positioned as per the GPS from the field. Then, suitable polygons and the patches on the Google image were identified. The details of the previous slides were also collected from secondary bases like Topographical Maps (78 A/4, 78 A/8 and 78 B/1) of a survey of India, government offices (Bijanbari and Sukhiapokhri block offices), Google Earth and conversations with the locals. Sixty landslide patches were identified and used to create an inventory map and model validation (Fig. 5.2).

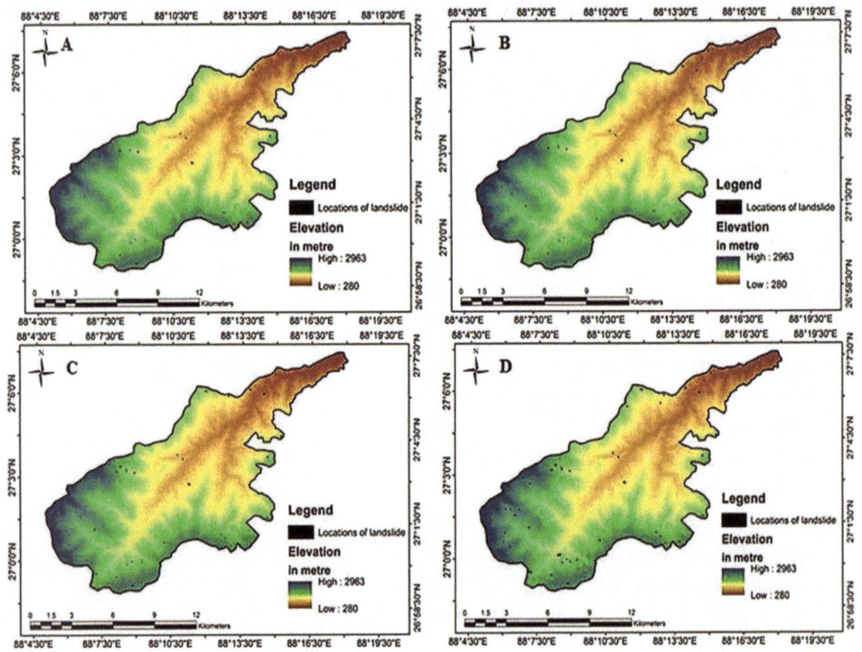

Fig. 5.2 Location of landslide patches on digital elevation model and landslide inventory map (**a**). 1991. (**b**). 2001. (**c**). 2011 and (**d**) 2022

5.3.3 Validation of the Model

The AHP model is validated based on the ROC. 'ROC (Receiver operating curve) is constructed by plotting the true positive rate (sensitivity) opposite the false positive rate (1-specificity) with the various cut-off thresholds' (Ghosh and Lepcha 2019). The area under the ROC is used to evaluate the models quantitatively. The value of AUC 1 means a perfect model, and when AUC equals 0, it means a non-perfect model.

5.3.4 Preparation of Factor Cluster

Five-factor cluster maps (Figs. 5.23, 5.24, 5.25, 5.26, and 5.27) have been prepared to see the impact of single-factor clusters in the landslide vulnerability in the Chhota Rangit river basin. To prepare the factor cluster maps, the 18 selected effective factors were first grouped into 5 different groups such as causing factors (precipitation, peak ground acceleration), ground factors (geology, soil texture, lineament density, gravity anomaly, distance from the lineament), surface causal factors (slope, elevation, slope curvature, slope aspect, SPI, distance from the stream, TWI), anthropogenic factors (LULC, road density, distance from the road) and protective factors (NDVI). The cluster diagrams were then prepared using a weighted linear combination model (Eq. 5.3).

$$\mathrm{WLC} = \sum_{n}^{j=1} a_{ij} w_j \tag{5.3}$$

The a_{ij} = ith rank of jth attribute; w_j = weightage of jth attribute (Ghosh and Lepcha 2019).

5.4 Results and Discussion

The spatial database mainly causes the landslide hazard and has been taken into consideration. Altogether, 18 factors are considered to make the landslide vulnerability map concerning LULC dynamicity of the Chhota Rangit basin since 1991 to 2022 using the AHP model in ArcGIS 10.3.1. Thus, 18 thematic data layers were prepared based on spatial geo-database such as geology, elevation, slope, NDVI, LULC, annual precipitation, lineament density, distance from the lineament, road density, distance from the road, soil texture, curvature, distance from the stream, SPI, earthquake magnitude, TWI, slope aspect and gravity anomaly.

5.4.1 Geology

Geology is an essential factor for landslide vulnerability studies (Dai and Lee 2002; Ayalew and Yamagishi 2005). Darjeeling district's geological data was extracted from NATMO and GSI (Geological Survey of India), and from there, the geology map of the Chhota Rangit river basin was extracted in raster format utilising ArcGIS 10.3.1. Two geological groups, the Damuda series and Daling and Darjeeling gneiss, are existent in the region (Fig. 5.3). The Damuda series belongs to the Permian Age bearing the lithological composition of quartzitic sandstones with slates, carbonaceous shales, seams of graphitic coal and marginal strips of limestone. Another geological group is Daling and Darjeeling gneiss, which belongs to the Precambrian period. Precambrian lithological characteristics, such as unclassified crystals, mainly gneisses, are present in this group of geological formations.

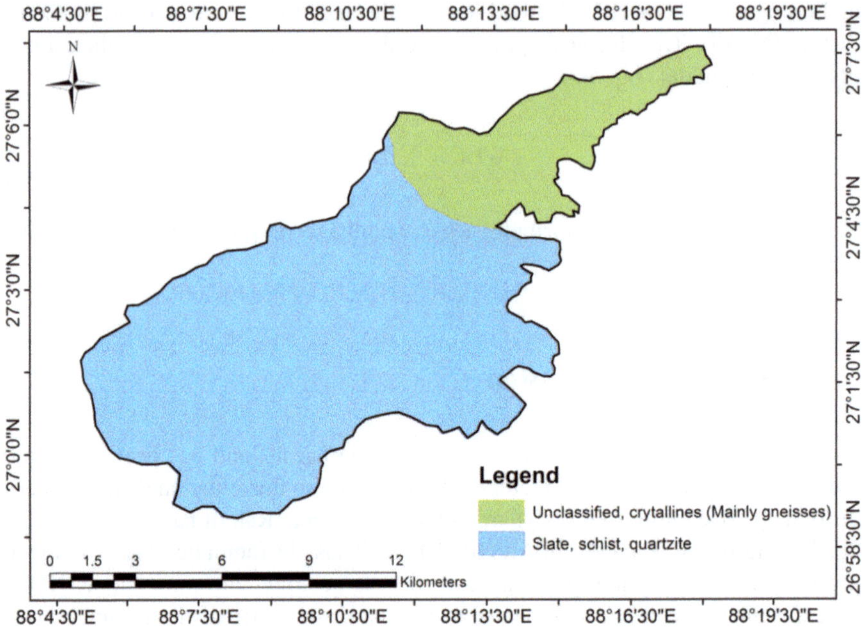

Fig. 5.3 Geological map of the Chhota Rangit river basin

5.4.2 *Elevation*

Elevation represents the local relief, including the highest and the lowest points of an area relative to mean sea level. Mostly, atmospheric phenomena such as temperature and rainfall are influenced by elevation, which impacts the growth of plants, and these conditions probably influence slope instability (Vivas 1992). The elevation map of the region is extracted from ASTER-DEM using GIS 10.3.1. Elevation of the study area (Fig. 5.4) is grouped into five classes using quantile classification method, and it varies between 280 m and 2963 m above the mean sea level and acts as the important parameter for landslide vulnerability in the region.

5.4.3 *Slope*

The slope is a key factor for determining the landslide vulnerability zones as it has a direct impact on the landslide occurrences (Lee and Min 2001). After the precipitation events, the slope steepness affects subsurface flow velocity, surface runoff rate and water content of the soil. Typically, the shear stress of soil, especially in the unconsolidated soil layer, tends to rise as the slope increases. A slope map of the study area has been created from ASTER-DEM in GIS 10.3.1 environment. The slope map of the Chhota Rangit river basin (Fig. 5.5) has been categorised into ten

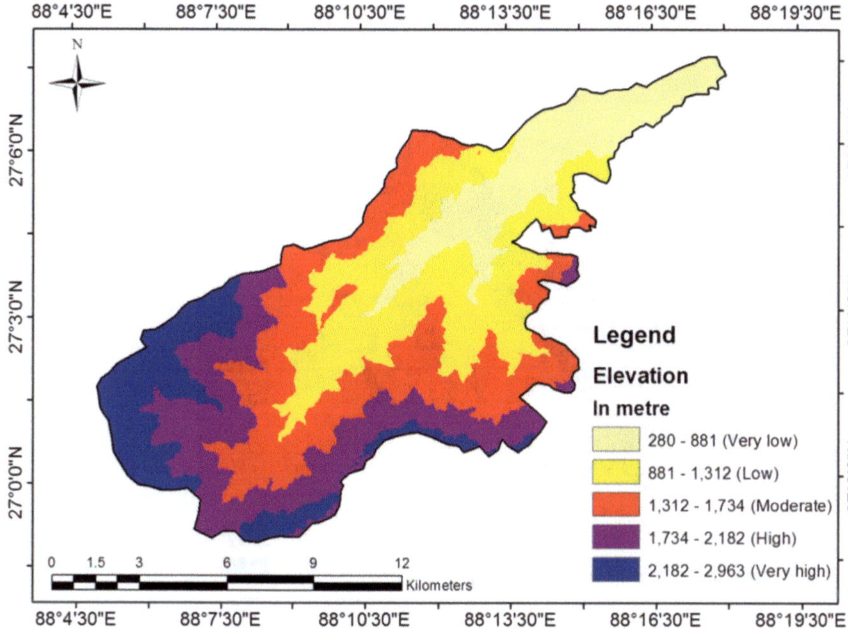

Fig. 5.4 Elevation map of Chhota Rangit river basin

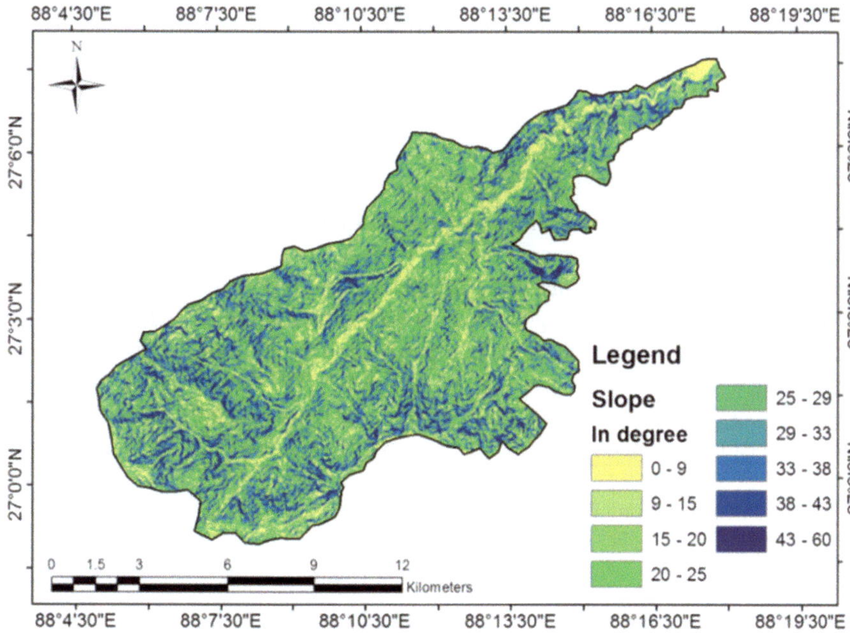

Fig. 5.5 Slope map of Chhota Rangit river basin

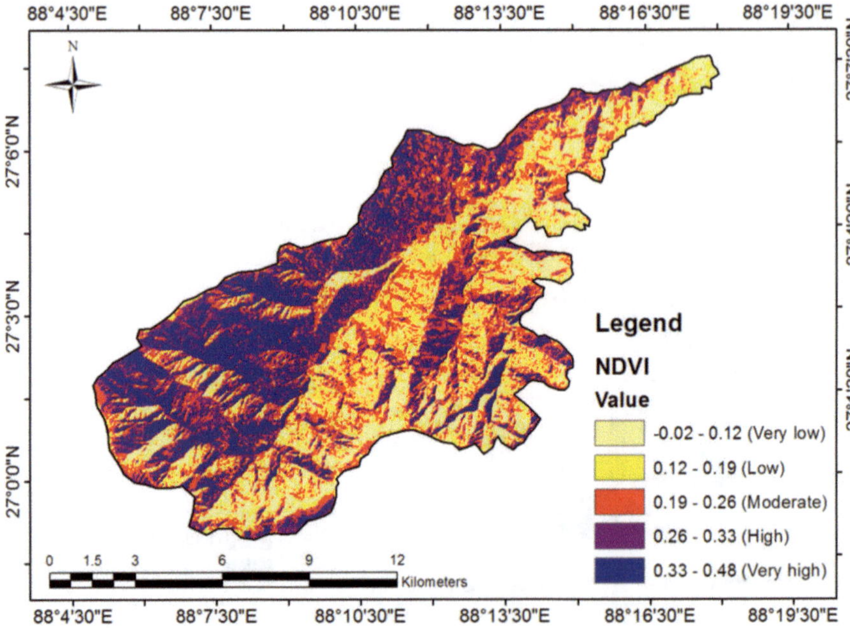

Fig. 5.6 NDVI map of Chhota Rangit river basin

classes utilising the quantile classification approach. The slope steepness of the Chhota Rangit river basin ranges from 9° to 60°.

5.4.4 NDVI

The NDVI is an efficient and widely used technique for understanding vegetation's health and quality. A healthy forest cover protects the area from soil erosion and landslide occurrences (Sharma and Ram 2014). The NDVI map (Fig. 5.6) of the river basin was generated using satellite images from Landsat-5 TM data for 1991 and 2011, Landsat-7 for 2001 and Landsat-8 OLI for 2022 employing the raster calculator in GIS software with Eq. 5.4. Higher NDVI values indicate healthier vegetation, whereas lower values suggest poorer vegetation health. The NDVI value for river basin is grouped into five classes such as very low, low, moderate, high and very high.

For NDVI extraction, Townshend and Justice (1986) method is used:

$$\text{NDVI} = \frac{(\text{IRband} - \text{Rband})}{(\text{IRband} + \text{Rband})} \tag{5.4}$$

where IR is the near-infrared band and R is the red band.

5.4.5 (Land Use/Land Cover)

The stability of the slopes is significantly affected by the nature of LULC (Ocakoglu et al. 2002). Often land cover serves as the protection that lowers the likelihood of soil erosion. Recently, the dynamic changes in land use in hilly regions have had a substantial impact on the mechanisms of soil erosion and landslides. To create the LULC map for the Chhota Rangit river basin, satellite images from Landsat-5 TM for 1991 and 2011, Landsat-7 for 2001 and Landsat-8 OLI for 2022 were processed using image classification tools in the GIS software. The LULC in the area under study has been categorised into seven distinct classes such as water bodies, settlement, forest, social forestry, agriculture (single crop), agriculture plantation (tea) and agriculture (fallow) (Fig. 5.7). Spatial distribution and patterns of different LULC in Chhota Rangit is shown in Table 5.6. According to the analysis, the forest cover has progressively reduced, which was 23.97% in 1991, 23.82% in 2001, 23.14% in 2011 and only 14.25% in 2022.

5.4.6 Rainfall

Precipitation is a critical factor contributing to landslides in the hilly area of Darjeeling. Review of previous literature and conversations with the residents indicate that every past landslide has been triggered by the intense rainfall during the

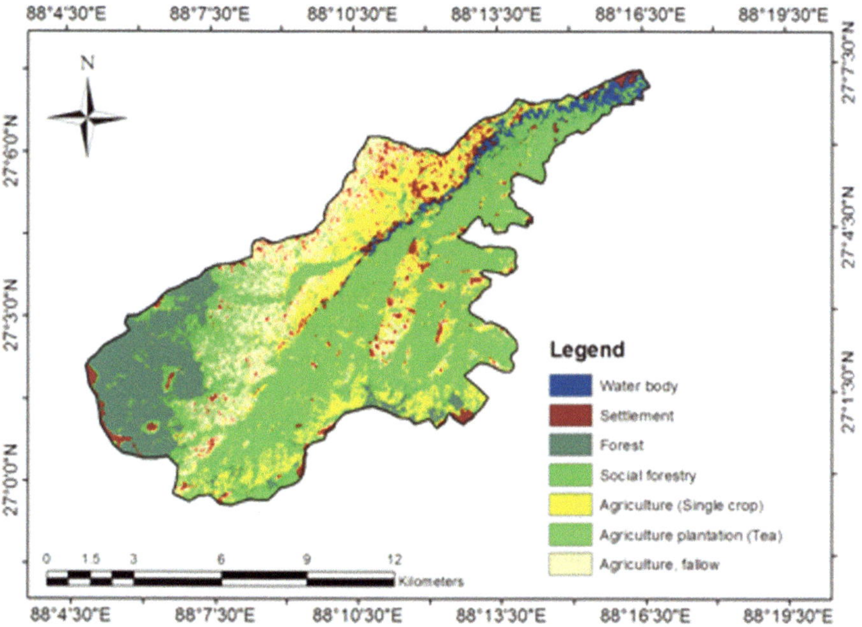

Fig. 5.7 LULC of Chhota Rangit river basin

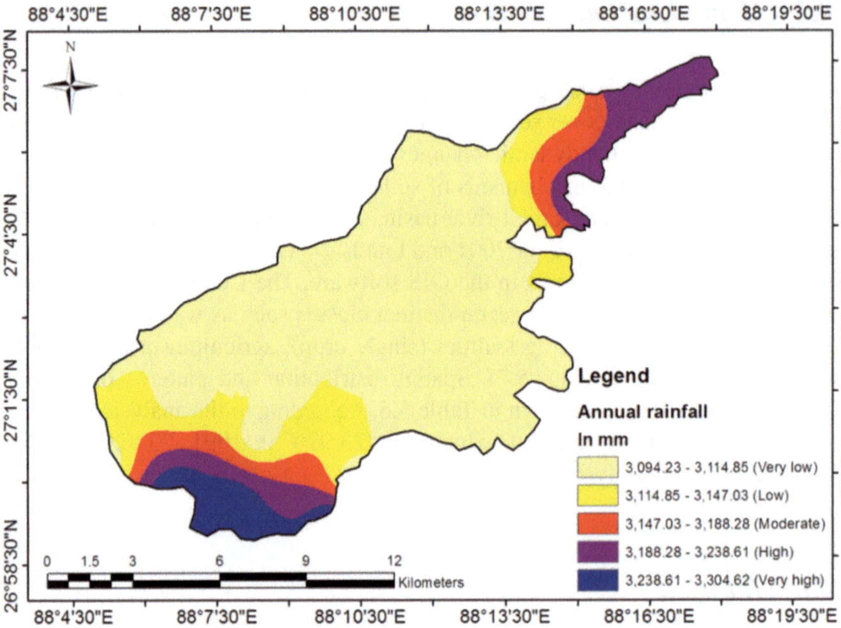

Fig. 5.8 Annual rainfall map of the Chhota Rangit river basin

rainy season. Typically, the Darjeeling region receives heavy and persistent rainfall during the monsoon season, which leads to dangerous landslides. The rainfall distribution map (Fig. 5.8) of the river basin is prepared based on TRMM data obtained from the mirador.gsfc.nasa.gov website employing the IDW method and categorised into five classes with the assistance of the natural break classifier method in ArcGIS 10.3.1.

5.4.7 Lineament Density

Lineaments are linear or curvilinear features that are influenced by structural patterns, representing the surface features associated with the underlying structures. High lineament density indicates greater jointed and fractured rocks, indicating more significant landslide chances (Mezughi and Akhir 2011). The lineament map for the study area was obtained from bhuvan.nrsc.gov.in (ISRO), and the lineament density of the river basin was generated utilising the line density analyst extension tool within ArcGIS software. It has been categorised into five levels: very low, low, moderate, high and very high (Fig. 5.9). A significant part of the study area is characterised by very low lineament density.

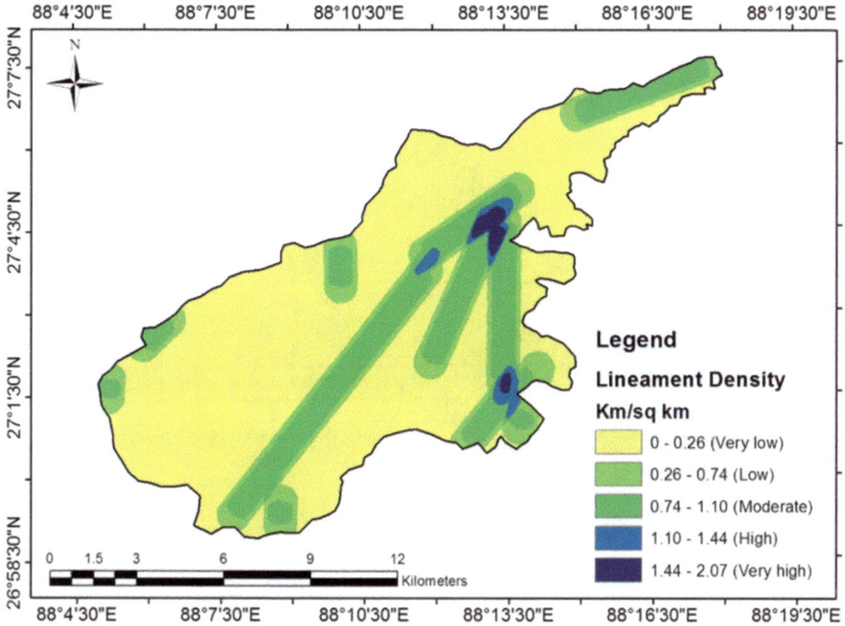

Fig. 5.9 Lineament density of the Chhota Rangit river basin

5.4.8 *Distance from the Lineament*

Lineament indicates the underlying structural character of the surface topography. It represents the weak zone of surface topography and its effects on slope stabilisation. A lineament map for the region was acquired from bhuvan.nrsc.gov.in (ISRO). The nearest distance from the lineament represents a high risk for landslide, and the far distance signifies a lower probability of a landslide. The distance from the lineament map is generated by utilising a spatial analyst's tool within the GIS software (Fig. 5.10).

5.4.9 *Road Density*

The construction of roads on the steep slopes is the critical cause for reoccurring landslides in Darjeeling. Construction of roads disturbs the stability of the hill regions, which initiates a huge number of landslides especially along the roads (Sujatha et al. 2012). Landslides along the road during the rainy season have caused a vast socio-economic hindrance to the region in recent times. To prepare the road density map (Fig. 5.11) of the Chhota Rangit river basin, firstly, all roads were digitised from Google Earth (2022), and density map was created using the line density analyst tool in ArcGIS 10.3.1 software environment.

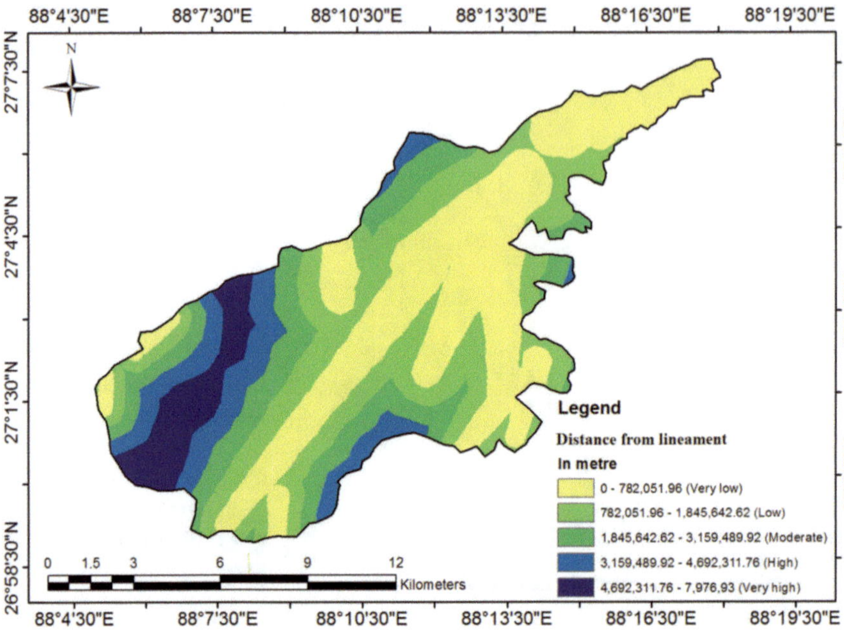

Fig. 5.10 Distance from the lineament map of Chhota Rangit river basin

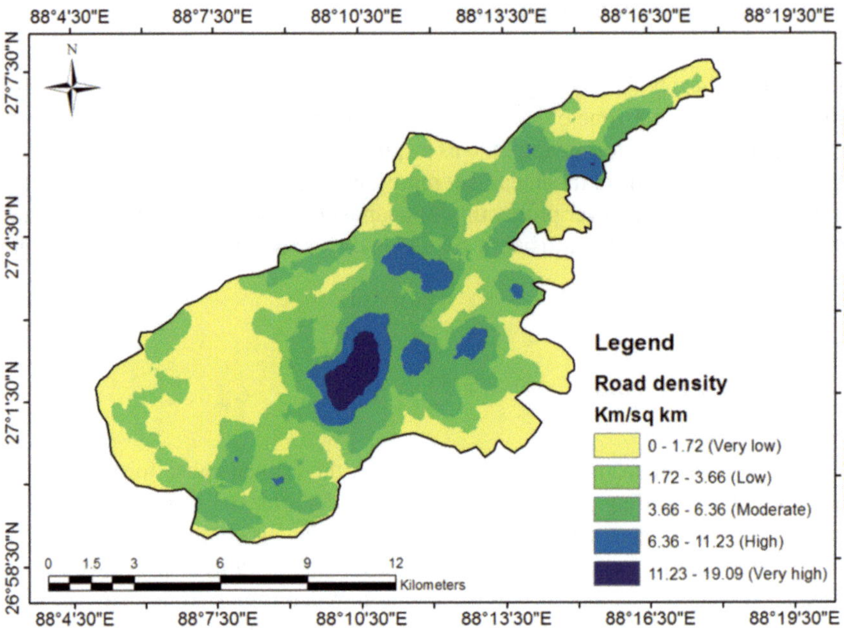

Fig. 5.11 Road density map of Chhota Rangit river basin

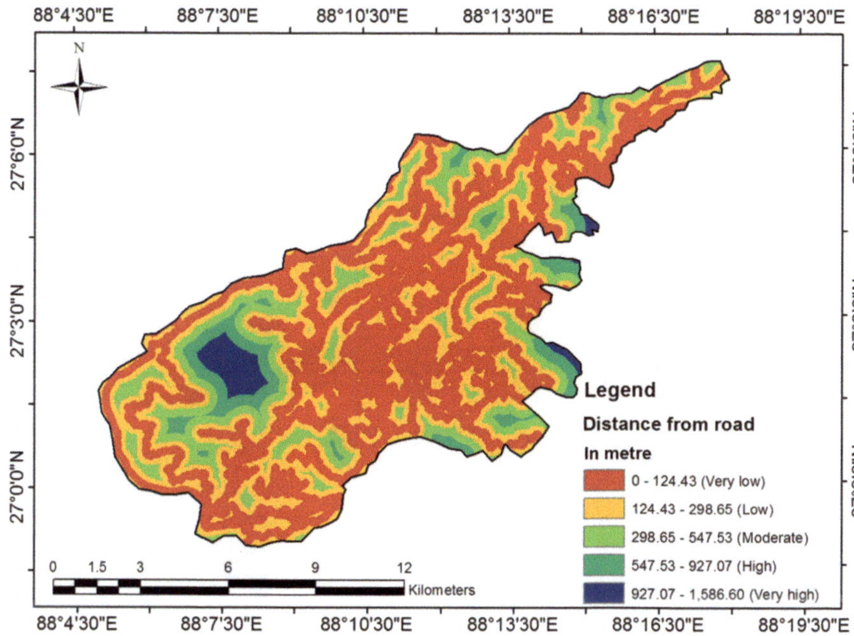

Fig. 5.12 Distance from the road map of the Chhota Rangit river basin

5.4.10 Distance from the Road

The position of the road highly impacts the likelihood of landslides (Sujatha and Rajamanickam 2012). The areas closest to the road represent the high probability of landslides, and the areas far from the road indicate a minimal probability. Distance from the road map of the Chhota Rangit river basin was generated using a spatial analyst's tool within the GIS environment (Fig. 5.12).

5.4.11 Soil Texture

Soil texture is a critical factor for preparing landslide susceptible zones. Thirty soil samples from the entire river basin were collected from 30 sites. Then, the sieving method was applied for the textural segregation of the accumulated samples. Lastly, a soil texture map was generated through IDW methods in GIS software. The soil texture of the area varies from 44.45% to 73.85% (Fig. 5.13).

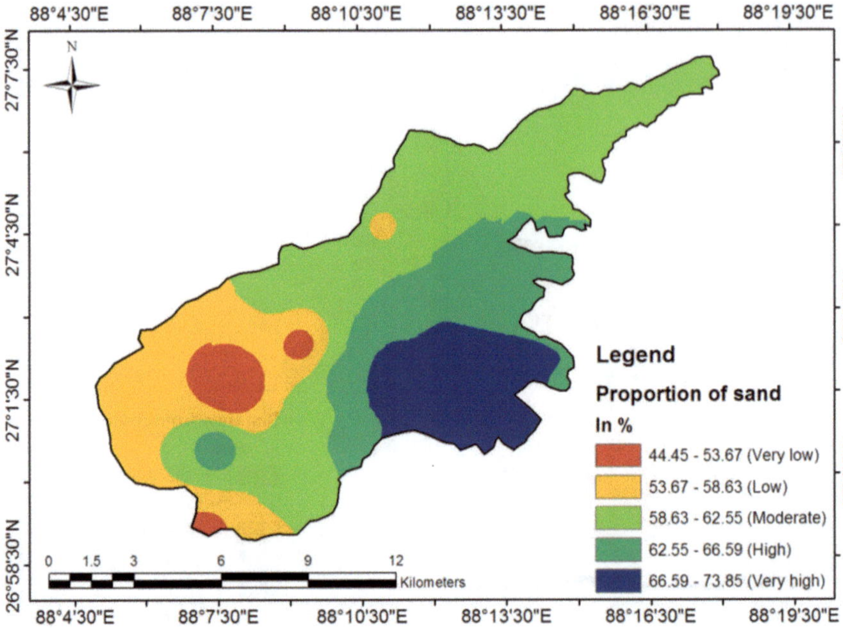

Fig. 5.13 Soil texture of Chhota Rangit river basin

5.4.12 Slope Curvature

Slope curvature helps to understand the changing landform character (Gilbert 1909). Slope curvature is a topographic attribute for landslides (Lee and Sambath 2006), affecting the soil cover's hydrological condition. After precipitation, the concave slope holds maximum water and preserves it for extended time than the convex slope. As a result, the concave slopes have a higher probability of landslide occurrences. A slope curvature map of the region was created from ASTER-DEM using surface analyst's tools. The slope curvature of the study area is classified into three categories: concave, flat and convex (Fig. 5.14).

5.4.13 Distance from the Stream

Distance from the stream influences the stability of a slope (Sar et al. 2016). The areas closer to rivers have a greater likelihood of landslide incidents. The Chhota Rangit river system drains the entire Chhota Rangit river basin. Distance from the river for the study area is grouped into five classes: very low, low, moderate, high and very high (Fig. 5.15).

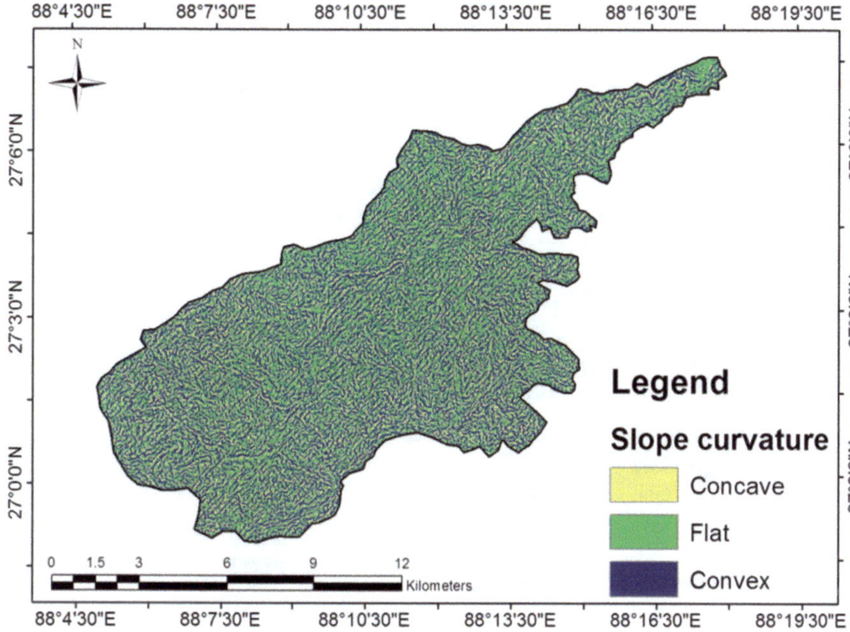

Fig. 5.14 Slope curvature map of Chhota Rangit river basin

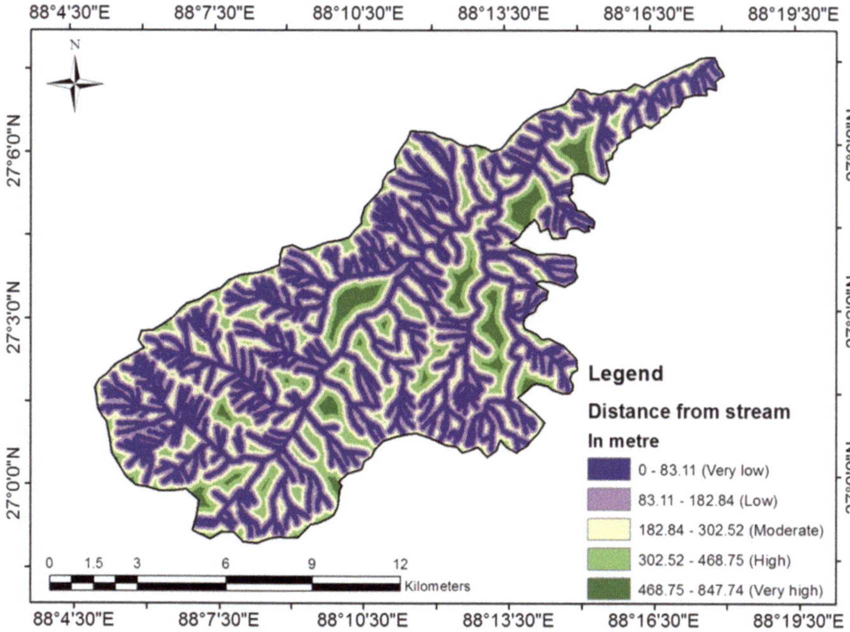

Fig. 5.15 Distance from the stream map of the Chhota Rangit river basin

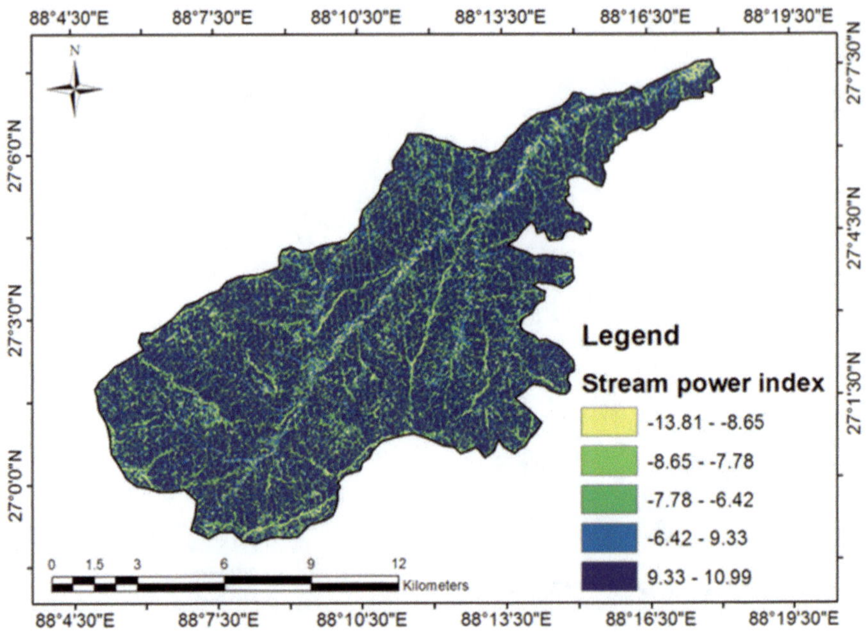

Fig. 5.16 Stream power index map of Chhota Rangit river basin

5.4.14 Stream Power Index

SPI is a multiplex topographic feature (Moore et al. 1991). It is a useful indicator for measuring the erosive power of water and slope erosion (Nefeslioglu et al. 2008). The stream power index of the study area is generated from ASTER-DEM using Eq. 5.5 used by Moore et al. (1991) through the raster calculator within the ArcGIS environment (Fig. 5.16):

$$\mathrm{SPI} = \mathrm{As} \times \tan \beta \tag{5.5}$$

where As is the upslope contributing region per unit contour length and β is the local slope gradient.

5.4.15 Peak Ground Acceleration

PGA may be described as the utmost ground acceleration at the time of earthquake. Peak ground acceleration may also be an important landslide-causing factor for geologically fragile areas like hills and mountains. For this study, a peak ground acceleration (PGA) map is prepared from a secondary source (http://www.wbdmd.gov.in/writereaddata/Chapter-4.pdf). Peak ground acceleration (PGA) of the Chhota Rangit river basin is classified into five categories, and the value of PGA ranges from 0.301 g to 0.324 g (Fig. 5.17).

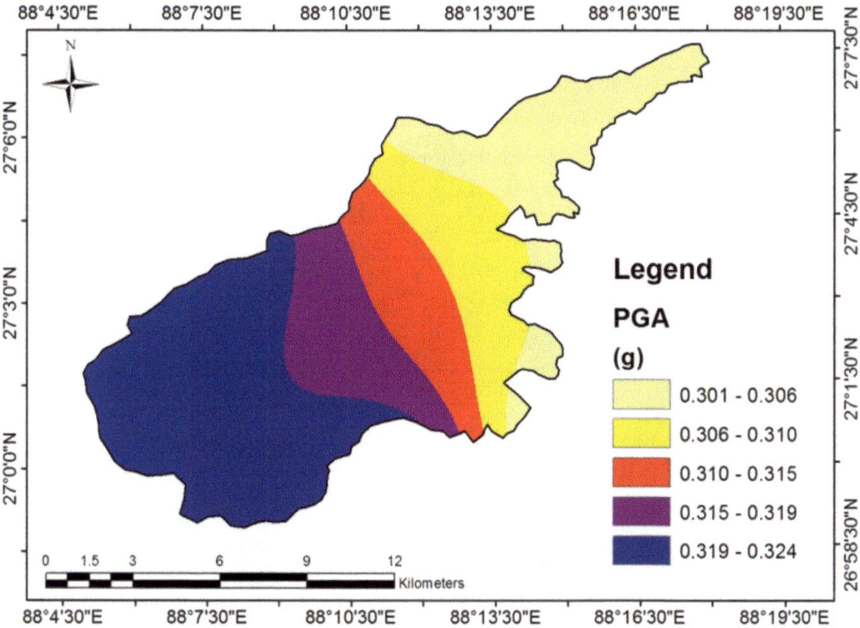

Fig. 5.17 Peak ground acceleration map of the Chhota Rangit river basin

5.4.16 Topographic Wetness Index

The TWI helps to understand and quantify hydrological processes and topographic control (Sörensen et al. 2006). It explains the role of topography about the positioning and size of the saturated source regions of runoff production (Pradhan and Kim 2014). The TWI map of the Chhota Rangit river basin was created from ASTER-DEM using Eq. 6 through the raster calculator within ArcGIS software, and it is classified into five classes (Fig. 5.18). The high TWI values are mainly observed in valley bottoms, terrace surfaces and moderate slopes.

$$\text{TWI} = \ln\left(\alpha / \tan\beta\right) \tag{5.6}$$

where α is the upslope contributing area and β is the local slope gradient.

5.4.17 Slope Aspect

The slope aspect is another prime parameter for preparing landslide susceptibility maps (Guzzetti et al. 1999; Saha et al. 2002; Lee 2005). Slope aspect elements like exposures to sunlight, winds, rainfall, etc. determine the amount of moisture on the ground that affects the likelihood of landslides in the area (Komac 2006; Magliulo

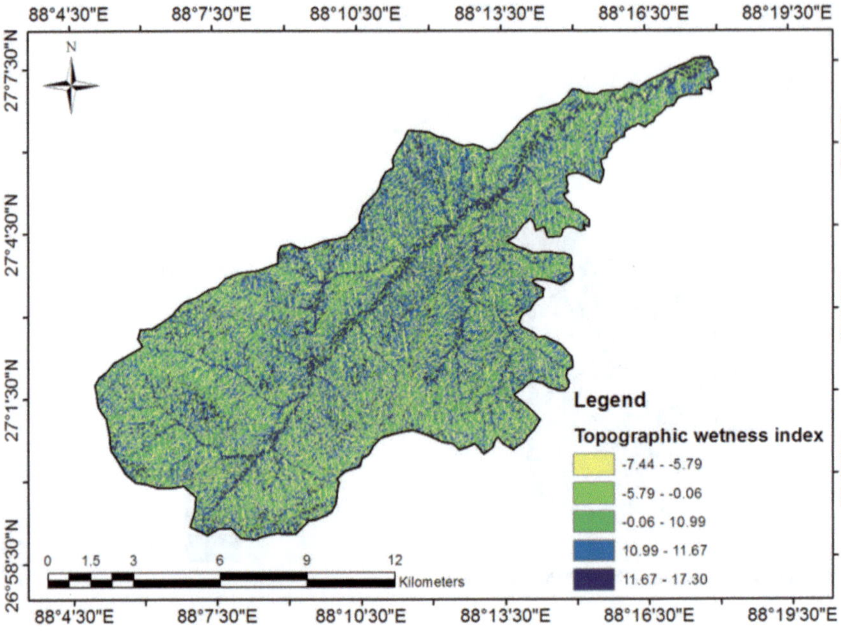

Fig. 5.18 TWI map of Chhota Rangit river basin

et al. 2008). The slope aspect map of the Chhota Rangit river basin is prepared from ASTER-DEM using 3D analyst tools within ArcGIS software. The slope aspect of the region is divided into ten categories (Fig. 5.19).

5.4.18 Gravity Anomaly

Gravity anomaly represents the variation in the density of rock. Gravity anomaly is low where the rock density is low, and the higher value of gravity anomaly reflects more rock density than the surrounding areas, showing the higher possibility for landslides. The gravity anomaly map for Chhota Rangit basin is generated from secondary sources (Mahadevan 2003) and it varies from −67.3645 milligals to −69.8684 milligals (Fig. 5.20).

5.4.19 Landslide Susceptibility Model

The sudden and frequent landslides in and around the Chhota Rangit river basin in recent times due to increased human interventions through continuous LULC changes including the other natural factors have become one of the significant

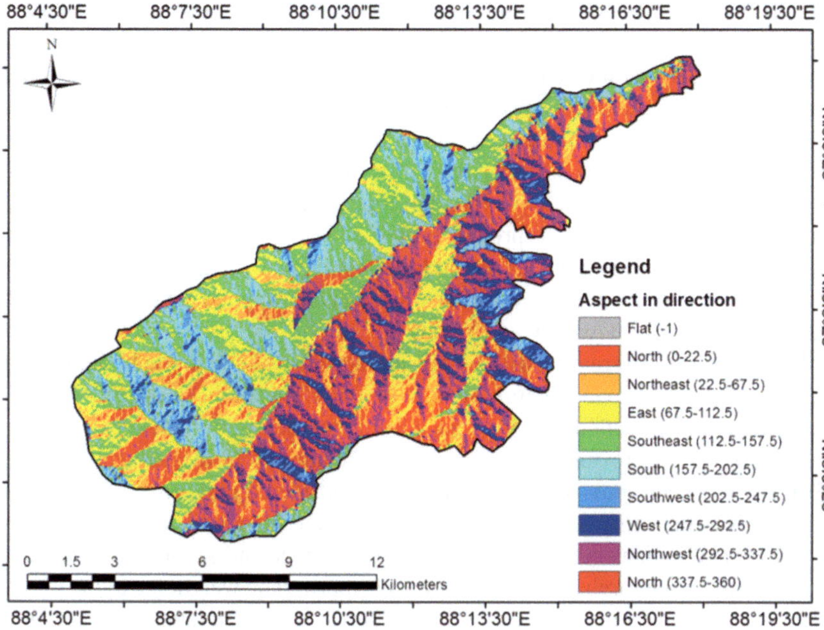

Fig. 5.19 Slope aspect map of Chhota Rangit river basin

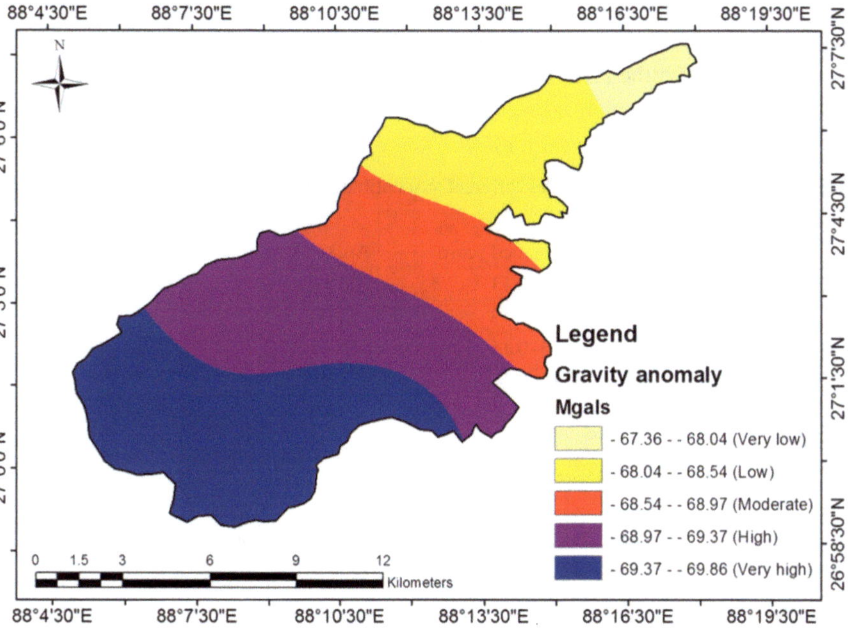

Fig. 5.20 Gravity anomaly map of Chhota Rangit river basin

reasons for socio-economic disturbances, especially during the rainy seasons. The region has also experienced increments in the regional and temporal distribution of landslide threats over the years, making many places and people susceptible to landslide hazards. The landslide susceptibility study, thus, can be the best step to identify and delineate the probable areas of landslide hazards of different magnitudes in the river basin. The landslide susceptibility analysis helps identify the areas at higher risk, which would help planner developers and authorities reduce the damage caused by landslide hazards through proper and appropriate mitigating measures (Solaimani et al. 2013). Various methods are available to analyse landslide susceptibility, and the AHP is one of the most extensively applied models for landslide susceptibility studies (Ayalew and Yamagishi 2005).

Eighteen (18) different landslide-causing factors, which include both the natural factors and the man-made factors, such as geology, elevation, slope, NDVI, LULC, precipitation, lineament density, distance from the lineament, road density, distance from the road, soil texture, slope curvature, distance from the stream, SPI, PGA, TWI, slope aspect, gravity anomaly, etc., are considered and ranked for preparing landslide susceptibility zones of the study area. They are ranked according to the hypothetical assumption and based on field experience and secondary information. Saaty's analytic hierarchy process assigns the appropriate weightage to the different landslide-causing factors in the river basin (Table 5.4). Each landslide-causing factor is further classified into subclasses, which are also given weight based on the analytic hierarchy processes. Table 5.5 represents the weights of the 18 landslide-causing factors, their subclasses, and the calculated consistency ratio (CR). The calculated consistency ratio of each selected parameter in the Chhota Rangit river basin ranges from 0.01 to 0.09 (Table 5.5). Finally, the landslide susceptible map is generated through the AHP model employing Eq. 7 in the raster calculator in GIS software.

Table 5.4 Percentage of area under different susceptible zones in the Chhota Rangit river basin

Soil erosion susceptible zone	Classified score	Area in km^2	% of area
Very low	0.05–0.11	26.16	18.55
Low	0.11–0.13	32.04	22.72
Moderate	0.13–0.16	34.45	24.43
High	0.16–0.18	34.91	24.76
Very high	0.18–0.26	13.40	9.50
Total		140.98	100

Source: Compiled by the authors

Table 5.5 Connection between landslide vulnerability layers and various factor clusters

Factors	Triggering	Lithological	Surface	Anthropogenic	Protective
Calculated values	0.749796	0.987436	0.953286	0.957537	0.953286

Source: Compiled by the authors

$$
\begin{aligned}
&(\text{Road density} \times 0.10) + (\text{LU / LC} \times 0.17) + (\text{Slope} \times 0.13) \\
&+ (\text{NDVI} \times 0.09) + (\text{Elevation} \times 0.07) + (\text{Rainfall} \times 0.09) \\
&+ (\text{Lineament density} \times 0.04) + (\text{Distance from} \\
&\text{the lineament} \times 0.04) + (\text{Geology} \times 0.04) \\
&+ (\text{Distance from the road} \times 0.03) + (\text{Soil texture} \times 0.02) \\
&+ (\text{Slopecurvature} \times 0.01) + (\text{Distance from the stream} \times 0.04) \\
&+ (\text{SPI} \times 0.01) + (\text{PGA} \times 0.01) + (\text{TWI} \times 0.01) \\
&+ (\text{Slope aspect} \times 0.009) + (\text{Gravity anomaly} \times 0.009)
\end{aligned}
\tag{5.7}
$$

To gain a clearer understanding of the landslide risks in the river basin, the susceptibility map for the Chhota Rangit river basin is divided into five levels: very low (0.05–0.11), low (0.11–0.13), moderate (0.13–0.16), high (0.16–0.18) and very high (0.18–0.28) (Fig. 5.21).

It is observed that very low and low susceptible zones are mostly present on the left bank of the basin, particularly in the lower part of the upper catchment, a large part of the middle catchment and some scattered parts on the lower and right sides of the river. These zones are primarily located in areas covered either by the natural forest or man-made forest coverage. Nearly 26.72 sq. km (18.95%) (Table 5.4) of the study area falls under a very low susceptible zone, and about 33.05 sq. km

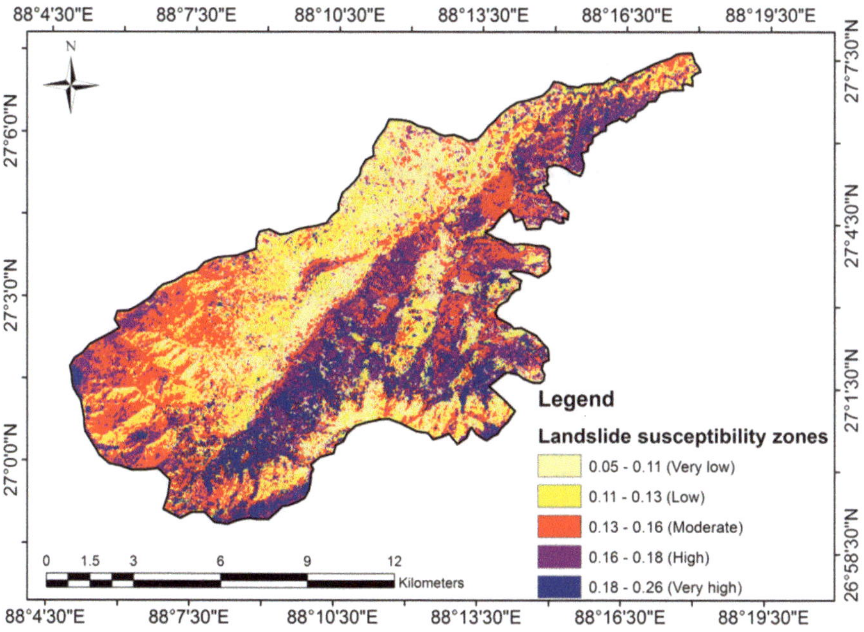

Fig. 5.21 Landslide susceptible map of Chhota Rangit river basin

(23.44%) (Table 5.4) area falls under a low susceptible zone. The most significant portion, 35.40 sq. km (25.11%) of the river basin, falls under the moderate landslide susceptible zone. This zone is scattered throughout the basin, but the more significant portion is in the upper catchment, mainly on the right bank of river. They are scattered on moderate to steep slopes and highly intervened by the people, like the plantation areas.

High landslide susceptible zone covers nearly 28.53 sq. km (20.24%) of the river basin (Table 5.4). The areas under the high landslide susceptibility are in small patches in the upper catchment, and the maximum areas under this zone are in middle and lower catchment, especially on the right bank. The high landslide susceptible zone mainly falls in areas with high road density, primarily covered by human settlement and plantation areas, low natural vegetation coverage and varying slope conditions from very gentle to steep slopes.

A very high susceptible zone covers around 17.25 sq. km (12.24%) of the Chhota Rangit river basin. This zone is present in the small patches in upper catchment, a large portion of middle catchment and scattered portions in the lower catchment. It is known from the susceptibility map (Fig. 5.21) that very high susceptible areas are present, mainly on the right bank river, which is covered mainly by tea plantations, human settlements, high road density, building of new routes and slopes varying from gentle to steep. The detail analysis of the susceptibility zonation of the study area shows that a major portion of the basin is under threat of moderate to high landslide susceptibility, and a considerably high part of the area remains under a very high susceptible zone. These landslide susceptible zones seem to be highly influenced by natural and anthropogenic factors. Thus, it can be rightly said that the location of susceptibility zones of the basin area is highly influenced by LULC dynamicity and other factors that cause landslides.

5.4.20 Validation of Landslide Susceptible Model Through ROC Curve

The ROC model (Fig. 5.22) has been developed using 60 landslide spots derived after the inventory map to assess the validation of susceptibility. The area under the ROC curve is calculated for qualitative comparison of the models. The AUC values can be categorised into five classifications: excellent (0.90–1.00), good (0.80–0.90), fair (0.70–0.80), poor (0.60–0.70) and fail (0.50–0.60), indicating varying levels of accuracy (Rasyid et al. 2016). In this study, the AUC is found to be 0.90, which qualifies as excellent class (Rasyid et al. 2016). Therefore, it can be concluded that the landslide susceptibility map generated through the AHP model is reliable for identifying the landslide susceptibility zones within the river basin.

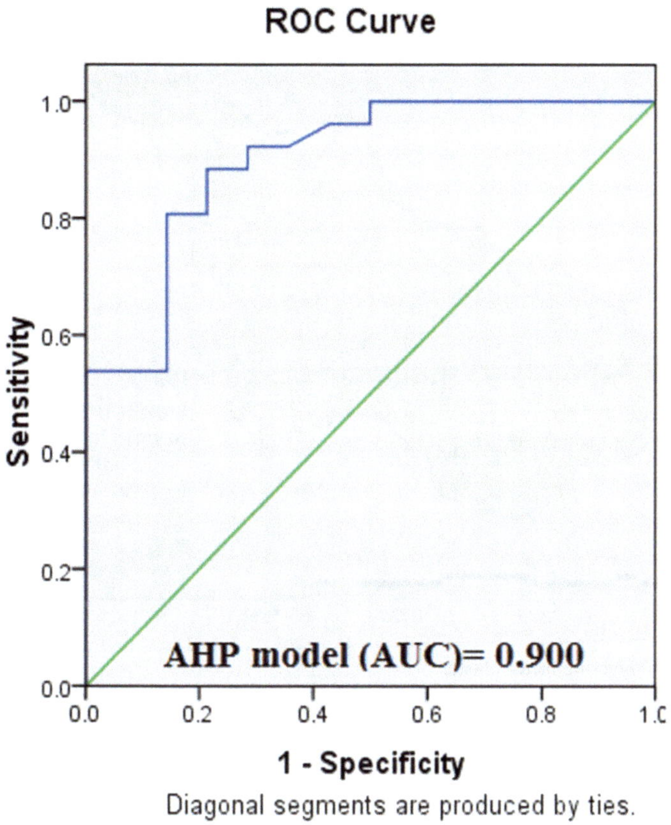

Fig. 5.22 Receiver operating curve (ROC)

5.4.21 Responsible Factor Clusters for Landslide Vulnerability in the Chhota Rangit Basin

According to historical recording of the landslide sites and the generated factor cluster maps, it is known that the triggering factor (Fig. 5.23) exhibits a low relationship with landslide susceptibility (Table 5.5). In contrast, the lithological factor cluster (Fig. 5.24) demonstrates the highest relationship with landslide exposure in the researched area (Table 5.5). The surface causal factor cluster (Fig. 5.25) and the anthropogenic factor cluster (Fig. 5.26) seem to share an equal relationship with the landslide susceptibility (Table 5.5). Even though maximum of the landslides in the region happens in the monsoon time and the whole region experiences substantial rainfall, areas with highest landslide susceptibility are not located within the area of highest rainfall. The protective factor cluster (Fig. 5.27) shows a positive relationship with the landslide susceptibility zones (Table 5.5), i.e. low landslide susceptibility zones are in areas with high vegetation cover and vice versa.

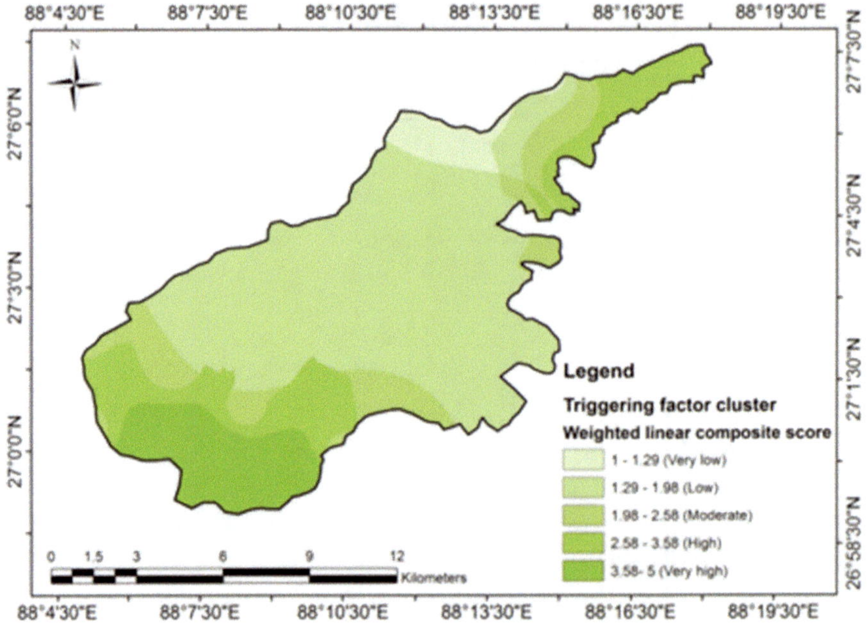

Fig. 5.23 Triggering factor cluster

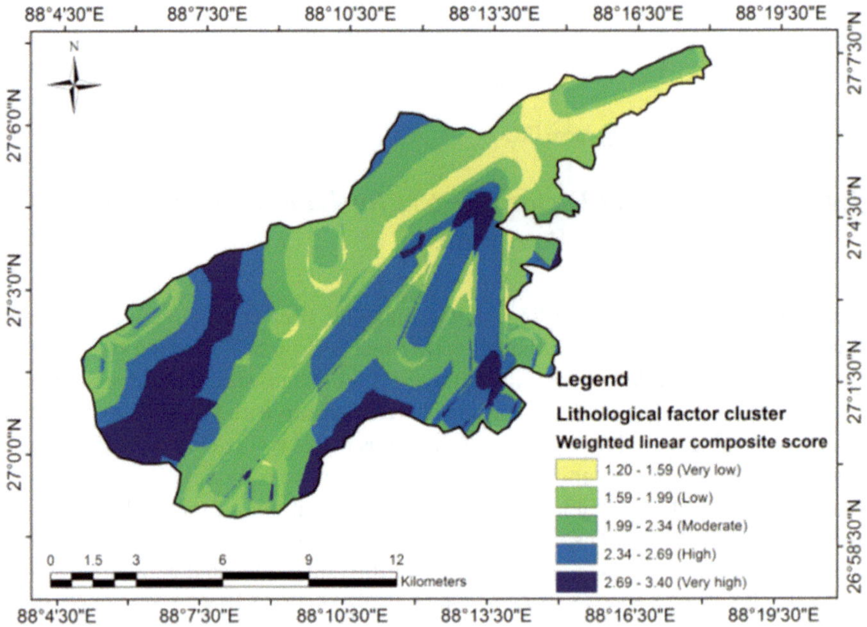

Fig. 5.24 Lithological factor cluster

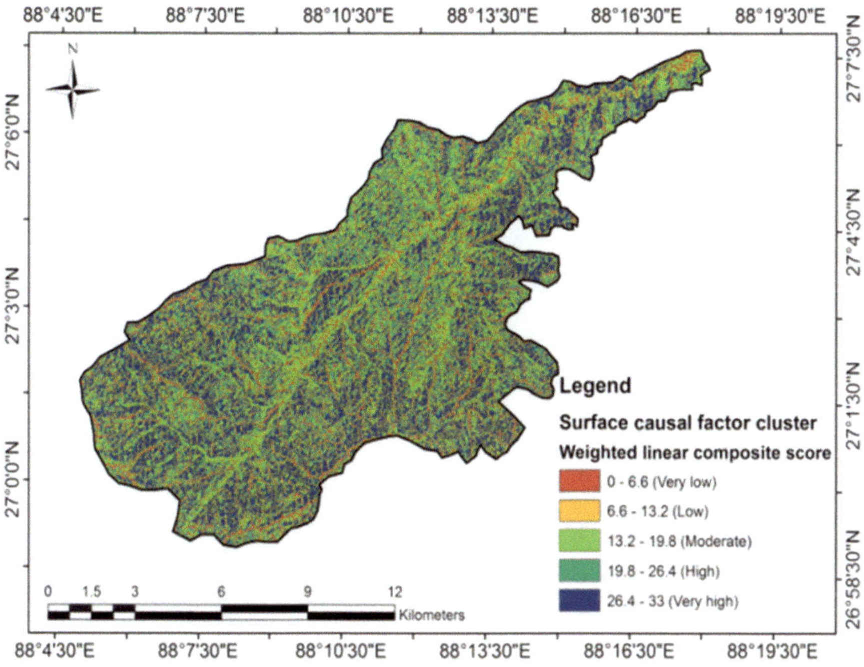

Fig. 5.25 Surface causal factor cluster

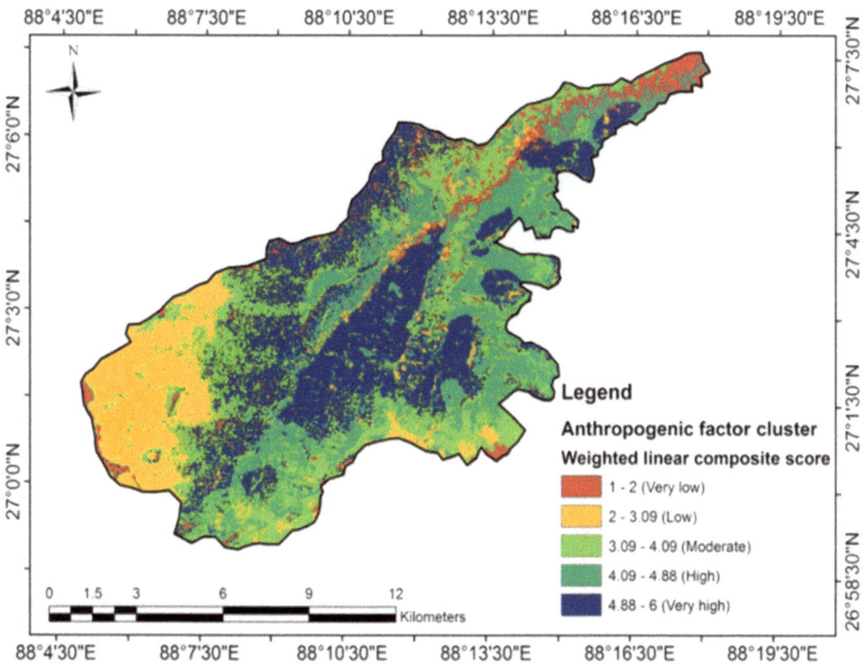

Fig. 5.26 Anthropogenic factor cluster

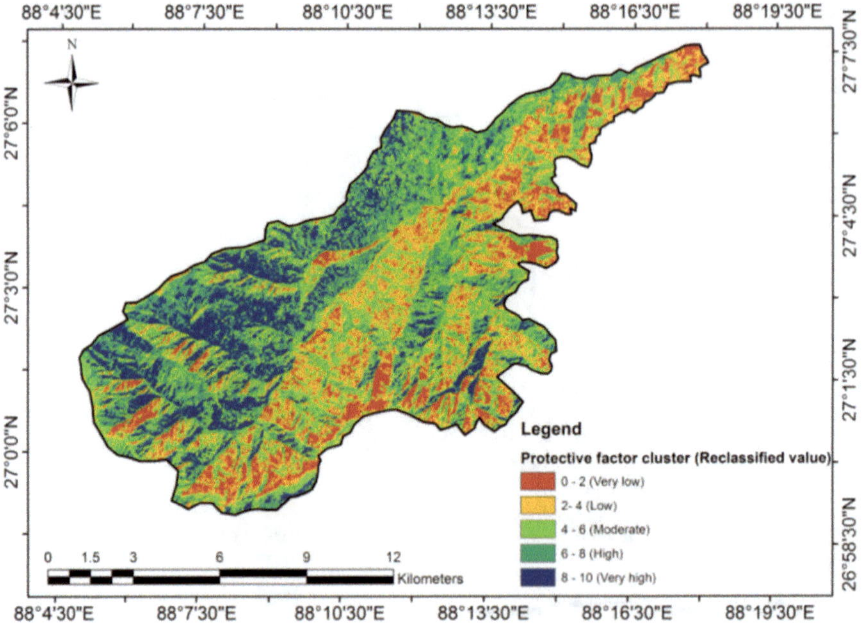

Fig. 5.27 Protective factor cluster

5.4.22 Landslide Distribution in Different LULC from 1991 to 2022

The details of the spatiotemporal pattern of landslides in Chhota Rangit river basin have been compiled in Table 5.6 as per the historical records, field studies and other secondary sources. Table 5.6 shows that along with the changes in LULC characteristics, area under landslide has also changed over the period. The places occupied with water bodies was 6.60 sq.km in 1991, 7.12 sq.km in 2001, 6.74 sq.km in 2011 and 4.69 sq.km in 2022, while the areas under landslide in water bodies were 0.47 sq.km in 1991, 0.91 sq.km in 2001, 0.92 sq.km in 2011 and 2.05 sq.km in 2022. The area under settlement was 6.21 sq.km in 1991, 8.06 sq.km in 2001, 8.20 sq.km in 2011 and 8.75 sq.km in 2022, and the area under landslide in settlement area was 1.61 sq.km in 1991, 2.53 sq.km in 2001, 2.55 sq.km in 2011 and 2.83 sq.km in 2022. Likewise, the area covered by forest was 33.80 sq. km in 1991, 33.59 sq.km in 2001, 32.62 sq.km in 2011 and 20.09 sq.km in 2022, and the landslide area under forested area was 1.71 sq. km in 1991, 2.37 sq.km in 2001, 2.33 sq. km in 2011 and 1.32 sq.km in 2022. The area under social forest was 8.49 sq.km in 1991, 12.33 sq.km in 2001, 21.41 sq.km in 2011 and 18.44 sq.km in 2022, and the landslide under social forestry was 2.35 sq.km in 1991, 1.72 sq.km in 2001, 1.27 sq. km in 2011 and 2.35 sq.km in 2022. The area under agriculture single crop was 35.52 sq.km in 1991, 10.54 sq.km in 2001, 9.61 sq.km in 2011 and 25.98 sq.km in 2022, and landslide area was 0.88 sq.km in 1991, 1.4 sq.km in 2001, 1.29 sq.km in

Table 5.6 Landslides in different LULC during 1991 to 2022 in Chhota Rangit river basin

Land use/land cover classes	1991 Area/km^2		2001 Area/km^2		2011 Area/km^2		2022 Area/km^2	
	LU/LC	Landslide	LU/LC	Landslide	LU/LC	Landslide	LU/LC	Landslide
Waterbody	6.60	0.47	7.12	0.91	6.74	0.92	4.69	2.05
Settlement	6.21	1.61	8.06	2.53	8.20	2.55	8.75	2.83
Forest	33.80	1.71	33.59	2.37	32.62	2.33	20.09	1.32
Social forestry	8.49	2.35	12.33	1.72	21.41	1.27	18.44	2.35
Agriculture (single crop)	35.52	0.88	10.54	1.4	9.61	1.29	25.98	1.37
Agriculture plantation (tea)	42.68	1.23	50.80	1.25	42.24	1.91	42.03	2.0
Agriculture (fallow)	7.65	2.74	18.51	0.75	20.11	0.73	20.96	1.09
Total	140.98	10.99	140.98	10.93	140.98	11	140.98	11.91

Source: Compiled by the authors

2011 and 1.37 sq.km in 2022. Meanwhile, area covered by plantation was 42.68 sq.km in 1991, 50.80 sq.km in 2001, 42.24 sq.km in 2011 and 42.03 sq.km in 2022, and the landslide area was 1.23 sq.km in 1991, 1.25 sq.km in 2001, 1.91 sq.km in 2011 and 0.9 sq.km in 2022. Lastly, the area under agriculture (fallow) was 7.65 sq. km in 1991, 18.51 sq.km in 2001, 20.11 sq.km in 2011 and 20.96 sq.km in 2022, and the landslide area under agriculture (fallow) was 2.74 sq.km in 1991, 0.75 sq. km in 2001, 0.73 sq.km in 2011 and 1.09 sq.km in 2022, respectively. Thus, Table 5.6 depicts the changes in landslide area coverage and LULC in the researched area.

5.5 Major Findings

The creation of landslide vulnerability map and developing the factor cluster maps with a GIS-based AHP model have revealed the following major findings in the research area.

- Landslide vulnerability zone of Chhota Rangit river basin varies from a very low (0.05–0.11) zone to a very high (0.18–0.28) zone.
- Nearly 26.72 sq. km (18.95%) of the basin falls in the category of very low susceptibility, and around 33.05 sq. km (23.44%) are in a low susceptible area. Majority of these zones are located on the left bank, where the NDVI values are high (0.26–0.33) to very high (0.33–0.48).
- The largest portions of the river basin, around 35.40 sq. km (25.11%) and 28.53 sq. km (20.24%), are under moderate to high landslide susceptibility zones. These zones mostly engulf the parts of upper and middle catchment, on the right bank of the river.
- Plantation areas, settlement areas and high-road-density areas cover the areas with moderate and high landslide vulnerability.
- Highest susceptibility zones cover 17.25 sq. km (12.24%). They are present in small patches in the upper catchment, where the slopes are very steep, in a large portion of the middle catchment, where human interventions are maximum, and in scattered patches in the lower catchment.
- Among all the factor clusters, the lithological factor cluster shows the highest relationship with the landslide susceptibility.
- Besides the lithological factor cluster, the anthropogenic and surface factor clusters are equally accountable for landslide susceptibility for research area.
- The triggering factor cluster shows a low relationship with landslide susceptibility in the study area.
- The protective factor cluster positively correlates with landslide susceptibility. The places covered with healthy vegetation are the least affected by the landslide susceptibility.

- The analysis clearly shows that both physical and anthropogenic factors are equally responsible for the frequent and sudden landslide hazards in the Chhota Rangit river basin of the Darjeeling Himalayas.
- Records of past landslides in the river basin (Table 5.5) show that the area under the landslides has changed along with the LULC dynamics of the region.
- Finally, it is observed that the area under the influence of landslides increased mostly from 1991 to 2022 in settlement areas, social forestry, plantation areas and water bodies.

5.6 Recommendations

Landslides are destructive and frequently occurring hazards that have recently impacted people's lives in the Chhota Rangit river basin of the Darjeeling Himalayas. Even though landslides are a natural phenomenon in this region, anthropogenic activities have significantly increased their severity and frequency over the years (see Figs. 5.28, 5.29, and 5.30). This work attempts to create landslide susceptibility zones for Chhota Rangit basin employing GIS-based analytic hierarchy process (AHP) model that analyses 18 key factors contributing to landslides.

Fig. 5.28 Landslides due to road construction near Majua. (Authors 2018)

Fig. 5.29 Road washed away by a landslide in Barnesbeg tea garden. (Authors 2022)

Fig. 5.30 Landslide near Bijanbari health centre. (Authors 2022)

The study identifies five landslide susceptibility zones. Areas classified as moderate to highly vulnerable are primarily in the upper and middle catchments of the research area. Areas, dominated by tea plantations and settlements, have seen rapid land use changes and increased road density, exacerbating the risk of landslides. Major risk areas include steep slopes and rocky outcrops, gully-eroded, incised riverbanks, roadside areas and fallow lands within tea plantations.

To address these challenges, the study recommends immediately implementing mitigation measures, focusing on an ecosystem-based approach (EbA). This approach has proven promising results in controlling soil loss and landslides in other mountainous areas (Hishamunda et al. 2024). EbA interventions include cost-effective and easy-to-adopt measures like vegetative measures, soil and water conservation interventions, reforestation of barren lands and sustainable land use planning, which are highly recommended for the study area. Interventions related to vegetative measures include the plantation of ecologically and economically important native grasses and trees, which can be undertaken along the first- and second-order streams and steep slopes, particularly in the upper part of the basin and in and around fallow lands of tea plantations and settlements and roadside areas.

Besides vegetative measures, infrastructure and community mobilising policy measures are important for mitigating landslide risks. These key recommendations include avoiding road construction in geologically weak areas or on steep slopes, as these are particularly prone to instability. Proper drainage systems must be incorporated into road designs to prevent water accumulation, which can trigger landslides. To ensure sustainable development, assessing the environmental impact of forest clearing and unplanned settlements is crucial, especially for tourism-related projects. High-risk zones should be managed with sustainable land use planning to reduce vulnerabilities. Raising community awareness about natural hazards and sustainable land management practices is also essential. Collaborative efforts involving research institutions, educational organisations, government bodies, NGOs and community-based organisations (CBOs) are needed to promote awareness and build resilience among local populations.

5.7 Conclusion

Addressing landslides in the Chhota Rangit river basin requires a multifaceted approach that combines ecosystem-based solutions, sustainable LULC planning and most importantly community involvement. Implementing these measures can decrease the likelihood of landslides and enhances the livelihood of the local population.

References

Ayalew L, Yamagishi H (2005) The application of GIS based logistic regression for landslide susceptibility mapping in the Kakuda Yahiko Mountains Central Japan. Geomorphology 65(1):15–31

Basu SR, Ghatowar L (1988) Landslides and soil erosion in the Gish drainage basin of the Darjeeling Himalaya and their bearing of North Bengal floods. Stud. Geomorphol. Carpatho-Balcanica

Basu T, Pal S (2017) Exploring landslide susceptible zones by analytic hierarchy process (AHP) for the Gish River basin, West Bengal, India. Spat Inf Res 25:665–675. https://doi.org/10.1007/s41324-017-0134-2

Basu SR, Sarkar S (1985) Some considerations on recent landslides at Tindharia and their control. Indian J Power River Valley Dev:188–194

Basu SR, Sarkar S (1987) Ecosystem vis-a vis landslide: a case study in Darjeeling Himalaya, proce. Impact of development on environment. Geog Soc India 2:45–53

Bera A, Mukhopadhyay BP, Das D (2019) Landslide hazard zonation mapping using multi-criteria analysis with the help of GIS techniques: a case study from eastern Himalayas, Namchi, South Sikkim. Nat Hazards 96:935–959. https://doi.org/10.1007/s11069-019-03580-w

Bhattacharya SK (2016) A comparison of the predicted vulnerability zones with the data based on hazard zones of landslide in the Kurseong hill subdivision, Darjeeling district, West Bengal, India. Environ Earth Sci 75:923. https://doi.org/10.1007/s12665-016-5729-8

Chen Y, Khan S, Paydar Z (2010) To retire or expand? A fuzzy GIS-based spatial multi-criteria evaluation framework for irrigated agriculture. Irrig Drain 59(2):174–188. https://doi.org/10.1002/ird.470

Dai FC, Lee CF (2002) Landslide characteristics and slope instability modelling using GIS, Lantau Island, Hong Kong. Geomorphology 42:213–228

Dutta KK (1966) A landslip in Darjeeling and Neighbouring Hills slopes in June. 1950. Bull Geol Surv India Series B 15(1):7–30

Feizizadeh B, Blaschke T (2013) GIS-multicriteria decision analysis for landslide susceptibility mapping: comparing three methods for the Urmia lake basin, Iran. Nat Hazards 65(3):2105–2128

Froehlich W, Starkel L (1987) Normal and extreme monsoon rains: their role in the shaping of the Darjeeling Himalaya. Stud Geomorphol Carpatho-Balc 21:129–160

Ghosh P, Lepcha K (2019) Weighted linear combination method versus grid based overlay operation method—a study for potential soil erosion susceptibility analysis of Malda district (West Bengal) in India. Egypt J Remote Sens Space Sci 22(1):95–115. https://doi.org/10.1016/j.ejrs.2018.07.002

Gilbert GK (1909) The convexity of hill tops. J Geol 17:344–350

Glade T (2003) Landslide occurrence as a response to land use change: a review of evidence from New Zealand. Catena 51(3–4):297–314. https://doi.org/10.1016/S0341-8162(02)00170-4

Gunarathna BWGID, Jayasinghe AB, Abenayake CC, Mahanama PKS (2018) Assessing the impact of land use on triggering landslides: a case of Sabaragamuwa Province, Sri Lanka Bhumi. Plann Res J 6(2):15–28. https://doi.org/10.4038/bhumi.v6i2.42

Guzzetti F, Carrara A, Cardinali M, Reichenbach P (1999) Landslide hazard evaluation: a review of current techniques and their application in a multi-scale study, Central Italy. Geomorphology 31:181–216

Hishamunda S, Fashaho A, Uwihirwe J et al (2024) Controlling soil erosion and landslides through ecosystem-based adaptation interventions in the hilly landscape of western Rwanda. Soil Adv 2. https://doi.org/10.1016/j.soilad.2024.100020

Kolat C, Ulusay R, Suzen ML (2012) Development of geotechnical microzonation model for Yenisehir(Bursa, Turkey) located at a seismically active region. Eng Geol 127:36–53. https://doi.org/10.1016/j.enggeo.2011.12.014

Komac M (2006) A landslide susceptibility model using the analytical hierarchy process method and multivariate statistics in perialpine Slovenia. Geomorphology 74(1–4):17–28

Lee S (2005) Application of logistic regression model and its validation for landslide susceptibility mapping using GIS and remote sensing data. Int J Remote Sens 7:1477–1491
Lee S, Min K (2001) Statistical analysis of landslide susceptibility at Yongin, Korea. Environ Geol 40:1095–1113
Lee S, Sambath T (2006) Landslide susceptibility mapping in the Damrei Romel area, Cambodia using frequency ratio and logistic regression models. Environ Geol 50:847–855. https://doi.org/10.1007/s00254-006-0256-7
Magliulo P, LisioA D, Russo F, Zelano A (2008) Geomorphology and landslide susceptibility assessment using GIS and bivariate statistics: a case study in southern Italy. Nat Hazards 47:411–435
Mahadevan TM (2003) 4D-modeling of Indian continental evolution through a geophysical window: geological constraints. J Virtual Explor 12:1–29
Malczewski J (1999) GIS and multicriteria decision analysis. Wiley, New York
Mandal S, Maiti R (2015) Geo-spatial variability of physiographic parameters and landslide potentiality. In: Semi-quantitative approaches for landslide assessment and prediction. Springer Natural Hazards, Singapore. https://doi.org/10.1007/978-981-287-146-6_2
Mezughi TH, Akhir JM (2011) Landslide susceptibility assessment using frequency ratio model applied to an area along the E-W highway (Gerik-Jeli). Am J Environ Sci 7(1):43–50
Moore ID, Grayson RB, Ladson AR (1991) Digital terrain modelling: a review of hydrological, geomorphological, and biological applications. Hydrol Process 5(1):3–30
Mugagga F, Kakembo V, Buyinza M (2012) Land use changes on the slopes of mount Elgon and the implications for the occurrence of landslides. Catena 90:36–46. https://doi.org/10.1016/j.catena.2011.11.004
Nefeslioglu HA, Duman TY, Durmaz S (2008) Landslide susceptibility mapping for a part of tectonic Kelkit Valley (eastern Black Sea region of Turkey). Geomorphology 94:401–418
Ocakoglu F, Gokceoglu C, Ercanoglu M (2002) Dynamics of a complex mass movement triggered by heavy rainfall: a case study from NW Turkey. Geomorphology 42(3):329–341
Pisano L, Zumpano V, Malek Ž et al (2017) Variations in the susceptibility to landslides, as a consequence of land cover changes: a look to the past, and another towards the future. Sci Total Environ 601–602:1147–1159. https://doi.org/10.1016/j.scitotenv.2017.05.231
Pradhan AM, Kim YT (2014) Relative effect method of landslide susceptibility zonation in weathered granite soil: a case study in Deokjeok-Ri Creek, South Korea. Nat Hazards 72:1189–1217. https://doi.org/10.1007/s11069-014-1065-z
Rasyid AR, Bhandary NP, Yatabe R (2016) Performance of frequency ratio and logistic regression model in creating GIS based landslides susceptibility map at Lompobattang Mountain, Indonesia. Geoenviron Disasters 3:19. https://doi.org/10.1186/s40677-016-0053-x
Rehman A, Song J, Haq F, Mahmood S, Ahamad MI et al (2022) Multi-Hazard susceptibility assessment using the analytical hierarchy process and frequency ratio techniques in the Northwest Himalayas, Pakistan. Remote Sens 14:554. https://doi.org/10.3390/rs14030554
Reichenbach P, Busca C, Mondini AC et al (2014) The influence of land use change on landslide susceptibility zonation: the Briga catchment test site (Messina, Italy). Environ Manag 54:1372–1384. https://doi.org/10.1007/s00267-014-0357-0
Saaty TL (1980) The analytic hierarchy process. McGraw Hill, New York
Saaty TL (1990) The analytic hierarchy process: planning, priority setting, resource allocation, 1st edn. RWS Publications, Pittsburgh, p 502
Saaty TL (1994) Fundamentals of decision making and priority theory with analytic hierarchy process, 1st edn. RWS Publications, Pittsburgh, p 527
Saaty TL (1996) Decision making with dependence and feedback, the analytic network process. RWS Publications, Pittsburgh
Saaty TL, Vargas LG (2001) The seven pillars of the analytic hierarchy process. In: Models methods, concepts and applications of the analytic hierarchy process. Springer, New York, pp 27–46
Saha AK, Gupta RP, Arora MK (2002) GIS-based landslide Hazard zonation in the Bhagirathi (ganga) valley, Himalaya. Int J Rem Sens 23:357–369
Samy IE, Morghany MM, Mohamed MM (2014) Landslide modeling and analysis using remote sensing and GIS: a case study of Cameron highland, Malayasis. J Geom 8(2):140–147

Sar N, Khan A, Chatterjee S, Das A, Mipun BS (2016) Coupling of analytical hierarchy process and frequency ratio based spatial prediction of soil erosion susceptibility in Keleghai river basin, India. Int Soil Water Conserv Res:1–21. https://doi.org/10.1016/j.iswcr.2016.09.004

Sharma A, Ram S (2014) A review on effects of deforestation on landslide: hill areas. Int J Sci Res Dev 2(7):1–10

Solaimani K, Mousavi SZ, Kavian A (2013) Landslide susceptibility mapping based on frequency ratio and logistic regression models. Arab J Geosci 6(7):2557–2569

Sörensen R, Zinko U, Seibert J (2006) On the calculation of the topographic wetness index: evaluation of different methods based on field observations. Hydrol Earth Syst Sci Discuss 10(1):101–112

Starkel L, Basu SR (eds) (2000) Rains, landslides and floods in the Darjeeling Himalaya. Indian National Science Academy, New Delhi, p 168

Sujatha ER, Rajamanickam GV (2012) Analysis of landslide susceptibility to different land use patterns in a part of Kodaikkanal, Tamil Nadu. Bonfring Int J Ind Eng Manag Sci 2(1):53–55

Sujatha ER, Rajamanickam GV, Kumaravel P (2012) Landslide susceptibility analysis using probabilistic certainty factor approach: a case study on Tevankarai stream watershed, India. J Earth Syst Sci 121(5):1337–1350

Townshend JR, Justice CO (1986) Analysis of the dynamics of African vegetation using the normalized difference vegetation index. Int J Remote Sensing 7(11):1435–1445

Vanacker V, Vanderschaeghe M, Govers G et al (2003) Linking hydrological, infinite slope stability and land-use change models through GIS for assessing the impact of deforestation on slope stability in high Andean watersheds. Geomorphology 52(3–4). https://doi.org/10.1016/S0169-555X(02)00263-5

Vivas L (1992) Los Andes Venezolanos. Academia Nacional de la Historia, Caracas

Chapter 6
Assessment of Spatiotemporal Variation of Vegetation and Its Relationship with Precipitation and Temperature in a Changing Climate in Nepal Mountain Ecosystems

Shiva Pokhrel and Chungla Sherpa

Abstract This chapter examines vegetation dynamics and climate trends in Nepal's Gandaki Province, highlighting implications for ecosystem-based resilience in Himalayan landscapes. Integrating satellite-derived NDVI, ERA5-Land climate data, and field-based forest biomass measurements, we identify pronounced seasonal differences in vegetation growth. Spring exhibits a strong greening signal, with NDVI increasing significantly at 0.0018 yr^{-1} (p 0.001), while summer shows only a modest, non-significant rise (0.0007 yr^{-1}) despite notable warming (+0.032 °C yr^{-1}; p 0.001) and a significant increase in precipitation. Autumn and the annual growing season also display significant greening, accompanied by warming and wetter conditions. Correlation analyses reveal a strong positive relationship between temperature and precipitation ($r = 0.631$; $p < 0.001$), whereas NDVI shows only weak associations with these climate drivers. This suggests that non-climatic factors such as community forestry initiatives, national conservation policies, and rural out-migration that facilitate regeneration can be primary drivers of vegetation recovery in parts of the province. Sustaining these positive trends requires integrated strategies, including climate-resilient afforestation using native and locally useful species combined with invasive-species control, strengthened community forest governance, incentive-driven conservation to promote human–nature coexistence, and science-based land-use planning. These approaches are essential for building

S. Pokhrel (✉)
Central Department of Botany, Tribhuvan University, Kirtipur, Nepal

Research Centre for Applied Science and Technology (RECAST), Tribhuvan University, Kirtipur, Nepal

Center for Environmental and Sustainable Agriculture Research, Pokhara, Kaski, Nepal
e-mail: shiva1.815702@iost.tu.edu.np

C. Sherpa
Forest Research and Training Center Gandaki Province, Pokhara, Nepal

Y. D. I. Khan et al. (eds.), *Ecosystem-based Approaches for Resilience Building in Himalayan Landscapes*, Disaster Resilience and Green Growth,
https://doi.org/10.1007/978-981-95-2007-7_6

ecological resilience and supporting livelihoods that depend on these ecosystems amid ongoing ecological and socio-economic change.

Keywords Annual growing season (AGS) · Normalized difference vegetation index (NDVI) · Vegetation dynamic · Gandaki Province

6.1 Introduction

Vegetation is considered one of the decisive components of the terrestrial ecosystem and plays a critical role in regulating the local, regional, and global climate systems (Chen et al. 2015; Cortes et al. 2021). Being an important component of the ecosystem, vegetation is a primary bond that connects soil, water, and atmosphere and has a crucial role in ecosystem functioning and the generation of various ecosystem goods and services (Lhomme et al. 2020). Vegetation health and distribution face mounting stress from anthropogenic pressures, climate change, geomorphological factors, and natural influences globally (Wang et al. 2022). As there is an interconnected relationship between vegetation and precipitation, precipitation is a key driver of soil water, runoff, and evapotranspiration, where vegetation plays an important role. Vegetation has the ability to enhance soil water by influencing infiltration, percolation, and interception, overall impeding surface runoff (Shi et al. 2022). Environmental degradation is a global challenge, but its impacts are disproportionately severe in developing countries, making them more vulnerable to its consequences. As many studies have shown (Liu et al. 2022; Wang et al. 2022), surface vegetation is a sensitive indicator of environmental changes that can directly respond to the overall ecological situation (Liang et al. 2017).

In such scenarios, understanding vegetation dynamics at a long-term spatial and temporal scale has been a concern of many ecological studies (Rees et al. 2001; Ndayisaba et al. 2016; Montoya et al. 2018; Quamar et al. 2023; Zhang et al. 2024a). Similarly, with growing climate change impacts throughout the world and their adverse consequences, it is urgent and important for researchers to better understand and develop better models to predict how these changes have affected vulnerable ecosystems (Gallopín et al. 2006; Jiang et al. 2018; Karki et al. 2018; Nguyen et al. 2019; Xue et al. 2019; Wei et al. 2020; Boori et al. 2021; Zhang et al. 2022). This could help better understand the role of climate in ecosystem functioning, co-evolution, and the interaction between the climate system and vegetation. As the impacts of climate change are highly noticeable in the hydro-meteorological cycles, temperature and precipitation are crucial climate components that play a vital role in the growth and development of vegetation and also influence the ecosystem goods and services generated during their interaction from different land with vegetation cover (Ndayisaba et al. 2016; Afuye et al. 2021; Sun et al. 2021; Zhang et al. 2024a, b).

On another note, it's important to understand the linkages between climate components and vegetation for the development of adaptations and mitigation strategies. Likewise, several studies have been carried out to better understand the relationship between climate components and vegetation worldwide, indicating varying degrees of relationship, ranging from strong to weak to no relation (Lawler et al. 2013; Zhang et al. 2019; Anderson et al. 2020; Ugarković et al. 2022).

A study conducted by Du et al. (2015) in China recommended that vegetation growth and development patterns are influenced by precipitation. Similarly, a study using AVHRR NDVI from 1989 to 1997 has found a strong relationship between NDVI and precipitation, with a positive correlation with temperature on the Great Plains in the Central USA during the early growing season (Zhuang et al. 2020). Likewise, another study carried out in Inner Mongolia using a SPOT VGG NDVI found a higher correlation between growing season NDVI and precipitation (2020; Wang et al. 2021; Zhang et al. 2024b).

However, the correlation analysis between NDVI response to precipitation and temperature at the global level was insignificant (Ichii et al. 2002). With the advancement and easy, open availability of decades of Earth observation data, it offers many advantages for understanding vegetation phenology at temporal and spatial continuity without being limited to time and space over local, regional, and global scales (Du et al. 2015; Yuan et al. 2015; Baniya et al. 2018; Zhu et al. 2021).

Earth observation data facilitate access to various vegetation phenological parameters at different spatial and temporal scales using the normalized difference vegetation index (NDVI). Myneni et al. (1997) and Tucker et al. (2001) were pioneering scientists who utilized Earth observation-derived data to investigate vegetation dynamics across multiple spatial and temporal scales. Numerous studies in recent decades have examined the application of vegetation indices, temperature, precipitation, and other climatic factors to assess their relationships with surface vegetation conditions, growth, development, and overall monitoring of vegetation changes over various timescales (Bengang and Shu 2000; Rees et al. 2001; Buitenwerf et al. 2015; Ndayisaba et al. 2016; Myers-Smith et al. 2020; Zhuang et al. 2020; Wang et al. 2021; Quamar et al. 2023).

The study of vegetation dynamics requires consistent and repetitive observations, where geographic information systems (GIS) and remote sensing play crucial roles in understanding ecosystems and their current states (Chang et al. 2023; Matyukira and Mhangara 2024). Remote sensing provides proxy measurements derived from satellite data, which are essential for monitoring vegetation health and distribution. The majority of satellite-derived vegetation studies rely on medium- to coarse-resolution sensors, such as Moderate Resolution Imaging Spectroradiometer (MODIS), Landsat, and Advanced Very High Resolution Radiometer (AVHRR) (Zhang et al. 2023). These sensors allow researchers to fuse imagery from different sources to enhance both temporal and spatial resolutions, thereby improving the analysis of vegetation patterns. Spectral vegetation indices, like the normalized difference vegetation index (NDVI), are frequently employed to quantify vegetation cover and health (Guan et al. 2021; Balata et al. 2022; Priya and Vani 2024).

Nepal is situated in the middle part of the greater Himalayan range with exclusive mountain ranges that have formed due to geological events. Being the youngest mountain range on Earth, it supports a diversity of flora, which may be due to the overlap of Eastern and Western Himalayan floral elements. However, the vegetation of Nepal is vulnerable to the impacts of climate changes (NAPA 2010) since they are important indicators of ecosystem dynamics (Walther et al. 2002). The impact of climate change has been noticeable due to increased temperature and rainfall variability, causing a wide range of impacts that lead to alterations in the structure, function, productivity, and seasonal phenology of vegetation (Higgins et al. 2023). As there is spatial and temporal variation in precipitation observed in Nepal, the summer monsoon dominates the total precipitation, with only 3% during winter (Sigdel et al. 2020). So far, very few studies have been carried out on trends in vegetation analysis in relation to precipitation and temperature in Nepal. A study was carried out on the trend and correlation of NDVI, temperature, and precipitation in the Manaslu Conservation Area from 2000 to 2008 (Mainali et al. 2015). Another study carried out in the Koshi River basin region by Zhang et al. (2013) found a significant increasing trend in the warm season NDVI for 1982–2006, however, a decline between 1994 and 2000. Similarly, Baniya et al. (2018) found an overall increasing trend in NDVI for Nepal from 1981 to 2015. Hence, temperature and precipitation have a profound impact on the growth and development of vegetation. Several studies suggest that Nepal is also one of the vulnerable countries where climate change has heavily altered the ecosystem productivity and land surface processes, and they are noticeable in various sectors (NAPA 2010; MoFE 2021; Pokhrel et al. 2025).

In this study, we used MODIS and ECMWF ERA5-Land climate data from 2000 to 2021. Given the unique location and topography of the Himalayan region, studies utilizing resolutions as fine as 1 km are especially advantageous. This high resolution is crucial for accurately modelling complex terrains like the Himalayas in Nepal, where local weather patterns are significantly affected by the intricate landscape (Norris et al. 2020). The main objectives were (i) to analyze trends in average monthly NDVI, surface temperature, and precipitation over a decade and (ii) to quantify changes in vegetation patterns within the study area. This book chapter could provide a baseline for understanding vegetation dynamics under a global change scenario at the provincial level in a mountainous country.

6.2 Material and Method

6.2.1 Study Area

Gandaki Province is located in western Nepal between 27°20′ N to 29°20′ N and 82°52′ E to 85°12′ E, covering 21,733 km^2 (Fig. 6.1), which is about 14.66% of the total country's area. It consists of major three physiographic zones, with 5819 km^2

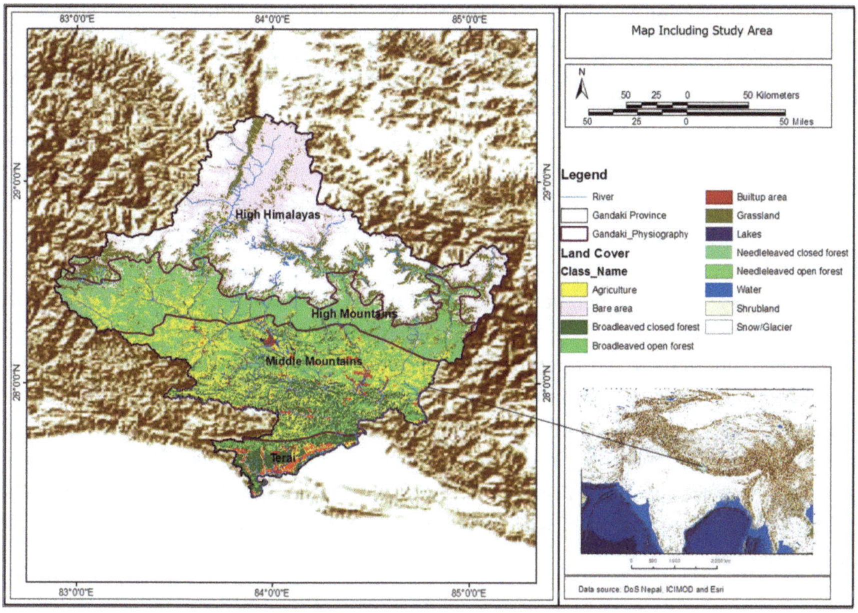

Fig. 6.1 Study area map with different vegetation types in Gandaki Province

(26.8%) as mountain, followed by 14,604 km^2 (67.2%) as hilly region, and 1210 km^2 (6%) as plain. The elevation of the province ranges between 7761 and 103 m above sea level. It consists of a total of 11 administrative districts with 85 local governments. The majority of people, about (72.4%), have their livelihoods in the province based on agro-based activities, such as agriculture, farming, and livestock. We have chosen Gandaki Province for the study as the province has a unique landscape and vegetation, with various protected areas, namely, Annapurna Conservation Area (CA), Manaslu CA, Dhorpatan Hunting Reserve (HR), and a small part of Chitwan National Park (NP). The province is home to a rich diversity of forest types, each shaped by variations in altitude, climate, and ecological conditions. In the lower plains, particularly in the Bhabar and Terai regions, sal (*Shorea robusta*) dominates as the primary canopy species, forming extensive sal forests. Moving to higher elevations, hill sal forests are found up to around 1300 m and beyond, showcasing the adaptability of this species to varying terrains. Along the streams of the Bhabar and Dun valleys, tropical deciduous riverine forests thrive, with prominent species such as *Bombax ceiba* and *Adina cordifolia*, accompanied by a variety of associated flora. Below around 900 meters, tropical evergreen forests flourish in damp and shaded areas, featuring species like *Eugenia jambolana* and *Phoebe lanceolata*, alongside other associated plants.

Between 900 and 1600 m, subtropical evergreen forests take prominence, characterized by species such as *Eugenia tetragona* and *Castanopsis indica*. Other significant forest types in the region include *Terminalia* forests, *Dalbergia sissoo-Acacia catechu* forests, and *Schima-Castanopsis* forests, each distinguished by unique canopy and understory species compositions. At higher altitudes, the landscape transitions to temperate and alpine forests, where species like *Quercus*, *Rhododendron*, and conifers such as *Abies spectabilis* and *Betula utilis* dominate. Near the timberline, alpine scrub and steppes feature hardy, resilient species like *Juniperus*, *Rhododendron*, and *Betula*, which are well-adapted to the harsh conditions of the upper regions. Similarly, the province links the ecologically unique eco-region between the north and south of Nepal through terrestrial connectivity via existing different forest regimes: government-managed, community, protection, leasehold, and buffer zone community forest and agricultural land (Pokhrel and Sherpa 2019). The province has varied climatic conditions due to the diverse nature of topography and altitude. Elevation less than 1500 m consisting of bhabar, chure, inner tarai, and valley has a subtropical climate, elevation from 1500 m to 3000 m has temperate types of climate, while the Himalayan region ranging from above 3000 m to 4500 m has alpine climate and tundra types of climate in the high Himalayas. In terms of precipitation, the province receives 250–2500 mm of average rainfall, with the lowest being 146 mm in the Mustang district and the highest being 5284 mm in the Lumle Kaski district, and with an average annual temperature of 21.5 °C.

6.2.2 *Data Sources*

6.2.2.1 NDVI Data

Datasets from 2000 to 2021 of the MODIS NDVI product (250 m, 16 days) were used for this study. The product, MOD13Q1, was obtained from the Land Processes Distributed Active Archive Center (LPDAAC) and downloaded from https://modis.gsfc.nasa.gov/ (Didan 2015). The NDVI datasets underwent geometric and atmospheric correction and a maximum value composite (MVC) was carried out to further eliminate interference from atmosphere, solar altitude, and clouds (Wu et al. 2014). The formula for NDVI is (NIR − RED)/(NIR + RED). The NDVI quantifies vegetation by measuring the differences between the near-infrared and red regions, which vegetation strongly reflects and red light vegetation absorbs.

6.2.2.2 Climate Data

For studies on vegetation response to climate variation, spatiotemporal data are very important. In many cases, in situ-measured datasets are used; however, the data do not fulfill the necessary standards and spatial density (Ebert et al. 2007; Karger et al. 2017). In the case of a least developed country like Nepal, where uniformly distributed situ-measurement stations are not available, various sources of climate data have been used for ecological studies. These data utilize situ-station interpolated data in combination with remote sensing algorithms and simulated products, which

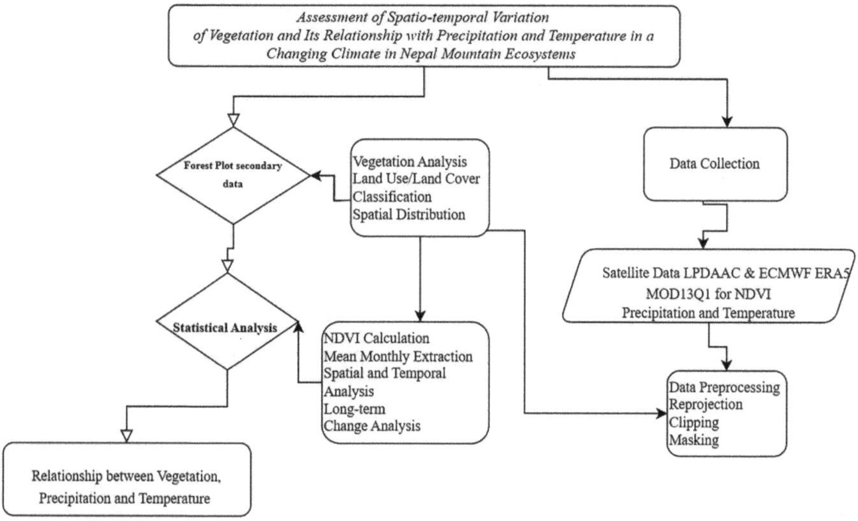

Fig. 6.2 Flowchart of study methodology

are available globally (Harris et al. 2014). The study utilizes temperature and precipitation data from the European Centre for Medium-Range Weather Forecasts (ECMWF) ERA5 dataset, which offers a 31 km reanalysis resolution with hourly temporal granularity (Hersbach et al. 2020).

6.2.3 Methods

A methodological flowchart of the study is shown in Fig. 6.2, and the following parameters were used in this study: mean monthly NDVI, mean monthly precipitation, and temperature. The overall changes in NDVI, temperature, and precipitation at the pixel level were analyzed based on the annual growing season, considering April to October. Similarly, seasonal classifications were carried out based on a study conducted in a similar ecological zone (Zhang et al. 2013), considering spring (April–May), summer (June–August), and autumn (September–October). Likewise, different vegetation distribution data were obtained from International Centre for Integrated Mountain Development (ICIMOD) land use/land cover data (Uddin et al. 2015). After that, we extracted the NDVI, precipitation, and temperature data for the growing season for forests and grasslands during the period 2000–2021.

6.2.4 Historical Long-Term Changes

Duration, annual and average NDVI, temperature, and total precipitation were estimated to detect variations in vegetation and climatic variables in the study area from 2000 to 2021. The linear regression method was applied to establish the trend using Eq. 6.1, with positive slope values indicating an increasing trend and negative values indicating a decreasing trend (Fensholt and Proud 2012; Piao et al. 2015; Baniya et al. 2018). R software was used to calculate and show the slope of NDVI from 2000 to 2021.

$$\text{Slope} = \frac{n \times \sum_{i=1}^{n} i \times \text{NDVI}_i - \sum_{i=1}^{n} \times \sum_{i=1}^{n} i\ \text{NDVI}_i}{n \times \sum_{i=1}^{n} i^2 - \left(\sum_{i=1}^{n} i\right)^2} \tag{6.1}$$

where the slope is the trend of NDVI, n is the number of years in the study period, i is the year, and NDVIi is the ith year.

6.2.5 The Amount of Changes in NDVI, Precipitation, and Temperature

The total changes in vegetation that occurred during an annual or growing season were determined using the slope and mean NDVI values, which indicate whether there was an increase or decrease. Equation 6.2 was used for the calculation.

$$\text{Amount of changes in values}(\%) = \frac{\text{Slope}}{\text{Mean}} \times n \times 100 \tag{6.2}$$

R = Linear regression slope of annual or monthly average, i.e., NDVI.
MNDVI = Average NDVI over study period.
N = The length of study period, i.e., 2000–2021.

6.2.6 Multivariate Regression Analysis

To better understand the relationship between NDVI and climate in our study area, we performed a multivariate regression analysis using ordinary least squares (OLS) techniques. OLS is a statistical method that examines the linear relationship between a dependent variable and one or more independent variables. This regression analysis also employs appropriate methods to assess the spatial effects of climatic variability on vegetation growth (Ding et al. 2007; Hou et al. 2015; Ghebrezgabher et al. 2020). For this study, the dependent variable was NDVI, while the independent variables were precipitation and temperature. The coefficients of the independent variables in the model helped us determine the direction and strength of the relationship between the variables. A positive coefficient for precipitation indicated that an increase in precipitation was associated with an increase in NDVI, while a negative coefficient for temperature indicated that an increase in temperature was associated with a decrease in NDVI.

The model was represented as follows:

$$\mathrm{NDVI} = \beta 0 + \beta 1^{*}\, x1 + \beta 2^{*}\, x2 + \varepsilon \tag{6.3}$$

where NDVI is the dependent variable; x1 and x2 are the independent variables representing precipitation and temperature, respectively; β0 is the y-intercept of the model; β1 and β2 are the coefficients representing the effect of precipitation and temperature on NDVI, respectively; and ε is the error term that represents the difference between the observed NDVI value and the value predicted by the model. It is important to note that the coefficients (β_1 and β_2) can be either positive or negative, reflecting a positive or negative relationship between NDVI and the independent variables. This indicates that NDVI may increase or decrease with changes in precipitation and temperature, respectively. The analyses were conducted within a GIS environment.

6.2.7 Mann-Kendall Trend Test

The Mann-Kendall Trend Test, as outlined in Eq. 6.4, was utilized to identify monotonic trends over time (Nyeko-Ogiramoi et al. 2013). This test helps determine whether the variables have shown an increasing or decreasing trend during the study period.

$$\mathrm{S} = \sum_{n-1}^{i=1} \sum_{j=i+1}^{n} \mathrm{sgn}(\mathrm{xj} - \mathrm{xi}) \tag{6.4}$$

where:

SS = Mann-Kendall test statistic.
nn = number of data points in the time series.
xixi and xjxj = data values at times ii and jj, respectively.
sgn(xj − xi)sgn(xj − xi) = the sign function, defined as

$$\mathrm{sgn}(x) = \{ \begin{matrix} 1 \; if\, x > 0 \\ 0 \; if\, x = 0 \\ -1 \;\; x < 0 \end{matrix}$$

A positive SS suggests an increasing trend.

A negative SS suggests a decreasing trend.

S = 0S = 0 indicates no trend.

6.2.8 *Comparative Correlation Coefficient Analysis*

To enhance our understanding of the satellite-derived vegetation index, we conducted a comparative analysis by correlating it with approximately 200 field-measured aboveground biomass data points. These data were collected from forests distributed across various geographic regions, as documented in previously published research (Pokhrel and Sherpa 2019). The biomass data were calculated at the plot level, and corresponding NDVI values were extracted from raster data for the same plot locations. Statistical analysis was then performed to determine the coefficient of determination (R^2) and assess the agreement level between MODIS-derived NDVI and the field-measured biomass.

6.3 Result and Discussion

6.3.1 *Spatial Variation of NDVI, Precipitation, and Temperature*

The spatial distribution of the decades from 2000 to 2021 mean forest NDVI (Fig. 6.3) revealed high values, greater than 0.5, in dense forest areas. These areas are geographically situated in the central, eastern, western, and southern regions, where deciduous and evergreen forests dominate. In terms of climatic zones, these

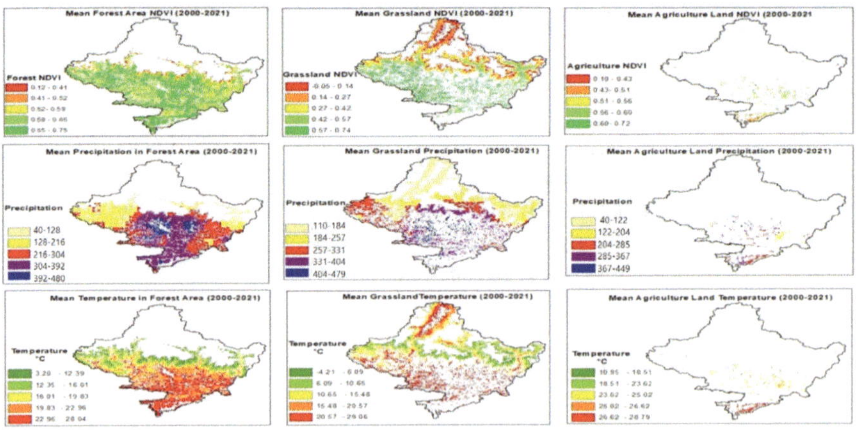

Fig. 6.3 Spatial variations of NDVI, precipitation, and temperature of forest, grassland, and agriculture area

regions consist of the tropical to subtropical and the edge between the subtropical and temperate zones. Similarly, low mean NDVI (less than 0.4) was observed in forest areas near the timberline and nearby areas influenced by anthropogenic activities, such as around agricultural land, roads, and development projects. In terms of the forest areas receiving decade's years of mean precipitation (Fig. 6.3), the central-southern forest region received the highest amount of precipitation (304–480 mm), which may be due to the presence of the foothill barrier of the Himalayas, creating a subtropical to temperate rainforest in the northwestern region.

While the forest areas' mean 22-year temperature (Fig. 6.3) was highest in the central and southern forest areas, ranging from 19 °C to 28 °C, the northern area near or around the timberline received lower temperatures ranging from 3 °C to 12 °C. In the grassland area, a high mean NDVI value greater than 0.4 was observed in healthy grassland largely distributed in the central western and central eastern regions, with the least observed in the northern region (Fig. 6.3). Similarly, the grassland in the central northern and southeastern regions (Fig. 6.3) received the greatest amount of precipitation (367–449 mm), while the least amount (less than 122–204 mm) was observed in the northwestern region. At the same time, the grassland in the far-northwestern and central southern regions received the highest mean temperature ranging from 15 °C to 29 °C, while the least mean temperature was observed in the central northeast to central northwest region (−4 °C to 6 °C). On the other hand, mean annual NDVI for the study period I in the agriculture/crop area (Fig. 6.3) showed a varied pattern depending on the geographic location. In the central and far southern regions, crop lands exhibited both heterogeneous and homogeneous patches of NDVI values. The greater recorded NDVI values ranged from 0.6 to 0.7, while the least values ranged from 0.1 to 0.4, with high precipitation in the central and far southern regions (404–479 mm) and the least precipitation in the central northern region (184–257 mm). Likewise, greater temperature values between 20 °C and 28 °C were observed in the far southern region, with the least temperatures (10–18 °C) observed in the central northern regions.

6.3.2 Interannual Variation of Growing Season NDVI and Climatic Factors

Analysis results indicate that the highest average NDVI during the growing season occurred in September (0.55), October (0.53), and August (0.46), reflecting peak vegetation activity in late summer and early autumn (Fig. 6.4). Seasonal NDVI trend analysis reveals a pronounced greening in spring, with a slope of 0.0018 $year^{-1}$ ($p < 0.01$), a minor and nonsignificant increase in summer (0.0007 $year^{-1}$), and a moderate rise in autumn (0.0017 $year^{-1}$; $p < 0.05$). These patterns suggest that vegetation health improves most strongly during spring, continues moderately in autumn, and remains relatively stable in summer. Mann–Kendall test results support this interpretation, showing a robust positive NDVI trend in spring ($\tau = 0.581$, p =

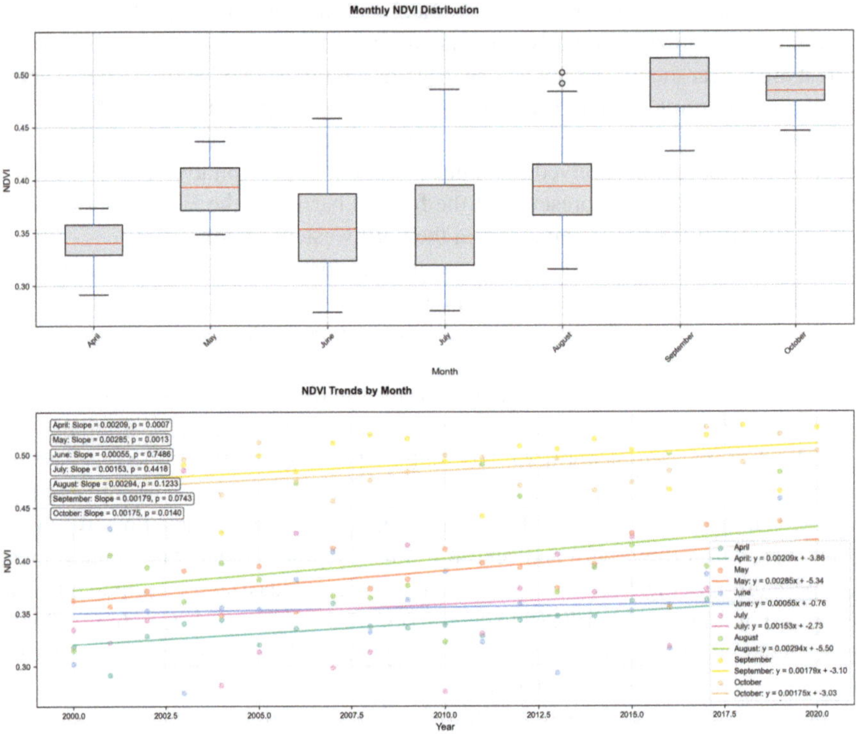

Fig. 6.4 Comparison of Annual Growing season (AGS) NDVI d analysis, showing trends across Gandaki Province from 2000 to 2021 during growing months

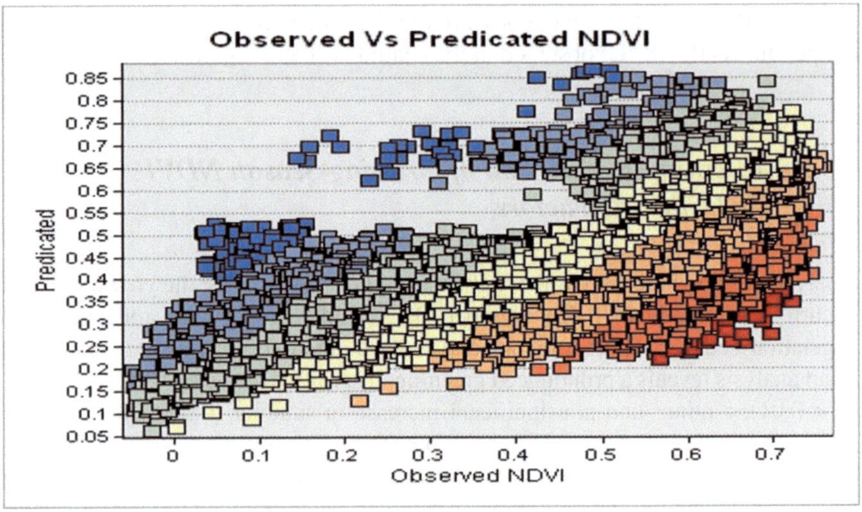

Fig. 6.5 Observed versus predicted NDVI values based on OLS

0.001), a weaker but significant increase in autumn ($\tau = 0.387$, $p = 0.013$), and no significant change in summer ($\tau = -0.113$, $p = 0.480$), highlighting the seasonal dynamics of vegetation across Gandaki Province between 2000 and 2021. Temperature trends also exhibited seasonal variation, with the most pronounced warming occurring in summer (slope = 0.032 °C year^{-1}; $p < 0.001$) and autumn (0.045 °C year^{-1}; $p < 0.01$). Changes in spring temperature were negligible and statistically nonsignificant. These findings indicate that summers and autumns are

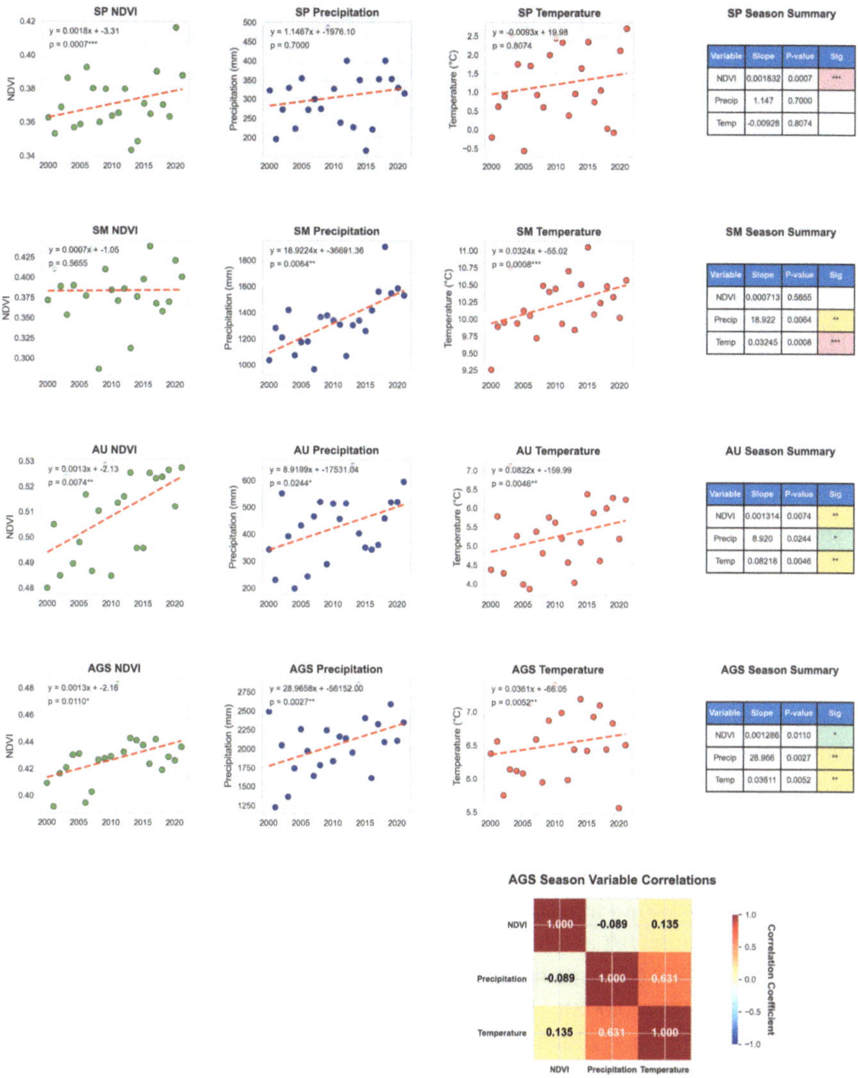

Fig. 6.6 Growing season (autumn, spring, and summer) and annual, the overall relationship between mean annual NDVI, precipitation, and temperature

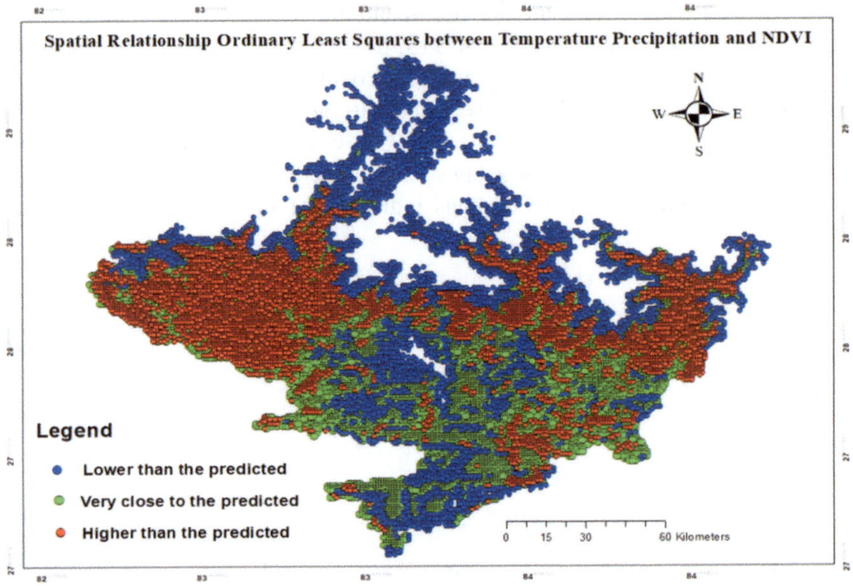

Fig. 6.7 Map showing OLS-derived standard residual

Table 6.1 Annual and AGS NDVI slope, test statistic, and change ratio (CR) in Gandaki Province, 2000–2021

Timescale	Linear slope	*P*-value	Mann-Kendall Z	Change (%)
Spring	0.0018	0.000692*	3.158	10.804
Summer	0.000713	0.566	0.733	4.102
Autumn	0.001314	0.007355	2.256	5.666
Annual	0.001286	0.011012	2.256	6.707

warming most rapidly, likely interacting with vegetation growth and precipitation dynamics (Figs. 6.5–6.6). Precipitation displayed complex seasonal patterns. Spring precipitation showed a weak and nonsignificant trend, while summer rainfall increased markedly (slope = 18.9 mm $year^{-1}$; $p < 0.01$) and autumn precipitation rose moderately (9.5 mm $year^{-1}$; $p < 0.05$). Mann–Kendall tests confirmed significant positive trends in summer ($\tau = 0.351$, $p = 0.024$) and autumn ($\tau = 0.342$, $p = 0.028$), with the spring trend remaining nonsignificant ($\tau = -0.021$, $p = 0.910$). At the annual scale, NDVI exhibited a gradual yet consistent greening trend (slope = 0.0009 $year^{-1}$; $p < 0.01$), with the Mann–Kendall test confirming a moderate positive trend ($\tau = 0.449$, $p = 0.008$). This indicates steady vegetation improvement over the study period (Fig. 6.7; Table 6.1). Collectively, these results suggest that while spring and autumn contribute most to vegetation recovery, summer experiences stronger climatic intensification, particularly in rainfall and temperature, without corresponding increases in vegetation activity.

The overall relationship between mean NDVI, precipitation, and temperature is shown in Fig. 6.6. Various studies at the national, regional, and global levels also suggest that vegetation is significantly experiencing the influences of climate changes (Baniya et al. 2019; Li et al. 2022; Hu et al. 2023; Dahal et al. 2024; Wang et al. 2024). A study carried out by Zhang et al. (2013) from 1997 to 2011 in the Koshi River basin showed that there is a significant increasing trend ($R^2 = 0.19$, $P = 0.03$). Likewise, another study carried out by Baniya et al. (2019) from 1982 to 2015 for Nepal also indicated a significant increasing rate of NDVI ($R^2 = 0.33$, $P = 0.0003$).

6.3.3 Analysis Between NDVI and Climate Variables

Based on the analysis of precipitation, temperature, and NDVI over the study period, Fig. 6.6 shows a significant increasing trend in vegetation greenness (NDVI), particularly during the spring and autumn seasons. The relationship between precipitation and NDVI is weak, as is the relationship between temperature and NDVI. Overall, the influence of climatic factors on vegetation dynamics is limited, with precipitation exhibiting a marginally stronger though still weak correlation with NDVI compared to temperature.OLS (ordinary least squares) multivariate regression analysis was used to model the relationship between multiple dependent and independent variables. In this case, the dependent variable is NDVI (normalized difference vegetation index) and the independent variables are temperature and precipitation. The analysis aims to determine the impact of temperature and precipitation on NDVI. Based on the standard residual values from −2.5 to 2.5 standard deviations, the OLS multivariate regression analysis map (Fig. 6.7) values are divided into categories. The lower than predicted values having negative standard residual values from −2.5 to −0.5 std. deviation indicate that the observed NDVI values are slightly lower than the predicted values, which are within the expected range of the predicted values by the model. The model may need to be refined to capture these deviations. Standard residual values from −0.5 to 0.5 std. deviation indicate that the observed NDVI values are very close to the predicted values, suggesting that the model is likely to be a good fit.

Similarly, standard residual values greater than 1 to 2.5 std. deviation indicate that the observed NDVI values are much higher than the predicted values and may indicate an issue with the model or the underlying data. Hence, the OLS multivariate regression analysis map (Fig. 6.7) shows values between NDVI, temperature, and precipitation in central western, central eastern, and some southern parts of forest areas. NDVI is more likely to be influenced by temperature and precipitation in these areas. In other areas, the underlying data may need further investigation with other influencing factors, such as site quality, microclimate, and soil.

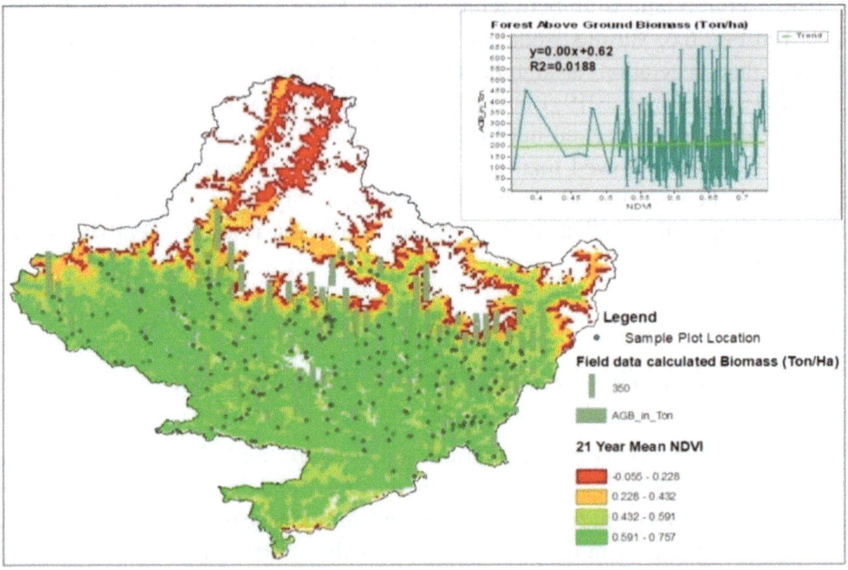

Fig. 6.8 Map showing study period year mean annual NDVI and ground-measured forest aboveground biomass with correlation between them

6.3.4 *Comparative Correlation Analysis Between AGB and NDVI*

The correlation coefficient measures the strength and direction of the linear relationship between two variables. This study used forest inventory data from 2012 to 2015, which covered a major part of the study area, to investigate the relationship between forest aboveground biomass (AGB) and MODIS-derived NDVI (Pokhrel and Sherpa 2019). The aboveground biomass data used in this study were taken from previous research where plot data, along with GPS locations, were used to calculate biomass values for specific plots. Each point in the dataset represents biomass at that point. In this study, these biomass values are related to NDVI data measured at the same point. The correlation coefficient value obtained was 0.0186, hence indicating a very weak positive correlation between AGB and NDVI in this complex mountain terrain.

This means that as the NDVI value increases, the value of AGB increases, but shows very low correlation and therefore more variability in this relationship. This might be for reasons of poor spatial and temporal resolution of the MODIS-derived NDVI or its incapacity to take small-scale variation in biomass into account that could otherwise be detected using field measurements. Again, a certain degree of heterogeneity among the types of forests present within Gandaki Province might contribute to the reason why such weak correlation is being reported in Fig. 6.8.

6.4 Discussion and Conclusion

This chapter provides a comprehensive perspective on the spatiotemporal dynamics of vegetation in Gandaki Province, illustrating how climatic and non-climatic drivers jointly shape Himalayan ecosystems. The analysis highlights strong spatial and seasonal heterogeneity in vegetation responses, consistent with broader evidence that the sensitivity of vegetation to climate drivers is context-dependent (Zhang et al., 2016). In Gandaki, precipitation, especially monsoon rainfall, emerges as the primary driver of interannual NDVI variability, while temperature plays an important but secondary role. These findings reinforce patterns identified in prior research: Krakauer et al. (2017) reported a positive correlation between NDVI and precipitation, while Baniya et al. (2018) found positive correlations between NDVI and temperature in Nepal. Similarly, Mu et al. (2021) showed that vegetation cover in China has been increasing in response to both precipitation and temperature, underscoring the broader regional importance of these drivers.

The greening trend observed across Gandaki Province aligns with findings from other parts of Nepal and the Himalaya. Zhang et al. (2013) documented increasing NDVI in the Koshi River basin between 2000 and 2016, although they observed distinct sub-periods with differing trends. Mishra and Mainali (2017) likewise reported widespread greening since 2000, and Wu et al. (2020) confirmed a greening trajectory in the Koshi basin from 2000 to 2018. Yet, as Krakauer et al. (2017) noted, the long-term NDVI increase across Nepal from 1981 to 2015 is not primarily explained by temperature or precipitation but rather correlates most strongly with rising atmospheric CO_2. This chapter adds further evidence that interannual NDVI variability is climate-driven, particularly linked to monsoon rainfall, while long-term greening may be due to non-climatic factors such as CO_2 fertilization in agricultural areas and the expansion of community forestry and conservation initiatives. These trends illustrate the effectiveness of community-based management as an Ecosystem-based Adaptation (EbA) measure, demonstrating how local stewardship can directly enhance vegetation recovery, carbon sequestration, and ecosystem services. At the same time, this apparent greening masks critical vulnerabilities. Steady warming, especially in alpine and high-elevation zones, is driving shifts in vegetation distribution. Evidence of potential upward migration of plant species (Sigdel et al. 2020) points to risks of habitat contraction, biodiversity loss, and disruption of water regulation services. Such ecological changes carry profound socio-economic consequences. Communities reliant on forests, pastures, water resources, and tourism (Shrestha et al. 2012) face growing challenges to their livelihoods, illustrating the limits of adaptive capacity in climate-sensitive areas. Thus, the vegetation dynamics of Gandaki Province represent a dual narrative: on the one hand, management-driven greening that builds resilience, and on the other, climate-driven fragility in precipitation-dependent and alpine ecosystems. The implications of these findings extend beyond ecological observations. They underscore the need for climate adaptation strategies that are spatially targeted and context specific. In areas where greening is reinforced by community forestry and conservation, policies should support and scale these EbA successes by strengthening institutional capacity and ensuring sustained local participation. Conversely, in regions where vulnerability is primarily climate-driven, adaptation must prioritize resilience building through livelihood diversification, the development of early warning systems, and the protection of ecological corridors to facilitate species migration. This chapter demonstrates the importance of understanding vegetation dynamics in the Himalaya through an integrated lens that connects ecological processes with human management and climate change. It contributes to ongoing debates on the drivers of greening in mountain landscapes while providing a province-level case study with global relevance. Looking forward, future research must integrate biophysical monitoring with socio-economic assessments to better quantify the cascading impacts of vegetation change on ecosystem services and human well-being. Such interdisciplinary work is critical for developing robust, evidence-based, and equitable adaptation strategies capable of securing both ecological integrity and community resilience in the Himalayas under a rapidly changing climate.

Conflicts of Interest The authors declare no conflicts of interest.

References

Afuye GA, Kalumba AM, Orimoloye IR (2021) Characterisation of vegetation response to climate change: a review. Sustainability 13(13). https://doi.org/10.3390/su13137265

Anderson K et al (2020) Vegetation expansion in the Subnival Hindu Kush Himalaya. Glob Change Biol 26(3):1608–1625. https://doi.org/10.1111/gcb.14919

Balata D et al (2022) Using satellite NDVI time-series to monitor grazing effects on vegetation productivity and phenology in Heterogeneous Mediterranean Forests. Remote Sens. https://doi.org/10.3390/rs14102322

Baniya B, Tang Q, Huang Z, Sun S, Techato K (2018) Spatial and temporal variation of NDVI in response to climate change and the implication for carbon dynamics in Nepal. Forests 9(6). https://doi.org/10.3390/F9060329

Baniya B et al (2019) Spatial and temporal variation of drought based on satellite derived vegetation condition index in Nepal from 1982–2015. Sensors (Switzerland) 19(2):430. https://doi.org/10.3390/s19020430

Bengang L, Shu T (2000) Correlation between AVHRR NDVI and climate factors. Acta Ecol Sin:898–902. http://europepmc.org/abstract/CBA/533697

Boori MS et al (2021) Spatiotemporal ecological vulnerability analysis with statistical correlation based on satellite remote sensing in Samara, Russia. J Environ Manag [Preprint]. https://doi.org/10.1016/j.jenvman.2021.112138

Buitenwerf R et al (2015) Three decades of multi-dimensional change in global leaf phenology. Nat Clim Chang [Preprint]. https://doi.org/10.1038/nclimate2533

Chang C-T, Chiang J-M, Dai J (2023) Remote sensing of climate-vegetation dynamics and their effects on ecosystems. Remote Sens. https://doi.org/10.3390/rs15215097

Chen Z et al (2015) Roles of climate, vegetation and soil in regulating the spatial variations in ecosystem carbon dioxide fluxes in the Northern Hemisphere. PLoS One 10(4):1–14. https://doi.org/10.1371/journal.pone.0125265

Cortés J et al (2021) Where are global vegetation greening and browning trends significant? Geophys Res Lett 48(6):1–9. https://doi.org/10.1029/2020GL091496

Dahal NM et al (2024) Spatiotemporal assessment of drought and its impacts on crop yield in the Koshi River Basin, Nepal. Theor Appl Climatol 155(3):1679–1698. https://doi.org/10.1007/s00704-023-04719-3

Didan K (2015) MOD13Q1 MODIS/Terra vegetation indices 16-Day L3 global 250m SIN Grid V006. NASA EOSDIS land processes DAAC https://doi.org/10.5067/MODIS/MOD13Q1.006

Ding M et al (2007) The relationship between NDVI and precipitation on the Tibetan Plateau. J Geogr Sci 17:259–268

Du JQ et al (2015) Dynamic changes in vegetation NDVI from 1982 to 2012 and its responses to climate change and human activities in Xinjiang, China. Ying Yong Sheng Tai Xue Bao J Appl Ecol 26(12):3567–3578

Ebert EE, Janowiak JE, Kidd C (2007) Comparison of near-real-time precipitation estimates from satellite observations and numerical models. Bull Am Meteorol Soc 88(1):47–64

Fensholt R, Proud SR (2012) Evaluation of Earth observation based global long term vegetation trends – comparing GIMMS and MODIS global NDVI time series. Remote Sens Environ 119:131–147. https://doi.org/10.1016/j.rse.2011.12.015

Gallopín GC, Gallopín GC, Gallopin GC (2006) Linkages between vulnerability, resilience, and adaptive capacity. Glob Environ Change [Preprint]. https://doi.org/10.1016/j.gloenvcha.2006.02.004

Ghebrezgabher MG et al (2020) Assessment of NDVI variations in responses to climate change in the Horn of Africa. Egypt J Remote Sens Space Sci 23(3):249–261. https://doi.org/10.1016/j.ejrs.2020.08.003

Guan X et al (2021) Fusing MODIS and AVHRR products to generate a global 1-km continuous NDVI time series covering four decades. Earth Syst Sci Data Discuss 2021:1–32. https://doi.org/10.5194/essd-2021-156

Harris I et al (2014) Updated high-resolution grids of monthly climatic observations – the CRU TS3.10 dataset. Int J Climatol 34(3):623–642. https://doi.org/10.1002/joc.3711

Hersbach H et al (2020) The ERA5 global reanalysis. Q J R Meteorol Soc [Preprint]. https://doi.org/10.1002/qj.3803

Higgins SI, Conradi T, Muhoko E (2023) Shifts in vegetation activity of terrestrial ecosystems attributable to climate trends. Nat Geosci 16(2):147–153. https://doi.org/10.1038/s41561-022-01114-x

Hou W et al (2015) Interannual variations in growing-season NDVI and its correlation with climate variables in the Southwestern Karst Region of China. Remote Sens:11105–11124. https://doi.org/10.3390/rs70911105

Hu B et al (2023) Spatio-temporal evolution of cropland ecosystem services value and its spatially varying dominate over the past two decades: a case study in Jiangxi Province in southern China. J Clean Prod 427:139228. https://doi.org/10.1016/j.jclepro.2023.139228

Ichii K, Kawabata A, Yamaguchi Y (2002) Global correlation analysis for NDVI and climatic variables and NDVI trends: 1982-1990. Int J Remote Sens 23(18):3873–3878. https://doi.org/10.1080/01431160110119416

Jiang L et al (2018) Method for evaluating ecological vulnerability under climate change based on remote sensing: a case study. Ecol Indic [Preprint]. https://doi.org/10.1016/j.ecolind.2017.10.044

Karki G, Paudel B, Uprety BK (2018) Integrating forests and biodiversity in Nepal's National Adaptation Plan: a review and synthesis of knowledge stock on opportunities and way forward. Banko Janakari 27(2):21–31. https://doi.org/10.3126/banko.v27i2.21220

Krakauer NY, Lakhankar T, Anadón JD (2017) Mapping and attributing normalized difference vegetation index trends for Nepal. Remote Sens 9(10). https://doi.org/10.3390/RS9100986

Lawler JJ et al (2013) Mitigation and adaptation strategies to reduce climate vulnerabilities and maintain ecosystem services. Clim Vulnerability:315–335. https://doi.org/10.1016/B978-0-12-384703-4.00436-6

Lhomme P et al (2020) The wild bees (Hymenoptera: Apoidea) of Morocco. Zootaxa. https://doi.org/10.11646/zootaxa.4892.1.1

Li Y et al (2022) Quantitative assessment for the spatiotemporal changes of ecosystem services, tradeoff–synergy relationships and drivers in the semi-arid regions of China. Remote Sens. https://doi.org/10.3390/rs14010239

Liang L, Schwartz MD, Fei S (2017) Validating satellite phenology through intensive ground observation and landscape scaling in a mixed seasonal forest. Ecosphere 8(3):e01919. https://doi.org/10.1002/ecs2.1919

Liu B et al (2022) The sensitivity of vegetation dynamics to climate change across the Tibetan Plateau. Atmos 13(7). https://doi.org/10.3390/atmos13071112

Karger DN, Conrad O, Böhner J, Kawohl T, Kreft H, Soria-Auza RW, Zimmermann NE, Linder HP, Kessler M (2017) Climatologies at high resolution for the earth's land surface areas. Scientific Data 4(1):170122. https://doi.org/10.1038/sdata.2017.122

Mainali J, All J, Jha PK, Bhuju DR (2015) Responses of Montane Forest to climate variability in the Central Himalayas of Nepal. Mt Res Dev 35(1):66–77. https://doi.org/10.1659/MRD-JOURNAL-D-13-00101.1

Matyukira C, Mhangara P (2024) Advances in vegetation mapping through remote sensing and machine learning techniques: a scientometric review. Eur J Remote Sens 57(1):2422330. https://doi.org/10.1080/22797254.2024.2422330

Mishra NB, Mainali KP (2017) Greening and browning of the Himalaya: spatial patterns and the role of climatic change and human drivers. Sci Total Environ 587–588:326–339. https://doi.org/10.1016/j.scitotenv.2017.02.156

Ministry of Forests and Environment (MoFE) (2021) Vulnerability and risk assessment and identifying adaptation options. Kathmandu, Nepal: Government of Nepal

Montoya E et al (2018) Long-term vegetation dynamics in a megadiverse hotspot: the ice-age record of a pre-montane forest of Central Ecuador. Front Plant Sci 9. https://www.frontiersin.org/journals/plant-science/articles/10.3389/fpls.2018.00196

Mu B et al (2021) Vegetation cover change and its attribution in China from 2001 to 2018. Remote Sens. https://doi.org/10.3390/rs13030496

Myers-Smith IH et al (2020) Complexity revealed in the greening of the Arctic. Nat Clim Chang [Preprint]. https://doi.org/10.1038/s41558-019-0688-1

Myneni RB et al (1997) Increased plant growth in the northern high latitudes from 1981 to 1991. Nature 386(6626):698–702. https://doi.org/10.1038/386698a0

NAPA(National Adaptation Programme of Action) (2010) National adaptation programme of action to climate change. Government of Nepal, Ministry of Environment

Ndayisaba F et al (2016) Understanding the spatial temporal vegetation dynamics in Rwanda. Remote Sens. https://doi.org/10.3390/rs8020129

Nguyen KA et al (2019) Global mapping of eco-environmental vulnerability from human and nature disturbances. Sci Total Environ [Preprint]. https://doi.org/10.1016/j.scitotenv.2019.01.407

Norris J et al (2020) Warming and drying over the central Himalaya caused by an amplification of local mountain circulation. npj Clim Atmos Sci 3(1):1. https://doi.org/10.1038/s41612-019-0105-5

Nyeko-Ogiramoi P, Willems P, Ngirane-Katashaya G (2013) Trend and variability in observed hydrometeorological extremes in the Lake Victoria basin. J Hydrol 489:56–73. https://doi.org/10.1016/j.jhydrol.2013.02.039

Piao S et al (2015) Detection and attribution of vegetation greening trend in China over the last 30 years. Glob Chang Biol 21(4):1601–1609. https://doi.org/10.1111/gcb.12795

Pokhrel S, Sherpa C (2019) Analysis of land use land cover in Annapurna Conservation Area in Gandaki Province, Nepal using vegetation index and land use land cover datasets. J Geogr Stud 2(2):87–99. https://doi.org/10.21523/gcj5.18020204

Pokhrel S, Kanta Subedi C, Thakuri S, Mardan Kunwar R, Prasad Sharma K, Prasad Chaudhary R, Ghimire SK (2025) Evaluating the ecological vulnerability and adaptive capacity of forest vegetation under climate stress across Nepal's mountain landscapes. Geocarto International 40(1). https://doi.org/10.1080/10106049.2025.2551806

Priya RS, Vani K (2024) Vegetation change detection and recovery assessment based on post-fire satellite imagery using deep learning. Sci Rep 14(1):12611. https://doi.org/10.1038/s41598-024-63047-2

Quamar MF et al (2023) Vegetation dynamics and hydro-climatic changes during the Middle Holocene from the Central Himalaya, India. Quaternary. https://doi.org/10.3390/quat6010011

Rees M et al (2001) Long-term studies of vegetation dynamics. Science (New York, NY) 293(5530):650–655. https://doi.org/10.1126/science.1062586

Shi S et al (2022) Soil infiltration properties are affected by typical plant communities in a semi-arid desert grassland in China. Water (Switzerland) 14(20):1–14. https://doi.org/10.3390/w14203301

Shrestha UB, Gautam S, Bawa KS (2012) Widespread climate change in the Himalayas and associated changes in local ecosystems. PLoS One 7(5):e36741

Sigdel SR et al (2020) Tree-to-tree interactions slow down Himalayan treeline shifts as inferred from tree spatial patterns. J Biogeogr 47(8):1816–1826. https://doi.org/10.1111/jbi.13840

Sun H et al (2021) Vegetation change and its response to climate change in Yunnan Province, China. Adv Meteorol 2021(1):8857589. https://doi.org/10.1155/2021/8857589

Tucker CJ et al (2001) Higher northern latitude normalized difference vegetation index and growing season trends from 1982 to 1999. Int J Biometeorol 45(4):184–190. https://doi.org/10.1007/s00484-001-0109-8

Uddin K, Shrestha HL, Murthy MSR et al (2015) Development of 2010 national land cover database for the Nepal. J Environ. Manage 148:82–90. https://doi.org/10.1016/j.jenvman.2014.07.047

Ugarković D et al (2022) Climatic relationship of vegetation in forest stands in the Mediterranean Vegetation Belt of the Eastern Adriatic. Atmos. https://doi.org/10.3390/atmos13101709

Walther G-R et al (2002) Ecological responses to recent climate change. Nature 416(6879):389–395. https://doi.org/10.1038/416389a

Wang H et al (2021) Response of NDVI of natural vegetation to climate changes and drought in China. Land. https://doi.org/10.3390/land10090966

Wang X et al (2022) An overview of vegetation dynamics revealed by remote sensing and its feedback to regional and global climate. Remote Sens 14(20):2–5. https://doi.org/10.3390/rs14205275

Wang Z et al (2024) Biodiversity conservation in the context of climate change: facing challenges and management strategies. Sci Total Environ 937:173377. https://doi.org/10.1016/j.scitotenv.2024.173377

Wei W et al (2020) Regional-scale assessment of environmental vulnerability in an arid inland basin. Ecol Indic [Preprint]. https://doi.org/10.1016/j.ecolind.2019.105792

Wu C et al (2014) Modeling growing season phenology in North American forests using seasonal mean vegetation indices from MODIS. Remote Sens Environ 147:79–88. https://doi.org/10.1016/j.rse.2014.03.001

Wu X et al (2020) Vegetation changes and their response to global change based on NDVI in the Koshi river Basin of central Himalayas since 2000. Sustainability (Switzerland) 12(16). https://doi.org/10.3390/su12166644

Xue L et al (2019) Spatiotemporal analysis of ecological vulnerability and management in the Tarim River Basin, China. Sci Total Environ [Preprint]. https://doi.org/10.1016/j.scitotenv.2018.08.321

Yuan X et al (2015) Effects of precipitation intensity and temperature on NDVI-based grass change over Northern China during the period from 1982 to 2011. Remote Sens:10164–10183. https://doi.org/10.3390/rs70810164

Zhang Y et al (2013) NDVI-based vegetation changes and their responses to climate change from 1982 to 2011: a case study in the Koshi River Basin in the middle Himalayas. Glob Planet Chang 108:139–148. https://doi.org/10.1016/J.GLOPLACHA.2013.06.012

Zhang T et al (2016) Spatial-temporal patterns of water use efficiency and climate controls in China's Loess Plateau during 2000-2010. Sci Total Environ 565:105–122. https://doi.org/10.1016/j.scitotenv.2016.04.126

Zhang Y et al (2019) Vegetation change and its relationship with climate factors and elevation on the Tibetan Plateau. Int J Environ Res Public Health 16(23). https://doi.org/10.3390/ijerph16234709

Zhang X et al (2022) Spatiotemporal evolution of ecological vulnerability in the Yellow River Basin under ecological restoration initiatives. Ecol Indic [Preprint]. https://doi.org/10.1016/j.ecolind.2022.108586

Zhang J et al (2023) Coarse spatial resolution remote sensing data with AVHRR and MODIS miss the greening area compared with the Landsat data in Chinese drylands. Front Plant Sci 14:1129665. https://doi.org/10.3389/fpls.2023.1129665

Zhang C et al (2024a) Spatial and temporal dynamics of ecological parameters in various land use types in China during the first 20 years of the 21st century. Land. https://doi.org/10.3390/land13050572

Zhang H et al (2024b) Changes in vegetation NDVI and its response to climate change and human activities in the Ferghana Basin from 1982 to 2015. Remote Sens. https://doi.org/10.3390/rs16071296

Zhu M, Zhang J, Zhu L (2021) Article title variations in growing season NDVI and its sensitivity to climate change responses to green development in mountainous areas. Front Environ Sci 9. https://www.frontiersin.org/journals/environmental-science/articles/10.3389/fenvs.2021.678450

Zhuang Q et al (2020) Analysis and prediction of vegetation dynamics under the background of climate change in Xinjiang, China. PeerJ 8:e8282. https://doi.org/10.7717/peerj.8282

Chapter 7
Current and Future Possible Potentiality of *Rauvolfia serpentina* an IUCN Listed Endangered Species Using Maxent Modelling

Neelam Rawat, Saurabh Purohit, Vikas Painuly, Govind Singh Negi, and Mahendra Pratap Singh Bisht

Abstract *Rauvolfia serpentina (L.) Benth. ex Kurz* is a globally significant species important for people's livelihoods. The unsustainable and unscientific over-harvesting of this plant has threatened its continuous survival. The IUCN has listed it as endangered. The present study was done in the Dehradun district of Uttarakhand Himalaya to assess the current and potential future distribution of *R. serpentina* based on two Representative Concentration Pathways (RCP) (RCP 2.6 and RCP 8.5) for the 2050s and 2070s using maximum entropy (MaxEnt) modelling. The model showed good accuracy with an area under the curve (AUC) value of 0.870, with the major contributions of the bioclimatic variables to the model being the precipitation of the mean diurnal range (Bio 2; 76.8%) and precipitation of the coldest quarter (Bio 19; 11.9%). The current distribution showed that 184.2 km^2 (5.96%), 683.21 km^2 (22.12%), and 525.3 km^2 (17.01%) were under high potential, good potential, and moderate potential habitats of *R serpentina* in the study area, respectively. However, the total area of the high-potential and good-potential habitat decreased under the RCPs (RCP 2.6 and RCP 8.5) in the 2050s and 2070s. The present study will help the decision makers make appropriate plans for conserving *R. serpentina* while considering future climate scenarios.

Keywords *Rauvolfia serpentina* · Endangered · MaxEnt · Representative Concentration Pathway

---

N. Rawat (✉) · V. Painuly · G. S. Negi · M. P. S. Bisht
Uttarakhand Space Application Centre, Dehradun, India

S. Purohit
WOTR Centre for Resilience Studies (W-CReS), Watershed Organisation Trust (WOTR), Pune, Maharashtra, India

Y. D. I. Khan et al. (eds.), *Ecosystem-based Approaches for Resilience Building in Himalayan Landscapes*, Disaster Resilience and Green Growth,
https://doi.org/10.1007/978-981-95-2007-7_7

7.1 Introduction

Research on medicinal plants continues to garner substantial national and global attention worldwide, mainly due to its multifaceted roles in supporting healthcare and alleviating poverty (Anmol et al. 2024; Dalir et al. 2025). These are one of the most valuable non-timber forest products (NTFP) (Astutik et al. 2019; Singh et al. 2024) and form an integral part of traditional healthcare systems around the world (Alemu et al. 2024; Dal Cero et al. 2023; Gaur 2024; Ogata et al. 2024; Painuli et al. 2021; Saising et al. 2022).

Rauvolfia serpentina (L.) Benth. ex Kurz generally known as "sarpagandha" is one of the significant therapeutic herbs native to Indian subcontinent and Southeast Asia (Sugimura et al. 2024). French botanist Charles Plumier proposed the generic name *Rauvolfia* to honour German physician and botanist Leonhard Rauwolf (1535–1596) who documented the plant in his extensive travels (Surendran et al. 2021). *R. serpentina* is one of the most important and economically valuable medicinal plants used in traditional as well as modern systems of medicine (Ayyappan et al. 2025; Gadkari et al. 2024) because it is rich in bioactive secondary metabolites of the terpenoid indole alkaloid type including reserpine, ajmaline, ajmalicine, serpentine, yohimbine, etc. (Dey et al. 2022; Lata and Gond 2025; Ray et al. 2024). These alkaloids contribute to the plant's diverse pharmacological effects, such as antihypertensive, antiarrhythmic, sedative, antioxidant, antifungal, and antipsychotic activities (Alshahrani et al. 2021; Zhang et al. 2025). Due to its declining status, *R. serpentina* currently classified as an endangered species by the International Union for Conservation of Nature (IUCN) is listed under Appendix II of the Convention on International Trade in Endangered Species of Wild Fauna and Flora (CITES). Recognising its high demand and conservation concerns, the National Medicinal Plants Board (NMPB) in India have given financial subsidies under various programs for its sustainable management, cultivation, and conservation (Pidigam et al. 2023).

Climate change (alterations in temperature, precipitation patterns, and increased frequency of extreme weather events) poses a significant threat to the survival, distribution, and phytochemical composition of medicinal plants worldwide (Mykhailenko et al. 2025; Wani et al. 2024). These climatic stressors can also influence the biosynthesis of secondary metabolites, often resulting in variations in the concentration and efficacy of bioactive compounds (Sun and Fernie 2024). Species potential distribution is directly related to the suitability of the climatic conditions in the landscape. As climatic variables such as temperature and precipitation change, the potential distribution of a species also changes. Areas that were once suitable may become inhospitable, leading to local population declines or extinction (Jha and Dev 2024). Conversely, new regions may emerge as climatically suitable, but the ability of a species to migrate, establish, and thrive in these areas depends on its dispersal capacity, reproductive strategy, and landscape connectivity (Simson 2024). Species distribution modelling (SDM) plays a vital role in predicting potential future habitats, enabling researchers to forecast shifts in habitat ranges under future climate scenarios (Liu et al. 2024; Maphanga et al. 2025). The results of SDM can help develop targeted conservation actions to ensure the long-term survival and sustainable use of medicinal plant species. Furthermore, by forecasting shifts in species ranges, SDMs aid policymakers and resource managers in formulating adaptive strategies to ensure medicinal plants' long-term availability and sustainable use (Singh et al. 2023).

The maximum entropy (MaxEnt) model (Elith et al. 2011; Phillips et al. 2006) is a widely used species distribution modelling (SDM) tool for identifying suitable habitats of medicinal plants in the Himalayan region, especially when only presence data is available (Boral and Moktan 2021; Chauhan et al. 2025; Dhyani et al. 2021; Paul and Samant 2024; Rawat et al. 2021, 2022; Singh et al. 2022; Wani et al. 2022; Wani et al. 2021). The MaxEnt model estimates species distribution by identifying the most probable distribution pattern that aligns with environmental conditions at recorded presence locations. It produces response curves to show the influence of each environmental variable and includes a jackknife analysis to evaluate the contribution of each predictor (Zhang et al. 2024). This approach is particularly advantageous for medicinal plant studies, where absence data is often unavailable or unreliable. Compared to other SDM techniques, such as generalized linear models (GLM) or BioClim, MaxEnt performs robustly with limited sample sizes and provides high predictive accuracy (Kaky et al. 2020; Remya et al. 2015). Its ability to handle complex interactions between predictor variables and produce clear probability maps makes it especially useful for guiding conservation planning, sustainable harvesting, and habitat restoration efforts for ecologically and economically valuable medicinal species.

The MaxEnt model utilised in this study aims to achieve the following objectives: (1) to assess the effects of climate change on the current and projected distribution of *R. serpentina* within Dehradun district, Uttarakhand, and (2) to identify the primary environmental variables influencing the species' potential distribution.

7.2 Materials and Methods

7.2.1 Study Area

The present study focuses on the Dehradun district of Uttarakhand, a part of the western Himalayas in India. The latitude and longitude of Dehradun are 30° 15′ N and 79° 15′ E, respectively (Fig. 7.1). The geographical area of the district is 3088 km^2, out of which nearly 53.12% is under forest cover (1640.30 km^2) (ISFR 2023). Dehradun district has different forest types which varies with the altitudinal gradient. The lower altitudes are predominantly covered by moist and dry deciduous forests, which include species like *Shorea robusta* (Sal), *Terminalia* spp. and *Mallotus philippensis (Rohini)* species. As the elevation increases, the forests transition with species such as *Pinus roxburghii* (chir pine), *Quercus* (oak), and *Cedrus*

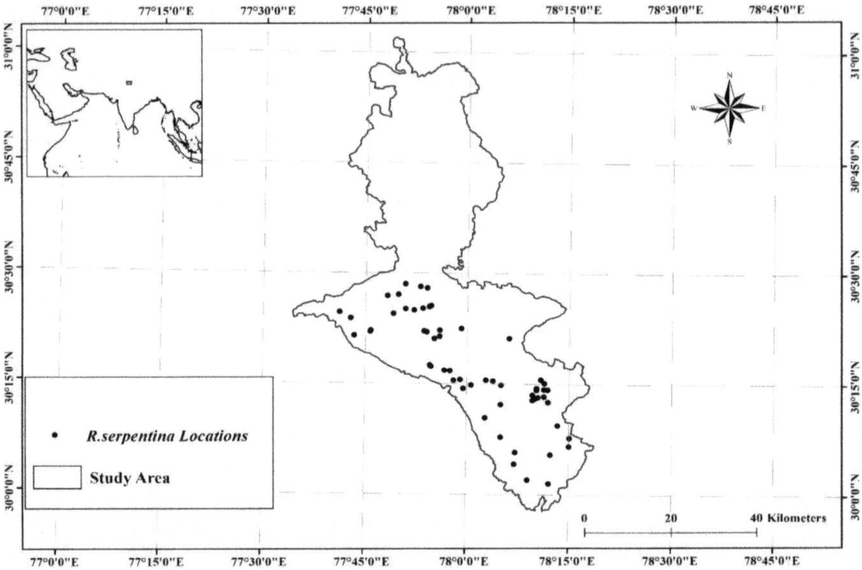

Fig. 7.1 Map showing the study area in the Western Himalaya region of India

deodara (Deodar). At higher elevations, the landscape is dominated by coniferous forests and alpine meadows, supporting a mix of species like *Abies pindrow* (Himalayan fir) and *Picea smithiana* (spruce). This variety of forest types contributes to the region's rich biodiversity and ecological significance, providing habitat for a range of flora and fauna (Champion and Seth 1968).

7.2.2 Species Occurrence and Environmental Data

R. serpentina is an erect woody shrub usually found in moist, damp, and shady places in moist deciduous forests especially associated with *Shorea robusta*, *Terminalia alata*, *Dalbergia sissoo*, and *Adina cordifolia* (Jiang et al. 2022). It favours deep, fertile soil rich in organic matter (Gadkari et al. 2024). For this study, sixty-three (63) non-overlapping occurrence points were collected by field surveys using a GPS device from June to November 2019 in the study area to be used for species distribution modelling using the MaxEnt model.

7.2.3 Data Used

Bioclimatic variables represent the primary determinants in delineating the environmental niches of species, as they encapsulate the critical climatic parameters influencing species distribution and ecological dynamics. The set of 19 bioclimatic variables is derived from long-term climatological records of monthly precipitation and temperature, providing ecologically meaningful indicators of annual trends, seasonality, and extreme or limiting environmental factors (1950–2000) (Hijmans et al. 2005). The dataset, obtained from the WorldClim database, possesses a spatial resolution of approximately 1 km^2 (30 arc-seconds), enabling fine-scale ecological and biogeographical analyses (Table 7.1) (www.worldclim.org). To test its habitat suitability ranges in the 2050s (2041–2060) and 2070s (2061–2080), RCP 2.6 (the minimum scenario for GHG emissions), and RCP 8.5 (the highest scenario for GHG emissions), Community Climate System Model version 4 (CCSM4) (Gent et al. 2011), a type of general circulation model (GCM), was used. The eight bioclimatic variables were drawn out based on a correlation matrix ≥0.90, viz. mean diurnal temperature range (Bio 2), isothermality (Bio 3) temperature seasonality (Bio 4), mean temperature of driest quarter (Bio 9), precipitation of wettest month (Bio 13), precipitation of driest month (Bio 14), precipitation of warmest quarter (Bio 18), and precipitation of coldest quarter (Bio 19) (Table 7.1). ArcGIS 10.5 was used to convert the bioclimatic layers and reclassify the final outputs.

Table 7.1 Different bioclimatic variables used for modelling the spatial distribution of *R. serpentina* (variables shown in bold were used for the final analysis)

Data source	Category	Variable	Abbreviation	Units
WorldClim	Bioclimatic	Annual mean temperature	Bio 1	°C
		Mean diurnal range	**Bio 2**	**°C**
		Isothermality	**Bio 3**	**%**
		Temperature seasonality	**Bio 4**	–
		Maximum temperature of warmest month	Bio 5	°C
		Minimum temperature of coldest month	Bio 6	°C
		Temperature annual range	Bio 7	°C
		Mean temperature of wettest quarter	Bio 8	°C
		Mean temperature of driest quarter	**Bio 9**	**°C**
		Mean temperature of warmest quarter	Bio 10	°C
		Mean temperature of coldest quarter	Bio 11	°C
		Annual Precipitation	Bio 12	mm
		Precipitation of wettest month	**Bio 13**	**mm**
		Precipitation of driest month	**Bio 14**	**mm**
		Precipitation seasonality	Bio 15	–
		Precipitation of wettest quarter	Bio 16	mm
		Precipitation of driest quarter	Bio 17	mm
		Precipitation of warmest quarter	**Bio 18**	**mm**
		Precipitation of coldest quarter	**Bio 19**	**mm**

7.2.4 Data Analysis Using MaxEnt Species Distribution Model

MaxEnt model (version 3.4.4) was used for data analysis. For model calibration, the occurrence records were divided into a training set (75%) and a test set (25%) to evaluate its accuracy. Model performance was assessed using the area under the curve (AUC) of the receiver operating characteristic (ROC) curve, a widely used model performance metric ranging from 0 to 1, where a value of 1 indicates the perfect discriminatory capability of the model (Fielding and Bell 1997; Phillips et al. 2006). Model performance is classified as failing (0.5–0.6), bad (0.6–0.7), reasonable (0.7–0.8), good (0.8–0.9), or great (0.9–1) (Swets 1988). The relative importance of bioclimatic predictors influencing the species' potential distribution was evaluated using a jackknife test, which systematically excludes each variable to assess its contribution to model performance. Species response curves were constructed to quantify the functional relationship between environmental gradients and habitat suitability for the focal species, facilitating the interpretation of ecological tolerances and thresholds. The resulting potential species distribution map displayed values ranging from 0 to 1. The values are classified into four categories based on their potential: high (> 0.6), good (0.4–0.6), moderate (0.2–0.4), and low (< 0.2) (Yang et al. 2013).

7.3 Results

The model accuracy for *R. serpentina* was found to be good (AUC = 0.870, Fig. 7.2). The jackknife test indicates that the distribution of *R. serpentina* was mainly influenced by two bioclimatic variables, the mean diurnal range (Bio 2) and coldest quarter precipitation (Bio 19) which contributed 76.8% and 11.9% to the MaxEnt model, respectively (Fig. 7.3).

Based on permutation importance analysis, precipitation during the coldest quarter (Bio19) emerged as the most influential variable in the habitat model, accounting for 37.2% of the total contribution. This was followed by mean diurnal temperature range (Bio 2) and isothermality (Bio 3), which contributed 35.3% and 15.8%, respectively, to the model's predictive performance (Table 7.2). Species response curves illustrate the relationship between environmental variables and the likelihood of species occurrence, revealing the target species' biological tolerances and habitat preferences. The response curves depict the relationship between the predicted probability of *R. serpentina* occurrence and the four most influential environmental variables: (a) mean diurnal range (Bio 2), (b) precipitation of the coldest quarter (Bio 19), (c) precipitation of the warmest quarter (Bio 18), and (d) precipitation of the driest quarter (Bio 14). These variables were identified as the top contributors to the species distribution model, and their respective response curves highlight the sensitivity of *R. serpentina* to changes in climatic conditions (Fig. 7.4).

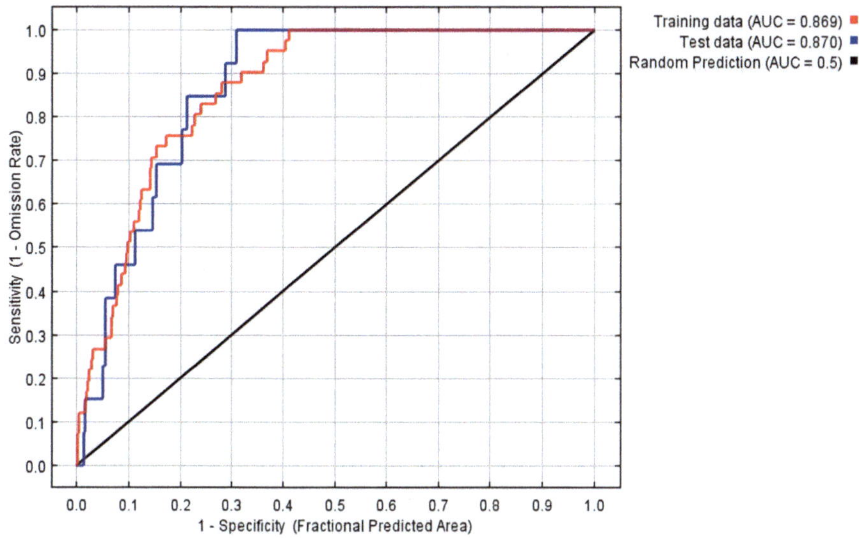

Fig. 7.2 Reliability (sensitivity) test of the distribution model

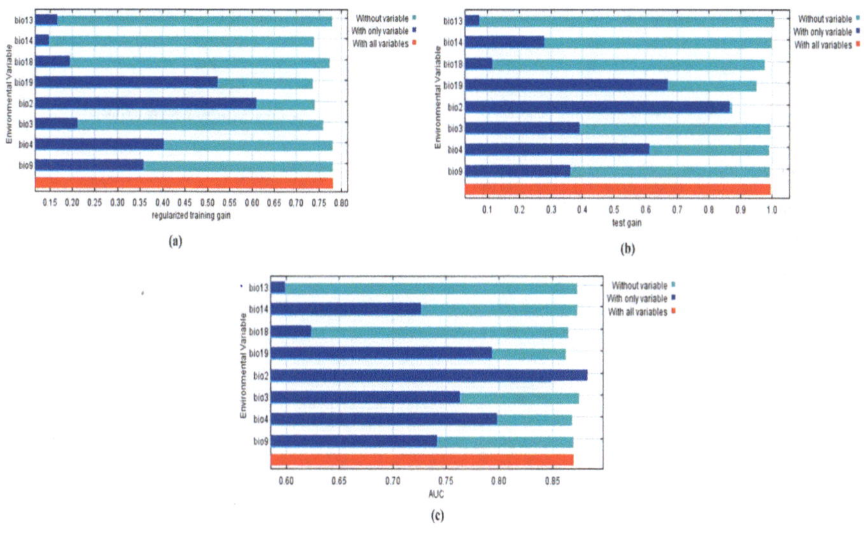

Fig. 7.3 The results of jackknife test for evaluating the relative importance of bioclimatic variables. (**a**) Regularised training gain. (**b**) Regularised test gain. (**c**) AUC for *R. serpentina*

Table 7.2 Percentage contribution and permutation importance for different bioclimatic variables in the model

Variable	Percent contribution	Permutation importance
Bio 2	76.8	35.3
Bio 19	11.9	37.2
Bio 18	4.3	0.5
Bio 14	3.4	10.4
Bio 3	2.2	15.8
Bio 4	0.5	0
Bio 9	0.5	0
Bio 13	0.4	0.9

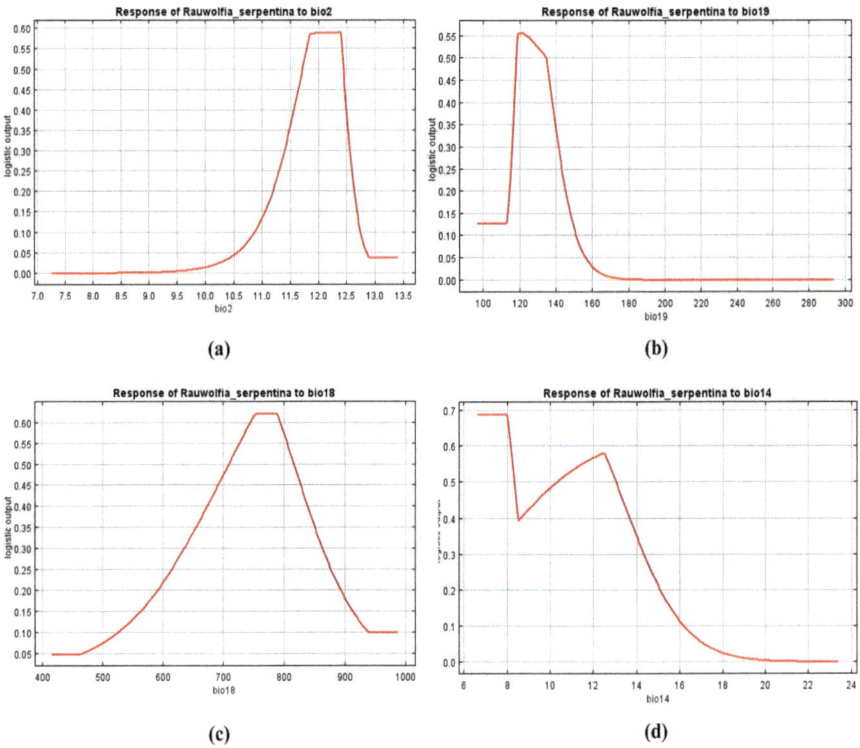

Fig. 7.4 Response curves illustrating the relationship of predicted probability of *R. serpentina* occurrence to top four contributing environmental variables: (**a**) mean diurnal range (Bio 2), (**b**) precipitation of the coldest quarter (Bio 19), (**c**) precipitation of warmest quarter (Bio 18), and (**d**) precipitation of driest quarter (Bio 14)

7.3.1 Current Potential Distribution of R. serpentina

Out of the total geographical area (3088 km^2), 1626.3 km^2 (52.67%) is found to be a less probable habitat and 184.2 km^2 (5.96%) as a high-potential habitat (Table 7.3). The Vikasnagar zone was found to have a high potential (110.0 km^2), followed by the Dehradun zone (73 km^2) (Table 7.4).

Table 7.3 Area under the current potential distribution of *R. serpentina* in present climatic conditions

Category		Least potential (<0.2)	Moderate potential (0.2–0.4)	Good potential (0.4–0.6)	High potential (>0.6)
Current	Area (km^2)	1626.3	525.3	683.21	184.2

Table 7.4 Predicted potential distribution areas under different zones for *R. serpentina* in current climatic conditions

Number	Zone name	Moderate potential (km^2)	Good potential (km^2)	High potential (km^2)
1	Tyuni	–	–	–
2	Chakrata	–	–	–
3	Kalsi	51.59	5.75	–
4	Vikasnagar	233.41	214.07	110.0
5	Dehradun	103.25	253.0	73.0
6	Rishikesh	137.05	210.39	1.2

7.3.2 Model Evaluations and Its Importance to Variables Under Different Scenarios

The model for *R. serpentina* predicted for 2050 in RCP 2.6 shows the high-potential habitat area as 161.02 km^2 (5.21%), slightly less than in the current potential scenario, i.e. 184.2 km^2 (5.96%). Similarly, for 2070 (RCP 2.6), habitat has decreased to 66.0km^2 (2.13%).

Due to climate warming under RCP 8.5, a reduction of 22.02 km^2 (0.71%) for 2050 and 53.52 km^2 (1.73%) for 2070 was recorded. The high and good suitability areas will be reduced in both climatic scenarios (RCP 2.6 and RCP 8.5) (Table 7.5). By 2070, the suitable areas, although reduced, will shift to the newer areas.

Considering different zones, it was found that the high-potential habitat will decrease in Vikasnagar and increase in Dehradun and Rishikesh under RCP 2.6 (2050) (Table 7.6). In RCP 2.6 (2070), a new high-potential habitat will emerge in the Kalsi zone. The area under the high-potential habitat will considerably decrease in the Dehradun and Rishikesh zones, but slight improvement was observed in the Vikasnagar area (Table 7.7).

Under the RCP 8.5 (2050) scenario, the high-potential habitat will be shrunk only to the Vikasnagar and Dehradun zones (Table 7.8). In RCP 8.5 (2070), Kalsi will have the highest area under the high-potential habitat zone, followed by Vikasnagar and Dehradun zones (Table 7.9).

Table 7.5 Predicted suitable areas for *R. serpentina* under future climatic conditions

Category		Least potential (<0.2)	Moderate potential (0.2–0.4)	Good potential (0.4–0.6)	High potential (>0.6)
RCP 2.62050	Area (km^2)	1751	639	468.01	161.02
RCP 2.62070	Area (km^2)	2040	640.02	273.01	66.0
RCP 8.52050	Area (km^2)	2002	680	315.01	22.02
RCP 8.52070	Area (km^2)	1955	473.31	537.20	53.52

Table 7.6 Predicted potential distribution areas for *R. serpentina* under RCP 2.6 (2050) scenario

Number	Zone name	Moderate potential (km^2)	Good potential (km^2)	High potential (km^2)
1	Tyuni	–	–	–
2	Chakrata	66.0	1.44	–
3	Kalsi	48.0	34.39	–
4	Vikasnagar	288.0	168.27	10.40
5	Dehradun	190.0	214.29	110.02
6	Rishikesh	47.0	49.62	40.60

Table 7.7 Predicted potential distribution areas for *R. serpentina* under RCP 2.6 (2070) scenario

Number	Zone name	Moderate potential (km^2)	Good potential (km^2)	High potential (km^2)
1	Tyuni	–	–	–
2	Chakrata	79.10	15.0	–
3	Kalsi	39.45	50.0	12.90
4	Vikasnagar	214.27	38.0	16.50
5	Dehradun	256.14	140.01	22.22
6	Rishikesh	51.06	30.0	14.38

Table 7.8 Predicted potential distribution areas for *R. serpentina* under RCP 8.5 (2050) scenario

Number	Zone name	Moderate potential (km^2)	Good potential (km^2)	High potential (km^2)
1	Tyuni	–	–	–
2	Chakrata	69.43	0.72	–
3	Kalsi	71.19	8.63	–
4	Vikasnagar	244.17	198.13	2.0
5	Dehradun	239.12	105.37	20.02
6	Rishikesh	56.09	2.16	–

Table 7.9 Predicted potential distribution areas for *R. serpentina* under RCP 8.5 (2070) scenario

Number	Zone name	Moderate potential (km^2)	Good potential (km^2)	High potential (km^2)
1	Tyuni	–	–	–
2	Chakrata	62.48	20.13	–
3	Kalsi	59.48	39.25	24.45
4	Vikasnagar	182.37	377.14	15.51
5	Dehradun	145.25	89.17	12.22
6	Rishikesh	23.73	11.51	0.72

7.4 Discussion

R. serpentina is an endangered plant whose population in its native regions has been consistently decreasing due to various human-induced threats, mainly habitat fragmentation and illegal harvesting (Roopashree et al. 2024). The AUC value of 0.870 is considered within the acceptable range (Ali et al. 2024; Araújo et al. 2005; Li et al. 2022). Abolmaali et al. (2018) used MaxEnt modelling to predict the suitable habitat for *Daphne mucronata* and found that Bio 19 is one of the most important contributing factors to the model. Li et al. (2019) used MaxEnt modelling to identify suitable planting areas for tuber mustard in China and found that Bio 2 was the most important factor in determining the distribution of tuber mustard. Li et al. (2020) used MaxEnt modelling to identify three Coptis herbs' present and future potential distribution in China. They found the mean diurnal range (Bio 2) to be the major contributing factor to the model. Gebrewahid et al. (2020) analysed the distribution of *Oxytenanthera abyssinica* in Northern Ethiopia and found that precipitation during the coldest quarter (Bio 19) was one of the dominant contributing factors in the model. Bhandari et al. (2020) used the MaxEnt model to predict *R. arboreum* distribution in Uttarakhand and got an AUC value of 0.886. Rawat et al. (2021) used MaxEnt modelling to predict the potential distribution of *Valeriana jatamansi* Jones. in the Chakrata Forest Division of Garhwal Himalaya and got an AUC value of 0.89. Purohit and Rawat (2021) used MaxEnt modelling to predict the current and future distribution of *C. infortunatum* L. in the Dehradun district, India, and got an AUC of 0.837. Soilhi et al. (2022) used MaxEnt modelling to predict the current and future distribution of *Mentha pulegium* L. in Tunisia and found that Bio 19 was the most critical factor in shaping its distribution. Zhao et al. (2023) used MaxEnt modelling for suitable habitat distribution of different forest species in China and found that Bio 2 is the top contributor to the model.

7.4.1 Changes in the Distribution of R. serpentina *in the Future*

The model predicted that global warming will result in habitat expansion in most areas for moderate suitable areas, but these suitable habitats will be effectively reduced to a few particular places. Both climatic scenarios, i.e. RCP 2.6 and RCP 8.5, show a reduction in the high-potential habitat (Fig. 7.5). Previous studies have

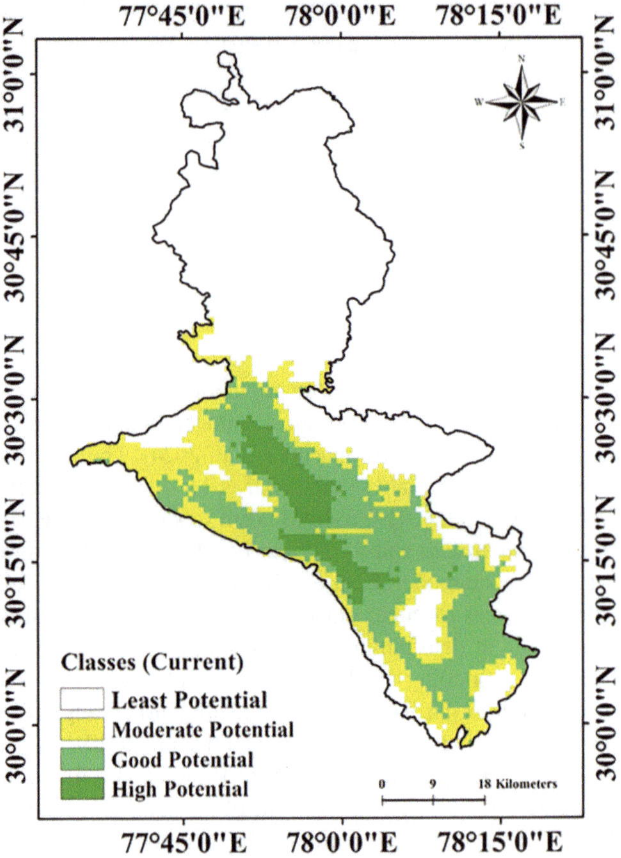

Fig. 7.5 Current distribution for *R. serpentina* in the study area

also shown that global warming would greatly influence species distributions by reducing the species' geographic distribution. Boral and Moktan (2021) used MaxEnt modelling for predicting the suitable habitat for the *Swertia bimaculata* in the Darjeeling-Sikkim region of Eastern Himalaya under current and future climatic scenarios. They found a considerable decline in the suitable habitat in the future climatic conditions. Chen et al. (2022) used MaxEnt modelling to identify the present and future habitat distribution of *Pomatosace filicula* under climate change. They found a decrease in the suitable habitat. Bao et al. (2024) used MaxEnt modelling for habitat suitability for *Vaccinium mandarinorum* (wild blueberry) in China and found that the suitable habitats will decline under the different climatic scenarios. Hosseini et al. (2024) used MaxEnt model to identify the current and future geographical distribution of *Thymus* species in Iran using RCP 4.5 and 8.5. They found that there is a negative area change in its suitable habitat. Meena et al. (2024) used MaxEnt modelling to identify the potential habitat of *Thamnocalamus spathiflorus* (alpine bamboo) in the Western Himalayas and found a decline in the distribution area of the species (Fig. 7.6).

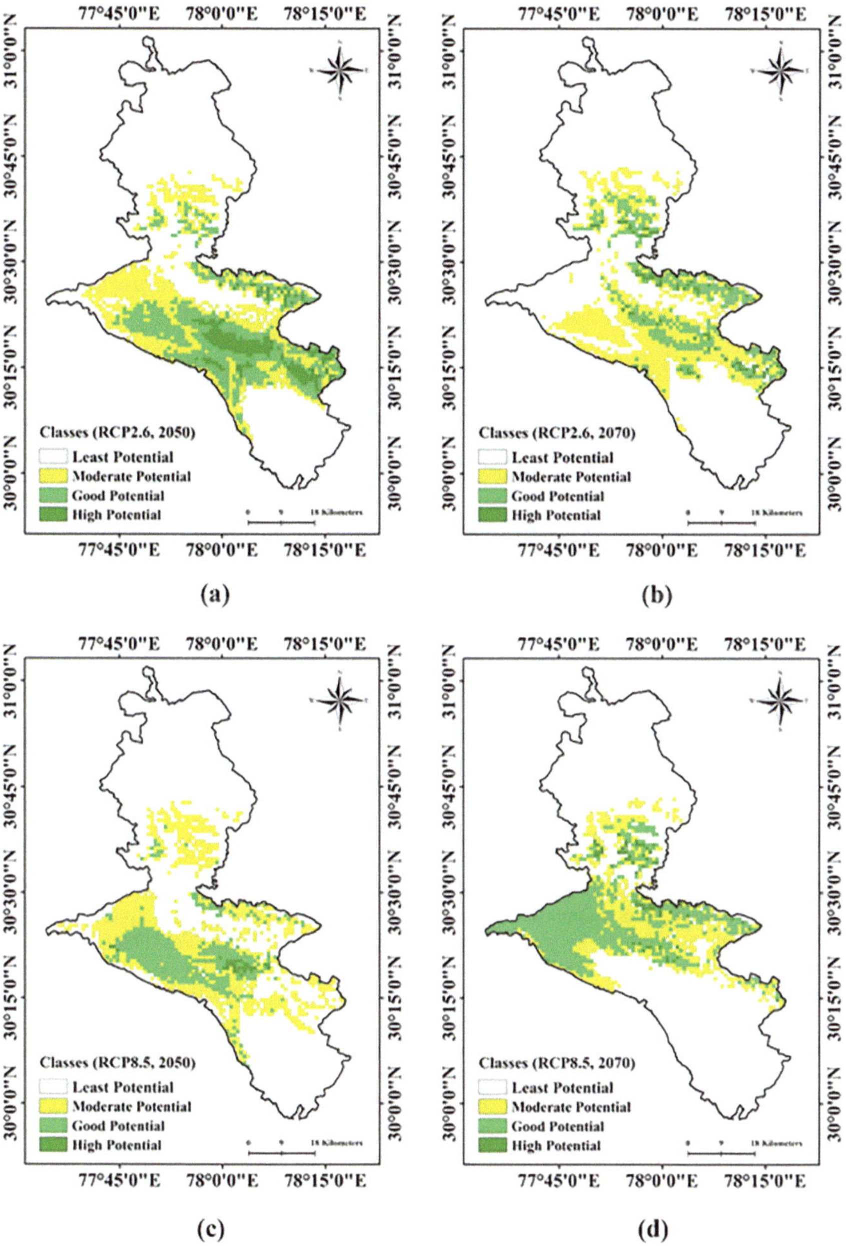

Fig. 7.6 Future distribution for *R. serpentina*: (**a**) RCP 2.6, 2050, (**b**) RCP 2.6, 2070, (**c**) RCP 8.5, 2050, and (**d**) RCP 8.5, 2070 in study area

7.5 Conclusion

The present study provided the current and future distribution map of *R. serpentina* for the Dehradun district of Uttarakhand, India. The findings suggest that the future habitat suitability of *R. serpentina* will decline under climate warming in both RCP (2.6 and 8.5) scenarios. These results will assist the relevant agencies in identifying and prioritising suitable areas for the species' reintroduction and conservation efforts. To decrease the pressure on the wild population of *R. serpentina*, the government should encourage the farmers to cultivate *R. serpentina* in private lands.

Acknowledgement The authors gratefully acknowledge the financial support provided by the National Medicinal Plants Board (Z.18017/187/CSS/R&D/UK-03/2015-16-NMPB). Additionally, the authors appreciate the cooperation of the Uttarakhand Forest Department in granting permission to collect field data.

References

Abolmaali SM-R, Tarkesh M, Bashari H (2018) MaxEnt modeling for predicting suitable habitats and identifying the effects of climate change on a threatened species, Daphne mucronata, in central Iran. Eco Inform 43:116–123. https://doi.org/10.1016/j.ecoinf.2017.10.002

Alemu M, Asfaw Z, Lulekal E, Warkineh B, Debella A, Sisay B, Debebe E (2024) Ethnobotanical study of traditional medicinal plants used by the local people in Habru District, North Wollo Zone, Ethiopia. J Ethnobiol Ethnomed 20(1):4. https://doi.org/10.1186/s13002-023-00644-x

Ali S, Umair M, Makanda TA, Shi S, Hussain SA, Ni J (2024) Modeling current and future potential land distribution dynamics of wheat, rice, and maize under climate change scenarios using MaxEnt. Land. https://doi.org/10.3390/land13081156

Alshahrani MY, Rafi Z, Alabdallah NM, Shoaib A, Ahmad I, Asiri M et al (2021) A comparative antibacterial, antioxidant, and antineoplastic potential of Rauwolfia serpentina (L.) leaf extract with its biologically synthesized gold nanoparticles (R-AuNPs). Plants. https://doi.org/10.3390/plants10112278

Anmol, Aggarwal G, Sharma M, Singh R, Shivani, Sharma U (2024) Ethnopharmacologically important highly subsidised Indian medicinal plants: systematic review on their traditional uses, phytochemistry, pharmacology, quality control, conservation status and future prospective. J Ethnopharmacol 320:117385. https://doi.org/10.1016/j.jep.2023.117385

Araújo MB, Pearson RG, Thuiller W, Erhard M (2005) Validation of species–climate impact models under climate change. Glob Chang Biol 11(9):1504–1513. https://doi.org/10.1111/j.1365-2486.2005.01000.x

Astutik S, Pretzsch J, Ndzifon Kimengsi J (2019) Asian medicinal plants' production and utilisation potentials: a review. Sustainability. https://doi.org/10.3390/su11195483

Ayyappan N, Raju R, Devairakkam EGWJ (2025) Ethnobotany, bioactive compounds and therapeutical values of Rauvolfia tetraphylla L. – a review. Curr Pharmacol Rep 11(1):21. https://doi.org/10.1007/s40495-025-00400-7

Bao X, Zhou P, Zhang M, Fang Y, Zhang Q (2024) MaxEnt-based habitat suitability assessment for Vaccinium mandarinorum: exploring industrial cultivation opportunities. Forests. https://doi.org/10.3390/f15122254

Bhandari MS, Meena RK, Shankhwar R, Shekhar C, Saxena J, Kant R et al (2020) Prediction mapping through maxent modeling paves the way for the conservation of Rhododendron arboreum in Uttarakhand Himalayas. J Indian Soc Remote Sens 48(3):411–422. https://doi.org/10.1007/s12524-019-01089-0

Boral D, Moktan S (2021) Predictive distribution modeling of Swertia bimaculata in Darjeeling-Sikkim Eastern Himalaya using MaxEnt: current and future scenarios. Ecol Process 10(1):26. https://doi.org/10.1186/s13717-021-00294-5

Champion HG, Seth SK (1968) A revised forest types of India. Government of India, Delhi

Chauhan J, Purohit VK, Patni B, Charles B, Prasad P (2025) Impact of climate change on the distribution range and niche dynamics of Himalayan Aconites: a highly important medicinal plant species of the higher Himalayan range. Environ Monit Assess 197(4):496. https://doi.org/10.1007/s10661-025-13848-3

Chen K, Wang B, Chen C, Zhou G (2022) MaxEnt modeling to predict the current and future distribution of Pomatosace filicula under climate change scenarios on the Qinghai–Tibet Plateau. Plants. https://doi.org/10.3390/plants11050670

Dal Cero M, Saller R, Leonti M, Weckerle CS (2023) Trends of medicinal plant use over the last 2000 years in Central Europe. Plants. https://doi.org/10.3390/plants12010135

Dalir M, Choobchian S, Abbasi E (2025) Identifying the relative importance of cultivating medicinal plants in improving rural livelihoods. Environ Dev 54:101159. https://doi.org/10.1016/j.envdev.2025.101159

Dey A, Roy D, Mohture VM, Ghorai M, Rahman MH, Anand U et al (2022) Biotechnological interventions and indole alkaloid production in Rauvolfia serpentina. Appl Microbiol Biotechnol 106(13):4867–4883. https://doi.org/10.1007/s00253-022-12040-8

Dhyani A, Kadaverugu R, Nautiyal BP, Nautiyal MC (2021) Predicting the potential distribution of a critically endangered medicinal plant Lilium polyphyllum in Indian Western Himalayan Region. Reg Environ Chang 21(2):30. https://doi.org/10.1007/s10113-021-01763-5

Elith J, Phillips SJ, Hastie T, Dudík M, Chee YE, Yates CJ (2011) A statistical explanation of MaxEnt for ecologists. Divers Distrib 17(1):43–57. https://doi.org/10.1111/j.1472-4642.2010.00725.x

Fielding AH, Bell JF (1997) A review of methods for the assessment of prediction errors in conservation presence/absence models. Environ Conserv 24(1):38–49. https://doi.org/10.1017/S0376892997000088

Gadkari P, Daharwal SJ, Umakant S, Nagendra Singh C, Khushbu S, Suman S, Devangan N (2024) Method development and validation for quantification of Yohimbine in root of Rauwolfia serpentina by HPTLC. Anal Chem Lett 14(4):549–560. https://doi.org/10.1080/22297928.2024.2385814

Gaur R (2024) A brief history: traditional Chinese medicinal system. Pharmacol Res Mod Chin Med 10:100387. https://doi.org/10.1016/j.prmcm.2024.100387

Gebrewahid Y, Abrehe S, Meresa E, Eyasu G, Abay K, Gebreab G et al (2020) Current and future predicting potential areas of Oxytenanthera abyssinica (A. Richard) using MaxEnt model under climate change in Northern Ethiopia. Ecol Process 9(1). https://doi.org/10.1186/s13717-019-0210-8

Gent PR, Danabasoglu G, Donner LJ, Holland MM, Hunke EC, Jayne SR et al (2011) The community climate system model Version 4. J Clim 24(19):4973–4991. https://doi.org/10.1175/2011JCLI4083.1

Hijmans RJ, Cameron SE, Parra JL, Jones PG, Jarvis A (2005) Very high resolution interpolated climate surfaces for global land areas. Int J Climatol 25(15):1965–1978. https://doi.org/10.1002/joc.1276

Hosseini N, Ghorbanpour M, Mostafavi H (2024) Habitat potential modelling and the effect of climate change on the current and future distribution of three Thymus species in Iran using MaxEnt. Sci Rep 14(1):3641. https://doi.org/10.1038/s41598-024-53405-5

ISFR (2023) Indian state of forest report. Dehradun

Jha MK, Dev M (2024) Impacts of climate change BT. In: Azrour M, Mabrouki J, Alabdulatif A, Guezzaz A, Amounas F (eds) Smart internet of things for environment and healthcare. Springer, Cham, pp 139–159. https://doi.org/10.1007/978-3-031-70102-3_10

Jiang L, Gaire D, Bhattarai S, Adhikari B, Panthi S (2022) Predicting the potential distribution, trade, and conservation of Rauvolfia serpentina in Nepal. Appl Ecol Environ Res 20(6)

Kaky E, Nolan V, Alatawi A, Gilbert F (2020) A comparison between Ensemble and MaxEnt species distribution modelling approaches for conservation: a case study with Egyptian medicinal plants. Eco Inform 60:101150. https://doi.org/10.1016/j.ecoinf.2020.101150

Lata R, Gond SK (2025) Antibacterial and antioxidant potentials, detection of host origin compounds, and metabolic profiling of endophytic Bacillus spp. isolated from Rauvolfia serpentina (L.) Benth. ex Kurz. Sci Rep 15(1):2094. https://doi.org/10.1038/s41598-024-84893-0

Li HQ, Liu XH, Wang JH, Xing LG, Fu YY (2019) Maxent modelling for predicting climate change effects on the potential planting area of tuber mustard in China. J Agric Sci 157(5):375–381. https://doi.org/10.1017/S0021859619000686

Li J, Fan G, He Y (2020) Predicting the current and future distribution of three Coptis herbs in China under climate change conditions, using the MaxEnt model and chemical analysis. Sci Total Environ 698:134141. https://doi.org/10.1016/j.scitotenv.2019.134141

Li Y, Shao W, Huang S, Zhang Y, Fang H, Jiang J (2022) Prediction of suitable habitats for Sapindus delavayi based on the MaxEnt model. Forests. https://doi.org/10.3390/f13101611

Liu S, Li T, Cong B, Yang L, Zhang Z, Zhao L (2024) Unveiling the suitable habitats and future conservation strategies of Tridacna maxima in the Indo-Pacific core area based on species distribution model. Ecol Evol 14(9):e70187. https://doi.org/10.1002/ece3.70187

Maphanga T, Shoko C, Sibanda M, Kavhu B, Coetsee C, Dube T (2025) Bush encroachment with climate change in protected and communal areas: a species distribution modelling approach. Ecol Model 503:111056. https://doi.org/10.1016/j.ecolmodel.2025.111056

Meena RK, Bhandari MS, Thakur PK, Negi N, Pandey S, Kant R et al (2024) MaxEnt-based potential distribution mapping and range shift under future climatic scenarios for an Alpine Bamboo Thamnocalamus spathiflorus in Northwestern Himalayas. Land. https://doi.org/10.3390/land13070931

Mykhailenko O, Jalil B, McGaw LJ, Echeverría J, Takubessi M, Heinrich M (2025) Climate change and the sustainable use of medicinal plants: a call for "new" research strategies. Front Pharmacol 15. https://www.frontiersin.org/journals/pharmacology/articles/10.3389/fphar.2024.1496792

Ogata R, Hakariya A, Okada S, Uneda K, Suzuki T, Suzuki T, Takayama S (2024) Encounters with Kampo medicine and its use in current position. Tradit Kampo Med 11(3):268–280. https://doi.org/10.1002/tkm2.1435

Painuli S, Semwal P, Cruz-Martins N, Bachheti RK (2021) Medicinal plants of Himalayan Forests BT. In: Husen A, Bachheti RK, Bachheti A (eds) Non-timber forest products: food, healthcare and industrial applications. Springer, Cham, pp 175–212. https://doi.org/10.1007/978-3-030-73077-2_8

Paul S, Samant SS (2024) Population ecology and habitat suitability modelling of an endangered and endemic medicinal plant Meconopsis aculeata Royle under projected climate change in the Himalaya. Environ Exp Bot 225:105837. https://doi.org/10.1016/j.envexpbot.2024.105837

Phillips SJ, Anderson RP, Schapire RE (2006) Maximum entropy modeling of species geographic distributions. Ecol Model 190(3):231–259. https://doi.org/10.1016/j.ecolmodel.2005.03.026

Pidigam S, Geetha A, Pandravada SR, Khan S, Thuraga V, Sivaraj N, Nagaraju K (2023) Conservation strategies for Indian medicinal plants BT. In: Máthé Á, Khan IA (eds) Medicinal and aromatic plants of India, vol 2. Springer, Cham, pp 279–302. https://doi.org/10.1007/978-3-031-44914-7_14

Purohit S, Rawat N (2021) MaxEnt modeling to predict the current and future distribution of Clerodendrum infortunatum L. under climate change scenarios in Dehradun district, India. Model Earth Syst Environ. https://doi.org/10.1007/s40808-021-01205-5

Rawat N, Mishra AP, Chandra N, Bisht MPS, Purohit S (2021) MaxEnt modeling for predicting the potential Distribution of Valeriana jatamansi Jones in Chakrata Forest Division of Garhwal Himalaya. Indian J Ecol 48(2):367–372

Rawat N, Purohit S, Painuly V, Negi GS, Bisht MPS (2022) Habitat distribution modeling of endangered medicinal plant Picrorhiza kurroa (Royle ex Benth) under climate change scenarios in Uttarakhand Himalaya, India. Eco Inform 68:101550. https://doi.org/10.1016/j.ecoinf.2021.101550

Ray S, Roychowdhury D, Majumder A (2024) Chapter 9 – recent trends in in vitro production of alkaloids from Rauvolfia serpentina. In: Kishor PBK, Pullaiah T, Suprasanna P, Rao AR, A. B. T.-B. P. of B. P. of M. V. Romano (eds). Elsevier, pp 221–243. https://doi.org/10.1016/B978-0-443-21818-7.00022-8

Remya K, Ramachandran A, Jayakumar S (2015) Predicting the current and future suitable habitat distribution of Myristica dactyloides Gaertn. using MaxEnt model in the Eastern Ghats, India. Ecol Eng 82:184–188. https://doi.org/10.1016/j.ecoleng.2015.04.053

Roopashree S, Anitha J, Challa S, Mahesh TR, Venkatesan VK, Guluwadi S (2024) Mapping of soil suitability for medicinal plants using machine learning methods. Sci Rep 14(1):3741. https://doi.org/10.1038/s41598-024-54465-3

Saising J, Maneenoon K, Sakulkeo O, Limsuwan S, Götz F, Voravuthikunchai SP (2022) Ethnomedicinal plants in herbal remedies used for treatment of skin diseases by traditional healers in Songkhla Province, Thailand. Plants. https://doi.org/10.3390/plants11070880

Simson A (2024) The potential for climate change and the resulting urban heat island effect in cities to assist the northerly migration of tree species. Arboricult J 46(3):229–234. https://doi.org/10.1080/03071375.2024.2386828

Singh L, Kanwar N, Bhatt ID, Nandi SK, Bisht AK (2022) Predicting the potential distribution of Dactylorhiza hatagirea (D. Don) Soo-an important medicinal orchid in the West Himalaya, under multiple climate change scenarios. PLoS One 17(6):e0269673. https://doi.org/10.1371/journal.pone.0269673

Singh H, Kumar N, Singh R, Kumar M (2023) Assessing the climate change impact on the habitat suitability of the range-restricted bird species (Catreus wallichii) in the Indian Himalayan ecosystem. Environ Sci Pollut Res 30(57):121224–121235. https://doi.org/10.1007/s11356-023-30789-x

Singh PA, Dash S, Choudhury A, Bajwa N (2024) Factors affecting long-term availability of medicinal plants in India. J Crop Sci Biotechnol 27(2):145–173. https://doi.org/10.1007/s12892-023-00219-y

Soilhi Z, Sayari N, Benalouache N, Mekki M (2022) Predicting current and future distributions of Mentha pulegium L. in Tunisia under climate change conditions, using the MaxEnt model. Eco Inform 68:101533. https://doi.org/10.1016/j.ecoinf.2021.101533

Sugimura K, Katsuki S, Arai R, Kamiya H, Kawasaki T, Iida O et al (2024) Relationship between the root processing method and quality of herbal medicine of Rauvolfia serpentina cultivated in Japan. Fitoterapia 178:106132. https://doi.org/10.1016/j.fitote.2024.106132

Sun Y, Fernie AR (2024) Plant secondary metabolism in a fluctuating world: climate change perspectives. Trends Plant Sci 29(5):560–571. https://doi.org/10.1016/j.tplants.2023.11.008

Surendran S, Raju R, Prasannan P, Surendran A (2021) A comprehensive review on ethnobotany, phytochemistry and pharmacology of Rauvolfia L. (Apocynaceae). Bot Rev. https://doi.org/10.1007/s12229-021-09262-2

Swets JA (1988) Measuring the accuracy of diagnostic systems. Science 240(4857):1285–1293. https://doi.org/10.1126/science.3287615

Wani IA, Verma S, Mushtaq S, Alsahli AA, Alyemeni MN, Tariq M, Pant S (2021) Ecological analysis and environmental niche modelling of Dactylorhiza hatagirea (D. Don) Soo: a conservation approach for critically endangered medicinal orchid. Saudi J Biol Sci 28(4):2109–2122. https://doi.org/10.1016/j.sjbs.2021.01.054

Wani BA, Wani SA, Magray JA, Ahmad R, Ganie AH, Nawchoo IA (2022) Habitat suitability, range dynamics, and threat assessment of Swertia petiolata D. Don: a Himalayan endemic medicinally important plant under climate change. Environ Monit Assess 195(1):214. https://doi.org/10.1007/s10661-022-10773-7

Wani ZA, Pant S, Bhat JA, Shukla G (2024) Distribution and survival of medicinal and aromatic plants is threatened by the anticipated climate change. Trees For People 16:100549. https://doi.org/10.1016/j.tfp.2024.100549

Yang X-Q, Kushwaha SPS, Saran S, Xu J, Roy PS (2013) Maxent modeling for predicting the potential distribution of medicinal plant, Justicia adhatoda L. in Lesser Himalayan foothills. Ecol Eng 51:83–87. https://doi.org/10.1016/j.ecoleng.2012.12.004

Zhang X-F, Nizamani MM, Jiang C, Fang F-Z, Zhao K-K (2024) Potential planting regions of Pterocarpus santalinus (Fabaceae) under current and future climate in China based on MaxEnt modeling. Ecol Evol 14(6):e11409. https://doi.org/10.1002/ece3.11409

Zhang Y, Rehman H, Khattak F, Tariq M, Khan BN, Chaman S et al (2025) Immunomodulatory and growth-promoting effects of Rauwolfia serpentina root powder in broiler chicks challenged with Salmonella Gallinarum. Front Vet Sci 12. https://www.frontiersin.org/journals/veterinary-science/articles/10.3389/fvets.2025.1534347

Zhao X, Zheng Y, Wang W, Wang Z, Zhang Q, Liu J, Zhang C (2023) Habitat suitability evaluation of different forest species in Lvliang Mountain by combining prior knowledge and MaxEnt model. Forests. https://doi.org/10.3390/f14020438

Part II
Traditional Practices, Local Knowledge and Adaptation Strategies

Chapter 8
People's Perceptions, Practices, and Traditional Knowledge on Climate Change in Meghalaya

Joram Beda, Albert Chiang, Vivek Lyngdoh, Marbakor Mary Lynrah, Amica L. Nongrang, Evenstone Wahlang, and Nivanaliza Wahlang

Abstract Climate change is a reality that is experienced globally. However, climate response planning requires an understanding of local and sector-specific vulnerabilities and the extent of their impacts. Recording people's perceptions of climate change and their traditional knowledge of resources and resource utilisation for their day-to-day activities is important in understanding past climate experiences and finding ways to tackle their impacts at the local and sectoral levels. The tribal communities in Meghalaya, through long-term trial-and-error experiments, have developed an elaborate, functional, and generally democratic system of conservation and management of forests and associated natural ecosystems, now offering a rich source of Indigenous traditional knowledge. The indigenous tribes of Meghalaya can be classified into three main groups—*Khasi*, *Pnar* or *Jaintia*, and *Garo*. The *Khasi* and *Jaintia* are believed to be the descendants of the Proto-Australoid Monkhmer race. In contrast, the Garo are descendants of the Tibeto-Burman race, who migrated from Tibet. Their cultural traits and ethnic origins remain distinctive due to their geographical isolation. The State of Meghalaya is governed under the Sixth Schedule of the Indian Constitution, which provides spe-

J. Beda
Meghalaya Climate Change Centre, Government of Meghalaya, Shillong, India

Meghalaya Basin Development Authority, Government of Meghalaya, Shillong, India

Planning, Investment Promotion and Sustainable Development Department, Government of Meghalaya, Shillong, India

A. Chiang (✉)
Meghalaya Climate Change Centre, Government of Meghalaya, Shillong, India

BRIC-Institute of Bioresources and Sustainable Development, Imphal, India
e-mail: Albert.IBSD@gov.in

V. Lyngdoh · M. M. Lynrah · A. L. Nongrang · E. Wahlang · N. Wahlang
Meghalaya Climate Change Centre, Government of Meghalaya, Shillong, India

Meghalaya Basin Development Authority, Government of Meghalaya, Shillong, India

Y. D. I. Khan et al. (eds.), *Ecosystem-based Approaches for Resilience Building in Himalayan Landscapes*, Disaster Resilience and Green Growth,
https://doi.org/10.1007/978-981-95-2007-7_8

cial provisions for administering tribal areas in the northeastern states, including Meghalaya. The Sixth Schedule empowers local governance and ensures that tribal interests are safeguarded. The functions included in the Sixth Schedule are forest conservation, land use regulation, water resources management, and biodiversity conservation to help maintain the region's ecological integrity and ensure that natural resources are used sustainably. The present study adopts the bottom-up approach to understand the climate impacts, identify best practices for adaptation, and look into indigenous practices that have served as sustainable living techniques with the power to overcome challenges. The tribal communities in Meghalaya have preserved and used traditional knowledge for generations as an integral part of life. Sacred groves, living root bridges, bamboo drip irrigation, etc., are some of the unique practices that showcase the local forest-related knowledge and practices and associated social institutions developed over generations under changing environmental conditions by Indigenous communities. These indigenous practices represent an important source of adaptive capacity for local forest-dependent communities in the face of climate change impacts on the ecosystems.

Keywords Climate change · Traditional knowledge · Indigenous · Adaptation · Mitigation

8.1 Introduction

Climate change is one of the biggest challenges faced by the world today. Evidence shows that temperature and rainfall patterns have modified hydrological cycles and biological systems across the globe (Bolan et al. 2024). The Intergovernmental Panel on Climate Change (IPCC) gives convincing evidence that climate change is real and highlights that developed and underdeveloped countries will be the worse affected (IPCC AR5 2013). Further, long-term shifts in annual averages and seasonal patterns of precipitation, temperature, and humidity, as well as erratic and extreme weather events leading to increased risk of floods, drought, and fire, are anticipated for the future (Coumou and Rahmstorf 2012; Hatfield et al. 2011).

IPCC Fourth Assessment Report (2007) noted that Indigenous knowledge is 'an invaluable basis for developing adaptation and natural resource management strategies in response to environmental and other forms of change'. This was reaffirmed at the 32nd Session of the IPCC in 2010, which stated that Indigenous or traditional knowledge may prove useful for understanding the potential of certain adaptation strategies that are cost-effective, participatory, and sustainable. Therefore, documenting traditional knowledge is essential for understanding the local-level impacts of climate change and the options a community might consider for adaptation appropriate to its ecological and sociocultural environments (UNFCCC 2013). There are two recognised pathways to tackle the issue of climate change, viz. mitigation and adaptation. Adaptation has been recognised as the most workable pathway that can be taken up at all levels, from international and national to local.

Policies are meaningful only when accompanied by local, micro-level adaptation initiatives that help to face the challenges of the changing climate.

Like other parts of the world, Meghalaya is also experiencing the impacts of climate change. Important sectors like agriculture, water resources, forests, human health, sanitation, and rural development are facing the effects of climate change and, given the trend, are likely to be affected in the future (Meghalaya State Climate Change Action Plan 2014). Within the State, the rural population, especially the farmers and marginalised sections, are particularly vulnerable as they often lack adaptive capacity in terms of resources and opportunities to adapt to climate change (Meghalaya State Climate Change Action Plan 2014).

Considering the essentiality of integrating peoples' understanding of climate change and their traditional knowledge (potential adaptation) in adaptation planning, this chapter draws on the peoples' perceptions, traditional knowledge, behaviours, and policy preferences associated with climate change, with particular references to rural areas. It also provides an understanding of climate change vulnerability and the level of awareness of the rural population.

8.2 Study Area

Meghalaya, 'the abode of clouds', is located between 24° 58′ N–26° 07′ N and 89° 46′ E–92° 51′ E and spreads over an area of 22,429 sq. km (Fig. 8.1). Meghalaya, with a boundary length of 1239.61 km, is a landlocked State surrounded by Bangladesh (South and West) and Assam (North and East). The state has 12 administrative districts and 46 blocks, comprising 6459 notified census villages (Directorate of Economics & Statistics 2018). The State has three Autonomous District Councils, viz. Khasi Hills Autonomous District Council, Jaintia Hills Autonomous District Council, and Garo Hills Autonomous District Council give special provisions with respect to the application of laws made by the Parliament and the State Legislature to the tribal areas in Meghalaya.

Meghalaya is a part of the Indo-Myanmar biogeospherical region, one of the world's mega biodiversity regions. Nokrek Biosphere Reserve; Balpakram National Park; Nongkhyllem, Siju, and Baghmara Wildlife Sanctuaries; and a large number of sacred groves found in different parts of the State are the main preserves of biodiversity (State of Environment Report, Meghalaya 2005).

Meghalaya is inhabited by three major tribes: the *Khasi*, *Jaintia,* and *Garo*. Other tribes include *Hajong*, *Koch*, *Karbi*, *Rabha,* and *Bodo*. As per the Census Report, 2011, Meghalaya has a total population of 2,966,889, of which male and female are 1,491,832 and 1,475,057, respectively. With a population growth of 27.95% this decade, the State constitutes 0.25% of India's total population (Directorate of Economics & Statistics 2023). Of the total population, 595,450 (20.07%) live in urban regions, of which 297,572 are males and 297,878 are females. On the other hand, around 2,371,439 (79.93%) constitute the rural population, of which males and females were 1,194,260 and 1,177,179, respectively. The sex ratio in Meghalaya

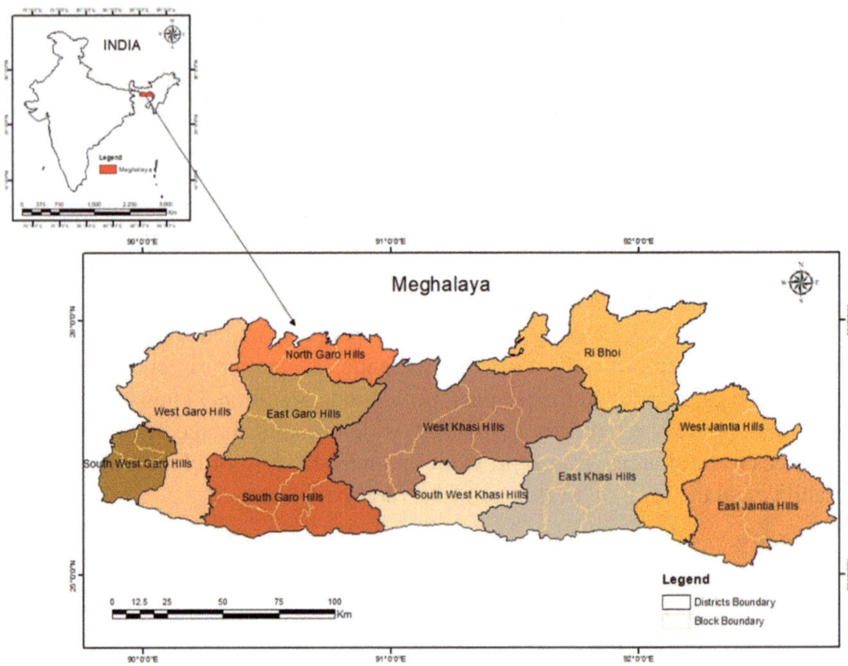

Fig. 8.1 Location map of Meghalaya

is relatively high, with 989 females per 1000 males. The urban region has a higher average literacy rate of 90.79% compared to the rural regions of 69.92%, where the literacy rate of males is 75.95%, and females are 72.89% (Directorate of Economics & Statistics 2023).

8.3 Methodology/Study Approach

As the study is Statewide, it is targeted to cover all rural areas, including 46 blocks under all 12 districts of Meghalaya. So far, the study has covered 163 villages in 10 districts, comprising around 700 cross-sectional participants/respondents. The primary data on people's perception of climate change and its impacts on different sectors, viz. water resources, agriculture, forest, livestock, fishery and human health, and socio-demographic profile, was collected through a well-designed questionnaire. The individual questionnaire was designed and translated into local vernaculars for easy comprehension of the questions. The questionnaires covered all related climate change issues at the local level and allowed the participants to respond based on their knowledge and experiences.

Traditional knowledge and practices are an integral part of the day-to-day life and activities of the rural population. As lives and livelihoods have been impacted by climate change, it is important to document how people have been using traditional knowledge in their everyday livelihood activities, which can be adopted as an adaptation. The standard approach of focus group discussion (FGD) was adopted to document traditional methods used for weather forecasts and decision-making to protect resources and livelihoods. Besides, traditional knowledge in different sectors like agriculture, livestock, water, forest conservation, and management was also captured. The documentation involved the usage of a semi-structured questionnaire as well as audio-visual devices.

8.4 Results/Findings

8.4.1 People's Perceptions Align with Scientific Predictions Regarding the Impact of Climate Change

The study encompasses a cross-sectional survey of participants at the village level. The target group for the study was the rural population of Meghalaya. Given the study's objectives and the availability of major resources (forest and water) upon which the livelihood activities of the large population depend, an arbitrary selection of survey villages was made. The target group for the survey comprised heterogeneous participants from different age groups, genders, education, and occupations.

The bottom-up approach for identifying local vulnerabilities has been recognised as an important tool globally. These bottom-up and top-down approaches together provide a complete and comprehensive understanding of climate and climate-induced impacts in an area or region. The vulnerability, coping and adaptive capacity, and resilience to climate change could differ from one place to another. International and macro policies are meaningful only when accompanied by local, micro-level initiatives that help them to innovate and adapt to face the challenge posed by the changing climate. Documenting people's perceptions, traditional knowledge, and understanding of climate change and its impacts will play an important role in identifying sectoral vulnerabilities and framing adaptation strategies at local levels.

The study revealed that climate change is already felt in Meghalaya and is expected to continue for a longer period. Crucial sectors like agriculture, water resources, forests, and health are likely to be affected by climate change. The rural population, especially farmers and marginalised sections, tend to be particularly vulnerable as they often lack the resources, power, and opportunities to adapt to climate change.

The findings from the present study validate what the IPCC Fifth Assessment Report (2013) confirmed: climate change is real and affects every aspect of livelihood and natural resources in the state.

8.4.2 Observed Climate Variabilities

Although people are not directly aware of climate change phenomena, they have observed changes in weather patterns over the years through their experiences. The study reveals an increase in temperature, irregular rainfall, an increase in hot days, and a decrease in rainy days over the years. The findings also indicate an increase in the frequency of extreme climate events such as landslides, floods, storms, hailstorms, and drought-like situations in certain parts of the State. These variabilities have significantly impacted their livelihood activities and resources, and the extent of the impact is not limited to a specific sector.

8.4.3 Sectoral Impacts of Climate Change

8.4.3.1 Water Resources

Climate change variability impacts the state's major water sources, such as rivers, streams, springs, bore wells, and ponds. The quality and quantity of water sources, underground and surface, are declining/deteriorating throughout the State.

8.4.3.2 Human Health

Human health is a serious concern in the State, mainly vector and waterborne diseases, which may emerge as a more significant threat due to climate change. People have perceived an increase in the occurrences of malaria and dengue in the State over the last two decades. Random occurrences of cholera have also been perceived as an important health concern in the State.

8.4.3.3 Agriculture

People's livelihoods largely depend on agriculture, the most climate-sensitive sector. As the State is running a deficit in food grains required to meet the food needs of 2.69 million populations, the sector needs extra attention and effective adaptation measures to ensure food security. Farmers' perceptions of climate change reflect a clear recognition of its impacts, particularly evident in shifting sowing periods.

Most of the farmers are aware of climate variability, and in response to that, they have adjusted their crop sowing time by a fortnight to a month. They have also observed increased occurrences and severity of pest attacks compared to the past. The State has faced significant crop damage, mainly by heavy hailstorms followed by heavy storms and drought. The study further shows that several factors have caused a reduction in crop production, and rainfall variability is seen as the most important potential detriment for crop or food production. Other noticeable factors responsible are variable temperatures, soil quality reduction, and increased pest and weed invasions.

8.4.3.4 Forests

Climate change and forests are closely linked. Meghalaya has forest and tree cover of almost 76% of its geographical area and 0.57 ha per capita of forest and tree cover, which is higher than the national average (ISFR 2021). However, over the last 30 years, communities have perceived a decline in forest cover. Major factors for reducing forest cover are logging, infrastructure/industrial development, forest fires, and agricultural expansion. A significant percentage of respondents also perceived climatic variability as an important factor responsible for the decline in forest cover. There has also been a noticeable decrease in the availability of timber-based forest products like log wood, fuel wood, bamboo, cane, and non-timber forest products like wild vegetables, fruits, potatoes, honey, etc. The study shows that no single factor is responsible for reducing forest cover. Still, a combination of anthropogenic and climatic factors (temperature and rainfall variability) has led to the decline in the forest cover and the supply of timber and non-timber forest products. This may lead to habitat loss of wildlife, causing potential increases in incidences of man-animal conflict, damage to property, and loss of life. Reducing forest and ecosystem services will impact the communities dependent on the forest for their livelihoods.

8.4.3.5 Livestock

The health and productivity of livestock are also perceived to be influenced largely by climate variability (extreme heat and cold) and associated factors like disease outbreaks and water and fodder scarcity. In the last 30 years, there has been a declining trend in livestock rearing, mainly due to diseases and financial constraints. Livestock disease outbreaks have become frequent in the last few years, making the herders reluctant to continue with the traditional activity. The major livestock diseases observed were plague, flu, and seizure in poultry and swine, the main livestock-based activity.

8.4.3.6 Fisheries

The state's fishery sector is in its nascent stage, and much impetus is being given to its development. Though pisciculture has the potential to contribute to food and livelihood security, this practice has not achieved its status of viable economic activity. Fishery, being a climate-sensitive sector, is vulnerable to climate variability. The farmers engaged in fish farming believe climate change influences this sector. Extreme heat has been seen as a significant factor affecting the health and productivity of fish.

8.4.4 Adaptive Strategies and Local Responses

The rural community is aware of climate variability and has used their traditional knowledge to cope with climate vagaries. However, with the changed climate scenario, traditional methods require supplementation with scientific climate adaptation methods. Farmers and labourers constitute the major working group in rural areas engaged in agriculture and allied activities. The study's findings suggest that agriculture emerges as the most climate-vulnerable sector, which will impact State economic development and farmers' livelihoods in the face of climate change.

8.4.5 Impacts of Climate Change: Local Ecological Indicators

Scientists and other observers of the natural environment are now seeing many impacts on the land that appear to be linked to climate change. They have noted climate-induced changes in at least 420 physical processes, species, and communities. They can now say with some certainty that it is affecting different species in different locations in different ways. Microlevel indicators at the local level might be more sensitive to climate change and could serve as early warning systems. Therefore, it is required to identify specific tangible, easily observable, and technically feasible indicators of local climate conditions for a given geographic area. Such indicators for specific regions need to be identified and recorded in the coming years to report changes in the region's climate and the impact those changes have on the region's environment, biodiversity, economy, and quality of human life. The people have provided important clues about climate change based on ecological indicators. Some of the micro-level indicators observed by the village communities are given below:

- One of the most striking indicators, as perceived by the people, is the reduction in nut size and productivity of *Areca catechu.*
- Black rot incidence in *Piper betle* is increasing in many villages in the State.
- Increased incidence of late blight in potatoes recently, causing severe yield losses to the farmers.
- Early ripening of certain plant species can also be a marking indicator of climate change. Notably, guava, wild pepper, and pitcher plant have recently changed their ripening period.
- Enhanced 'gall' formation in *Cinnamomum tamala*, leading to drying of the plant.
- The farmers have changed the sowing time of maize and potatoes.
- The orange fruit growers have noticed a drastic reduction in the quality of the fruit. The fruits look good visually, but the insides are usually infected or rotten.

- In Jaintia Hills, a plague locally called *Kjut Khlam*[1] occurs twice a year during March and December, causing the death of livestock.
- An increased attack of white insects was observed on flowers like *chrysanthemum* and marigold.
- Small insects have been seen attacking the planted paddy seeds in the field.

8.5 Traditional Knowledge: A Coping Mechanism and Adaptative Strategies

IPCC AR 4 (2007) states that Indigenous knowledge is 'an invaluable basis for developing adaptation and natural resource management strategies in response to environmental and other forms of change'. Traditional knowledge has mainly been used to supplement scientific observations of climate change and its impacts. Traditional knowledge has been recognised as a contributor to a knowledge base, alongside science, for promoting effective adaptation and understanding the impacts of climate change. With the increased global recognition and acceptance of traditional knowledge in observation, assessment, and adaptation planning, it is now seen as a locally available effective measure for climate change adaptation in agriculture, post-harvest grain management, water, health, livestock, forest, soil management, etc. The involvement of communities in decision-making and adaptation planning at the local level provides low-cost, effective, and sustainable adaptation measures.

The rural communities in Meghalaya have preserved and used traditional knowledge for generations as an integral part of life. The focus group discussions reveal much of the traditional knowledge utilised in climate-sensitive sectors such as agriculture, livestock, and human health. Albeit its former relevance, some of the traditional practices have lost their effectiveness. The traditional knowledge documented in the study is provided in the following table (Table 8.1).

Rural communities in Meghalaya are always seen as a huge repository of traditional knowledge and beliefs. These beliefs and practices vary from village to village and community to community, symbolising the State's richness and diversity of knowledge. Rural communities interact closely with nature and can easily observe any behavioural changes in plants and animals.

Some of the interesting observations of rural communities are listed below.

Behavioural Indicators in Predicting Weather Phenomena

- The nesting behaviour of an insect (locally called Jongsok) indicates the number of flood occurrences. One protruding nest twig would indicate one flood and so on. Interestingly, the twig's length indicates the flood's severity and scale.
- The ants coming out from the burrows indicate the onset of monsoon.

[1] "Kjut Khlam" or plague is a generic term used by the Pnar (Jaintia) community in Meghalaya. It translates to "plague-like situation", causing illness or diseases in the livestock population.

Table 8.1 Traditional knowledge and adaptation: sector-wise

Region	Traditional knowledge and adaptations		
	Agriculture	Livestock	Human health
Garo Hills	i. Pomelo (*Citrus maxima*), locally called *jambura*, and a flower locally called *Dosimah* are used as pesticides ii. Crop seeds are stored in earthen pots covered with straw and kept in a particular place called *Ongagre* iii. To keep pests and insects away, the locals would pierce a stick through a crab and place it in the fields iv. The farmers are opting for crop diversification with a growing interest in cash crops like pineapples, rubber, and cashew nut v. In the North Garo Hills region, numerous rice varieties are produced, like Sali, Black Johai, Malbok, Duria, Keju, and Minal (sticky rice). Intercropping of rice variety is practised to protect crops from a mass infestation of insects and pests. Varieties like Keju, Minal, and Malbak are prone to pest attacks, while Black Johai is highly pest resistant (interestingly, despite Black Johai being pest resistant, the locals do not prefer to grow it because of its low yield) vi. The farmers rotate from Indigenous rice variety to hybrid variety after 3 years of cultivation since they noticed a reduction in the yield	i. The cows that are selected for breeding must have no marks located at the back of the navel ii. The juice of the pomelo fruit is mixed with the feed to cure livestock diseases. The fruit is also dried and can be stored for later usage. The tree is widely available and each house has 3–4 trees iii. The root of a plant (locally called Nengribengri budu) is wrapped around the leg of the cow to treat fracture iv. The juice of *Citrus indica* and juice of *Citrus macroptera* are used as medicines to cure the common diseases of cattle. *Citrus macroptera* is mixed with the feed to cure livestock diseases and also to treat the loss of appetite in the animals v. Bamboo shoot is given to pigs to keep them healthy	i. There are herbal medicines that are available and used for menstruation problems, like *Dillenia pentagyna* (Agatchi budu), chram budu, *Careya arborea* (Gimbil budu), and Sokchon budu (least in quantity) ii. Tembil budu is used to treat bone fractures but is not readily available anymore

(continued)

Table 8.1 (continued)

	Traditional knowledge and adaptations		
Region	Agriculture	Livestock	Human health
Jaintia Hills	i. Pest and insect diseases are usually cured by spraying tobacco water on the plants ii. Lime is applied at the tree trunk to prevent disease attacks iii. Tomato is seen to be resistant to pest and insect attacks and can also survive in water-deficit areas	i. A mixture of onion, salt, and turmeric/rice, onion, and turmeric is used for treating poultry diseases ii. Ginger and turmeric are given to livestock to treat cough and cold iii. Sugar syrup is given during dysentery	i. Collection and usage of water from pre-monsoon showers are generally avoided, considering the high concentration of impurities risking human health
Khasi Hills	i. Branches of *Semecarpus anacardium* (Dieng-Sohbhala) and *Pinus kesiya* (Dieng-Kseh) are used for the prevention of pest infestation (white rot) in paddy. These branches are kept near the water inlet to the paddy fields ii. Lime and tobacco mixed with water are used as a pesticide iii. Seeds are visually checked for any deformity and are stored in a dry place iv. Ginger is stored underground v. People prefer to cultivate ginger over *Colocasia*	i. A ginger sp. *Zingiber zerumbet* (local nomenclature Sying blei) is given to the reared animals as part of their feed to prevent heat stress and skin diseases ii. The leaves of *Schima* sp. are tied around the neck of the animals to prevent insect attacks	i. Leaves of *Colocasia* and roots of lily are used to treat fever but are no longer effective ii. A local herb (Miniraid) is used to heal bone fracture iii. Tubers of *Flemingia vestita* (Sohphlang) is eaten as a natural control for intestinal worms iv. Bamboo shoot is considered unhealthy due to its properties to revive illness v. The fluid of an unopened pitcher plant flower (*Nepenthes khasiana*) is used as an eye drop to cure cataracts and night blindness. It is also used in treating gastro-intestinal problems, diabetes, and gynaecological problems

- Birds nesting on top of trees indicate no occurrence of cyclones/thunderstorms, while nesting on big branches indicates heavy rainfall.
- The laying of eggs by insects on cow dung indicates heavy rainfall.
- Spider webs weaved on grass indicate calm weather.
- When the *Chengari* insect makes noise, it symbolises the time to sow paddy.
- An increase in the population of small insects like ants indicates an increase in temperature.
- Coming out of an insect called *Ah Skhang* from its burrow predicts rainfall.
- The arrival of flying ants indicates the time for vegetation to bloom; when they do not, it symbolises delayed blooming.

8.6 Knowledge and Practices Developed Over Generations Under Changing Environmental Conditions by Indigenous Communities

8.6.1 Sacred Groves: Sanctified Solutions for Carbon Sequestration/Sink

The adaptive and mitigation strategies that can be taken up to combat climate change in the State can be drawn by efficient management and utilisation of rich biological diversity and forest wealth. The prevalence of community-conserved forests such as sacred groves can be tapped to adapt and mitigate the changing climate. The hill region has a time-tested local governance mechanism through traditional institutions and is predominantly inhabited by ecosystem people. Traditionally, the tribal communities in Meghalaya have been preserving small patches of virgin forest based on their religious belief since immemorial. Sacred groves represent a traditional way of protecting forests and biodiversity from human interference on grounds of religious beliefs and are considered a key reservoir of biodiversity. Traditional rules support conservation by limiting activities inside the sacred forests. These forests have been reported to be relict forests and may be the only remaining climax vegetation of an area (Gadgil and Vartak 1976). The sacred groves in the State enjoy legal support as they are covered by the United Khasi and Jaintia Hills Autonomous District Council (Management and Control of Forests) Act, 1958. The sacred groves in the State are known as *Law Lyngdoh, Law Niam,* and *Law Kyntang*, depending on where they are located. Around 133 sacred groves have been documented in the state, spread over approx. 2507 ha area across the State (Forest and Environment Department 2023–2024). According to several reports, sacred groves hold tremendous species diversity and richness, even far more than disturbed forests. Besides, sacred groves provide ecosystem services through carbon sequestration, soil erosion control, water quality, and underground water balance, maintaining the hydrological cycle and ecosystem balance (Tiwari et al. 1998). Several perennial streams originate from 58 of these groves, providing water to the nearby community. The sacred groves also help maintain the ecosystem's health, reduce habitat destruction, conserve the viable population of pollinators and predators, and serve as the potential source of propagules required for colonising wastelands and fallows. The sacred groves still possess a great heritage of diverse gene pools of many forest species, many of which are endemic.

The tribal communities in Meghalaya, through long-term trial-and-error experiments, have developed an elaborate, functional, and generally democratic system of conservation and management of forests and associated natural ecosystems. Several forest and natural resource management lessons can be learned from the institutional structure and decision-making system of the evolving and dynamic institutions of tribal communities of the region.

8.6.2 Living Root Bridges: Weaving Resilience Over Generations

Climate change should not be solved with the same approach that created the problem. Indigenous technological innovations such as the living root bridges, locally known as jing kieng jri, are a testament to the ingenuity of the *Khasi* and *Jaintia* tribes and offer valuable lessons in sustainable living and climate resilience. They exemplify how traditional knowledge and practices can contribute to modern environmental challenges.

Meghalaya receives the highest rainfall in the country. The state houses Mawsynram and Cherrapunjee, which are ranked as the wettest places in the world. The average rainfall at Cherrapunjee during the last 35 years has been 11,952 mm (470 inches), and there were several years when it was substantially more than this. In the past 35 years (1970–2005), the State has experienced increased precipitation. Naturally, the monsoon comes with the risk of periodic floods, storms, and cyclones that destroy all the bridges, except for the state's unique and exclusive living bridges developed by the indigenous people of Meghalaya (Paka 2022). The *Khasi* tribe of Meghalaya has developed this ingenious 'biotechnology' using their 'bio-engineering skills' (Mathew 2005) in dense forests and wet areas prone to environmental disturbances.

The building process involves tweaking the aerial roots of *Ficus elastica*, commonly known as rubber plants, across deep river valleys and gorges (Shankar 2015). These 'ethno-bio-engineered bridges' grow up to 50–100 feet, with a carrying capacity of more than 50 people at once and a life span of around 500–600 years (Chaudhuri et al. 2016), which only get stronger with time, unlike any artificial bridges. They offer potential eco-technology ideas and options for mitigating commutation and other problems in rural areas (Chaudhuri et al. 2016). The *Khasi* applying this technology to damaged steel suspension bridges shows its superior nature, success, and relevance while offering long-term sociological resilience and sustainable community-based infrastructure solutions (Shankar 2015).

Living root bridges are effective adaptive measures without much spending but a community action in dealing with climatic extremes like floods and storms. This local-level approach ensures the sustainability of these structures for future generations. They are incredibly resilient to extreme weather conditions such as heavy rainfalls and flash floods, which are becoming more frequent due to climate change. Unlike conventional bridges, they grow stronger over time and can withstand the harsh climatic conditions of Meghalaya (Paka 2022). They also provide other ecosystem services that mitigate climate change impacts, such as *Ficus elastica*, which absorbs the devious greenhouse gas carbon dioxide from the atmosphere. They aid in soil stabilisation and the prevention of landslides. In addition, they support a diverse range of plant and animal life, contributing to the ecosystem's overall health.

8.6.3 Bamboo Drip Irrigation System: An Inhouse 'More Drop per Crop' Technique

Under present climate variability, water stress is already high, which may exacerbate an anticipated climate change scenario. Although the State is fortunate to have abundant water resources, with Cherrapunjee and Mawsynram being the wettest spots on the earth, it is prone to seasonal water scarcity. Water resources in the State are under tremendous pressure to meet the increasing demand for water for domestic usage, agriculture, energy production, manufacturing, and industrial processes. Climate change impacts are likely to adversely affect the existing water resources, which will further exacerbate the water challenges in the State. Since the scope of these effects is unknown and unpredictable, responsive and flexible adaptation measures are the need of the hour. New and innovative approaches to water resources management and its development are vital to reduce vulnerability and livelihood insecurity among the poorest and facilitate adaptation to the uncertain effects of a changing climate.

Meghalaya has an ingenious system of tapping streams and springwater using bamboo pipes to irrigate plantations. About 18–20 litres of water entering the bamboo pipe system per minute gets transported over several hundred meters and reduced to 20–80 drops per minute at the plant site. The tribal farmers of the Khasi and Jaintia Hills of the Meghalaya Plateau have been practising this system for more than 200 years. However, this system is more prevalent in the 'war' Jaintia Hills than in the 'war' Khasi Hills region (Ryngnga 2018). The tribal farmers use an age-old tradition to drip irrigate crops such as betel leaf and black pepper planted in areca nut orchards or mixed orchards, with the help of locally available bamboo (Jasmine et al. 2018). In this system, water is carried out with the help of different forms of bamboo culms and further distributed into different bamboo water channels for irrigation of cropland (CSE 1998). The drip irrigation system is suitable for water scarcity conditions in the state since soils have poor water-holding capacity and the topography is rocky and undulating (Ryngnga 2018). In hills, natural perennial streams are the primary water source for domestic and irrigation purposes. The most convenient way is to divert the water from the upper reaches of the hill slope to the lower reaches through gravity flow. This concept is fully utilised, diverting water from the source to the desired lines.

Bamboo drip irrigation practice prevents leakage and loss of water on the way. It also increases crop yield with less water and uses natural, local, and inexpensive materials. The farmers lay out the bamboo networks proficiently to keep the sites productive. The practice involves not cutting trees and shrubs to clear the land to make channels through the forest areas on hills (Jasmine et al. 2018). The farmer goes for settled cultivation when bamboo drip irrigation practice is followed and shifting cultivation is reduced. The bamboo drip irrigation practice helps conserve and preserve the prestigious natural resources in the hilly terrain of Meghalaya (Mishra and Sharma 2001).

8.7 Conclusion

Most adaptations to local climate conditions serve multiple purposes, aligning closely with the ecosystem-based approach (EbA) principles. Local forest-related knowledge and practices and associated social institutions developed over generations by Indigenous and local communities under changing environmental conditions represent a critical source of adaptive capacity for local forest-dependent communities facing climate change impacts. By emphasising the integration of EbA, these practices strengthen ecological resilience and ensure the sustainability of local resources. People's perceptions of climate change and adaptive strategies are valuable for local and regional policy formulation. They can significantly contribute to a macro-level policy framework for global adaptation efforts and goals. Most responses to climatic variability have primarily originated from affected communities rather than external agencies or interventions. Thus, the lessons learned from these communities must be communicated to policymakers and farmers, creating a positive feedback loop that enhances long-term vulnerability reduction efforts through ecosystem-based, community-driven solutions.

References

Bolan S, Padhye LP, Jasemizad T et al (2024) Impacts of climate change on the fate of contaminants through extreme weather events. Sci Total Environ 909:168388. https://doi.org/10.1016/j.scitotenv.2023.168388

Chaudhuri P, Bhattacharyya S, Samal AC (2016) Living root bridge: a potential no cost eco-technology for mitigating rural communication problems. Int J Exp Res Rev (IJERR) 5:33–35

Coumou D, Rahmstorf S (2012) A decade of weather extremes. Nat Clim Chang 2:491–496

CSE (1998) Bamboo drip irrigation of the Northeastern Hills- rainwater harvesting, Northeastern Hills, Centre for Science and Environment, New Delhi

Directorate of Economics & Statistics (2018) Directory of Block-wise location codes of villages of all the 46 development blocks of the Meghalaya

Directorate of Economics & Statistics (2023) Statistical abstract Meghalaya. Government of Meghalaya

Forest and Environment Department (2023–2024) Forest inventory of Sacred groves in Meghalaya Volume III. Forest Resources survey division working plan, Research and Training and district council affairs wing. Published by Office of the Principal Chief Conservator of Forests, Working Plan, Research & Training and District Council Affairs, Forests and Environment Department, Government of Meghalaya Sylvan House, Lower Lachumiere

Gadgil M, Vartak VD (1976) Sacred groves of India – a plea for continued conservation. J Bombay Natl Hist Soc 73:623–647

Hatfield JL, Ort D, Thomson AM et al (2011) Climate impacts on agriculture: implications for crop production. Agron J 103:351–370

IPCC (2013) The physical science basis. Working group I contribution to the fifth assessment report of the intergovernmental panel on climate change

ISFR (2021) India State of Forests Report, Forest Survey of India, Dehradun. Assessed on 01.10.24. https://fsi.nic.in/forest-report-2021

Jasmine TS, Larihun J, Michelle K (2018) The contribution of indigenous knowledge of the Khasis in ecosystem management. In: Mawlong BL (ed) Climate change and developing countries. Cambridge Scholars Publishing, pp 210–218

Mathew R (2005) The living root bridges of Meghalaya. Curr Sci 89(1):10–11

Meghalaya State Climate Change Action Plan (2014) Government of Meghalaya

Mishra AK, Sharma UC (2001) Traditional wisdom in range management for resource and environment conservation in northeastern region of India. Envis Bull Himal Ecol Dev 9(1):31–36

Paka B (2022) A bottom-up approach to fusion of indigenous knowledge systems and digital technological systems for enhanced conservation of ecosystems (line break) case study on Meghalaya State of India

Ryngnga P (2018) Traditional irrigation system: bamboo dripping system in Meghalaya. Int J Sci Res 7:1532

Shankar S (2015) Living root bridges: state of knowledge, fundamental research and future application. In: IASBSE conference- structural engineering: providing solutions to global challenges. Geneva

State of Environment Report (2005) Department of Environment and Forests, Government of Meghalaya

Tiwari BK, Barik SK, Tripathi RS (1998) Biodiversity value, status, and strategies for conservation of sacred groves of Meghalaya, India. Ecosyst Health 4(1):20–32

United Nations Framework Convention on Climate Change (2013) Best practices and available tools for the use of indigenous and traditional knowledge and practices for adaptation, and the application of gender-sensitive approaches and tools for understanding and assessing impacts, vulnerability and adaptation to climate change. Technical paper, FCCC/Tp/2013/11

Chapter 9
Jhum Practices for Livelihood in North-Eastern Himalayan Regions of India: Current Status, Challenges, and Future Direction

Nabajit Kalita, Rupak Kr. Sarma, and Hemen Deka

Abstract *Jhum* cultivation or slash and burn agriculture is a common traditional practice of nearly 64 countries across the globe. This is a common farming practice in the hilly terrain of Northeast Indian states, including Assam. However, the rapid increase in human population reduces the *jhum* cycle (fallow period), hindering the ecological and economic sustainability of the primaeval agricultural practice. Overall, the livelihoods of the *jhumias* are adversely affected.

Jhum agriculture is intricately associated with socioeconomic and cultural fabric of many ethnic communities across the globe, serving as the sole source of their sustenance. Besides, the jhum fields play a pivotal role in sequestering carbon, offsetting the release during the preparation of the field. By maintaining the carbon neutrality, the mountain agriculture protects the hydrological features of the landscape, reduces soil erosion, and improves soil organic carbon and nutrient profiles. Therefore, safeguarding *jhum* agriculture is important to protect the dependent livelihoods and hill ecosystems. Accordingly, the present chapter outlines a focused overview of the current livelihood status of the *jhumias* of Northeast India. This chapter also offers insights into the various challenges of *jhumming* and possible strategies for improving this traditional practice.

Keywords Shifting cultivation · Fallow period · *Jhumias* · Sustainability

---

N. Kalita · H. Deka (✉)
Ecology and Environmental Remediation Laboratory, Department of Botany, Gauhati University, Guwahati, India
e-mail: hemendeka@gauhati.ac.in

R. Kr. Sarma
Department of Botany, Nalbari College, Nalbari, Assam, India

Y. D. I. Khan et al. (eds.), *Ecosystem-based Approaches for Resilience Building in Himalayan Landscapes*, Disaster Resilience and Green Growth,
https://doi.org/10.1007/978-981-95-2007-7_9

9.1 Introduction

The *jhum* or shifting agriculture is an ancient farming practice of Neolithic origin (Kingwell-Banham and Fuller 2012). Globally, more than 250 million people are engaged in shifting cultivation that covers approximately 410 million hectares of land (Giardina et al. 2000; Li et al. 2014; Dagar and Tewari 2016; Sarkar 2024). The northeast Himalayan region is renowned for this mountain agriculture (Mishra et al. 2021). The ethnic populations of the Northeast Indian hilly states exclusively practised the jhum agriculture, including more than 0.44 million families cultivating nearly 0.45 million hectares of land in a calendar year (Thong et al. 2022; Sarkar et al. 2021). The farmers practising this form of agriculture are known as *jhumias*. The tribal communities do not merely practice *Jhum* for their sustenance; it is a traditional livelihood method, utilising the indigenous farming techniques. This age-old agricultural practice evolves to the present status due to the region-specific interactions between the environmental and cultural factors (Tripura and Mondal 2016). The socio-cultural lives of the *jhumias* and their rituals/customs are intricately woven around the calendar of *jhum* cultivation, which primarily consists of site selection, clearing and burning forest vegetation, sowing of seeds, plant protection, and harvesting.

The sustainability of the *jhum* agriculture in the purview of ecological and economic status always depends on the low population density and long fallow period to restore the soil fertility and promote the regeneration of native flora (Saha et al. 2012). However, increasing demand for cultivable land due to the rapid rise in human population reduces the *jhum* cycle (fallow period) from 10–15 years to 2–3 years. The overall scenario adversely affects the soil health and ultimately destabilises the mountain ecosystems (Nunisa et al. 2023). In the long run, it will seriously impair the food and livelihood security of the *jhumias*.

The livelihoods and rural economy of the developing countries are totally dependent on the agriculture. Mountain agriculture plays a crucial role in maintaining the economy of the hilly terrain of Northeast Indian states. Despite this, the jhum cultivation has received minimal attention in national agricultural policies even though jhum cultivators face terrible livelihood challenges daily. The gradual depletion of primary forest cover due to climate change adversely affects the productivity of jhum lands as the forest ecosystems are crucial for the sustenance of the organic carbon contents in the *jhum* agricultural fields. Even a minor change in the mountainous environment can threaten the livelihood of the economically vulnerable *jhumias*. Looking ahead, the current writing focuses on the *jhum* system and the livelihood pattern of the *jhumias* associated with it. The study further addresses the challenges faced by *jhumias* in practising this mountain agriculture and the strategies that can be employed to improve this indigenous cultivation method.

9.2 *Jhum* Cultivation System: A Brief Overview

The principal process of *jhum* cultivation across the globe is more or less similar, with minor differences based on geographical and other environmental conditions prevailing in the region. Accordingly, the time of cultivation and harvesting of products usually varies from region to region. A common pattern reflecting the process of *jhum* cultivation has been presented in Fig. 9.1. In brief, it involves site selection, clearing and burning of vegetation, sowing of seeds of selected crops, weeding, protection, harvesting cultivated crops, and storage (Dey and Laila 2017). At each stage of cultivation, crucial decisions regarding site selection, timing of cultivation, crop selection, and labour allocation are essential. This decision-making process relies heavily on agroclimatic, environmental, social, and cultural factors. The plots for *jhum* cultivation are overseen by a system of structured local governance led by the village elders and the headman. The land access and tenure is under the regulation of traditional land access norms of the local governance. A family's right on land is inherited to their next generation and cannot be transferred outside the lineage (Chanu et al. 2023). However, nowadays, many *jhumias* share their lands with private parties (Punitha et al. 2018). This practice varies from place to place. Each tribe involved in *jhum* possesses some unique, traditional, and region-specific calendar of events for jhum, which they strictly follow while practising this cultivation. In the north-eastern Himalayan region, land selection for jhumming usually starts in January or February (Sitlhou 2020). *Jhumias* select the land considering their sacred

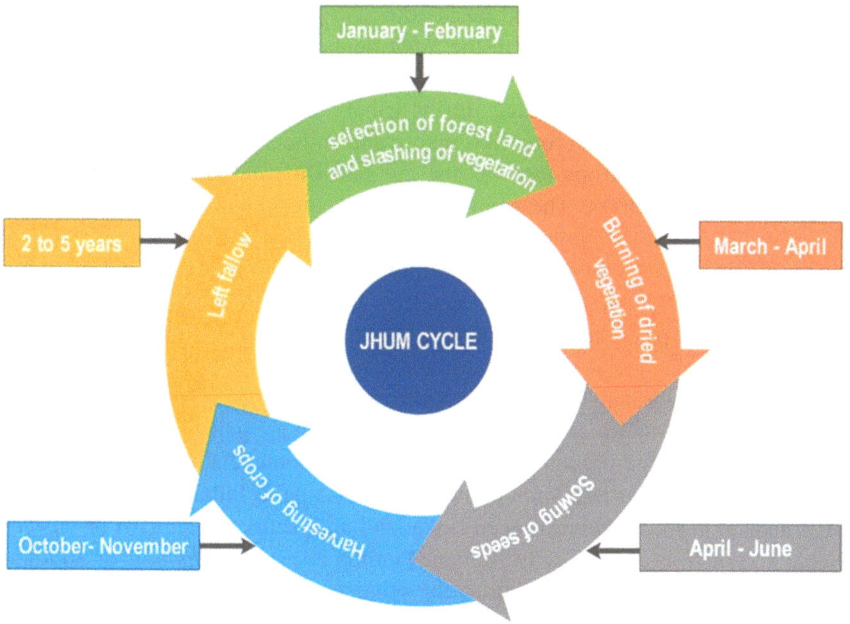

Fig. 9.1 Common pattern of *jhum* cultivation system in the north-eastern Himalayan region of India

beliefs due to the presence of rare vegetation and cursed locations. Tree cutting is strictly prohibited in those locations. The selection of land is primarily based on the region's soil type and the crop to be cultivated there. The felling of trees follows selection. Some tribes, before cutting down the trees, seek permission from deities, asking for mercy for felling the trees and offering gratitude for their long nurture and presence. Furthermore, while clearing the forest, certain tree species are deliberately untouched out of respect for local beliefs and traditions. Jhumias follow a distinct schedule of clearing the field. The felled trees are left to dry from February to March, followed by burning at a specific day by the end of March and April (Singh et al. 2017). The resulting ash is evenly distributed across the field. Before the onset of pre-monsoon rain, i.e. during April–June, sowing operations are carried out. The paddy is sown using dibbling technique where *jhumias* use pointed stick to make small holes and drop two to three seeds of desired crop into those holes with the skill they gain through years of practice. Women are primarily responsible for sowing (Kalita et al. 2017). The *jhumias* closely monitor the fields regularly to prevent weeds and other disturbances. The harvesting is generally done in October and November (Saha et al. 2023; Kalita et al. 2017). Before using the newly harvested grains for food, they set aside a portion of the grains separately to use as seed for future cultivation. After 2–3 years of cultivation, the cropland left fallow to naturally regain the fertility. By this time, the cultivation is shifted to another virgin forestland (Das et al. 2017). In earlier days, the whole village relocate to a new cultivation site and come back to the original site once the previous site naturally gained fertility after 10–15 years. Thus, the cropping cycle continues. However, in present days, the original jhum cycle is destabilised and shorten to 2–3 years due to the increasing population pressure (Tripathi and Barik 2003).

The primary crop cultivated in the *jhum* fields of the north-eastern Himalayan region is rice. Paddy is the staple food of north-eastern people and is also used to prepare their traditional beverages in the tribal communities. In addition to rice, *jhumias* also grow other crops like maize and millet, as traditional alternatives. Apart from the staple crops, jhumias also cultivate different vegetables, fruits, tuber crops, as well as spices. The examples of the commonly grown vegetables are tomato, brinjal, chilli, pumpkin, cucumber, cabbage, and others. The mountain slopes are occupied by root crops like potato, sweet potato, root potato, yam, radish, turnip, cassava, etc. These are sown near tree stumps, a strategy to prevent soil erosion. Creeper crops are also used as margin crop to demarcate the filed boundaries. Besides the crops, the *jhum* lands are also occupied by fruits plants like banana, apple, pineapple, orange, kiwi, and passion fruit. In some fields, cultivation of some cash crops like turmeric, ginger, and large cardamom are gaining importance.

Moreover, the age-old agriculture has different nomenclature within the north-east Himalayan countries. *Jhum* is the common name in India and Bangladesh. In Myanmar, it is called *taungya* or hill cropland and, similarly, *khoriya* or *basme* in Nepal and *tseri* or *phangzing* in Bhutan (Punitha 2017). The *jhumias* of South and Southeast Asian countries belong to the ethnic populations, often categorised as ethnic minorities and hill tribes (Erni 2015). A detailed overview of the various ethnic groups engaged in shifting cultivation in these countries has been presented in Table 9.1.

Table 9.1 Showing information on shifting cultivation in South and Southeast Asian countries

Country	Local name	Tribes/ethnic groups involved	Area (m ha)	References
India	*Jhum* cultivation	Angami, Chakhesang, Kuki, Lotha, Sumi, Lusei, Hmar, Khasi and Jaintia, Bondo, Didayi, Koya, Gadaba, Paroja, Soura, KutiaKondha, DongariaKondha, Kandha, Parenga, Jatapur, Juang, Paudi Bhuyan, Erenga Kolha, Reang, etc.	4.37	Sahu et al. (2005), Shikhu (2017), and Panda and Sharkar (2017)
Myanmar	Shwe Pyaung Taungya	Karen, Chin	15.41	Thet and Tokuchi (2020) and Win (2005)
Nepal	Khoriya or Bhasme	Chepang, Magar, Sherpa, Rai, Limbu, Tamang and Gurung	–	Kafle (2011) and Sapkota (2023)
Malaysia	Huma or ladang	Iban	0.67	Ngah and Kamarudin (2019), Wasli et al. (2009), and Silva et al. (2011)
Laos	Slash and burn agriculture or swidden agriculture	Lao, Khamu, Hmong	7.89	Hett and Lu (2014), Sodarak (1999), and Chen (2022)
Bangladesh	*Jhum* cultivation	Marma, Mro, and Bawm	1.51	Osman et al. (2013), Thapa and Rasul (2005), Nath et al. (2005), and Silva et al. (2011)
Indonesia	Cadong	Dayak	4.70	Siahaya et al. (2016) and Silva et al. (2011)
Bhutan	Tsheri and phangzing	Tshangla	0.2	Limbu et al. (2020) and Roder et al. (1992)
Thailand	Tam-rai	Lawa, H tin, Khmu, Karen, Hmong, Yao, Akha, Lahu, Lisu	6.66	Rerkasem (1998), Schmidt-Vogt (2001), and Silva et al. (2011)
Sri Lanka	Chena	Vedda	0.79	Verma et al. (2017), Mihiranie et al. (2023), and Silva et al. (2011)
China	Shifting cultivation	Bulang, Jinou, Dulong	6.77	Yin (2015), Yin et al. (2020), Long and Zhou (2001), Shi-Cai et al. (2017), and Silva et al. (2011)

(continued)

Table 9.1 (continued)

Country	Local name	Tribes/ethnic groups involved	Area (m ha)	References
Cambodia	Shifting cultivation	Khmer, Bunong, Stieng, etc.	3.23	Swift and Cock (2015), Chassagne et al. (2016), Mouy (2010), and Silva et al. (2011)
Vietnam	Lua	Hmong, Tay, Nung, Thai, and Muong	9.93	Verma et al. (2017), Tugault-Lafleur and Turner (2009), Huy (2023), and Silva et al. (2011)
Philippines	Kaingin	Igorot, Lumad, etc.	–	Mertz and Bruun (2017) and Mialhe et al. (2014)

'–' Information not available

As *jhum* is a no-till agricultural practice, it minimises soil disturbance and loss of nutrients. Similarly, it is an environmentally sustainable farming method as it avoids using chemical fertilisers to grow crops in nutrient-poor soil (Panda and Sharkar 2017). The overall practice of the mountain agriculture sustains greater biodiversity in comparison to the conventional lowland farming system. The agro-biodiversity is effectively maintained by upholding traditional cultivation system and local seed, knowledge exchange networks. The traditional seed preservation system for next generations is one of the key identities of the *jhum* agriculture. The unique seed conservation techniques of the local crop varieties are highly acknowledged (Pandey et al. 2019). The conservation of the germplasms of the wild crop plays a crucial role in ensuring the resilience and sustainability in crop during the present situation of global climate change (Lane and Jarvis 2007). Besides, carbon sequestered by the long-fallow *jhum* fields during field preparation is useful in maintaining the carbon neutrality in the ecosystem. The carbon neutrality protects the landscape's hydrological features, reduces soil erosion, and improves soil organic carbon and nutrient profiles (Ziegler et al. 2009; Yuen et al. 2013).

9.3 Status and Trends of *Jhum* Cultivation in the Northeast Indian Himalayas

Jhum agriculture is a widely practiced method of livelihood in the rural mountain region of Northeast India. However, in recent years, *jhum* has experienced various transformations. Earlier, simple tools like sticks or hoes were employed for dibbling to sow crop seeds with long fallow periods. However, the gradual rise in population increases food demand, ultimately driving the adoption of modern technologies to the shortening the fallow periods. Over time, this leads to soil degradation, rapid spread of invasive weeds and harmful pest, and low crop yields. Thus, the land availability for *jhumming* is steadily declining, disrupting the ecological balance. In addition, the changing aspirations of the younger generation, rising educational attainment, rural to urban migration, low productivity, low income from *jhumming*, lack of labour force, and limited government support are some of the significant reasons behind the periodic reduction of *jhum* lands. Additionally, the natural disasters, especially landslides and frequent earthquake, are a recurring problem for wrecking the jhum agricultural land. The change in *jhumming* area in the hill states of north-eastern Himalayan regions, including the hilly regions of Assam, from 2005–2006 to 2015–2016, has been presented in Fig. 9.2.

Arunachal Pradesh, as the largest hill state in NE India, has approximately 80% of rural population, practising *jhumming* as their primary way of livelihood (Buragohain 2022). The farmers predominantly cultivate varieties of paddy for their own consumption and prepare traditional wine, called Apong (Bhuyan 2019). Besides rice, millet and maize are the other important crops they cultivated. Apart from these staple crops, farmers also cultivate different cash crops as mentioned in the earlier section of the chapter. Being organic, the agricultural products have high demand in the national market. Although *jhumming* is the main livelihood practice in the state, the area used for *jhum* cultivation has been declining gradually. According to a survey by Sarkar (2024), there is a reduction of approximately 34% of the jhum land between 2003–2004 and 2015–2016. During 2003–2004, the jhum cultivation area was approximately 56% (111,691 ha) of the net area, which was reduced to about 22% (50,911 ha) by 2015–2016.

In Manipur, approximately 300,000 rural inhabitants practise *jhum* cultivation, locally known as Pam-Lou (Singh et al. 2017; Akoijam and Mahongnao 2018). The

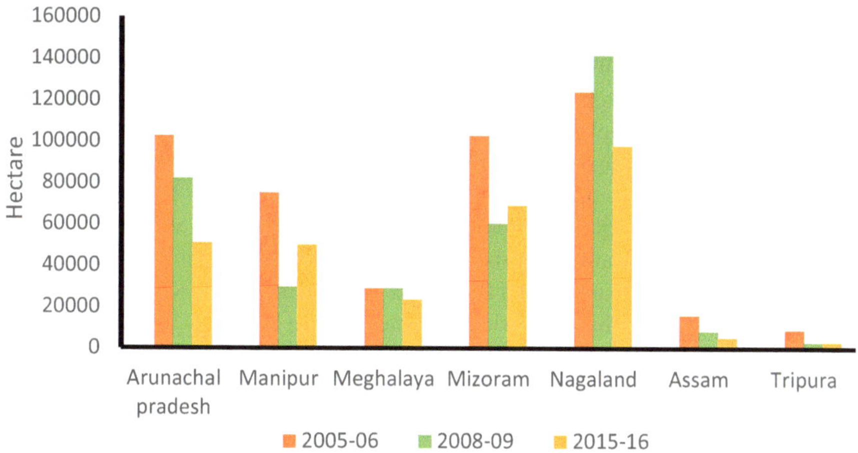

Fig. 9.2 Showing changes in the area under *jhum* cultivation in Arunachal Pradesh, Manipur, Meghalaya, Mizoram, Nagaland, Assam, and Tripura from 2005–2006 to 2015–2016. (Buragohain 2022)

two principal crops cultivated in the *jhum* lands are rice and maize, together with a variety of fruits, including pineapple, banana, lemon, guava, orange, peaches, and so on. Churachandpur and Tamenglong are the two major districts of Manipur where shifting cultivation is predominantly practised. In the Churachandpur district, over 80% of the population relies on *jhum* cultivation for sustenance (Sarkar 2024). According to Buragohain (2022), in Manipur, the extent of land used for *jhumming* decreased from 111,954 ha in 2003 to around 29,668 ha in 2008–2009; However, by 2015–2016, it had increased again to 49,996 ha. Sarkar (2024) mentioned that the shifting cultivated area is decreasing continuously in the Chandel and Ukhrul districts. Despite this increase in the jhum cultivation area, the net sown area dedicated to jhum decreased significantly from 51.6% in 2003 to 13.1% in 2016.

In Meghalaya, an estimated 257,140 people are involved in *jhum* cultivation (Jeeva et al. 2006). It is particularly prevalent in the East Garo and West Garo Hills, where over 70% of the land is under this cultivation (Sarkar 2024). Buragohain (2022) reported that in 2003, *jhum cultivation* in Meghalaya covered an area of 62,721 ha, which decreased to 23,787 ha by 2015–2016. The *jhumias* of Meghalaya primarily grow rice and maize for subsistence. The state is also famous for its Khasi mandarin oranges, pineapple, jackfruits, bananas, plums, pears, and peaches. Additionally, commercial broom cultivation is often practised on abandoned *jhum* lands by the local inhabitants of Meghalaya (Tiwari and Kumar 2008).

In Mizoram, about 75% of the rural population primarily depend on *jhum*, cultivated in around 69,155 ha of land (Buragohain 2022). According to a study by Maithani (2005), approximately 58,000 families were involved in *jhum* cultivation in Mizoram by 2000. However, a report by Sati (2019) at G.B. Pant Institute of Himalayan Environment and Development placed the estimate at 50,000 households. The main crop they grow is rice. However, they also cultivate maize, beans, cucumber, mustard, ginger, sesame, and pulses like cowpea, French beans, rice beans, etc., under shifting cultivation (Soni et al. 2020).

In Nagaland, around 190,000 households are engaged in *jhum* agriculture, spanning an area of approximately 94,380 ha (Rajkhowa et al. 2017). It is particularly prevalent in the Mokokchung, Mon, Tuensang, Wokha, Zunheboto, and Kohima districts. The Naga people mix rice, pulses, cereals, vegetables, and spices under *jhum*. Nearly 20,000 ha of forests are slashed and burned yearly in the state for practising this mountain agriculture (Sarkar 2024). The Nagaland Government has also implemented several initiatives to increase the productivity of *jhum*.

In Assam, *jhum* is mainly practised in Karbi Anglong and Dima Hasao districts. The Karbis refer it as 'Inglong Arit Sai Katiki'. The farming is practised by approximately 54,500 families in the district, with annual land use of 70,000 ha (Bey et al. 2023). According to the data documented by DAO, about 65% of the Dima Hasao population relies on *jhum* for livelihood (Nunisa et al. 2023). The *jhum* fields of Assam include rice, corn, cotton, ginger, garlic, turmeric, and vegetables of the local variety.

Tripura, having a population of 36,73,917, is the second most populated state in NE India (Census of India 2011). Out of 55,049 *jhumia* households in the state, only

21,677 (39.3%) rely solely on *jhumming*, and the remaining 60.7% are partly dependent. Buragohain (2022) estimated that approximately 28,489 ha of the land was under shifting cultivation in 2003 and reduces to 3062 ha by 2015–2016. This reduction can be attributed to government initiatives promoting settled cultivation among the *jhum* cultivators. In Tripura, 17 tribal communities still practise *jhumming* (Gupta 2000), among which the Tripuri and Reang tribes are the two major *jhumia* communities. Most (70.8%) of *jhumias* have only 0.8–2 ha of *jhumming* land per family, and only 6.1% have more than 2 ha (Gupta 2000). The jhum crops practised in the state include staple crops like rice, wheat, and maize, together with oilseeds, pulses, fruits and vegetables, rubber, tea, etc.

9.4 The Livelihood Dimensions of *Jhum* Cultivation

A livelihood designates the use of skills and resources for basic individual needs of people for food, income, and assets. An environmentally sustainable livelihood should have positive impact to the future generations (Erni 2015). *Jhum* cultivation is the primary source of livelihood among a significant portion of population of Northeast India. In *jhum,* both men and women are involved. Throughout the jhum cultivation process, the men folk handle the more laborious tasks such as slashing and burning the vegetation, removal of big and straight logs, and constructing fences. Women take the responsibilities of field preparation by clearing the undergrowth, big creepers, and climbers and managing the prime activities like sowing, weeding, harvesting, husking, thrashing, and the storage of crops (Kalita et al. 2017). Paddy is the chief crop considered for *jhum* cultivation. Apart from paddy; various vegetables and cash crops are also practised in the *jhum* fields. Crops are cultivated for self-consumption and for sale in the local market. They generate income by selling rice and locally brewed beer at the nearby weekly market to cover the household expenditures (Tamuli and Bora 2017). *Jhum* cultivation can fulfil the overall consumption needs of the *jhumias* households throughout the year, provided low population density and the fallow period is long enough to restore soil fertility. Thus, anyone can consider the *jhumias* as self-sufficient. However, rapid climate change due to human anthropogenic activities has led to uncertainty in rainfall and reduced soil fertility, ultimately affecting the jhum agriculture and reducing crop yields. The scientific feasibility of jhum-based livelihood in the North Garo Hills district was assessed by Tamuli and Bora (2017). According to the survey, 23% of the respondents believe on fulfilment of their economic needs from jhum agricultural system, whereas 77% of the respondents did not agree that the economic needs are covered from the jhum cultivation. Datta et al. (2015) did a substantial amount of work on the livelihoods of the *jhumias* of Tripura. According to their report, 54.29% of farmers self-sustained by *jhumming*. Additionally, livestock rearing raises additional income in 72.86% of the respondents. Furthermore, the benefit of using of common property resources was accessed by 100% of the farmers. The entire survey states that there was dissatisfaction of 82.14% of the population regarding the returns from the forest and non-timber forest products. According to a case study by Tripura and Chandn (2015), *jhumia* tribes of Tripura change their occupation in recent days. The *jhumias* choose the occupations like wage labourers (50%), bari agriculture (31%), government jobs (10%), and weaving, *jhumming*, business, and herding (2%). The study also revealed some interesting information about the secondary occupation among the *jhumias* of the state. It has been reported that among the *jhumias* of the state, about 12% of the population choose jhum as a secondary occupation. The authors highlighted the reasons for occupational changes. These include education, shortage of family labour, less availability of forest products, and shortage of food grain.

9.5 Challenges Encountered by *Jhumias*

The livelihood of a large proportion rural family of north-eastern Himalayan hilly terrain depends on *jhum* agriculture. However, while practising traditional jhum agriculture, the *jhumias* face numerous challenges that threaten the very survival of their way of life. One of the primary challenges faced by the *jhumias* in recent years is the gradual scarcity of forest land to perform shifting cultivation due to an increase in population. Increasing demand of food decreases the jhum cycle, one of the main reasons for declining soil fertility of the jhum land and ultimately gradual drop in agricultural productivity (Fig. 9.3). The remote location of jhum fields is another

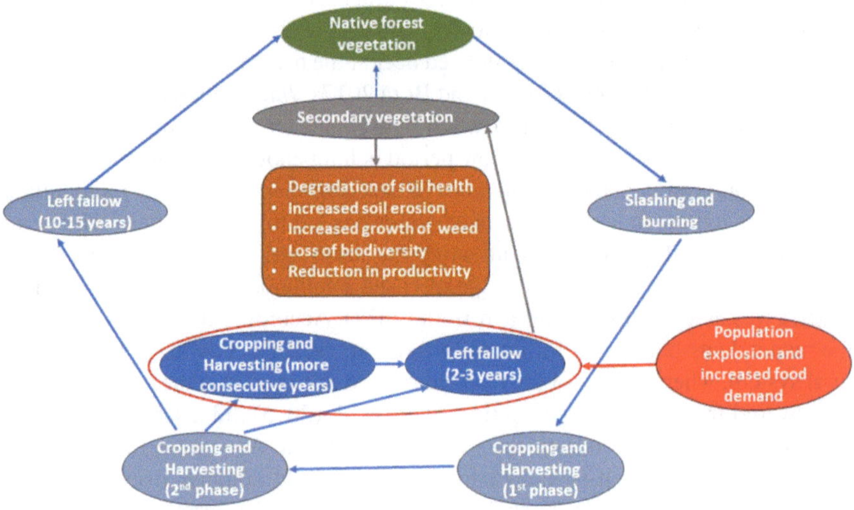

Fig. 9.3 Schematic representation of the effect of reduction of fallow period due to increased population density and food demand

major challenge faced by the jhumias. Farmers often have to shift from one hill area to another in search of new fertile land, which may not always be nearer to their homes. This increases the physical burden on the farmers (Tripura and Mondal 2016). Furthermore, *jhum* cultivation relies mainly upon the monsoon season for irrigation (Yadav 2013), making the timing of crop sowing critical and limited. The unpredictability of the monsoon, compounded by climate change, further complicates this dependency, making it more difficult for the *jhumias* to achieve consistent crop yields. The risk of soil erosion in the jhum lands due to climate change is another major challenge in the Northeast Indian Himalayan region (Schröder et al. 2024). Additionally, *jhumias* often lack knowledge about pest and disease management due to limited education and low to medium cosmopolitanism, resulting in higher crop losses and reduced productivity (Bhadana et al. 2013; Patra et al. 2023). Wild animal threats also pose a significant issue, as *jhum* fields are typically located in forest areas (Gouda et al. 2020). These animals often destroy crops and sometimes attack the *jhumias*.

Economically, the *jhumias* face substantial challenges due to the region's limited market access and inadequate infrastructure. Additionally, the inadequate roads and transportation facility hinder farmers from accessing larger markets, limiting their income opportunities and economic sustainability. Therefore, the *jhumias* often rely on middlemen who purchase their produce at very low prices (Tiwari and Pant 2018). Furthermore, the *jhumias* are often marginalised and excluded from the mainstream economy, which limits their access to education, healthcare, and other essential services. They are also vulnerable to social and cultural discrimination, leading to isolation and disconnection from the rest of society. There is a regular conflict between the governmental land use policies and forest conservation, which impose an adverse impact on the overall jhum agricultural pattern. The lack of representation and participation of the *jhumias* in policy-making processes also means that the *jhumias*' voices are often not heard. According to the experts, an underlying objective of several governmental initiatives is to gradually eliminate the overall agricultural practice (Erni 2015).

9.6 Strategies for the Improvement of *Jhum* Cultivation

Rapid increase in population and drastic climate change impose a serious threat on the sustainability of jhum agriculture. Around the world, many Indigenous tribes residing in hilly or undulated regions solely depend on *jhumming* for their livelihood. Therefore, identifying alternative practices or production systems to enhance the sustainability of the *jhum* cultivation is of utmost importance. Various scientific investigations and approaches have been employed in various regions, focusing on production and conservation perspectives. Some of the viable approaches are discussed as follows.

An integrated farming system (IFS) is vital in improving *jhum* cultivation. The IFS believes on the principle of utilising the by-products of one component as inputs for another, promoting resource efficiency and internal recycling within the system (Kalita et al. 2017). This is a sustainable approach based on technical skills and location-specific action, which plays a multifaceted role in terms of resource mobilisation and opening of new employment opportunities. Overall, the approach is considered as the combination of agriculture, horticulture, fisheries, and livestock farming to make self-sustained the economically vulnerable population (Prakash et al. 2015).

Agroforestry is a strategic approach to improve *jhum* cultivation in Northeast India by integrating trees and crops in a single system. Tree plantation has both ecological and economical importance. It helps in controlling soil erosion and improvement of soil micronutrients and carbon sequestration. Similarly, the use of plants as fuel woods, timber, and fodder is the economic benefit of tree plantation. Trees on the farm help to control soil erosion and improve the microclimatic condition of the soil; supply fuel wood, fodder, timber, and fibre; fix atmospheric N; mitigate global warming through C sequestration; and enhance the aesthetic and recreational values of the landscape (Haokip et al. 2021). Crops like maize, pineapple, ginger, coffee, and vegetables are cultivated alongside tree species such as *Alnus nepalensis*, *Pyrus communis*, *Pinus kesiya*, *Areca catechu, Schima wallichii*, *Prunus domestica*, etc. The selection of tree species and intercrops is influenced by the region's climate and the economic value of the chosen species (Mohapatra and Chandra 2017). If possible, the adoption of an alder-based jhum system can be considered as an excellent model for sustainable land use for hill ecosystems (Kithan 2014). Alder-based jhum cultivation typically follows a 4-year cycle, with 2 years of cropping interspersed among alder trees, followed by 2 years of fallow to support soil regeneration and coppice growth.

The cropping system can be changed with crop diversification strategy. Traditional cropping system in *jhum* lands always favour the technique of crop diversification. Planned and well-documented research is required to encourage suitable cropping pattern. Cropping with leguminous vegetables and heavy nutrient feeder crops like ginger and turmeric proved beneficial for maintaining soil fertility (Kalita et al. 2017). Integrating leguminous cover crops into jhum fields can reduce soil erosion, enhance soil fertility, control weed growth, and supply food for humans and fodder for livestock while also generating additional income (Baishya et al. 2016).

Establishing bio-terraces in jhum fields using fast-growing hedgerow species such as *Tephrosia candida*, *Crotalaria tetragona*, *Crotalaria juncea*, *Indigofera tinctoria*, *Flemingia macrophylla*, and *Cajanus cajan* holds potential for reducing soil erosion and enhancing soil productivity (Punitha et al. 2018). Saha et al. (2012) reported that barrier crops reduced soil loss by 94% and run-off by 78%.

In present days, jhum farmers acknowledge the sustainability and soil management qualities of organic farming. It is a holistic system of agriculture that prevents nutrient depletion and enhances the organic matter content of the soil. As a result, soils of organic farm lands retain more amount of water than the soils under conventional cultivation. Moreover, organic farming helps make farmers more resilient to the impacts of climate change (Babu et al. 2015). The *jhumias* often use locally available organic materials for their cropping system and are reluctant to use chemical fertilisers and pesticides. Therefore, there is a need to standardise different organic fertilisers like farm yard manure, green manures, vermicompost, neem cake, and biofertilisers to enable farmers to use the products judiciously. Using botanicals to control insect pests and diseases should be encouraged and promoted. Vermicomposting unit is also a profitable enterprise accepted and established by the farmers (Kalita et al. 2017).

Despite heavy rainfall during the monsoon season in the north-eastern region, there is an acute water shortage in winter. The construction of low-cost micro water harvesting structures with minimum seepage and evaporative losses, depending on the availability of suitable space in sloppy land, will be a viable option for the life-saving irrigation of crops. Harnessing perennial spring water in *jhum* areas through diversion channels may hold promise in overcoming the water scarcity problem in *jhum* fields. In situ soil moisture conservation through organic mulch may be one of the ideal options (Hazarika 2017).

The traditional *jhumias* have extensive knowledge of medicinal plants from generations, and they are already in the habit of growing these plants/crops in their jhum field (Kalita et al. 2017). Research should focus on the identification and utility of such crops. Farmers should be encouraged and supported to grow such crops in commercial mode.

9.7 Future Directions and Conclusion

For decades, scientists, conservationists, economists, intellectuals, and policymakers have divided views on *jhum* cultivation. Some argue in favour of *jhum* for its traditional value, cultural identity, and food security and emphasise the improvement of the existing *jhumming*. According to many others, shifting cultivation is an unscientific, wasteful, and ecologically dysfunctional system, detrimental to forests and soil and hence needs to be eradicated by inducing the cultivators to adopt alternative livelihood (Halder and Sarkar 2024; Baishya et al. 2015). Government organisations have taken several initiatives that directly or indirectly aim to eradicate shifting cultivation. This traditional agricultural system, deeply intertwined with the cultural and ecological fabric of the area, has evolved over centuries to adapt to the rugged terrain and variable climate. Thus, instead of finding alternatives to shifting cultivation, which has often failed, there is a need for viable solutions to make it more productive and less degradative for the future.

The future direction for *jhum* cultivation should depend on adopting balanced approaches that uphold the Indigenous knowledge systems embedded in *jhum* practices while incorporating scientific advancements. Research and extension services should focus on improving the productivity and ecological sustainability of *jhum*, potentially through improved fallow management and soil conservation techniques. Scientists, policymakers, and administrators have often tried to impose development plans that they consider suitable for the people without trying to understand the processes that operate in the traditional ecosystems. Instead, strategies for development should incorporate traditional value systems as much as possible so that development packages not only find ready acceptance by tribal societies but also ensure the participation of the people in the development processes. Community participation is crucial in designing and implementing interventions that respect local knowledge and address the needs of *jhum* practitioners. Empowering communities through education, capacity building, and secure land tenure can enhance their resilience and adaptability. Furthermore, attempts should be made to decrease the relative importance of rice in *jhum* fields and increase the production of other crops such as pulses, vegetables, tubers, and rhizomes. Diversification of rice mixed farming with legumes and pulses has to be promoted among the *jhumias* to improve *jhum* land. The production of more cash crops and multipurpose fast-growing trees in the abandoned jhum fields should also be encouraged. The traditional shifting cultivation system has the potential to lead to a sustainable society and healthy environment. The usage of indigenous varieties of seeds proves to be more resilient to environmental changes than the newly introduced hybrid varieties.

In conclusion, *jhum* cultivation remains a critical livelihood strategy for many indigenous communities in the north-eastern Himalayan regions. Despite of the different prevalent challenges, the traditional mountain agricultural practices deeply twisted with the regional cultural heritage and ecological fabric of the area. It has evolved over centuries to adapt to the rugged terrain and variable climate. Thus, an in-depth study is more essential to understand the impact of jhum agriculture on socioeconomic and cultural dynamics, as well as role in food security and environmental sustainability. The possible reason of reducing the fallow period, followed by scientific planning to overcome the key issues, is significantly important for an ecologically sustainable jhum agroecosystem. Our prime focus needs to be on strengthening the food security and safeguarding the traditional jhum agricultural practices of the indigenous tribes of the region.

References

Akoijam AS, Mahongnao M (2018) The challenges of adopting innovative agricultural practices under shifting cultivation in Northeast India. NEHU J 15(1):31–52

Babu S, Singh R, Avasthe RK et al (2015) Organic farming: problems and prospects in north East India. In: Training manual on integrated farming system approaches for sustainable hill agriculture under changing climatic scenarios. ICAR research complex for NEH Region, West Tripura, India, 28 October–6 November 2015

Baishya LK, Singh R, Ansari MA et al (2015) *Jhum* rehabilitation for food security of tribal farmers through sustainable agriculture. In: A comprehensive manual on integrated farming system:

an approach towards livelihood security and natural resource conservation. ICAR Research Complex for NEH Region, Umiam, Meghalaya, September 2015, p 63–68

Baishya LK, Ansari MA, Sarkar D et al (2016) Productivity enhancement in shifting cultivated lands through biofertilizer and micro-dosing of NPK in eastern Himalayan region. Research on crops 17(2):268–275

Bey BS, Ahmed R, Singh R et al (2023) Crop diversification vis-á-vis shifting cultivation in Karbi Anglong district of Assam: an empirical analysis. Curr Sci 125(5):536

Bhadana VP, Sharma PK, Ansari MA et al (2013) Food legumes for livelihood and nutritional security in north eastern Himalayan region: prospects and constraints. Indian J Agric Sci 83(9):899–906

Bhuyan R (2019) Review note on shifting cultivation in Northeast India amidst changing perceptions. Dhaulagiri J Soc Anthropol 13:90–95

Buragohain T (2022) Crop diversification and economy of *jhum* cultivation in Northeast India. In: A tradition in transition. The Energy and Resources Institute (TERI), New Delhi, pp 78–105

Census of India (2011) National population register and socio economic and caste census. https://censusindia.gov.in/nada/index.php/catalog/42619

Chanu LK, De A, Chakraborty K et al (2023) Characterization of shifting cultivation, trends, and diversification of livelihood patterns: a case study from forest villages in Barak valley, Assam, Northeast India. Trees For People 14:100420

Chassagne F, Hul S, Deharo E et al (2016) Natural remedies used by Bunong people in Mondulkiri province (Northeast Cambodia) with special reference to the treatment of 11 most common ailments. J Ethnopharmacol 191:41–70

Chen S (2022) Satellite-based monitoring, attribution, and analysis of forest degradation. Dissertation. Boston University

Dagar JC, Tewari JC (2016) Agroforestry research developments: anecdotal to modern science. In: Agroforestry research developments. Nova Publishers, New York, pp 1–45

Das A, Layek J, Mohapatra KP et al (2017) Shifting cultivation: some options for sustainable development in Meghalaya. In: Prakash N, Roy SS, Ansari MA et al (eds) Jhum improvement for sustainable farm livelihood and natural resource conservation in north Eastern Hill region: vistas and frontiers. ICAR Research Complex for NEH Region, Meghalaya, pp 20–32

Datta J, Gangadharappa NR, Biradar GS (2015) Livelihood status of tribal people practicing shifting (jhum) cultivation in Tripura state of north-East India. Trop Agric Res 25(3):316–326

Dey S, Laila R (2017) Reconstruction of women's role in jhum cultivation and shift in the gendered division of labor among the GAROS. In: Gendered lives, livelihood and transformation. University Press Limited, Dhaka, pp 8–39

Erni C (ed) (2015) Shifting cultivation, livelihood and food security: new and old challenges for indigenous peoples in Asia. AIPP, Bangkok

Giardina CP, Sanford RL, Døckersmith IC et al (2000) The effects of slash burning on ecosystem nutrients during the land preparation phase of shifting cultivation. Plant Soil 220(1):247–260

Gouda S, Decemson HT, Parida A et al (2020) Impact of shifting cultivation on mammalian diversity and distribution in fringe areas of Dampa Tiger Reserve, Mizoram, India. Environ Conserv J 21:103–115

Gupta AK (2000) Shifting cultivation and conservation of biological diversity in Tripura, Northeast India. Hum Ecol 28:605–629

Halder O, Sarkar A (2024) Assessment of shifting cultivation in the context of anthropogenic environmental burden in eastern Himalaya. In: Borthakur A, Singh P (eds) The Himalayas in the Anthropocene. Springer Nature, Cham, pp 219–235

Haokip IC, Devi HM, Das H et al (2021) Shifting cultivation in North East India: sustainability issues and strategies for improvement. Harit Dhara 4(2):3–6

Hazarika S (2017) Options for natural resource conservation for jhum improvement in Neh region. In: Prakash N, Roy SS, Ansari MA et al (eds) Jhum improvement for sustaining farm, livelihood and natural resource conservation in north eastern hill region: vistas and frontiers. ICAR Research Complex for NEH Region, Meghalaya, pp 103–109

Hett C, Lu J (2014) A literature review on carbon stocks in shifting cultivation landscapes in Laos. Climate Protection through Avoided Deforestation (CliPAD), Bonn and Eschborn, Germany

Huy B (2023) Shifting cultivation and rehabilitation of natural forest ecosystems: in the central highlands of Vietnam. In: Cairns M (ed) Farmer innovations and best practices by shifting cultivators in Asia-Pacific. CABI, Wallingford, pp 360–378

Jeeva SRDN, Laloo RC, Mishra BP (2006) Traditional agricultural practices in Meghalaya, north East India. Indian J Tradit Knowl 5(1):7–18

Kafle G (2011) An overview of shifting cultivation with reference to Nepal. Int J Biodivers Conserv 3(5):147–154

Kalita H, Baruah MS, Datta D et al (2017) Status of shifting cultivation (*Jhum*) in Arunachal Pradesh, India. In: Prakash N, Roy SS, Ansari MA et al (eds) Jhum improvement for sustaining farm, livelihood and natural resource conservation in north eastern hill region: vistas and frontiers. ICAR Research Complex for NEH Region, Meghalaya, pp 1–12

Kingwell-Banham E, Fuller DQ (2012) Shifting cultivators in South Asia: expansion, marginalisation and specialisation over the long term. Quat Int 249:84–95

Kithan LN (2014) Indigenous system of Paddy cultivation in terrace and Jhum fields among the Nagas of Nagaland. Int J Sci Res Publ 4(3):1–4

Lane A, Jarvis A (2007) Changes in climate will modify the geography of crop suitability: agricultural biodiversity can help with adaptation. ICRISAT 4(1):1–12

Li P, Feng Z, Jiang L et al (2014) A review of swidden agriculture in Southeast Asia. Remote Sens 6(2):1654–1683

Limbu DK, Rai BK, Rai KK (2020) In: Shang Z, Degen A, Rafiq M, Squires V (eds) Carbon management for promoting local livelihood in the Hindu Kush Himalayan (HKH) region. Springer, Cham, pp 63–87

Long CL, Zhou Y (2001) Indigenous community forest management of Jinuo people's swidden agroecosystems in southwest China. Biodivers Conserv 10:753–767

Maithani BP (2005) Shifting cultivation in north-east India: policy issues and options. Mittal Publications, New Delhi

Mertz O, Bruun TB (2017) Shifting cultivation policies in Southeast Asia: a need to work with, rather than against, smallholder farmers. In: Carins M (ed) Shifting cultivation policies: balancing environmental and social sustainability. CABI, Wallingford, pp 27–42

Mialhe F, Walpole P, Bruno E et al (2014) Spatio-temporal migration patterns to and from an upland village of Mindanao, Philippines. Popul Environ 36:155–179

Mihiranie MKS, Jayasinghe JMJK, Wanasundara JPD et al (2023) Indigenous food practices of the Dambana Vedda Community in Sri Lanka: past and present scenario. Vidyodaya J Humanit Soc Sci 8(02):1–12

Mishra G, Giri K, Jangir A (2021) Understanding the effect of shifting cultivation practice (slash-burn-cultivation-abandonment) on soil physicochemical properties in the north-eastern Himalayan region. Investigaciones Geograficas (Espana) 76:243–261

Mohapatra KP, Chandra P (2017) Rejuvenation of *jhum* land through agroforestry interventions. In: Prakash N, Roy SS, Ansari MA et al (eds) Jhum improvement for sustaining farm, livelihood and natural resource conservation in north eastern hill region: vistas and frontiers. ICAR Research Complex for NEH Region, Meghalaya, pp 73–78

Mouy M (2010) Assessment of the impact of bamboo harvesting on livelihoods and bamboo resources in the Seima Protection Forest, Mondulkiri, Cambodia. Cambodian J Nat Hist 1:27–37

Nath TK, Inoue M, Chakma S (2005) Shifting cultivation (jhum) in the Chittagong Hill Tracts, Bangladesh: examining its sustainability, rural livelihood and policy implications. Int J Agric Sustain 3(2):130–142

Ngah I, Kamarudin KH (2019) Malaysia: the state of/in village agriculture. In: Thomson E, Gillen J, Rigg J (eds) Asian smallholders in comparative perspective. Amsterdam University Press, Amsterdam, p 145

Nunisa B, Tamuly D, Hojai P et al (2023) Soil fertility status in relation to fallow cycle in shifting cultivated areas of Dima Hasao. Int J Plant Soil Sci 35(7):120–128

Osman KS, Jashimuddin M, Haque SS et al (2013) Effect of shifting cultivation on soil physical and chemical properties in Bandarban hill district, Bangladesh. J For Res 24(4):791–795
Panda BK, Sharkar S (2017) Shifting cultivation in north-East India: growth and progress. Dera Natung Gov Coll Res J 2(1):69–90
Pandey DK, Adhiguru P, Devi SV et al (2019) Quantitative assessment of crop species diversity in shifting cultivation system of eastern Himalaya. Curr Sci 117(8):1357–1363
Patra NK, Rohith GV, Das S (2023) Socio-economic and psychological status of rice (Oryza sativa L.) growers and constraints in cultivation: evidence from Mizoram, India. J Commun Mobil Sustain Dev 18(3):1057–1065
Prakash N, Roy SS, Ansari MA et al (2015) A comprehensive manual on integrated farming system: an approach towards livelihood security and natural resource conservation. ICAR Research Complex for NEH Region, Meghalaya
Punitha P (2017) Jhumias of Manipur in North-East India: a livelihood analysis. Dissertation,. Central Agricultural University (Imphal), Umiam, Meghalaya
Punitha P, Ansari MA, Pandey DK et al (2018) Shifting cultivation in North East India: social dimension, cross cultural reflection and strategies for improvement. Indian J Agric Sci 88(6):811–819
Rajkhowa DJ, Baishya LK, Ray SK et al (2017) Challenges, scope, and opportunities of jhum rejuvenation in Nagaland. In: Prakash N, Roy SS, Ansari MA et al (eds) Jhum improvement for sustaining farm, livelihood and natural resource conservation in North Eastern Hill region: vistas and Frontiers. ICAR Research Complex for NEH Region, Meghalaya, pp 73–78
Rerkasem K (1998) Shifting cultivation in Thailand: land use changes in the context of national development. In: ACIAR proceedings. Australian Centre for International Agricultural, Sydney, pp 54–63
Roder W, Calvert O, Dorji Y (1992) Shifting cultivation systems practised in Bhutan. Agrofor Syst 19:149–158
Saha R, Chaudhary RS, Somasundaram J (2012) Soil health management under hill agroecosystem of North East India. Appl Environ Soil Sci 2012(1):1–9
Saha S, Hussain S, Deka N et al (2023) Traditional ecological farming practices in the eastern Himalayan mountain environment: case of a Naga Village, Nagaland (India). Agrobiodivers Agroecol 3(02):34–54
Sahu SK, Nayak RK, Sarangi D (2005) Sustainable soil and land management under shifting cultivation in Orissa. Orissa Rev 5:67–74
Sapkota KN (2023) Shifting cultivation practices in Rai Community of eastern Nepal. Res CAB 2(1):179–190
Sarkar S (2024) The historical perspectives of shifting cultivation in north-eastern states of India. Int J Manag Public Policy Res 3(2):58–63
Sarkar R, Bora AK, Sarma T (2021) Shifting cultivation in relation to slope pattern and elevation in Karbi Anglong District of Assam. Ecol Environ Conserv 27:239–243
Sati VP (2019) Shifting cultivation in Mizoram, India: an empirical study of its economic implications. J Mt Sci 16(9):2136–2149
Schmidt-Vogt D (2001) Secondary forests in swidden agriculture in the highlands of Thailand. J Trop For Sci 13(4):748–767
Schröder LS, Rasche L, Jantke K et al (2024) Combined effects of climate change and agricultural intensification on soil erosion in uphill shifting cultivation in Northeast India. Land Degrad Dev 35(2):670–686
Shi-Cai S, Di-Yu L, Fu-Dou Z et al (2017) Swidden agriculture and sloping land conversion in China's Dulong valley. In: Cairns M (ed) Shifting cultivation policies: balancing environmental and social sustainability. CABI, Wallingford, p 295
Shikhu AC (2017) Women and resources management: a study of shifting cultivation in Nagaland. Dissertation. Nagaland University

Siahaya ME, Hutauruk TR, Aponno HS et al (2016) Traditional ecological knowledge on shifting cultivation and forest management in East Borneo, Indonesia. Int J Biodivers Sci Ecosyst Serv Manag 12(1–2):14–23

Silva JMN, Carreiras JMB, Rosa I et al (2011) Greenhouse gas emissions from shifting cultivation in the tropics, including uncertainty and sensitivity analysis. J Geophys Res Atmos 116. https://doi.org/10.1029/2011JD016056

Singh IM, Punitha P, Ansari MA et al (2017) Status of shifting cultivation in Manipur: an overview. In: Prakash N, Roy SS, Ansari MA et al (eds) Jhum improvement for sustaining farm, livelihood and natural resource conservation in north eastern hill region: vistas and frontiers. ICAR Research Complex for NEH Region, Meghalaya, pp 13–19

Sitlhou H (2020) Sacred ecology and ritual practices of the Thadou Kukis of Manipur. East Anthropol 73(3–4):445–462

Sodarak H (1999) Shifting cultivation practices by Hmong, Khamu and Lao ethnic categories in the nam Nane watershed, Nane district, Luang Prabang province, Lao PDR. Dissertation. Swedish University of Agricultural Sciences, Department of Rural Development Studies, Uppsala

Soni JK, Hota S, Shakuntala I (2020) Jhum cultivation: the way of livelihood security for Mizoram. Indian Farm 69(9):20–23

Swift P, Cock A (2015) Traditional Khmer systems of forest management. J R Asiat Soc 25(1):153–173

Tamuli TM, Bora AK (2017) Socio-economic status and livelihood opportunities of *jhumias* in Resubelpara Block, Meghalaya, India. J Xi'an Shiyou Univ Nat Sci Ed 17(05):157–167

Thapa GB, Rasul G (2005) Patterns and determinants of agricultural systems in the Chittagong Hill tracts of Bangladesh. Agric Syst 84(3):255–277

Thet APP, Tokuchi N (2020) Traditional knowledge on shifting cultivation of local communities in Bago Mountains. Myanmar J For Res 25(5):347–353

Thong P, Thangjam U, Sahoo UK et al (2022) Socio-economic vulnerability assessment of shifting cultivators (Jhumias) amidst the changing climate in Mizoram, Northeast India. Appl Geogr 147:102790

Tiwari BK, Kumar C (2008) Forest products of Meghalaya. Regional Centre, National Afforestation and Eco-development Board, North-Eastern University, Shillong

Tiwari BK, Pant RM (2018) Shifting cultivation: towards transformation approach. National Institute of Rural Development & Panchayati Raj North Eastern Regional Centre, Khanapara, Guwahati

Tripathi RS, Barik SK (2003) Shifting cultivation in North east India. In: Bhatt BP, Verma ND, Sharma YP (eds) Approaches for increasing agricultural productivity in hill and mountain eco system. ICAR Research Complex for NEH region, Umiam, pp 317–322

Tripura CS, Chandn R (2015) Jhum cultivation and occupational change in Tripura tribe: a case study of jatindra roaza Para of manu block Tripura, north East India. Asian J Multidiscipl Stud 3(10):25–33

Tripura B, Mondal S (2016) Socio-economic and agronomical constraints of Jhumias in shifting cultivation. J Crop Weed 12(3):118–120

Tugault-Lafleur C, Turner S (2009) Of rice and spice: Hmong livelihoods and diversification in the northern Vietnam uplands. Canada Research Chair in Asian Studies, Université de Montréal, Montreal

Verma PK, Kumar V, Chandra A et al (2017) Alternatives of shifting cultivation in north-eastern region of India. Rep Opin 9(12):1–8

Wasli ME, Tanaka S, Kendawang JJ et al (2009) Vegetation conditions and soil fertility of fallow lands under intensified shifting cultivation systems in Sarawak. Malay Trop 18(3):115–126

Win S (2005) Investigation on shifting cultivation practices conducted by the hill tribes for the development of suitable agroforestry techniques in Myanmar. Proceedings of the annual research conference (forestry sciences), Yangon, Myanmar, 7–9 January 2005

Yadav PK (2013) Slash-and-burn agriculture in North-East India. Expert Opin Environ Biol 2:2–5

Yin S (2015) Shifting cultivation and its changes in Yunnan province, China. In: Carins MF (ed) Shifting cultivation and environmental change. Routledge, London, pp 122–133

Yin L, Zachary M, Zheng Y et al (2020) Traditional ecological knowledge of shifting agriculture of Bulang people in Yunnan, China. Am J Environ Prot 3(3):51–58

Yuen JQ, Ziegler AD, Webb EL et al (2013) Uncertainty in below-ground carbon biomass for major land covers in Southeast Asia. For Ecol Manag 310:915–926

Ziegler AD, Bruun TB, Guardiola-Claramonte M et al (2009) Environmental consequences of the demise in swidden cultivation in montane mainland Southeast Asia: hydrology and geomorphology. Hum Ecol 37:361–373

Chapter 10
Cultural Wisdom in Environmental Stewardship: Kashmir's Traditional Practices

Soha ul Islam, Tabina Manzoor, Mohd Sharjeel Sofi, and Sami Ullah Bhat

Abstract This study examines the legacy of environmentally sustainable practices in Kashmir, focusing on traditional methods of agriculture, architecture, water and waste management, and small-scale enterprises. Rooted in centuries of indigenous knowledge, these practices demonstrate a deep ecological wisdom and provide valuable insights for addressing pressing environmental challenges. By positioning these age-old approaches within a contemporary context, the study underscores their relevance and potential to mitigate modern issues such as climate change, resource scarcity, and ecosystem degradation. The study argues for the integration of traditional ecological knowledge with modern scientific innovations as a pivotal strategy for advancing sustainable development in the region. These practices promote a harmonious relationship with the natural environment and offer practical, low-impact solutions for environmental conservation. By illustrating the potential of these sustainable practices to inspire environmentally conscious lifestyles, this review calls for collaborative efforts among policymakers, environmental practitioners, and local stakeholders to revitalise and promote these traditional systems.

Keywords Traditional practices · Sustainable · Kashmir · Natural resources · Development

10.1 Introduction

Climate change is an undeniable reality that spares no corner of the globe from its impact (Nuţă et al. 2024; Pimenta and Kamruzzaman 2024). It is no longer a distant concern affecting others but a shared experience that touches every individual on the planet (Ramzan et al. 2022). The consequences of climate change threaten to disrupt environmental sustainability, particularly as greenhouse gas emissions and

S. ul Islam · T. Manzoor · M. S. Sofi · S. U. Bhat (✉)
Department of Environmental Science, University of Kashmir, Srinagar, India

Y. D. I. Khan et al. (eds.), *Ecosystem-based Approaches for Resilience Building in Himalayan Landscapes*, Disaster Resilience and Green Growth,
https://doi.org/10.1007/978-981-95-2007-7_10

biodiversity loss continue to rise unchecked (Malhi et al. 2020). For developing economies, the challenge is particularly acute. Achieving a balance between sustainable production and economic growth presents a significant challenge, necessitating innovative approaches and collaborative efforts across multiple sectors (Ramzan et al. 2022). The severity of environmental degradation underscores the urgent need to address this crisis. Indeed, curbing the damage caused by human activity has become a critical global necessity, overshadowing the priorities of previous decades (Nuţă et al. 2024). Since the Industrial Revolution, a new era has arisen, the Anthropocene (Crutzen 2002), in which human actions have become the main driver of global environmental change (Steffen et al. 2007). Policymakers worldwide focus on mitigating climate change drivers and adapting to its impacts. However, human activities risk pushing the Earth system beyond the Holocene's stable conditions, potentially causing severe or catastrophic consequences for global ecosystems and societies. This highlights the need for urgent, science-driven action to maintain planetary stability. Policies aimed at mitigating degradation often fall short due to fragmented governance and prioritisation of short-term economic growth over sustainability (Meadows et al. 2004).

The failure to enforce environmental protections emphasises the need for policies integrating ecological limits into development (Steffen et al. 2015). As environmental concerns dominate political agendas worldwide, the urgency to address pressing ecological issues has never been greater. Questions on preventing environmental degradation, preserving habitats, and protecting endangered species are central to sustainable development discussions. These issues are now central to development planning, reflecting the growing recognition that planetary health supports economic and social well-being (Clarke 1990).

In this context, it is important to integrate traditional knowledge into modern resource management for its rhetorical and political importance and intrinsic conservationist and social benefits. This perspective was solidified by the SPREP (South Pacific Regional Environment Programme) region's 1982 declaration, emphasising the adaptability of traditional practices, such as land and reef tenure, to contemporary management needs. Deeply rooted in cultural and ecological understanding, these practices are increasingly recognised as vital for assessing developmental impacts (Gadgil et al. 1993). Traditional knowledge systems, which have evolved over centuries, offer environmentally sustainable practices that have long ensured ecological balance (Hariram et al. 2023; Bühler et al. 2024). Connected to local ecological contexts, these systems highlight how communities can sustain resource management without exhausting the planet's natural cycles. Indigenous groups offer alternative knowledge and perspectives based on their locally developed resource use practices (Rist and Dahdouh-Guebas 2006). Traditional knowledge, as a way of knowing, is similar to Western science, based on an accumulation of observations. Still, it is different from science in some fundamental ways. Anthropologist Claude Lévi-Strauss (1962:269) argued that these two ways of knowing are parallel modes of understanding the universe, with one supremely concrete and the other supremely abstract. This observation highlights the unique strengths of traditional knowledge, rooted in practical, place-based understanding, often passed down through generations. Unlike Western science, which seeks universals, traditional knowledge centres on the ecosystem and community relationships, which are vital for local environmental management and cultural resilience. Integrating both systems offers a more holistic framework for addressing complex ecological challenges (Rist and Dahdouh-Guebas 2006; DeWalt 1994). Coevolving with ecological and social systems, traditional ecological knowledge (TEK) enhances society's ability to navigate disturbances and sustain ecosystem services amid uncertainty (Colding et al. 2003).

In Kashmir, for example, sustainable agricultural practices, water harvesting techniques, and the preservation of freshwater springs are testaments to how traditional knowledge can align with the need for modern environmental management. These time-tested methods are particularly relevant today, where climate change, resource depletion, and environmental degradation present complex challenges. Since the advent of modernity, particularly following the Industrial Revolution, TEK has experienced significant erosion in many parts of the world, including Kashmir. This region, rich in its cultural heritage and sustainable practices, has seen its traditional systems increasingly challenged by globalisation and industrialisation.

The erosion of TEK in Kashmir is driven by factors such as formal schooling prioritising external knowledge and the decline of local languages essential for passing on TEK. Additionally, shifts in land use, mainly due to agricultural mechanisation and changes in land ownership patterns, have further contributed to the decline of traditional practices. Market integration and urbanisation in Kashmir have transformed community resource management and local crafts, eroding traditional values. While aiming to protect biodiversity, conservation programs have sometimes restricted local communities' access to traditional resources, causing further disconnect (Gómez-Baggethun et al. 2010). The decline of traditional lifestyles in Kashmir by the late twentieth century led to concerns in the 1980s about the survival of TEK into the next millennium (Cox 2000). However, revisiting and revitalising these knowledge systems could provide crucial lessons in addressing contemporary. Despite these advancements, the value of the traditional system has not diminished (Agatha 2016). For instance, Kashmir's traditional agricultural practices—such as organic farming, mixed cropping, and using ancient irrigation systems like the Kul (water channels)—have sustained the region's agrarian economy for centuries while maintaining ecological balance. The Kul irrigation system, a community-managed resource, exemplifies how traditional methods can effectively optimise scarce natural resources, ensuring equitable water distribution and enhancing crop yields (Ahmad 2021). This system reflects an indigenous understanding of sustainable water management, which remains crucial in modern water scarcity challenges. Similarly, the traditional architecture of Kashmir is another testament to environmentally sound practices. The designs are functional and deeply integrated with the environment, highlighting a sustainable construction approach that predates modern eco-friendly housing trends (Rather, 2024). The focus on passive heating and cooling in traditional Kashmiri architecture highlights its enduring relevance as the world seeks sustainable housing solutions. Despite the shift towards

Fig. 10.1 Glimpses of the traditional architecture of Kashmir House

modern construction methods, there is a growing recognition of the value of revisiting these traditional approaches. Indigenous building techniques (Fig. 10.1) and community-managed agriculture provide key insights for sustainability, especially in climate-vulnerable regions.

In this context, this review provides an in-depth look into the traditional sustainable practices of Kashmir, such as organic farming, sustainable architecture, traditional water mills (Gharats), and indigenous methods of food preservation. These practices offer practical, eco-friendly solutions relevant to contemporary environmental policy and sustainable development (Roy et al. 2023; Upadhyay et al. 2024). However, many of these practices are now threatened by modernisation, consumerism, and changing social dynamics (Garcia 2023). This chapter explores Kashmir's traditional environmentally sustainable practices and assesses their potential for addressing contemporary environmental challenges. By examining agricultural techniques, architecture, and community-based resource management systems, this review underscores the relevance of these practices in modern-day policy-making and environmental management (Ghayoumi et al. 2023).

10.2 Traditional Agriculture in the Kashmir Himalayas: A Model for Sustainability

Traditional farming practices in the Kashmir Himalayas are a hallmark of sustainable agriculture, shaped by centuries of indigenous knowledge and cultural heritage. Still widely embraced, these practices harmonise crops, livestock, soil, and water management, fostering natural pest control, enhancing soil fertility, and supporting diverse crop production (Kunju 2013). Methods such as agroforestry, intercropping, composting, and integrated livestock farming are vital for ensuring food security and maintaining ecological balance. These techniques rejuvenate soil health, contribute to carbon sequestration, and minimise the environmental impact of farming. The Kul irrigation system exemplifies community-driven resource management, skillfully conserving water through naturally crafted channels (Ahmad 2021).

By balancing tradition with ecological foresight, these practices address modern challenges like resource scarcity and climate resilience. Deeply rooted in the region's socio-cultural and environmental ethos, they offer a timeless framework for sustainable development (Upadhyay et al. 2024). This enduring legacy of nature-aligned farming in the Kashmir Himalayas highlights the potential of traditional wisdom in shaping a sustainable future.

10.2.1 The Agrarian Wealth of Kashmir: Harnessing Traditional Practices for Sustainable Agriculture

Kashmir's agriculture thrives on the fertile soil of the region, abundant water resources, and favourable topography, creating an ideal environment for diverse crop cultivation (Maqbool and Jan 2017). The valley's agrarian system predominantly focuses on staple crops such as rice, maize, wheat, and pulses, which form the backbone of its food security. Traditional farming methods, deeply embedded in the region's cultural and ecological heritage, emphasise sustainability through organic fertilisers and crop rotation (Puttoo 2008). These practices rejuvenate soil fertility and mitigate nutrient depletion, ensuring long-term agricultural productivity.

A cornerstone of Kashmir's sustainable agrarian system is the "Kul" irrigation network—a community-managed marvel of traditional engineering. This intricate system of water channels efficiently distributes water across fields, supporting irrigation needs while conserving the region's vital water resources. By preventing over-extraction and minimising wastage, the *Kul* system exemplifies how traditional knowledge can offer innovative solutions to modern resource management challenges. Rooted in balance and resource conservation principles, they provide a replicable model for promoting food security and ecological harmony in another region.

10.2.2 Organic Farming and Tools

Traditional farming in Kashmir is characterised by its organic nature, where synthetic fertilisers and pesticides are avoided. Farmers use cow dung and compost as fertilisers, which enrich the soil and prevent the soil from harmful chemicals. Tools like wooden ploughs, hoes, and sickles are designed to minimise environmental damage and are sourced locally; tools like these also help farmers live healthy lives. Unlike modern mechanised agriculture, where there is the incessant threat of soil erosion and loss of soil microflora, traditional farming techniques exert minimal pressure on natural resources. The environment of Jammu and Kashmir, characterised by optimal weather conditions and favourable climatic patterns, presents significant potential for the widespread adoption of organic agriculture (Dar 2021).

10.2.3 Sustainability in Rice Culture

Rice (Daniye in Kashmiri), was cultivated in Kashmir in ancient times. It has a vital significance as it was the staple food for the Kashmir populace and the primary revenue source to the state. Various varieties of rice were grown in Kashmir during the Dogra Period. Hassan Khuihami, Har Gopal Khasta, and Dewan Kripa mention 96 rice varieties cultivated in Kashmir during that period (Bhat and Bhat 2022). Farmers had developed sound traditional knowledge about the parameters for good rice harvest and blast (rai) disease management. Sustained rice production by adopting organic practices coupled with indigenous farming implements and bullock power was invariably followed. Two approaches—dry (tao) and wet (kenul)—were adopted for growing paddy depending upon location and soil under broadcasting and transplantation systems. In rice cultivation, simple tools such as wooden ploughs and organic manure are used to prepare the fields. Rice saplings are transplanted manually, and weeding is done by hand, which not only helps protect the soil but also maintains the health of the crops. In addition to that, crops are free from hazardous contaminants. These practices reflect an environmentally friendly approach to agriculture that has sustained generations of Kashmiri farmers.

10.3 Unique Characteristics of Traditional Kashmiri Architect

The vernacular architecture of Kashmir reflects traditional wisdom suited to its climate, lifestyle, and natural hazards like earthquakes and floods (Beg 2016). These architectural practices use locally available materials such as wood and stone and construction techniques like taq and dhajjidewari, specifically designed to withstand seismic activities (Langenbach 2009). The resilience of vernacular architecture in Kashmir has long been a subject of scholarly interest, particularly regarding seismic events. Gosain and Arya (1967) particularly highlighted the role of timber-laced construction techniques such as dhajjidewari, observing that the ductility imparted by timber elements significantly reduced damage during earthquakes. Their study emphasised that multi-story structures employing more significant timber volumes demonstrated superior performance, with minimal damage even in the 1967 earthquake. This aligns with findings from Langenbach (2009), who emphasises the seismic benefits of timber-framed systems across similar contexts globally. Furthermore, Dar and Ahmad (2015) elaborate on these traditional methods' energy efficiency and environmental sustainability, making them exemplary for modern green building practices. Integrating indigenous materials and techniques addresses historical challenges and continues to inspire contemporary architectural resilience frameworks. Such insights illustrate the architectural creativity in Kashmiri traditions, reaffirming their relevance in sustainable and disaster-resilient construction discourse. The vernacular architecture of Kashmir Valley symbolises extraordinary environmental adaptability and disaster resilience. These architectural practices, created in response to frequent natural calamities such as earthquakes, floods, and fires, reflect an understanding of local climatic conditions, including long, frigid winters and humid summers. Historical accounts highlight the devastation caused by these hazards (Shah 2002), yet local builders demonstrated excellent creativity, employing materials like timber for their strength and flexibility. Techniques such as taq (timber-laced masonry) and dhajjidewari (timber-braced frames) evolved not only for structural efficiency but also for mitigating risks of seismic collapse (Langenbach 2009). Langenbach emphasises that timber's characteristic ductility is vital for withstanding dynamic loads, a principle long entrenched in Kashmiri designs. Integrating diagonal braces and lightweight infill ensures load redistribution during seismic events, limiting structural failure.

Post-disaster evaluations, particularly after the 2005 earthquake, affirm the durability of these practices, revealing how traditional techniques outperform many modern counterparts in seismic zones (Jigyasu and Hamdani 2006). Similarly, the catastrophic floods of 2015 exposed vulnerabilities in contemporary designs while underscoring the resilience of heritage structures (Chawla and Iqbal 2015). This aligns with sustainable construction principles, as these methods optimise the use of local resources while minimising ecological impact. In a broader context, vernacular architecture of Kashmiri demonstrates how traditional knowledge systems can inform modern resilience strategies. By integrating environmental harmony with disaster mitigation, these practices offer critical lessons for sustainable development in vulnerable regions. The preservation of Kashmiri architecture is thus vital, as it represents a distinct cultural and architectural identity on the global stage. Ensuring preservation safeguards a vital heritage and promotes resilience and adaptability for contemporary architectural needs (Shah 2020).

10.4 Textile Industry

Kashmir's textile industryembody a rich tradition of sustainable craftsmanship (Divya et al. 2025).This art form goes beyond decoration, embodying the rich cultural heritage and exceptional craftsmanship of Kashmiri artisans, with techniques carefully preserved and transmitted across generations (Bhat 2012). Artisans use natural fibres like wool and pashmina to create exquisite shawls, carpets, and wood carvings, reflecting cultural heritage and environmental stewardship. These traditional industries rely on locally sourced materials, minimal waste, and eco-friendly processes, preserving artisanal skills while promoting sustainability. By blending tradition with sustainability, these industries offer a blueprint for environmentally conscious craftsmanship in the modern world (Wani et al. 2024; Roy et al. 2023).

By the 1850s, "cashmere" shawls had gained immense popularity among the affluent middle class in France and across Europe, while women from diverse social backgrounds adopted shawls in varying styles and qualities. (Daly 2002). The famous Pashmina wool of the region, derived from the Changthangi goat, is spun into fine threads for weaving luxurious shawls and textiles. These practices have been passed down through generations and are deeply embedded in the cultural identity of the region. Furthermore, producing these textiles requires minimal energy and water, making them sustainable alternatives to mass-produced synthetic fabrics.

10.4.1 Handicrafts and Sustainable Practices

In addition to textiles, Kashmir's cottage industry is known for handicrafts like wood carving, shawl weaving, and carpet making. These crafts use local materials, support the economy, and promote sustainable practices. The manual processes used to craft these items ensure minimal environmental impact, unlike industrial manufacturing processes that are resource-intensive and polluting.

10.5 Traditional Water Management: Gharats

Gharats, traditional water mills in Kashmir, exemplify sustainable water management and eco-friendly energy use. Powered by the natural flow of rivers and streams, these mills efficiently grind grains without relying on external energy sources. Gharats are a low-impact, renewable energy solution and reflect the community's deep connection with nature. Reviving and integrating these traditional systems into modern practices can promote sustainable resource management and energy conservation (Wani et al. 2024; Upadhyay et al. 2024).

10.5.1 Economic and Environmental Benefits

Traditional water mills, or Gharats, found in the villages of Kashmir harness the kinetic energy of flowing water to grind grains. These mills are constructed using local materials such as wood and stone and rely on renewable energy sources. Gharats play a vital role in rural communities, providing a cost-effective and environmentally friendly alternative to electric mills (Slathia et al. 2018).Water mills are crucial to local economies, providing income to farming families and offering a sustainable alternative to energy-intensive grain milling methods. Moreover, flour produced in water mills retains its nutritional value better than that produced by modern mills. These mills exemplify sustainable energy use, which is essential in promoting environmental protection and economic stability (Slathia et al. 2018).

10.6 Traditional Solid Waste Management

Traditional solid waste management in Kashmir emphasises resource conservation through eco-friendly practices like household composting and recycling organic waste for animal feed. These methods minimise waste, reduce landfill use, and promote nutrient recycling, fostering a closed-loop system that aligns with modern sustainability goals. By harnessing natural processes, these practices reflect a deep understanding of waste as a resource, offering valuable insights for contemporary waste management strategies (Basyouni and Mahmound 2024; Wani et al. 2024).In traditional Kashmiri communities, waste management was a collective responsibility. Communitynal labour was employed to build sanitation infrastructure, manage waste, and maintain hygiene. Household composting plays a vital role in managing organic waste. Kitchen scraps, farm residue, and animal manure were buried in shallow pits to decompose into nutrient-rich compost for farming. This mimimized the waster generation and supported local agriculture. Recycling practices, such as repurposing metal and using organic materials for household items, were deeply ingrained in traditional Kashmiri life. This practice ensured that waste was disposed of in an environmentally sustainable manner while promoting community bonds.

10.7 Traditional Domestication of Fowls

The traditional domestication of fowls in Kashmir plays a key role in sustainable farming and biodiversity preservation. Traditional poultry-rearing methods align with organic, low-input agricultural systems crucial for soil enrichment and sustainable ecosystem management. Poultry manure, rich in nutrients, significantly enhances soil fertility, supporting crop productivity. This practice contributes to local food security and promotes biodiversity conservation, as it reduces dependence on chemical fertilisers (Upadhyay et al. 2024).

The holistic integration of livestock and poultry farming with crop cultivation reflects the sustainable agricultural ethos of Kashmir. These traditional practices, rooted in centuries-old wisdom, offer an environmentally friendly alternative to industrial farming systems, aligning with modern sustainability goals (Shiva 1996; Wani et al. 2024).

10.7.1 Backyard Poultry Farming

Backyard poultry farming has been a long-standing tradition in Kashmir. Indigenous hardy and disease-resistant chicken breeds are reared in rural households, providing both food security and additional income. The chickens consume household waste, contributing to waste management, and their manure is used as a natural fertiliser for local farms.

10.8 Traditional Pottery

Traditional pottery in Kashmir is a centuries-old craft that harmoniously blends artistry with sustainability. Using locally sourced, biodegradable clay, Kashmiri artisans produce functional and decorative items with minimal environmental impact. The reliance on natural materials and manual techniques ensures this craft remains an eco-friendly alternative to mass-produced goods. By preserving traditional pottery, Kashmir's rich cultural heritage is sustained and promotes environmentally conscious production methods (Basyouni and Mahmoud 2024; Wani et al. 2024).

For generations, the people of Kashmir have been deeply engaged in crafting hand-made utensils, clothing, and decorative items (Fig. 10.2). The meticulous craftsmanship showcases the region's cultural richness, with pottery being one of the most cherished art forms in the valley. Among the various styles, glazed pottery—referred to as **Dalgate pottery**—holds a special place in Kashmiri artistic traditions. Artefacts such as oil lamps, tea kettles, and kitchen stoves serve as fine examples of the skill and ingenuity of local potters, and today, many of these pottery products are repurposed as decorative pieces in urban settings.

Fig. 10.2 Handmade Kashmiri willow wicker work

The roots of Kashmiri pottery run deep. Archaeological sites such as Burzhama in Srinagar and Gufkral in Pulwama, dating back over 5000 years, have uncovered ancient pottery traditions. Additionally, medieval sites like Avantipura, Devsar, and Martand have revealed fragments of earthen vessels, including jars, *gharas*, *haandis*, jugs, and bowls, showcasing the region's historical mastery in pottery-making.

Numerous workshops and fairs are regularly organised throughout Kashmir to revive and enhance this timeless art. These events feature talks on the origins and techniques of pottery design and include live demonstrations by skilled potters from various parts of the state (Ministry of Home Affairs, Government of India 2014).

10.8.1 Eco-friendly Pottery Craft

In Kashmir, pottery occupies an indispensable place, valued for its domestic utility and ritual significance, while also exemplifying a traditional craft rooted in environmentally sustainable practices. Potters use locally sourced clay to create biodegradable kitchenware, storage containers, and decorative items. Unlike metalware and plastics, pottery does not contribute to pollution or resource depletion.

10.9 Conclusion

Kashmir's traditional environmentally sustainable practices offer invaluable insights for addressing today's environmental challenges. Rooted in a deep connection with nature, these practices emphasise harmony between human activities and the ecosystem, utilising local resources and community-driven management. From agriculture and architecture to textile production and waste management, the principles of sustainability embedded in these methods provide a robust framework for modern environmental stewardship. Reviving and adapting these time-honoured approaches promote sustainable development and safeguard rich cultural heritage of Kashmir. To build a resilient, ecologically balanced future, policymakers, environmentalists, and local communities must integrate these traditional practices with modern innovations. By adopting an ecosystem-based approach, we can ensure sustainable development that respects both the environment and cultural heritage. This integration is crucial for creating policies that support sustainable practices, fostering community involvement, and promoting environmental education. Strengthening this synergy between tradition and innovation will pave the way for a sustainable path forward for the region.

Acknowledgement The authors express their gratitude to the Department of Environmental Science, University of Kashmir, for its support in conducting this study.

References

Agatha A (2016) Traditional wisdom in land use and resource management among the Lugbara of Uganda: a historical perspective. SAGE Open 6(3):2158244016664562. https://doi.org/10.1177/2158244016664562

Ahmad I (2021) From Kul to modern canals: the evolution of water management in Kashmir. Greater Kashmir

Basyouni YA, Mahmoud H (2024) Affordable green materials for developed cool roof applications: a review. Renew Sust Energ Rev 202:114722. https://doi.org/10.1016/j.rser.2024.114722

Bhat BA, Bhat MA (2022) Rice cultivation in Kashmir during Dogra period – Methods, extent and production. Int J Creat Res Thoughts (IJCRT) 10(5):2320–2882

Bhat S (2012) The Art and Tradition of Kashmiri Embroidery. Kashmir Studies J 10(2):143–156

Bühler M, Hollenbach P, Köhler L, Armstrong R (2024) Unlocking resilience and sustainability with Earth-based materials: a principled framework for urban transformation. Front Built Environ 10:1385116. https://doi.org/10.3389/fbuil.2024.1385116

Chawla M, Iqbal S (2015) Damage assessment mission in Srinagar city, Kashmir. INTACH Kashmir in collaboration with Prince Claus Fund, unpublished report

Clarke WC (1990) Learning from the past: traditional knowledge and sustainable development. Contemp Pac:233–253

Colding J, Elmqvist T, Olsson P (2003) Living with disturbance: building resilience in social-ecological systems. In Navigating social-ecological systems: building resilience for complexity and change, pp 163–185

Cox PA (2000) Will tribal knowledge survive the millennium? Science 287(5450):44–45. https://doi.org/10.1126/science.287.5450.44

Crutzen P (2002) Geology of mankind. Nature 415:23. https://doi.org/10.1038/415023a

Daly S (2002) Kashmir shawls in mid-Victorian novels. Vic Lit Cult 30(1):237–256

Dar M (2021) Organic farming in Jammu and Kashmir: Progress, challenges and prospects. https://www.jkpi.org/organic-farming-in-jammu-and-kashmir-progress-challenges-and-prospects/. Accessed 11 Dec 2024

Dar MA, Ahmad S (2015) Traditional earthquake resistant systems of Kashmir. Int J Civil and Struct Eng Res 2(2):86–92

DeWalt BR (1994) Using indigenous knowledge to improve agriculture and natural resource management. Hum Organ 53(2):123–131. https://doi.org/10.17730/humo.53.2.ku60563817m03n73

Divya R, Yadav S, Rose NM, Rukhaya S (2025) Beyond the needle: the symbolism and significance of Kashida embroidery in Kashmiri culture. Vigyan Varta: An International E-Magazine for Science Enthusiasts 6(5)

Gadgil M, Berkes F, Folke C (1993) Indigenous knowledge for biodiversity conservation. AMBIO-STOCKHOLM 22:151–151

Garcia CC (2023) Contemporary changes and challenges in the practice of trend forecasting. Int J Fashion Stud 10(1):75–98. https://doi.org/10.1386/infs_00086_1

Ghayoumi R, Charles A, Mousavi SM (2023) A multi-level analysis of links between government institutions and community-based conservation: insights from Iran. Ecol Soc 28(2). https://doi.org/10.5751/ES-14217-280233

Gómez-Baggethun E, Corbera E, Reyes-García V (2010) Traditional ecological knowledge and global environmental change: research findings and policy implications. Ecol Soc 15(4):28. https://doi.org/10.5751/ES-06288-180472

Gosain N, Arya AS (1967) A report on Anantnag earthquake of February 20, 1967. Bulletin of the Indian Society of Earthquake Technology

Hariram NP, Mekha KB, Suganthan V, Sudhakar K (2023) Sustainalism: an integrated socio-economic-environmental model to address sustainable development and sustainability. Sustainability 15(13):10682. https://doi.org/10.3390/su151310682

Jigyasu R, Hamdani S (2006) Research report on disaster risk management of vernacular heritage in Kashmir. INTACH, J&K in collaboration with Research Centre for Disaster Mitigation of Urban Cultural Heritage, Ritsumeikan University, Rits-DMUCH, Kyoto, Japan, unpublished report

Kunju PCA (2013) Agricultural technology development and transfer in the socio-economic transformation of Agrarian economies: The Kerala experience with special reference to Kuttanad. Doctoral dissertation, Mahatma Gandhi University, Kottayam

Langenbach R (2009) Don't tear it down! Preserving the earthquake resistant vernacular architecture of Kashmir. UNESCO, New Delhi. Available at: http://www.traditional-is-modern.net

Levi-Strauss C (1962) La pensée sauvage. Paris: Librairie Plon (English translation 1966). The savage mind. Chicago: University of Chicago Press

Malhi Y, Franklin J, Seddon N, Solan M, Turner MG, Field CB, Knowlton N (2020) Climate change and ecosystems: threats, opportunities and solutions. Philos Trans R Soc B 375(1794):20190104

Maqbool O, Jan F (2017) The agricultural sector of Kashmir: between growth and mismanagement. Acad Res J Agric Sci Res 5:341–349

Meadows D, Randers J, Meadows D (2004) A synopsis: limits to growth: the 30-year update. Chelsea Green, Estados Unidos, p 381

Ministry of Home Affairs (2014) Government of India, December. Jammu and Kashmir update 6:12

Nuţă FM, Sharafat A, Abban OJ, Khan I, Irfan M, Nuţă AC et al (2024) The relationship among urbanization, economic growth, renewable energy consumption, and environmental degradation: a comparative view of European and Asian emerging economies. Gondwana Res 128:325–339

Pimenta A, Kamruzzaman LM (2024) Assessing the comprehensiveness and vertical coherence of climate change action plans: the case of Australia. J Environ Manag 369:122419

Puttoo BL (2008) Traditional rice culture in Kashmir. Asian Agri-History 12(4):285–297

Ramzan M, Raza SA, Usman M, Sharma GD, Iqbal HA (2022) Environmental cost of non-renewable energy and economic progress: do ICT and financial development mitigate some burden? J Clean Prod 333:130066

Rather (2024) Roots in the Roof: preserving Kashmir's architectural heritage. Kashmir times. Assessed on 09-09-2025. https://kashmirtimes.com/culture/art/preserving-kashmirs-architectural-heritage

Rist S, Dahdouh-Guebas F (2006) Ethnosciences – a step towards the integration of scientific and indigenous forms of knowledge in the management of natural resources for the future. Environ Dev Sustain 8:467–493. https://doi.org/10.1007/s10668-006-9050-7

Roy S, Rautela R, Kumar S (2023) Towards a sustainable future: Nexus between the sustainable development goals and waste management in the built environment. J Clean Prod 415:137865. https://doi.org/10.1016/j.jclepro.2023.137865

Beg MS (2016) Issues of conservation and adaptation in protecting Kashmir's vernacular heritage. Int J Environ Stud 73(4):524–532

Shah H (2002) Tarikh Hassan, vol. I, Translation by Ahmed, S. Srinagar: J&K Academy of Art, Culture and Languages

Shah M (2020) Architecture of Kashmir – through ages symbolises syncretism and rich heritage. J Crit Rev. ISSN-2394-5125. https://www.jcreview.com/admin/Uploads/Files/659438ed573b35.26669382.pdf

Shiva V (1996) Agricultural biodiversity, intellectual property rights, and farmers' rights. Econ Polit Wkly 22:60–67

Slathia PS, Kumar R, Paul N, Sharma BC, Peshin R, Gupta SK (2018) Traditional water mills (Gharats)-A source of rural livelihood in mountainous region of Jammu and Kashmir. Indian J Tradit Knowl 17(3):569–575

SPREP (South Pacific Regional Environment Programme) (1982) Report of the conference on the human environment in the South Pacific. Rarotonga, Cook Islands, 8–11 March 1982. Noumea: South Pacific Commission

Steffen W, Crutzen PJ, McNeill JR (2007) The Anthropocene: are humans now overwhelming the great forces of nature. Ambio-J Hum Environ Res Manag 36(8):614–621

Steffen W, Richardson K, Rockström J, Cornell SE, Fetzer I, Bennett EM et al (2015) Planetary boundaries: guiding human development on a changing planet. Science 347(6223):1259855

Upadhyay H, Vinke K, Weisz H (2024) "We are still here" climate change, gender and immobility in highly mobile Himalayan communities. Clim Dev 16(5):443–457. https://doi.org/10.1080/17565529.2023.2230176

Wani NR, Rather RA, Farooq A, Padder SA, Baba TR, Sharma S et al (2024) New insights in food security and environmental sustainability through waste food management. Environ Sci Pollut Res 31(12):17835–17857. https://doi.org/10.1007/s11356-023-26462-y

Chapter 11
Livelihood Sustainability and Pastoralism: Challenges and Adaptation of the Gaddis of the Himalayas

Dattatray Bhandalkar and Sangeeta Singh

Abstract Pastoralism is an age-old livelihood system mainly practised in resource-poor regions such as arid, semi-arid, mountain, and cold desert areas. Pastoralism refers to the domestication of livestock for subsistence. It is practised on rangelands, which are communally owned, privatised, and agricultural fallow land. It is also a socioeconomic, cultural, and ecological form of life. Pastoralism as a livelihood activity has been coping with temporal and spatial variabilities. Pastoralists have developed resilience to cope with such variabilities and withstand livelihood shocks. Since pastoralism is being operated in regions like the mountains of the Himalayas, it has been exposed to various challenges. Current literature highlights that in recent two decades, this pastoral livelihood system has been changing due to various factors, such as land use-land change, climatic factors, geopolitical factors, economic factors, education factors, and motives for upward mobility, that have created challenges for the sustainability of pastoralism. Due to land accusations and regional and local regulations, traditional pasture lands are no longer readily available. Due to several challenges, the new generation of pastoral communities is no longer interested in continuing the pastoral traditional way of life; thus, it will affect the sustenance of the pastoral livelihood system. Those factors have led to various livelihood-associated vulnerabilities for pastoralists of the Indian Himalayan regions such as the Gaddis. This further leads to the migration of herders in the urbanised region in search of uncertain non-pastoral opportunities. The government has proactively taken various initiatives to build pastoralists' resilience; however, this is not halting the present change. Therefore, this paper highlights factors affecting the sustainability of the pastoral livelihood system with special reference to the Gaddis from the Himalayas.

D. Bhandalkar (✉)
Amity Institute of Liberal Arts (AILA) Amity University Mumbai (AUM), Mumbai, Maharashtra, India

S. Singh
Department of Education, University of Allahabad, Allahabad, Uttar Pradesh, India

Y. D. I. Khan et al. (eds.), *Ecosystem-based Approaches for Resilience Building in Himalayan Landscapes*, Disaster Resilience and Green Growth,
https://doi.org/10.1007/978-981-95-2007-7_11

Keywords Pastoralism · Factors affecting the Himalayan region · Gaddi community

11.1 Introduction

The world faces the challenge of climate change and is coming to terms with the present problems caused by climate change. The communities residing and practising traditional livelihood are adversely impacted by the threat of climate change at the front of livelihood activity. Traditional pastoralist communities in the Himalayas are one such community that is losing on various fronts—ecological challenges impacting their traditional livelihood of maintaining livestock. Existing literature prominently highlights various challenges observed in this region, such as ecological challenges, challenges related to water sources, and the rising man vs wild conflict (Dadas and Akhup 2025).

The Himalayan region is adversely affected by climate change, which has led to further changes in ecosystem dynamics. However, the effects of climate change on agricultural activity, demographic changes, and environmental changes and the impact on the water resources in the Himalayas change the dynamics (Immerzeel et al. 2020; Kattel 2022). Ecological changes directly affect the dependent population in the region. Farming activities were most affected, which led to a reduction in agricultural production (Nautiyal et al. 2023).

Climate change is having an impact on livelihoods that are directly related to natural resources (Eriksson et al. 2009). Pastoralism is one of these types of livelihoods. Climate is one of the many factors that negatively impact traditional pastoral living. Climate change is one of the main reasons why pastoralism as a means of subsistence is at a crossroads. Communities that have been pastoralists for generations are struggling to maintain their way of life. Pastoralists of the younger generation are not prepared to remain in such demanding occupations. Apart from climate-related problems, pastoralist communities are negatively impacted by economic and geopolitical factors. Present generations of traditional pastoralists have stopped practising pastoralist livelihoods due to their aspirations and exposure to education (Dadas et al. 2025).

Past pastoralist generations struggle to maintain resilience to climate change and lack support from the new generations. Furthermore, external factors like land accusations and regional and local regulations that govern access to the pasture are defining factors for pastoralist communities. Traditional pasture lands are shrinking or are no longer available or accessible for grazing. In other words, the inaccessibility of traditional pastures forces them to migrate to urban or semi-urban areas due to the inaccessibility of customary grazing land (Sharifian et al. 2022; Sharaftmandrad et al. 2024).

The sustainability of the pastoral livelihood system is significantly impacted by both internal and external issues, such as the decline of the younger generation of pastoral communities in maintaining the traditional pastoral way of life and external variables like the climate and access to traditional pastures.

However, these factors create a number of threats to livelihood-associated vulnerabilities for pastoralists in the Indian Himalayan regions. The Gaddis are the best illustration of how internal and external forces influence and limit the livelihood activities of the traditional pastoralist community in the Himalayas (Dadas and Akhup 2025). It is difficult to study these factors in isolation since these factors are directly or indirectly connected. The resilience of the pastoralist lifestyle of this well-known pastoralist community, the Gaddis, is significantly impacted by internal factors that are imperceptible to outsiders, even while the consequences of climate change as an external factory are easily noticeable.

The government has implemented a number of programs for pastoralist communities, but these are insufficient to increase their resilience or prevent them from relocating to cities where they are forced to engage in non-pastoralist occupations. This paper is an attempt to bring the factors affecting pastoralism in the Indian Himalayan region, especially collecting evidence from the Gaddi community, to the forefront of the discussion.

11.2 Methodology

This paper is an attempt to provide an overview of existing literature. The journey starts with exploring literature related to pastoralism in general and the Gaddis of the Himalayas in particular. Eventually, the existing literature was divided into themes and subthemes highlighting the challenges and adaptations associated with pastoralism in the Himalayan region. Various challenges affect the livelihood of pastoralist communities in the Himalayan region, which were again divided into two major categories of internal and external factors. External factors such as climate change, geopolitical issues, and economic conditions, along with internal factors like rising education levels and aspirations for upward mobility, have impacted the sustainability of pastoralism.

External factors are interlinked and primarily include environmental, policy-level, and cultural elements. In contrast,internal factors pertain to communities' access to land for grazing, the evolving attitudes of the current generation, exposure to education, and the resulting impact on the decline of traditional pastoralist livelihoods. The paper highlights significant issues and gaps in the literature regarding pastoral communities. Based on the overview of the literature, the paper is divided into four main themes: (1) ecological shifts and their impacts, (2) grazing land degradation in the Himalayan region, (3) issues and challenges of traditional pastoralist communities, and (4) challenges and adaptations of the Gaddis.

11.3 Findings or Results

11.3.1 About the Gaddi Community

The Gaddi community belongs to a pastoralist group that is popularly known for its semi-nomadic lifestyle in the Himalayan region. Traditionally, the Gaddis are a shepherd community located mainly in the Himalayan region and some parts of Nepal. Their livelihood activities are dependent on natural resources. Climate change threatens natural resources, impacting their socioeconomic and traditional lifestyle. However, the review of their situation highlights that the Gaddis have faced environmental problems posed by climate change for many decades. Socioeconomic problems result from the climatic changes and loss of traditional pasture for their herds. The central problem of the Gaddi community is climate change, which is impacting their livelihood activities and their seriocomic situation, which further leads to various challenges in their day-to-day life. The absence of governmental policies that target their problems further makes their situation vulnerable.

The Gaddi community has practised a unique nomadic pastoralist lifestyle for many years in the region of the Himalayas. Rapid political, environmental, and economic changes significantly threaten their traditional way of life and cultural legacy. The corpus of information about the Gaddi community is compiled in this thorough literature evaluation, with special attention to their socioeconomic situation, means of subsistence, customs, and challenges they face. Using a detailed examination of scholarly papers, reports, and ethnographic studies, this study highlights areas for future research and policy interventions. It shows how adaptable the Gaddis are to changing environments. To progress, it is essential to understand the complexities of the Gaddi's livelihood and cultural dynamics. The Gaddis are pastoralists, and many of the families still practise a semi-nomadic lifestyle. They have co-existed with nature for centuries now.

The Gaddi community's socioeconomic standing results from the intricate interactions between their traditional ways of subsistence, available economic resources, and outside influences. Numerous studies have detailed the Gaddis' demographics, sources of income, educational attainment, and accessibility to essential services (Khanyari et al. 2022). Despite their tenacity and adaptability, the Gaddi communities frequently endure poverty, limited access to healthcare and education, and inadequate infrastructure in remote mountainous locations (Raina et al. 2018; Punam 2021).

Regarding survival, the typical characteristics of the Gaddi community members are seasonal changeability and diversification of livelihood strategies. The Gaddi community still relies heavily on livestock rearing and transhumance as food sources and cultural identity (Nehria and Ghosh 2024). Following customary migration patterns between high-altitude summer pastures (*bugyals*) and lower-altitude winter grazing fields (*shoras*), the Gaddi pastoralists engage in seasonal migration with their herds (Bhasin 2013). Their primary income source is livestock (Sharma et al. 2022). The Gaddi's lifestyle and the cultural heritage they represent are unique. Oral traditions and community activities transmit traditional knowledge systems about herbal treatment, livestock management, and the mountain environment (Bhasin 2013; Malhotra et al. 2021). Their culture and identity are threatened by the rapidly growing urbanisation and globalisation (Pandey 2015). The Gaddis find adjusting to urbanisation and modernisation and their impact on their culture challenging.

The Gaddi shepherds have many intricate facets, impacted by various variables such as shifting market demands, government policy, land encroachment, and climate change. Livelihood dependency is on the livestock, so the lack of support from government officials and local politicians leads them to lose customary and traditional rights on grazing land (Bhasin 2011). If they do not have access to these grazing lands, they must migrate in search of grazing land. The Gaddi shepherds need assistance and interventions that tackle these issues, encourage sustainable practices, and guarantee their access to grazing areas and equitable market opportunities to maintain their traditional way of life. Their situation becomes more vulnerable due to political marginalisation and their dependency on the availability of grazing lands. So, to improve their situation and adapt to the changing times, they need to be trained in new farming techniques, best practices for managing livestock, and market strategies. The Gaddi shepherds endure in the face of these difficulties and adjust to the shifting conditions.

They have strong ties with their ancestral homelands, which is suitable for their livelihood activities and survival. They face numerous challenges, including climate change, land grabbing, market needs, and government policies. To protect their cultural identity and traditional way of life, they need access to grazing land, awareness about sustainable livelihood practices, and market opportunities. The Gaddi shepherds' way of life is intricately linked to various opportunities and problems. It necessitates addressing government regulations, market needs, land encroachments, and climate change.

11.3.2 Ecological Shifts and Its Impacts

Pastoralist communities in the Himalayas are especially facing several challenges due to climate change because it has resulted in changing the traditional grazing practices, the status of the flora, and extreme weather conditions (Khanyari et al. 2022). Rising temperatures and irregular rains disturb the traditional livelihood activities essential for the pastoralist groups (Dadas et al. 2025). The region's population growth and rapid development led to habitat loss and deforestation. Deforestation further leads to many consequences in the Himalayan region. Infrastructure development and population also contribute to agricultural activities.

Due to natural habitat loss, it severely impacts animal species and the food chain altogether. On the one hand, the animals have lost their habitat; on the other hand, flora and wildlife are lost due to the damage done by increasing human interference.

The impact on the ecosystem also puts various varieties of plants and fauna in danger, directly impacting pastoralists, where they lose a traditional pasture. The decline in biodiversity left a huge impact on pastoralist activities. There is an urgent need to conserve and safeguard biodiversity.

11.3.3 Grazing Land Degradation in the Himalayan Region

Deforestation and land degradation make it harder for traditional pastoralists to find suitable grazing areas. Expanded human habitation and activities are the reason for degradation, which could be an important reason for human-wildlife conflict in the Himalayan region (Poudel et al.2022) . Human-wildlife conflicts are becoming challenges for humans and their livelihoods, apparently impacting their survival.

Pastoralist groups are concerned about human-animal conflict. In addition, what worries them more is the growth of wildlife reserves and protected areas (Sharma et al. 2021). Wildlife conflicts are impacting their financial losses as well as overall safety. Wildlife conflict leads to the loss of sheep, which leads to financial losses and safety (Khatiwada et al. 2017). The present literature emphasises how anthropogenic and natural forces interact to create a complex web of ecological issues in the Himalayan region. Climate change worsens preexisting vulnerabilities, putting ecosystems, biodiversity, and human health in danger. Unabated habitat loss and deforestation further fracture landscapes and endangered species' existence. The region has additional issues due to declining biodiversity, depletion of water resources, and conflict between humans and wildlife. Climate change today has adversely impacted the Himalayan region. To find the immediate solution to these problems, coordinated efforts with the locals, state authorities, and national levels are the need of the hour. Building resilience in the region is necessary and can be achieved with conservation, sustainable development efforts, and community involvement.

11.3.4 Issues and Challenges of Traditional Pastoralist Communities

Communities in the Himalayas have genuinely lived by traditional pastoralism for ages. However, this livelihood is facing numerous issues and challenges in the modern era. Traditional pastoralism in the Himalayas often faces land tenure insecurity, with unclear or insecure land rights, leading to conflicts and displacement (Kapruwan et al. 2024). Government policies and land use changes further exacerbate land tenure issues, threatening pastoralists' access to grazing lands and water sources (Nautiyal et al. 2003). Traditional pastoralism in the Himalayas faces serious problems from environmental degradation, which many observers see as involving soil erosion, deforestation, and biodiversity loss (Tiwari et al. 2020). In the Himalayan region, traditional pastoralists like the Gaddis are vulnerable to socio-economic marginalisation, and they have very little scope to access essential services like economic, medical, and educational (Malhotra et al. 2021). Due to poverty, the absence of infrastructure and social prejudices add up to the shocks and strains from the outside world, making them more vulnerable (Joshi and Rawat 2021). In addition, modernisation and globalisation rapidly add to the decline in traditional cultural practices and knowledge systems (Rao 2024). The efforts to build resilience in pastoralist communities are threatened by increasing migratory activities, changing demography, and causing loss of cultural identity (Pandey et al. 2017).

However, numerous challenging factors threaten the sustainability and resilience built by the traditional pastoralists in the region. To address the concern, there is a dire need to work on several fronts. Collaborative efforts from academicians, government authorities, and non-governmental groups, as well as the pastoralist communities themselves, are needed. Adoption of an inclusive strategy to ensure the preservation of culture and natural resources in the region is needed.

11.3.5 Challenges and Adaptations of the Gaddis

Property tenure instability, which results from unclear property rights and intrusion by outside players, is one of the main issues the Gaddi community faces (Mishra and Pandey 2019). Land grabs, infrastructural development, and government conservation regulations that limit access to grazing regions pose growing threats to the traditional grazing pastures necessary for the Gaddi pastoralism (Wagner 2013). Insecure land tenure affects the Gaddi pastoralists' ability to support themselves, their sense of cultural identity, and social cohesiveness (Thakur et al.2025).

The Gaddi pastoralism and ecosystem resilience in the Himalayas are seriously threatened by environmental degradation, which includes deforestation, soil erosion, and water scarcity (Bhasin 2013). Their food security worsens due to unsustainable land usage, overgrazing, and logging, which destroy the traditional grazing areas and lead to loss of fodder (Kotnala et al. 2016). In addition, climate change as a threat started showing symptoms like changing weather patterns and disasters that impact traditional livelihood practices (Dadas et al. 2025). As to their socioeconomic marginalised situation, the Gaddi community has less scope to access essential services like education, health, and economics (Sharma 2014). To make their situation more adverse, the Gaddi communities become the victims of poverty and the absence of infrastructure, impacting their socioeconomic growth and leading them to prolonged cycles of suffering (Sharma et al. 2003).

Socioeconomic gaps within the Gaddi community are further exacerbated by discrimination and marginalisation by government institutions and dominant social groupings (Pattanaik and Singh 2005). The Gaddi community's traditional knowledge, customs, and beliefs are threatened by the quick speed of industrialisation and cultural globalisation, which could result in cultural erosion and identity loss (Sharma et al. 2022). In pursuit of more excellent prospects, younger generations are moving to cities more frequently, undermining community bonds and causing a loss in traditional livelihoods (Singh et al. 2015). The degradation of cultural norms and values weakens the Gaddi community's resiliency and fuels social disintegration (Dhiman et al. 2022).

The Gaddi population in the Himalayas has intricate and varied problems that stem from environmental deterioration, socioeconomic inequality, and historical injustices. The intersection of land tenure insecurity, environmental degradation, socioeconomic marginalisation, and cultural decline threatens the Gaddi people's resilience and general well-being. Furthermore, these issues are made worse by outside forces like governmental regulations, development initiatives, and climate change, underscoring the necessity of comprehensive and inclusive strategies to address the root causes of vulnerability.

Several interrelated issues affect the Gaddi population in the Himalayas, endangering their traditional way of life, cultural legacy, and socioeconomic stability. Land tenure insecurity, environmental degradation, socioeconomic marginalisation, and cultural erosion undermine the Gaddi people's resilience and agency; hence, immediate action is required to protect their rights and advance sustainable development. Acknowledging the Gaddi community's rights, knowledge, and ambitions to promote fair and inclusive development that honours the Himalayan region's ecological integrity and cultural diversity is imperative.

They face numerous issues that affect the Gaddi community's well-being, livelihoods, and cultural resilience. The research has identified several significant difficulties, including insecure land tenure, environmental degradation, the effects of climate change, socioeconomic marginalisation, and cultural erosion (Dadas 2023; Mishara and Pandey 2019). These problems are worsened by restricted access to grazing areas, competition for natural resources, poor infrastructure, and government policies disregarding pastoral communities' requirements (Nautiyal et al. 2023).

The range of ecological issues that pastoralist communities in the Himalayas must deal with highlights the necessity of comprehensive and long-term solutions. Promoting the resilience of pastoralist people and the ecosystems they depend on requires integrated land management techniques, community-based conservation programmes, and climate-resilient livelihoods.

11.4 Discussion

The case of the Gaddis demonstrates the complex relationships among the Gaddi community's socioeconomic standing, livelihood strategies, cultural customs, and difficulties in the Himalayas. Despite their adaptability and tenacity, the Gaddis face several obstacles to cultural preservation and sustainable development. Significant risks to the Gaddi community members' identity and well-being include socioeconomic marginalisation, land tenure instability, environmental degradation, and cultural erosion. To address these issues, comprehensive strategies that incorporate community involvement, indigenous knowledge, and policy interventions targeted at advancing equitable and sustainable development in the Himalayan area are needed.

Pastoralist communities in the Himalayas confront intricate and diverse ecological issues from environmental, socioeconomic, and institutional factors. To find context-specific solutions that support the sustainability of ecosystems and livelihoods in the Himalayan region, researchers, government agencies, and pastoralist communities must work together to address these issues.

11.5 Conclusion

The Himalayan Gaddi community exemplifies a unique fusion of customary ways of life, cultural legacy, and adaptability to shifting environmental and socioeconomic circumstances. This comprehensive literature assessment offers insightful information about the Gaddi communities' socioeconomic situation, livelihood options, cultural customs, and difficulties. Acknowledging the Gaddis' rights, knowledge, and agency is imperative to advance sustainable development, protect cultural variety, and build resilience in the Himalayan region. Future research and policy interventions should prioritise their needs and ambitions to ensure the Gaddi community's continued well-being and cultural vibrancy in the face of ongoing transformations.

The issues that pastoralist communities face appear to be resolved by the ecosystem-based approach; this approach aims to strengthen societal and environmental resilience among vulnerable communities. The livelihood of pastoralist communities like the Gaddis is impacted by both internal and external factors, which are addressed by ecosystem-based approaches. By implementing various measures, including ecosystem improvement initiatives, forest management, restoration, and ecosystem maintenance.

Furthermore, effective environmental management policies, considering human and natural factors, and sustainable management techniques are crucial to addressing these issues and protecting the Himalayas' ecological integrity. To foster resilience and guarantee the long-term viability of the region's ecosystems and livelihoods, integrated approaches incorporating biodiversity conservation, climate change mitigation, and community engagement are required.

References

Bhasin V (2011) Pastoralists of Himalayas. J Hum Ecol 33(3)
Bhasin V (2013) Pastoralists of Himalayas. J Biodivers 4(2):83–113
Biswas MÁ, Rao RÁ (2016) Socio-economic status of gaddi tribe in Himachal Pradesh: a study. Int J Adv Res 4(8):159–167
Dadas DR, Khan YDI, Bhandalkar D, Indurkar C, Gokhale K (2025) Ecological disruption and its impact on Indian pastoralism: ecosystem-based adaptation is way forward. In Handbook of nature-based solutions to mitigation and adaptation to climate change. Cham: Springer Nature Switzerland, pp 1–22
Dadas DR (2023) Indian pastoralism amidst changing climate and land use: evidence from dhangar community of semi-arid region of Maharashtra. In The Palgrave Handbook of socio-ecological resilience in the face of climate change: contexts from a developing country. Singapore: Springer Nature Singapore, pp 85–97
Dadas D, Akhup A (2025) Pastoral livelihood vulnerabilities in the Himalayan region: negotiating for community resilient practice. In Social work research: methodologies in fields of practice. Singapore: Springer Nature Singapore, pp 81–99
Dhiman VK, Bandyopadhyay B, Singh J, Kumar A, Chaudhary A (2022) Himachal Pradesh's Gaddi Tribe: A Socio-Cultural Study. ECS Transactions 107(1):10061
Eriksson M, Xu JianChu XJ, Shrestha AB, Vaidya RA, Santosh Nepal SN, Sandström K (2009) The changing Himalayas: impact of climate change on water resources and livelihoods in the greater Himalayas, p 24
Immerzeel WW, Lutz AF, Andrade M et al (2020) Importance and vulnerability of the world's water towers. Nature 577, 364–369. https://doi.org/10.1038/s41586-019-1822-y
Joshi NC, & Rawat GS (2021) An integrated approach for the identification and prioritization of areas based on their livelihood vulnerability index: a case study of agro-pastoral community from Western Indian Himalaya. Mitig Adapt Strat Glob Chang 26(6):27
Kotnala S, Bhardwaj S and Cherlet J (2016) Pastoralists re-establish traditional ecosystems and customary rights. Case study of the ILC Database of Good Practices. Rome: ILC
Kapruwan R, Saksham AK, Bhadoriya VS, Kumar C, Goyal Y, Pandey R (2024) Household livelihood resilience of pastoralists and smallholders to climate change in Western Himalaya, India. Heliyon 10(2)
Khatiwada SP, Deng W, Paudel B, Khatiwada JR, Zhang J, Su Y (2017) Household livelihood strategies and implication for poverty reduction in rural areas of central Nepal. Sustain 9(4):612
Khanyari M, Robinson S, Milner-Gulland EJ, Morgan ER, Rana RS, Suryawanshi KR (2022) Pastoralism in the high Himalayas: Understanding changing practices and their implications for parasite transmission between livestock and wildlife. Pastor 12(1):44
Kattel GR (2022) Climate warming in the Himalayas threatens biodiversity, ecosystem functioning and ecosystem services in the 21st century: is there a better solution?. Biodivers Conserv 31(8):2017–2044
Malhotra A, Nandigama S, Bhattacharya KS (2021) Food, fields and forage: a socio-ecological account of cultural transitions among the Gaddis of Himachal Pradesh in India. Heliyon, 7(7)
Mishra H, Pandey BW (2019) Navigating the impacts of social and environmental changes to traditional lifestyle: A case study of Gaddi transhumance of Chamba district in Himachal Pradesh. Orient Anthropol 19(2):326–337
Nautiyal S, Prakash S, Goswami M, Premkumar A (2023) Impact of climatic variations on crop yield in Chamoli District, Uttarakhand, India. In The Palgrave handbook of socio-ecological resilience in the face of climate change: contexts from a developing country. Singapore: Springer Nature Singapore, pp 53–69
Nehria S, Ghosh A (2024) Gaddi: Himachal Pradesh's evolving pastoral tribe. Antrocom: Online J Anthropol 20(1)
Nautiyal S, Rao KS, Maikhuri RK, Saxena KG (2003) Transhumant pastoralism in the Nanda Devi Biosphere Reserve, India: A case study in the buffer zone. Mt Res Dev 255–262

Nautiyal S, Gupta AK, Goswami M, Khan YDI (2023) The Palgrave Handbook of Socio-ecological Resilience in the Face of Climate Change. Palgrave Macmillan Singapore. https://doi.org/10.1007/978-981-99-2206-2

Poudel TR, Aryal PC, Thapa-Parajuli R, Thapa A, Yadav SK, Prakash M (2022) Depredation loss drives human–wildlife conflict perception in the Trans-Himalayas. J Environ Manag 311:114763

Pandey R, Kumar P, Archie KM, Gupta AK, Joshi PK, Valente D, Petrosillo I (2018) Climate change adaptation in the western-Himalayas: household level perspectives on impacts and barriers. Ecol Indic 84:27–37

Pandey K (2015) Ethnography of food and food habits in a transhumant population of north India. Int J Med Res Prof 1(3):132–141

Punam S (2021) Social and economic status of mountainous communities of Himachal Pradesh. Sambodhi 44(1):48–54

Pandey A, Pradhan N, Chaudhari S, Ghate R (2017) Withering of traditional institutions? An institutional analysis of the decline of migratory pastoralism in the rangelands of the Kailash Sacred Landscape, western Himalayas. Environ Sociol 3(1):87–100

Pattanaik BK, Singh K (2005) Socio-economic conditions of Gaddi tribals: findings from a survey in Kangra district, Himachal Pradesh. Soc Chang 35(2):13–24

Raina SK, Singh M, Chander V, Raina S (2018) Prevalence and co-relates of hypertension among Gaddi (Agro Pastoralist) tribal's at high altitude in North-West India. Adv Hum Biol 8(3):159–163

Rao KK (2024) Cultural constraints on knowledge transmission and knowledge erosion: an indigenous community in India. Asian J Soc Sci 52(4):23–30

Sharma VP, Köhler-Rollefson I, Morton J (2003) Pastoralism in India: a scoping study. Indian Institute of Management and League of Pastoral Peoples, Ahmedabad, India

Sharma A, Parkash O, Uniyal SK (2022) Moving away from transhumance: the case of Gaddis. Trees, Forests and People 7:100193. https://doi.org/10.1016/j.tfp.2022.100193

Sharma P, Chettri N, Wangchuk K (2021) Human–wildlife conflict in the roof of the world: Understanding multidimensional perspectives through a systematic review. Ecol Evol 11(17):11569–11586

Sharma V (2014) Education and women empowerment among Gujjars, Bakerwals and Gaddis in Jammu region of Jammu and Kashmir. Int J Res (IJR) 1–4

Singh R, Sharma RK, Babu S (2015) Pastoralism in transition: livestock abundance and herd composition in Spiti, Trans-Himalaya. Hum Ecol 43(6):799–810

Sharaftmandrad M, Abedi Sarvestani A, Shahraki M, Hassanzadeh Nafooti M (2024) Uncovering the reasons behind the failure of pastoralists in adopting climate change adaptation strategies. Sci Rep 14(1):20602

Sharifian A, Fernández-Llamazares Á, Wario HT, Molnár Z, Cabeza M (2022) Dynamics of pastoral traditional ecological knowledge: a global state-of-the-art review. Ecol Soc 27(1):14

Tiwari KR, Sitaula BK, Bajracharya RM, Raut N, Bhusal P, & Sengel M (2020) Vulnerability of pastoralism: A case study from the High Mountainsof Nepal. Sustain 12(7):2737

Thakur SN, Kahlon S, Bhutani S, Sharma P (2025) Nomadic pastoralism in Western Himalaya: A journey through history, politics, development, and the fall of Gaddi pastoralists. Hum Ecol 1–14

Wagner A (2013) The Gaddi beyond pastoralism: making place in the Indian Himalayas. Berghahn Books

Chapter 12
An Ecosystem-Based Assessment of Resilient Traditional Restoration Practices Towards Biodiversity Conservation and Ecosystem Restoration in High Altitudes of Darma Valley in Uttarakhand Himalayan Region

Rishav Rawal, Harshit Pant, and Sanjay Ghildiyal

Abstract The ecosystem-based adaptation (EbA) is a strategical framework of actions for practices involving adapting towards climate change. It harnesses both the nature-based solutions and ecosystem services, which can be linked with forest restoration as it plays an important role in conserving the essence of traditional practices and people's benefit through environmental conservation. In this aspect, addressing the drivers affecting the restoration approaches is of prime importance, and the present ecosystem-based assessment of resilient traditional restoration practices towards biodiversity conservation and ecosystem restoration in the high altitudes of Darma Valley in the Uttarakhand Himalayan region can have an important path for understanding the process. The Darma Valley is located in the Kailash Sacred Landscape region of India, with significant ecological and cultural values. The present study evaluates the traditional and resilient restoration practices intended for biodiversity conservation in community-supported forest plantations in the Darma region. The study examines how local communities' interventions such as afforestation initiatives can facilitate the rehabilitation of degraded ecosystems, increase biodiversity, and enhance the economy of communities. Since this area has the integration of native species, it improves the adaptive capacity of ecosystems to changes and threats posed by climate and human activities. The outcome has demonstrated that the retention of local people's participation in the restoration provides long-term benefits. The results show that the incorporation of traditional knowledge

R. Rawal · S. Ghildiyal
Department of History D.S.B. Campus, Kumaun University, Nainital, Uttarakhand, India

H. Pant (✉)
Centre for Socioeconomic Development (CSED), G.B.Pant National Institute of Himalayan Environment (NIHE), Almora, Uttarakhand, India
e-mail: harshit.pantj@gbpihed.nic.in

Y. D. I. Khan et al. (eds.), *Ecosystem-based Approaches for Resilience Building in Himalayan Landscapes*, Disaster Resilience and Green Growth,
https://doi.org/10.1007/978-981-95-2007-7_12

with modern approaches to nature protection is quite efficient for the problem of biodiversity preservation and combating climate change impacts in these difficult mountain conditions.

Keywords Biodiversity Conservation · Community forest · Resilient restoration · Darma Valley · Kailash Sacred Landscape

Abbreviations

CC	Climate change
DRR	Disaster risk reduction
EbA	Ecosystem-based adaptation
EEK	Ethno-ecological knowledge
GIS	Geographical information systems
ICT	Information and communication technology
KSL	Kailash Sacred Landscape
PES	Payment for ecosystem services
SDG	Sustainable Development Goals

12.1 Introduction

With respect to the rural regions of the Himalayan landscape, the local communities have usually addressed and defined the utilisation of their natural resources in the near vicinity in accordance with their historical management practices or relevant conservation rituals adopted over the centuries. This makes them susceptible and resilient towards the unprecedented instances of macro- and micro-ecosystem-based anomalies, which arise due to various climatic and human influences (Gadgil et al. 2000; Berkes et al. 2000; Colding et al. 2003; Folke et al. 2003; Toledo et al. 2003; Tompkins and Adger 2004; Berkes 2007; Dimri et al. 2021; Negi et al. 2021, 2022; Rawat and Schickhoff 2022). Moreover, it has been noted that in the marginalised region's villages, the abrupt conditions of weather and climatic incongruities have impacted the socio-economical and livelihood generation sectors, with the prominent ones encompassing the agriculture, forest, and water sectors (Jugran et al. 2024).

Formal and informal institutions mediate local responses to socio-economic and ecological change, access to influence, and/or decision-making over land and natural resources, as well as by existing ecological and traditional knowledge (Holt-Gimenez 2002; Adger et al. 2005; Eakin 2005; Lebel et al. 2006; Janssen et al. 2007; Naess 2012; Ruiz-Mallén and Corbera 2013; Chopra 2014; Black-Lock and Rodehorst 2023).

The thrilling ecological conditions of the high-altitude terrain have shaped lifestyles and cultures since ages (Pant 1935) along with sustenance practices adopted by the communities. Similarly, the traditional crop varieties and races improved or acclimatised to the local high-altitude environment and circumstances not only meet the nutritional and food security needs of these traditional communities but also simplify the commodity exchange practices that are prevailing in the region. It has been noted that the combined effects of socio-economic change and changing climatic patterns have resulted in a decline in natural resource management and significant disenchantment towards traditional resource management practices which has further led to economic and socio-ecological uncertainties, augmented outmigration, and societal conflicts over scarcely available resources, impacting various socio-economic and ecological aspects of these highland mountain communities. In this aspect, the inclination of mountain-specific EbA needs to be explored in the aspect of that nature-based solution that sustainably harnesses the regional biodiversity and ecosystem services towards reducing ecological, environmental, and socio-economical vulnerability and building the resilience of human communities for adaptation towards climate change and changing socio-economic and socio-ecological dynamics.

12.1.1 Resilience in Climatic Adaptation

The focal community that has been focused on during the present study is the Bhotia community, an ethnologically and traditionally heterogeneous community that conventionally lived a nomadic life (Srichandan et al. 2021). While the Johari community reside in the Gori valley, the Rung community inhabits the other three valleys. Rungs further subdivide themselves according to the valley they live in: Darmani (Darma Valley), Chaudansi (Chaudans Valley), and Byansi (Byans Valley) as reported in earlier studies (Bergmann et al. 2011; Bergmann 2016, Rawal et al. 2012; Srichandan et al. 2021).

Rawat and Schickhoff (2022) reported that the migratory households of the Darma region practice assorted mountain agriculture, i.e. a mix of animal husbandry, crop farming, forest, and meadowland use for their sustenance (Ehlers and Kreutzmann 2000; Bergmann 2016), which is linked through periodic and recurrent migration (Nüsser 2006).

At the beginning of the winter season (October–November), the community usually migrates to warmer provinces in lower locations near Dharchula, and they typically used to return to Darma Valley in March to spend the summers with their livestock (Rawat and Schickhoff 2022), which is known as transhumant pastoralism or *Kuncha*. This seasonal migratory pattern of the community is entirely dependent on resource availability (nutritional fodder and water) and requirements for their livestock. This practice usually extends over several altitudinal zones of the Kumaon Himalaya (Nüsser 2006), and at diverse altitudinal belts, environmental conditions such as the duration of the snow cover, its abundance or the onset of vegetation growth necessitates location-specific potentials and limitations for pastoral land use by the transhumant community (Bergmann et al. 2008). This *Kuncha* practice also enables these stakeholders to gather rare medicinal herbs for trade and self use while providing time to utilise agricultural land in the summer villages for cultivating high-altitude crops and vegetables such as buckwheat (*Fagopyrum esculentum* Moench.), barley (*Hordeum vulgare* L.), cabbage (*Brassica oleracea* L.), and potatoes (*Solanum tuberosum* L.). Due to the testified decline in the practice of transhumance in this region by earlier studies (Rawat and Schickhoff 2022), collection of wild medicinal plants mainly caterpillar fungus (*Ophiocordyceps sinensis* (Berk.) G.H. Sung, J.M. Sung, Hywel-Jones & Spatafora, which is locally known as Keedajadi or Yarchagumba, and cultivation of native medicinal plants like Kutki (*Neopicrorhiza scrophulariiflora* (Pennell) D.Y. Hong), Atish (*Aconitum heterophyllum* Wall. ex Royle), Hathazari (*Dactylorhiza hatagirea* (D. Don) Soó), Kuth (*Dolomiaea costus* (Falc.) Kasana & A.K. Pandey), Chippi (*Pleurospermum angelicoides* (DC.) Benth. ex C.B. Clarke), Jambu (*Allium stracheyi* Baker), Kala Jeera (*Carum carvi* L.), etc. as a form of subsistence have become prevalent, which represents a considerable share of the household income and value chain development in the region.

According to Hoon (1996) and Negi (2009), the term *Kuncha* implies a setup of family and livestock caravan intermingled coupled with the rearing and guarding of animals that usually migrate to high-altitude villages (within a single river valley system) in summer to cultivate a variety of traditional crops and to provide grazing facilities for their animals (livestock usually containing cattle and horses), covering usually 60–90 km of distance during the movement. This upward migration of the Bhotiyas (also known as Shokas or Rung) to their summer habitations is governed by the prerequisite to arrive in time to their native village to sow their crops and to capitalise on the short growing season. Their migration begins at the end of April and continues throughout May. The crops are usually ready for harvest by the end of August (sometimes mid-September). September is also the period for harvesting medicinal plants, chiefly native *Kutki*. The return migration to their winter homes begins in October or in early November.

Negi (2009) reported another resilient adaptation factor, which described the details of the transhumance practice prevalent in the region. It has been described as a sustainable practice involving the utilisation of resources, the accumulation of day-to-day utilisable goods and minor commodities, and activities to enhance and improve their economic status.

12.1.2 Rationale of EbA in Darma Valley: Present Scenario

At present, ecosystem-based adaptation (hereby EbA) emerges as an approach to address multiple challenges like climate change (CC), biodiversity conservation, and restoration of degraded environments through ecosystem services. This adaptation method is crucial for the Himalayan region, where ecological surroundings are sensitive to changing climate and anthropogenic pressure. In such conditions, attaining EbA by maintaining ecological balance, supporting biodiversity, and preserving cultural heritage are essential to supporting sustainable development by mitigating adverse impacts that degrade the environment. The valley consists of rich biodiversity, including the variety of pervasive endemic plant and animal species. In addition, local inhabitants have maintained a harmonious relationship with the environment for decades. Like many other Himalayan communities, these residents still rely on their traditional ecological knowledge (hereby TEK), which has been conceded down from generation to generation. These knowledge and practices help sustainably preserve cultural identity and customs.

The traditional forest restoration methods in Darma Valley have evolved, shaped by the community's deep understanding of the local ecosystem and its dynamics. These methods include planting native tree species, conserving sacred groves, and maintaining natural regeneration processes, all of which contribute to the restoration of degraded lands and the enhancement of biodiversity. Importantly, these methods are often adapted to address the specific challenges of the high-altitude environment, where climate conditions and soil types can vary widely and topographic changes test the management scenarios. The integration of such traditional knowledge with modern conservation practices presents an opportunity to enhance the effectiveness of restoration efforts and build greater ecological resilience in the region.

The study shows how the integration of traditional restoration practices of the Darma Valley, with the present approach of conservation, may contribute to the conservation and maintenance of biodiversity and ecosystem restoration. The study highlights the crucial role of local communities in maintaining long-term ecological sustainability by focusing on community-driven afforestation and reforestation efforts. The participation of local inhabitants in these activities not only aids in rehabilitating the degraded ecosystem but also strengthens the sense of belongingness towards the land, ensuring the long-term sustainability of the natural environment in the region. This work also explores the linkages between traditional knowledge and contemporary approaches, highlighting the need to combine these two for addressing present-day concerns related to the environment. In the context of the Darma Valley, which has a long-driven tradition of human-environment relations, this combination of Indigenous ecological knowledge with modern techniques of restoration showcases ways to achieve ecological and sustainable goals for the benefit of the local landscape by underscoring the importance of such practices in high-altitude regions like Darma. Understanding these methods with EbA helps in enhancing knowledge about the conservation of biodiversity, ecosystem resilience, and the creation of models for addressing complex environmental challenges.

12.2 Description of the Study Area

The study area is located in the highlands of Bhot region of Pithoragarh district, comprising four valleys: Johar, Darma, Chaudans, and Byans valley. These valleys are situated within the Kali River system (Gori, Darma, Kuti Yankti, Kali River) and are inhabited by the Johari-Shauka and Rung-Shauka communities known as Bhotias. The present study was conducted in Darma Valley situated in the Kailash Sacred Landscape (KSL)-India region of the Pithoragarh district, Uttarakhand, India, consisting of 17 villages (14 migratory and 3 permanent settlement villages) distributed across an altitudinal range around 2400 to over 4000 m above sea level (asl) in the Western Himalayas.

The Darma Valley is located north of the Panchachuli Mountains (30°N and 80°E) in the Dharchula subdivision and contains 12 villages at altitudes between 2500 and 4800 m. The total population in 2001 was about 1000, containing a very small number of families in each village. Conventionally, the main professions of villagers are trade, sheep rearing, collection of medicinal herbs from the wild, and cultivation of crops, chiefly *Fagopyrum esculentum* and *F. tataricum* and, to a lesser extent, potatoes and other vegetables since ages. The literacy of these villages is quite high, with the highest recorded in Goe village (76.11%), followed by Dugtu (74.19%), and the lowest is around 60% in other villages (Census 2011).

As part of KSL-India, this region has significant cultural, religious, and socio-economic significance (Fig. 12.1) and rich biodiversity. Darma, along with the valleys of Byans and Chaudans, forms part of the Rung cultural landscape, which is renowned for its systematic and ritualistic approach founded on the depth of ethno-ecological knowledge (EEK), making a substantial contribution to conservation and subsistence sustenance. Given the high diversity, rarity, and endemism of floral and faunal species in the forests, communities' traditional knowledge-based practices are playing an important role in safeguarding this invaluable natural heritage in the landscape (Fig. 12.1).

12.3 Methodology

The purposive sampling method is used in this research work for selecting sources and respondents following the standard methodology by Kunwar et al. (2020), Castagnetti et al. (2021), and Srichandan et al. (2021). Through the purposive goal-directed sampling method, the first set of respondents for the interview was chosen based on prior experience in the region. A total of ten interviews were conducted with semi-structured, open-ended questionnaires that covered the historical background of the place, significant events that influenced the shaping of history, Bhotias and their worldview along with recent horizons, the process of seasonal migration, past trade activities, the influence of Mount Kailash and the religious beliefs pertaining to it, traditional knowledge of the region, and the adopted conservation and restoration methods for the resource management.

Fig. 12.1 Map of the study site describing the Rung cultural landscape and Darma Valley

As the study progresses, various methods, including interviews, questionnaire surveys, group discussions, case studies, and observations, are considered to enhance the quality of work. Qualitative findings through these methods are analysed as per the aims and objectives of the research. This work is supported by a review of historical sources, published materials, and research accounts to make the findings robust. These methods contribute to important insights into cultural diversity, ethno-ecological knowledge (EEK), and biodiversity conservation in the Darma Valley and ease the data interpretation (Fig. 12.2).

Fig. 12.2 (**a**, **b**) Natural resources and biodiversity of the Darma landscape

12.4 Results

This research recognises the participation of the local community as a focal point for sustainable efforts in ecosystem restoration. Integrating TEK in the restoration of forests with a combination of modern ecological techniques leads to a resilient, robust, and ecologically sustainable future for the valley by contributing to livelihood opportunities, enhancing food security, and providing ecosystem services. In the restoration process, native species play a significant part, as their restoration enhances the region's biodiversity. Still, on the other hand, it also strengthens the ability of ecosystem resistance against the challenges of climate change (Negi et al. 2017). In Darma Valley, native plants are also better adapted to the high-altitude environmental conditions, supporting improved ecosystem resilience. Native species in local restoration practices manage ecological functioning and connection of landscape. This work signifies the crucial role of the community in the forest management of Darma Valley, showcasing ecological and economic benefits for the community. These initiatives support residents' improved livelihoods by generating income sources, enhancing food security, and safeguarding them from climate change vulnerability. In Darma Valley, this integration of traditional and modern methods of restoration develops a holistic model, demonstrating enhanced effectiveness of biodiversity conservation and local resilience through the pairing of contemporary approaches in combination with traditional knowledge, and can help in covering the issues of addressing EbA for different resource management regimes (Bhatt et al. 2024a, b). Moreover, in the instances of market-based rationalisation of food security, a reckonable point of view for resource management due to their experience in the food and commodity exchange can be enhanced with improved connectivity to distant markets, which may bring additional prosperity to individual households, which in turn could improve the overall living condition of local families.

12.4.1 Landscape Diversity and Its Essence in Ecosystem Restoration: An EbA Approach in Darma Valley

It has been reported that there is a push towards constant, transparent, and replicable assessments of ecosystem features, which can be ascertained through diversity assessments. This includes using indices for ecosystem extent, biodiversity, abiotic characteristics, and land uses to monitor ecosystems' situation and support natural assets accounting from local to global levels and reinforcement of ecosystem service valuation through community perspectives (Bhatt et al. 2024a, b). In this aspect, landscape diversity plays a crucial role in ecosystem restoration as it is a more recognised and much-explored part of high-altitude regions due to its ecosystem vulnerability. With its rich eco-cultural significance, the community residing in Darma Valley incorporates various EbA methods for restoring forests, offering examples of biodiversity conservation. This EbA approach is based on integrating restoration with communities' socio-cultural practices. Our study focuses on age-long cultural traditions linked with natural surroundings in Darma Valley as the people in the valleys have been engaged in varied ecosystem management in the form of forest conservation and sustainable land use practices, which are equally important in democratic and socio-economic front (Fig. 12.3). Darma's forests, which are mostly community managed, are located in high-altitude ranges and are important for biodiversity, regulating water and mitigating climate, and have faced adverse effects of changing climatic conditions. This study of resilient restoration practices of Darma Valley also highlights the ways of addressing factors leading to ecosystem degradation, including climate change, deforestation, and unsustainable land use. Led by indigenous communities in the valley, the local afforestation practices provide examples of ecosystem recovery and are majorly adopted, as described in Table 12.1.

Fig. 12.3 Field-based focal-group discussion and questionnaire survey

Table 12.1 Case studies of villages with EbA management practices undertaken by community

Village	Region	Management practices	Sacred and protected status	Key biodiversity conservation/ restoration measures	Cultural and ecological importance
Tedang	Vallashe	Wood utilisation requires permission from the village authority, and a goat sacrifice is required for fulfilment	Restricted, but accessible with permission	Community management with traditional rituals (goat sacrifice)	Ecological and cultural integration, sacred rituals for resource use
Dakar	Betula (Bhoj patra)	Rotation system for resource utilisation, with complete restrictions on Betula forest (Larungdev) utilisation	Sacred and protected by the name of Larungdev (Shyankshay)	Sustainable resource extraction, traditional knowledge integration	Sacred status for Betula trees, balancing traditional and modern practices
Sipu	Mahdev Gufa	Restricted access, dedicated to Mahadev	Sacred, restricted forest	Preservation of sacred forest patch, no exploitation	Cultural and religious significance, conservation through spiritual belief
Dugtu	Sacred Bugyals (Sangsuti bugyal, Phatu bugyal, Gabbe bugyal)	Sacred grazing areas with limited access. Villagers have exclusive rights to Neppa bugyal	Sacred bugyals, community-managed	Controlled grazing and conservation through exclusive community access	Sacred and culturally significant grazing areas, protecting local biodiversity
Dugtu	Ringjangti Tal (Bholenathka Tal)	Sacred lake with conservation practices by the village	Sacred lake	Protection of the lake and its surrounding ecosystem as a sacred space	Cultural significance, ecosystem protection linked with spiritual beliefs

12.4.2 Key Features of Biodiversity Conservation Practices Being Undertaken in Darma Valley

In every village of the valley, traditional knowledge and spiritual practices are blended as biodiversity conservation practices (Fig. 12.4). Sacred sites such as forests, lakes, and bugyals as per cultural norms are protected as sacred sites. Local communities of the region play a significant part in resource management through

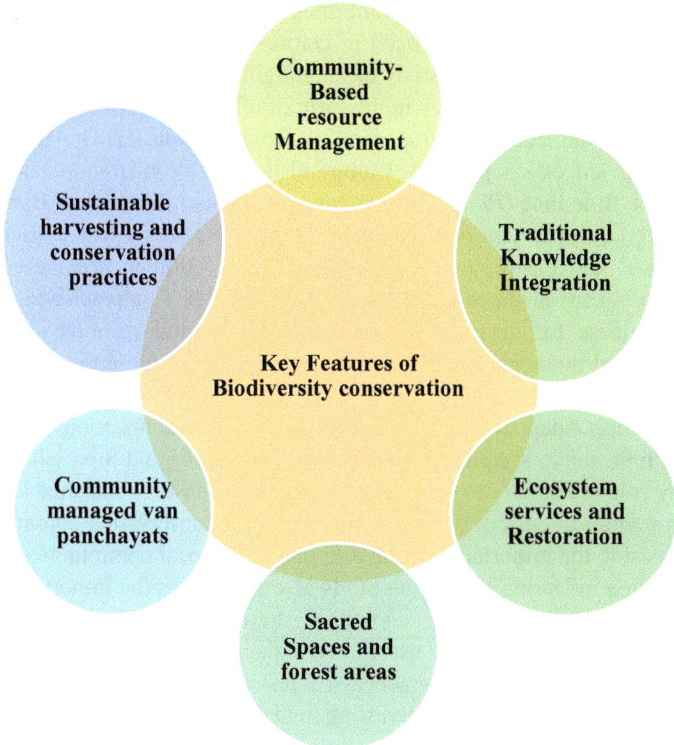

Fig. 12.4 Key features of biodiversity conservation practices in Darma Valley

practices like permission-based extraction of wood (Tedang), rotational harvesting (Dakar), and restricted access to sacred areas (Sipu and Dugtu), ensuring sustainable resource use. In these regions sacred forests, bugyals, and lakes (such as Mahadev Gufa, Velleshe, and Ringjangti Tal) play a significant role in reinforcing conservation efforts as religious and spiritual values strengthen the protection and management of these ecosystems, creating long-term conservation outcomes. This valley also rehabilitated the degraded ecosystem by afforestation and integrating native species.

12.5 Discussion

Darma Valley is an ecologically sensitive zone that lies within the Kailash Sacred Landscape and showcases an integrated model for traditional management practices and contemporary conservation strategies to promote the conservation of biodiversity and ecosystem resilience. Communities' sustainable resource management practices and deep understanding of local ecosystems enable them to effectively adapt against changing climate and anthropogenic pressures. Underscoring the significance of TEK, the EbA approach in Darma Valley addresses the present-day environmental challenges. The study has highlighted the significance of traditional restoration practices of the valley in the form of afforestation of native tree species such as Himalayan Cedar (*Cedrus deodara* Roxb. D.Don (G. Don)), Himalayan Oak (Quercus leucotrichophora A. Camus), Rhododendron (*Rhododendron arboreum* Smith), Blue Pine (*Pinus wallichiana* A. B. Jacks.), Himalayan Birch (*Betula utilis* D. Don), Juniper (*Juniperus communis* Linn.), etc. along with conservation of sacred groves and the promotion of natural regeneration in pristine conditions, while the recent advancement is inclusive of upgraded ICT outreach and simultaneous preparedness for disaster risk reduction for natural resource conservation (Fig. 12.5). Traditional knowledge combined with state-of-the-art conservation tactics enhances capacity to withstand changing climate by developing and reinforcing strong resilience. Adaptation of spiritual beliefs and deep connections with the land, which are reflected in sacred forests, like those in Tedang and Sipu villages, play a significant role in natural resource conservation. This ensures sustainable preservation and provides a sense of belonging through local conservation commitment.

Emphasising the importance of the role played by local communities in ecosystem restoration in Darma Valley, this study also showcases the importance of community participation through community-based activities in afforestation, reforestation, and long-term land use, rooted in the generation of passing Indigenous knowledge as key focal points for successful restoration. The involvement of locals in methods like rotational wood harvesting in Dakar and regulated grazing in Dugtu

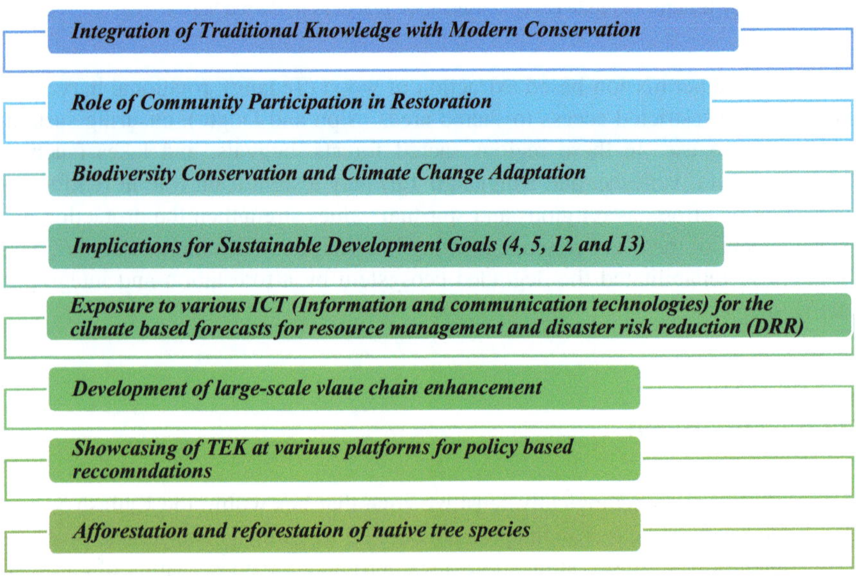

Fig. 12.5 Key focus point of discussion emerged after the focused group discussion with the community

shows that the work has advocated sustainable management practices of sustainable resource utilisation as a factor for enhancing and rehabilitating ecosystems. These steps can contribute to the mutually beneficial relationship between the environment and the community.

Restoration practices and their impact on improving biodiversity in the Darma Valley high-altitude ecosystem are highlighted in this study, where native plant species adapted to challenging climatic scenarios remain focal points for improving biodiversity by developing ecosystem resilience. This restoration effort at the grassroots level supports climate change mitigation in various forms, such as water flow regulation, prevention against soil erosion, and support for carbon sequestration (Bhatt and Jugan 2024). The ecosystem restoration approach which is being adopted by the community of Darma Valley is exemplary, setting a model for high-altitude regions across the globe and showcasing how the combination of TEK with community-driven conservation methods can achieve SDGs, improve biodiversity, and enhance livelihoods. Effective adoption of this model can be utilised and upgraded to resolve environmental issues in the Himalayas and beyond and can cover the uncertainty of the EbA scope in the region by interpreting various socio-economic nexuses envisaging community participation.

12.6 Conclusion

Ecological research on climate change mitigation to biodiversity loss in the valley can be addressed through EbA within this landscape of the Western Himalayas, which is culturally significant and complements the exercise of blending TEK with contemporary conservation strategies. By highlighting the significance of local communities in their community practices, this research emphasises restoration of ecosystems and climate resilience built through generations as in the past, Darma Valley has been famous for its traditional models of sustainable environmental management through espousing ancient traditions with contemporary ecological approaches in high-altitude locations which are vulnerable to climate change (Negi 2007). Based on the present observations, some suggestions which have been documented during the study for the successful EbA in Darma Valley are described below:

- **Stakeholder Engagement**: Engagement of local communities, government agencies, NGOs, and other stakeholders in the assessment process, as their knowledge and participation are crucial for accurate data collection and effective implementation of conservation measures.
- **Use of GIS and Remote Sensing**: Utilise geographical information systems (GIS) and remote sensing technologies to map and monitor ecosystem changes over time. This can help identify areas of concern and track the progress of conservation efforts.
- **Integration of Traditional Knowledge**: Incorporate traditional ecological knowledge from local communities into the assessment process. This can provide

valuable insights into sustainable practices and help design culturally appropriate conservation strategies.

- **Focus on Ecosystem Services**: Identify and prioritise ecosystem services critical for local communities, such as water supply, soil fertility, and biodiversity. Assess how human activities and climate change are impacting these services.
- **Adaptive Management**: Implement adaptive management practices that allow flexibility and adjustments based on monitoring results. This approach ensures that conservation strategies can be modified in response to new information or changing conditions.
- **Capacity Building**: Invest in capacity building for local communities and stakeholders to enhance their ability to participate in and benefit from EbA initiatives. This includes training in data collection, monitoring techniques, and sustainable land management practices.
- **Policy Integration**: Work towards integrating EbA findings into local, regional, and national policies. This can help ensure ecosystem conservation is considered in decision-making and development plans.
- **Involvement of Local Community Perception**: The articulation of the local community's perception of climate change and its adequacy can be crucial for policymakers to better manage plans and implement mitigation and adaptation strategies at regional, national, and global levels, as perception-based studies can substantially complement the observed climate change evidence for remote data-deficient locations.

By following these suggestions, one can develop a robust EbA framework that supports sustainable development and conservation efforts in the Darma Valley and other high-altitude regions; a proposed framework for the possible EbA approach is shown in Fig. 12.6. Moreover, it has been noted that the restoration of the ecosystem

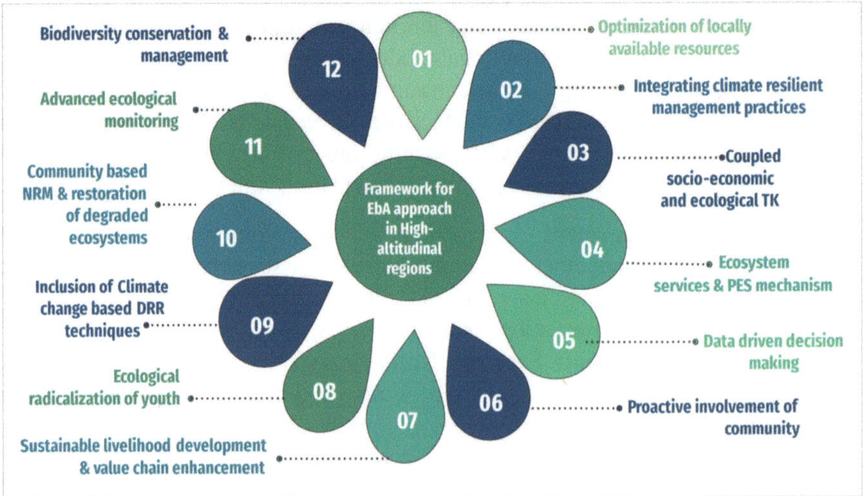

Fig. 12.6 Proposed framework for EbA approach in high-altitude regions

in Darma Valley has been a success with the local knowledge and inspiring spiritual practices, which include the preservation of sacred groves afforestation with native species and regulated extraction of resources. These practices are traditional but work well under high-altitude terrain, whose insurmountable difficulties include climate variability and soil erosion. Sacred sites of the valley, such as forests, grazing meadows (bugyals), and lakes, are important components of the biodiversity conservation strategy for the valley. They are sacred spaces managed with care and reverence, sometimes with stony restrictions on access that help sustain natural regeneration and add their part in restoring ecosystems that are otherwise degraded under anthropogenic pressures. Combining these TEKs with modern approaches to conservation has proved fruitful in restoring the ecological balance, safeguarding biodiversity, and enhancing specific ecosystem services like water regulation and soil conservation.

It can be of conclusive statement that the local, regional and national policies must formally recognize and empower highland communities as essential partners in ecosystem restoration. This requires legally securing their ancestral land tenure, which is the foundation of their stewardship. Funding mechanisms must be established to directly support community-led initiatives that integrate traditional knowledge with scientific tools. Furthermore, policies should mandate the inclusion of highlanders in all levels of environmental planning and decision-making. This rights-based approach ensures sustainable, culturally respectful, and cost-effective conservation, securing vital ecosystem services for both highland and downstream populations. Also, this study brings the importance of community participation in ecosystem restoration to the fore. Their active involvement by local peoples guarantees that the efforts towards restoration are ecologically sound, but culturally as well. Ecologically and socio-economically sustainable activities such as rotational wood harvesting, controlled grazing and socio-cultural gatherings, and folklore-based resource utilisation are indeed evidence of the profound realisation of communities on the environment and their intuitive desire for ecological balance. Such practices, therefore, promote ownership and responsibility towards the conservation endeavour, ensuring longevity in its course. Although traditional practices in the valley have been seen to bear fruits, great potential is available to augment this by using modern science. Advanced ecological monitoring techniques, climate modelling, and studies on the genetic resistance of native plants could even optimise restoration projects. Together with modern science, indigenous knowledge makes it possible to articulate and employ a broader adaptive approach for ecosystem rehabilitation to improve this valley's resilience to climate change and other environmental stressors and pave a well-trodden way for the ecosystem-based restoration, adaptation, and sustainable management approaches.

Acknowledgement The authors wish to dedicate this chapter to the encouraging memory of the late Dr. R.S. Rawal, former Director of GB Pant National Institute of Himalayan Environment (GBP-NIHE), for his immense support. They would also like to thank Dr. Shekhar Pathak and Darma Valley's people for their valuable input.

Note Photo credits are duly acknowledged for Fig. 12.2 a,b for which the copyright resides with Dr. Vikram .S. Negi (Assistant Professor, Department of Botany, HNBGU, Pauri-Garhwal, Uttarakhand). The remaining images are sourced by and copyrighted by the first author (RR).

Conflict of Interest The authors declare that the research was conducted without any commercial or financial relationships that could be construed as a potential conflict of interest amongst the authors.

References

Adger WN, Huges TP, Folke C, Carpenter SR, Rockström J (2005) Social-ecological resilience to coastal disasters. Science 309:1036–1039. https://doi.org/10.1126/science.1112122

Bergmann C (2016) The Himalayan border region: trade, identity and mobility in Kumaon, India. In: Advances in Asian human-environment research. Springer, Cham

Bergmann C, Gerwin M, Nüsser M, Sax WS (2008) Living in a high mountain border region. The case of the 'Bhotiyas' of the Indo-Chinese border region. J Mt Sci 5(3):209–217

Bergmann C, Gerwin M, Sax WS, Nüsser M (2011) Politics of scale in a high mountain border region: being mobile among the Bhotiyas of the Kumaon Himalaya, India. Nomad People 15:104–129. https://doi.org/10.3167/np.2011.150207

Berkes F (2007) Community-based conservation in a globalized world. Proc Natl Acad Sci 104(39):15188–15193

Berkes F, Colding J, Folke C (2000) Rediscovery of traditional ecological knowledge as adaptive management. Ecol Appl 10(5):1251–1262. https://doi.org/10.1890/1051-0761(2000)010[1251:ROTEKA]2.0.CO;2

Bhatt H, Jugran HP (2024) Community-managed forests and their effectiveness in SDG implications in the Western Himalayan Region. In: Warming mountains: implications for livelihood and sustainability. Cham, Springer, pp 435–458

Bhatt H, Jugran HP, Pandey R (2024a) Cultural ecosystem services nexus with Socio-Cultural attributes and traditional ecological knowledge for managing community forests of Indian western Himalaya. Ecol Indic 166:112379

Bhatt H, Verma D, Jugran HP, Kumar V, Pandey R (2024b) Community perception about ecosystem services assessment from forests managed under different regimes in Hindu Kush Himalaya. Environ Sustain Indic:100463

Black-Lock T, Rodehorst D (2023) Traditional knowledge and practices for resilience to climate change in Nepal's Mid-Hills: perspectives from Darma and Madi rural municipalities. Division of Risk Management and Societal Safety, Faculty of Engineering Lund University, Lund

Castagnetti F, Bhatta J, Greene A (2021) An offering of grain: the agricultural and spiritual cycle of a food system in the Kailash Sacred Landscape, Darchula, Far Western Nepal. Front Sustain Food Syst 5:646719

Chopra R (2014) Uttarakhand: development and ecological sustainability

Colding J, Elmqvist T, Olsson P (2003) Living with disturbance: building resilience in social-ecological systems. In: Berkes F, Colding J, Folke C (eds) Navigating social-ecological systems: building resilience for complexity and change. Cambridge University Press, Cambridge, pp 163–186. https://doi.org/10.1017/CBO9780511541957.011

Dimri AP, Allen S, Huggel C, Mal S, Ballesteros-Canovas JA, Rohrer M, Shukla A (2021) Climate change, cryosphere and impacts in the Indian Himalayan Region. Curr Sci 120(5):774–790

Eakin H (2005) Institutional change, climate risk, and rural vulnerability: cases from central Mexico. World Dev 33(11):1923–1938. https://doi.org/10.1016/j.worlddev.2005.06.005

Ehlers E, Kreutzmann H (2000) High mountain ecology and economy: potential and constraints. In: Ehlers E, Kreutzmann H (eds) High mountain pastoralism in Northern Pakistan, vol 132. Erdkundliches Wissen, pp 9–36

Folke C, Colding J, Berkes F (2003) Synthesis: building resilience and adaptive capacity in social-ecological systems. In: Berkes F, Colding J, Folke C (eds) Navigating social-ecological systems: building resilience for complexity and change. Cambridge University Press, Cambridge, pp 352–387. https://doi.org/10.1017/CBO9780511541957.020

Gadgil M, Rao PRS, Utkarsh G, Pramod P, Chhatre A (2000) New meanings for old knowledge: the people's biodiversity registers program. Ecol Appl 10(5):1307–1317

Holt-Gimenez E (2002) Measuring farmers' agroecological resistance after Hurricane Mitch in Nicaragua: a case study in participatory, sustainable land management impact monitoring. Agric Ecosyst Environ 93:87–105. https://doi.org/10.1016/S0167-8809(02)00006-3

Hoon V (1996) Bhotiyas of the Kumaun Himalayas. Living on the move series. Sage, New Delhi

Janssen MA, Anderies JM, Ostrom E (2007) Robustness of social-ecological systems to spatial and temporal variability. Soc Nat Resour 20:307–322. https://doi.org/10.1080/08941920601161320

Jugran HP, Bhatt H, Bisht DS, Bisht D (2024) Impact of changing climate pattern on forest resources and socio-economic dynamics of a marginalised village in Kumaon Himalayan region. In: Himalayan ecology: special focus on millets. EIACP- GBPNIHE, pp 63–66. ISBN 978-93-5967-660-9

Kunwar RM, Evans A, Mainali J, Ansari AS, Rimal B, Bussmann RW (2020) Change in forest and vegetation cover influencing distribution and uses of plants in the Kailash Sacred Landscape, Nepal. Environ Dev Sustain 22(2):1397–1412

Lebel L, Anderies JM, Campbell B, Folke C, Hatfield-Dodds S, Hughes TP, Wilson J (2006) Governance and the capacity to manage resilience in regional social-ecological systems. Ecol Soc 11(1):19. [online] http://www.ecologyandsociety.org/vol11/iss1/art19/

Naess LO (2012) The role of local knowledge in adaptation to climate change. WIRE's Clim Change 4(2):99–106. https://doi.org/10.1002/wcc.204

Negi CS (2007) Changing face of polyculture in the Darma and Johaar valleys, Pithoragarh, Kumaun Himalayas. Int J Sustain Dev World Ecol 14(4):428–436

Negi CS (2009) Habitat ecology, biochemical analysis and pharmacological tests of crude extracts of Yar Tsa Gumba (cordyceps sinensis Berk.). The Botanica 57:71–79

Negi VS, Maikhuri RK, Pharswan D, Thakur S, Dhyani PP (2017) Climate change impact in the Western Himalaya: people's perception and adaptive strategies. J Mt Sci 14(2):403–416

Negi VS, Thakur S, Dhyani R, Bhatt ID, Rawal RS (2021) Climate change observations of indigenous communities in the Indian Himalaya. Weather Clim Soc 13(2):245–257

Negi VS, Tiwari DC, Singh L, Thakur S, Bhatt ID (2022) Review and synthesis of climate change studies in the Himalayan region. Environ Dev Sustain 24(9):10471–10502

Nüsser M (2006) Ressourcennutzung und nachhaltige Entwicklungim Kumaon-Himalaya. Geogr Rundsch 58:14–22

Pant SD (1935) A social economy of the Himalayans. George Allen and Unwin, London

Rawal RS, Rastogi A, Palni LMS (2012) Journey through a sacred land. GB Pant Institute of Himalayan Environment and Development, Almora

Rawat D, Schickhoff U (2022) Changing climate scenario in high altitude regions: comparison of observed trends and perceptions of agro-pastoralists in Darma Valley, Uttarakhand, India. In: Mountain landscapes in transition: effects of land use and climate change. Springer, pp 429–447

Ruiz-Mallén I, Corbera E (2013) Community-based conservation and traditional ecological knowledge: implications for social-ecological resilience. Ecol Soc 18(4)

Srichandan S, Pasupuleti RS, Mishra AJ (2021) Environ Challenges. https://doi.org/10.1016/j.envc.2021.100291

Toledo VM, Ortiz B, Cortés L, Moguel P, Ordóñez MJ (2003) The multiple use of tropical forests by indigenous peoples in México: a case of adaptive management. Conserv Ecol 7:9

Tompkins EL, Adger WN (2004) Does adaptive management of natural resources enhance resilience to climate change? Ecol Soc 9(2):10. [online] http://www.ecologyandsociety.org/vol9/iss2/art10/

Part III
Ecosystem-Based Approaches and Resilience Building

Chapter 13
Ecosystem-Based Adaptation: Concepts, Principles, and Case Studies from Mountainous and Hilly Ecosystems of the World

Anandan Das, Y. D. Imran Khan, Reneema Hazarika, Mrinalini Goswami, Sunil Nautiyal, Niranjan Roy, Debojyoti Chakraborty, and Shubhadeep Roychoudhury

Abstract Ecosystem-based adaptations (EbA) are nature-based solutions that include a broad array of ventures for the management of ecosystems, such as environmental protection, restoration of ecosystems, and activities to lower the vulnerability and improve the resilience of communities under a changing climate. In fact,

A. Das
Department of Zoology, Pandit Deendayal Upadhyaya Adarsha Mahavidyalaya, Eraligool, India

Y. D. I. Khan
WOTR Centre for Resilience Studies, Watershed Organisation Trust, Pune, India

R. Hazarika
Institute of Silviculture, University of Natural Resources and Life Sciences (BOKU), Vienna, Austria

M. Goswami
Researcher in Sustainability Science, Bengaluru, Karnataka, India

S. Nautiyal
Centre for Ecological Economics and Natural Resources, Institute for Social and Economic Change, Bengaluru, Karnataka, India

G. B. Pant National Institute of Himalayan Environment, Kosi-Katarmal, Almora, Uttarakhand, India

N. Roy
Department of Economics, Assam University, Silchar, India

D. Chakraborty
Department of Forest Growth, Silviculture and Genetics, Austrian Research Centre for Forests, Vienna, Austria
e-mail: debojyoti.chakraborty@bfw.gv.at

S. Roychoudhury (✉)
Department of Life Science and Bioinformatics, Assam University, Silchar, India

Y. D. I. Khan et al. (eds.), *Ecosystem-based Approaches for Resilience Building in Himalayan Landscapes*, Disaster Resilience and Green Growth,
https://doi.org/10.1007/978-981-95-2007-7_13

adaptation measures can tackle the harmful aspects of climatic variations as well as assist vulnerable communities in preparing to face climatic changes in the long run. The concepts of EbA and resilience are inherently linked as they both address the capacity of ecosystems and communities to absorb disturbances and still be able to bounce back. EbA and resilience depend on thresholds, which are boundaries around a system state. If crossed, these thresholds represent a transition to a new system with a new set of components and relationships. The system's response will also depend on its adaptive capacity (i.e. the capacity to learn, cope, innovate, and adapt), which in turn is determined by the amount and diversity of social, economic, physical, and natural capital, as well as social networks, institutions, and entitlements to the distribution and use of such capital. The implementation of EbA can be envisaged independently or combined with engineered measures (e.g. the raising of dykes of reservoirs of water), blended approaches (e.g. artificial reefs), and measures that build up the capability of communities as well as institutions to adapt to climatic risks (e.g. the introduction of early warning systems). Therefore, effective implementation of EbA should include transformative action involving both climatic and non-climatic drivers of change. In this chapter, we define the concepts and principles of EbA and evaluate its application, using lessons learned from mountainous and hilly ecosystems worldwide.

Keywords Climate change · Resilience · Communities · Mountain · Hilly landscape

Abbreviations

ALPARC	Alpine Network of Protected Areas
BES	Biodiversity and ecosystem services
CO_2	Carbon dioxide
EbA	Ecosystem-based adaptation
EU	European Union
GIS	Global information system
PES	Payment for ecosystem services
RS	Remote sensing
TEK	Traditional ecological knowledge
UNFCCC	United Nations Framework Convention on Climate Change
UV	Ultraviolet

13.1 Introduction

Shifting rainfall patterns and increased frequency of extreme weather events due to climate change have threatened global concerns for decades. As a result, ecosystems are brutally damaged, resulting in forced migration of species, alteration of life processes, and reshaping the structure and functionality of ecosystems. Coral reefs are undergoing extensive bleaching as a result of rising sea temperatures, which is causing a decline in biodiversity and the disruption of marine ecosystems (Hoegh-Guldberg et al. 2007). Terrestrial ecosystems are similarly impacted, seeing increased occurrences of forest diebacks and alterations in vegetation zones (Chen et al. 2024). Subsequently, the subtle equilibrium of ecosystems is altered, leading to the disruption of vital ecosystem services and a reduction in carbon sequestering capacities (Reid et al. 2005).

Ecosystem services have been essential mitigators of climate change. Various ecosystem services provide natural solutions that increase resilience and reduce vulnerability to environmental challenges. Forests serve as important carbon sinks, meaning they absorb carbon dioxide (CO_2) from the atmosphere through the process of photosynthesis (Pan et al. 2011). Wetlands have a role in carbon sequestration and offer other advantages, including flood control and water purification (Mitsch et al. 2013). Mangroves and salt marshes in coastal regions have the dual purpose of safeguarding shorelines against erosion and storm surges and capturing sediments rich in carbon (Duarte et al. 2013). Mangroves and salt marshes protect shorelines against erosion and capture carbon-rich sediments (Gowdy et al. 2010). Moreover, ecosystem services help communities adapt to climate change by providing resources and means of livelihood, thus diminishing their susceptibility (Adger et al. 2005).

Ecosystem-based adaptation (EbA) is a sustainable strategy to minimise the impact of climate change by utilising the natural abilities of ecosystems to protect against climate change's impacts and aid human adaptation. EbA is 'the use of biodiversity and ecosystem services as part of an overall adaptation strategy to help people adapt to the adverse effects of climate change' (Convention on Biological Diversity 2007). The main goal of EbA is to decrease the vulnerability and increase the resilience of communities (Munang et al. 2013a). EbA addresses climate and non-climatic drivers of change by utilising ecosystem services, offering a holistic approach to sustainable development. Restoring degraded landscapes enhances water infiltration and diminishes soil erosion, thereby boosting agricultural productivity and securing the food supply (Mills et al. 2020).

EbA integrates engineered solutions with community-driven initiatives to make the ecosystem more resilient against evolving environmental conditions. This may be achieved by constructing artificial reefs to protect coastal areas and foster marine biodiversity, as well as developing early warning systems to alert communities of impending climate-related threats (Doswald et al. 2014). The success of EbA depends on the adaptive abilities of both ecosystems and human communities, which are influenced by the availability and diversity of social, economic, physical, and natural resources, alongside robust social networks and institutions (Folke et al. 2010).

This evidence-based study aims to provide a comprehensive outline of the concepts and principles of EbA, reviewing its ability to mitigate climate change and support community sustenance. Moreover, it delineates the characteristics of mountainous and hilly ecosystems and investigates various such ecosystems around the globe.

13.2 Ecosystem-Based Adaptation

13.2.1 Definition

The United Nations Environmental Programme defines EbA as 'Ecosystem-based adaptation is a strategy for adapting to climate change that harnesses nature-based solutions and ecosystem services' (Environment 2021). It essentially includes a wide array of efforts that can utilise the ecosystem's implicit capability to counter climate effects and may eventually prove to be decisive to protect communities. This approach encompasses the preservation, responsible administration, and rehabilitation of ecosystems in order to offer services that decrease susceptibility and improve adaptability (Munang et al. 2013a). EbA practices encompass activities such as reforestation, wetland restoration, and sustainable agriculture. These methods serve the dual purpose of preserving or reinstating ecological services and offering socio-economic advantages to communities (Paracchini et al. 2015). In addition, EbA combines the adaptive management of ecosystems with social and institutional strategies to address climate risks completely (Doswald et al. 2014).

13.2.2 Characteristics

EbA is notable for its ability to address climate-related threats by conserving, sustainably managing, and restoring ecosystems. It utilises ecosystems' complex characteristics to deliver a wide range of services that bolster resilience and decrease vulnerability. This includes providing ecosystem services such as carbon sequestration, flood control, and shoreline stability by entailing natural elements such as forests, wetlands, and mangroves (Munroe et al. 2012). EbA methods are highly important in preserving the health and variety of ecosystems (Scarano 2017).

Community engagement is another important aspect of EbA, which focuses on integrating indigenous knowledge in designing and executing adaptation strategies. This method ensures that adaptation measures are culturally appropriate, hence improving their efficacy and long-term viability (Reid 2016). EbA incorporates adaptive management principles, which include monitoring, learning, and contextual modification of approaches (Italian Development Corporation).

In addition, EbA provides other advantages beyond climate adaptation, such as enhanced preservation of biodiversity, improved water quality, and increased prospects for livelihood. It is a cost-effective strategy to simultaneously address several environmental and socio-economic concerns (Seddon et al. 2020).

13.2.3 Importance

EbA addresses climate change challenges through sustainable and cost-effective strategies. Key practices, including the restoration of wetlands and forest conservation, are vital to EbA. They offer effective means to lessen the vulnerability of communities to severe weather events and safeguard them against floods, storms, and droughts. This buffering effect helps attenuate such occurrences' impacts (Munang et al. 2013a). EbA has a crucial role in preserving and renewing ecosystem services, hence promoting biodiversity, which is vital for ecosystems' overall well-being and operation (Scarano 2017). In addition, EbA is an essential element of sustainable development that helps enhance food security, livelihoods, and human well-being (Doswald et al. 2014). Moreover, EbA assures long-term resilience by making provisions for adaptable and responsive management solutions to address climate change's uncertainties and transient nature (CARE Climate Change 2014). In a nutshell, EbA is an essential and comprehensive method that tackles the ecological and socio-economic aspects of climate adaptation, by offering holistic solutions to global environmental concerns (Seddon et al. 2020).

13.3 Concepts and Principles of Ecosystem-Based Adaptation

13.3.1 Overview of Ecosystem-Based Adaptation Concepts

EbA is a nature-centric, versatile, cost-efficient approach with broad applicability that has gained importance in worldwide climate policy (Jones et al. 2012; Cousiño and Penha-Lopes 2021). It can effectively deal with the uncertainty of climate change, along with having the capacity to provide diverse economic, social, and environmental benefits (Jones et al. 2012; Munang et al. 2013b). The efficacy of ecosystem-based approaches focused on recognising indigenous and traditional knowledge while also considering the requirements of marginalised community members, such as women, older people, and the impoverished (Girot et al. 2012).

EbA was first introduced into the global political arena through the United Nations Framework Convention on Climate Change (UNFCCC) in 2008. Since then, EbA approaches have gained importance in the global climate policy framework. A clear demonstration can be found in the European Union's (EU) climate adaptation strategies, which explicitly advocate for the implementation of these approaches (European Commission 2023).

Nature-based solutions are economically efficient strategies that derive inspiration and assistance from the natural world. They provide concurrent advantages to the environment, society, and economy while strengthening resilience. These solutions integrate elements of nature, such as natural features and processes, into urban areas, landscapes, and seascapes by implementing locally customised, resource-efficient, and systematic interventions (Scarano 2017). One such strategy is an adaptive transition, designed by integrating technological, ecological, economic, livelihood, and lifestyle shifts, thereby achieving sustainable change and boosting local adaptive capabilities (Pant et al. 2015). EbA typically integrates policy instruments and tools pertaining to biodiversity and ecosystem services (BES) and policies concerning socio-economic and development aspects. EbA is essential for promoting sustainability transitions, especially in developing nations prioritising global biodiversity conservation and maintaining healthy ecosystems (Scarano 2017).

13.3.2 Core Principles of Ecosystem-Based Adaptation

13.3.2.1 Ecosystem Services and Biodiversity Conservation

Ecosystem services have a fundamental interconnection with biodiversity conservation. These services are essential for ensuring environmental sustainability, which may be achieved by encompassing four primary categories: provisioning, regulating, supporting, and cultural. Provisioning services involve the constant supply of food and water; regulating services include facilities like climate regulation and disease control; supporting services include nutrient cycling and soil formation; and cultural services encompass recreational and spiritual benefits (Millennium Ecosystem Assessment (Program) 2005). Therefore, to guarantee the relentless provision of essential natural services, the preservation of biodiversity gains the utmost urgency (Cardinale et al. 2012). The intricateness of the correlation between biodiversity and ecosystem services is further established by the fact that biodiversity is crucial in providing stability to ecosystems and increasing their productivity (Tilman et al. 2012).

Studies have shown that biodiversity enrichment frequently improves ecosystem functioning, which, in turn, results in better delivery of ecosystem services (Hooper et al. 2012). For instance, the productivity and efficiency of varied plant communities in utilising nutrients have been demonstrated, which has been shown to upregulate primary production and further augment the recycling of nutrients (Isbell et al. 2011). Diverse life forms can help ecosystems withstand and recover from environmental changes and disruptions (Elmqvist et al. 2003).

To regain ecosystem resilience, efforts like creating protected areas, rehabilitation of damaged ecosystems, and encouragement of sustainable land-use practices may be considered (Foley et al. 2005). Moreover, wider policy frameworks and land management practices must accommodate biodiversity protection to preserve ecosystem services and promote human well-being (Naidoo et al. 2008).

13.3.2.2 Community Involvement and Participatory Approaches

Community engagement and participatory methodologies are essential elements of EbA. These approaches amalgamate indigenous knowledge and improve the effectiveness and durability of adaptation solutions. For the sake of designing, executing, and evaluating adaptation solutions to tackle climate change, active involvement of local stakeholders, such as community members, local governments, and non-governmental organisations, holds immense importance in achieving the goals of EbA along with ensuring effective tackling of the unique vulnerabilities of local populations (Reed et al. 2009). Active community participation also helps promote a sense of responsibility and care for natural resources, paramount in achieving EbA goals (Vignola et al. 2015).

Integrating traditional ecological knowledge (TEK) with scientific data improves the strength of EbA techniques, resulting in more efficient and culturally suitable adaptation measures (Berkes et al. 2000). In addition, participatory approaches have the potential to enhance the societal acceptability of adaptation strategies, mitigate tensions, and foster trust among stakeholders (Stringer et al. 2006). Active community participation in decision-making processes can enhance the long-term sustainability of EbA projects (Moser and Ekstrom 2010).

Transparent and inclusive governance mechanisms must be formulated to achieve effective community involvement. In this regard, promoting discourse, capacity enhancement, and fair distribution of benefits must be ensured (Ensor et al. 2019). Acknowledging and rectifying power disparities are crucial, guaranteeing the active inclusion of marginalised people in the process (Armitage et al. 2011). Participatory approaches in EbA not only improve ecological resilience but also promote social fairness and community empowerment, hence strengthening the ability of socio-ecological systems to adapt.

13.3.2.3 Sustainability and Resilience Building

Sustainability and resilience in EbA involve purposefully using biodiversity and ecosystem services to support communities in adapting to the harmful effects of climate change. This approach involves the management of ecosystems to enhance their efficiency, hence promoting human adaptability (Seddon et al. 2016). The EbA strategy reduces vulnerability to climate-related threats, enhances adaptive capacity, and provides several benefits, such as biodiversity conservation and improved livelihoods (Munang et al. 2013a). Restoring mangroves defends coastal communities against storm surges and the increasing levels of the sea. In addition, it can potentially enhance fisheries and sequester carbon (Temmerman et al. 2013). EbA strategies strive to maintain the long-term viability of ecosystem services, thus safeguarding the capacity of natural systems to meet human requirements without experiencing deterioration (Jones et al. 2018). Furthermore, these strategies increase the ability of ecosystems to absorb disturbances, recover from damages, and adapt to changing conditions, thereby enhancing their resilience (Chong 2014). By formulating a balanced concoction of indigenous knowledge and scientific research in EbA methods, interventions can be customised to address specific circumstances and to ensure their local applicability and efficacy (Nalau et al. 2015). However, to complete elimination of barriers to incorporate EbA into policy and practice, efforts such as creating stronger institutional structures, improving efforts to develop skills and knowledge, and the active participation of individuals and groups must be ensured (Doswald et al. 2014).

13.4 Characteristics of Mountainous and Hilly Ecosystems

13.4.1 Unique Features

Mountainous and hilly habitats are differentiated by a blend of distinctive features and traits that enhance their ecological intricacy and biodiversity. These ecosystems exhibit altitudinal zonation, meaning that the floral and faunal populations change as the elevation increases or decreases. This is primarily due to variations in temperature, humidity, and soil composition (Körner 2007). The variation in climatic conditions from the base to the summit results in the formation of unique ecological niches, which in turn support a significant number of species that are exclusive to the area and a high level of biodiversity (Rahbek 1995).

Due to differences in aspect, slope, and exposure, a wide range of microclimatic conditions can be frequently experienced over small distances in mountain locations (Barry 2008). These climatic variations bring about distinctions in ecological niches, thereby enabling the coexistence of species with diverse ecological needs (Grytnes and Vetaas 2002). Moreover, mountainous areas are characterised by the presence of distinct geographical features. These attributes impact the water cycle, creating distinct wetland environments such as alpine bogs and fens (Viviroli et al. 2007).

Diverse topographical conditions, such as steep slopes, act as natural obstacles to the free movement of species, resulting in genetic isolation and speciation (Bertuzzo et al. 2016). Glacial and deglaciation cycles have historically had impactful consequences on species distribution and community composition, which has worsened this isolation pattern (Hewitt 2004). Mountain ecosystems play a crucial role in controlling regional temperatures and acting as water sources for downstream ecosystems and human populations (Viviroli et al. 2003).

The biotic interactions in mountain ecosystems exhibit a high degree of specialisation due to challenging environmental conditions, including restricted growing seasons, intense ultraviolet (UV) radiation, and reduced oxygen levels at higher elevations (Körner 2012). These conditions favour the development of distinct adaptations in plants and animals, such as the ability of plants to withstand cold and the ability of animals to acclimatise to high altitudes (Körner 2012). Altitudinal gradients, microclimatic variation, and distinctive biotic interactions further influence the already complex ecology of mountainous and hilly regions. These characteristics give rise to high biodiversity, endemism, and specialised adaptations. This signifies the importance of these ecosystems in terms of ecological and climatic regulations (Viviroli et al. 2003; Körner 2007).

13.4.2 Importance for Biodiversity and Livelihood

Mountainous and hilly ecosystems are crucial in preserving biodiversity and supporting human livelihoods by acting as repositories of ecological, genetic, and cultural variation. Around 25% of the Earth's land area, these ecosystems contain a substantial amount of the world's biodiversity. These regions are pivotal in sustaining about one-third of mostly native terrestrial species adapted to their distinct habitats (Körner et al. 2005). Intricate terrains, variable microclimates, and secluded habitats characterise mountain ecosystems. These features are essential in fostering biological niches that sustain various plant and animal species (Körner and Ohsawa 2005).

Furthermore, mountain habitats offer vital ecological services. They help sustain human lifestyles by functioning as reservoirs providing freshwater to approximately 50% of the world's population (Viviroli et al. 2007). These areas govern the processes of water movement, impact the weather patterns, and help reduce the occurrence of natural disasters like floods and landslides (Dhyani and Thummarukuddy 2016). Moreover, mountainous ecosystems maintain agricultural diversity, especially in growing traditional crops and raising livestock. These ecosystems counter the challenges of ever-changing environmental conditions, which, in turn, helps ensure food security and support local economies (FAO 2022).

Their cultural significance further uplifts the prominence of mountainous regions. Indigenous and local populations living in these places have distinct knowledge systems and practices that help protect biodiversity and promote sustainable land use (Berkes 2012). The value of TEK for global conservation efforts is being increasingly acknowledged (Gadgil et al. 1993). Nevertheless, mountainous ecosystems are highly threatened due to their increased susceptibility to the impacts of climate change and excessive exploitation. Higher occurrence of extreme weather events, including rising temperatures and changes in precipitation patterns, poses a tremendous risk in sustaining the stability and functioning of mountain ecosystems (Pepin et al. 2015). Agricultural operations, deforestation, and infrastructure development worsen the already delicate surroundings (Meybeck et al. 2001).

Ultimately, conserving mountainous and hilly habitats is crucial to uphold worldwide biodiversity and sustain human livelihoods. Based on scientific research and traditional knowledge, sustainable management and conservation techniques must be employed with a sense of urgency to protect these indispensable ecosystems from increasing environmental threats (Grêt-Regamey et al. 2012). The inherent and practical worth of mountainous regions must be acknowledged along with promoting ecological resilience and socio-economic well-being during the formulation of policies by stakeholders.

13.5 Case Studies

13.5.1 Case Study I: The Himalayas

13.5.1.1 Description of the Ecosystem

The Himalayan mountainous region falls into the countries of India, Bhutan, Nepal, China, and Pakistan and houses the world's highest mountain peaks. This region has significant altitudinal variation and intricate topographical features, representing its distinct and diversified ecosystems (Singh). The transition of vegetation pattern is evident; the lower montane forests, which are mainly made up of broadleaved species like *Quercus* and *Rhododendron*, gradually change into coniferous forests having species such as *Pinus*, *Abies*, and *Picea* as the elevation increases (Körner 2003a).

At elevations above the tree line, the vegetation consists of resilient grasses, herbaceous plants, and dwarf shrubs well-adapted to challenging climatic circumstances (Gairola et al. 2008). Temperature, precipitation, and soil types are the main driving forces causing this zonation. These variations are leading factors behind the occurrence of endemic species (Bhattarai et al. 2004). To put it in perspective, the Himalayan blue poppy (*Meconopsis betonicifolia*) and the critically endangered snow leopard (*Panthera uncia*) are endemic to the Himalayan region because they have evolved to thrive in specialised ecological niches (Fox et al. 1992).

The hydrology of the Himalayas is crucial in determining the structure and function of its ecosystems. The region serves as the origin of numerous significant rivers, such as the Ganges, Brahmaputra, and Indus. These rivers supply water to millions of people living downstream (Immerzeel et al. 2010). The combination of glacial meltwater and monsoonal precipitation creates a variety of aquatic habitats and wetlands. These areas are essential for breeding several aquatic species and migrating birds (Xu et al. 2009).

Climate change presents a substantial risk to the complex interplay of biotic and abiotic components of the Himalayan ecosystem (Shrestha et al. 2012). Moreover, human activities such as deforestation, overgrazing, and infrastructure development worsen the susceptibility of these ecosystems (Pandit et al. 2007).

Preserving the Himalayas is essential for upholding ecological equilibrium and ensuring the continuation of ecosystem services that support regional biodiversity and human welfare. To develop sustainable management plans, it is essential to utilise integrated techniques that blend TEK with current scientific methods (Nautiyal et al. 2015). Furthermore, it is crucial to have cooperation among the countries that share the Himalayan terrain to effectively tackle the concerns of conservation and climate resilience (Bawa et al. 2010).

13.5.1.2 Specific Ecosystem-Based Adaptation Interventions

EbA in the Himalayan high region has emerged as a viable approach to enhance resilience against the impacts of climate change. At the same time, it has been crucial in preserving biodiversity and ecosystem services. An example of a crucial EbA strategy is sustainably restoring and effectively managing forests. Due to the inherent potential of trees functioning as natural barriers against landslides and soil erosion, this method can make EbA projects more effective when implemented. Afforestation and reforestation efforts in the forests of Nepal have resulted in the stabilisation of slopes and a decrease in downstream sedimentation. These initiatives have enhanced water quality and fewer floods (Sharma et al. 2021). Agroforestry is an essential method for protecting the goals of EbA, as this technique has enhanced soil fertility and agricultural productivity (Kumar et al. 2018).

Conservation and sustainable management of grasslands and rangelands are also significant EbA measures. Grasslands possess tremendous water retention capability, which can counter and reverse the impact of droughts. This approach was highlighted in a study conducted in the Indian Himalayas, in which researchers argued that traditional grassland management practices hold immense importance in maintaining livestock productivity and biodiversity. A study in the Indian Himalayas has underscored the significance of traditional grassland management in sustaining livestock productivity and biodiversity. Furthermore, the adoption of community-driven water management systems, including the revival of age-old water harvesting structures such as the 'Kuhl' and 'Zabo' in Himachal Pradesh, has bolstered water security and lessened the susceptibility of rural populations to water shortages (Thakur et al. 2022).

It is well-known that wetlands serve as natural water reservoirs and play a significant role in flood control and replenishing groundwater. Naturally, wetland protection is considered a crucial policy for EbA. A case study undertaken in Sikkim demonstrates that protecting and restoring high-altitude wetlands can improve the availability of local water supply and help conserve biodiversity (Choudhary et al. 2011). The importance of undertaking these measures gains further escalation considering the glacial melting and changing precipitation patterns in the Himalayas.

Furthermore, endeavours are underway to promote sustainable tourism practices to establish a harmonious equilibrium between economic advancement and ecological preservation. Promoting environmental awareness and generating stable income for the local communities is of utmost significance in fulfilling the desire to implement EbA approaches. Ecotourism is one such initiative that can play a vital role in achieving this initiative. Ecotourism practices undertaken in the Himalayan regions of Bhutan and Uttarakhand have provided positive results, generating environmental awareness and employment and consequently reducing the burden placed on natural resources (Mishra et al. 2018).

Watershed management systems have been another promising proposition in addressing the complex hydrological challenges in the Himalayan region. The Garhwal Himalayas have effectively executed watershed management initiatives, combining soil and water conservation, afforestation, and sustainable farming practices. This has enhanced water availability and decreased soil erosion (Tiwari and Joshi 2011).

13.5.1.3 Outcomes and Conclusions

Due to its unique socio-economic and ecological conditions, customised EbA solutions are required in the Himalayan region. Case studies demonstrate that the resilience of both Himalayan ecosystems and communities against climate change can be significantly enhanced by combining forest restoration, grassland management, wetland conservation, sustainable tourism, and integrated watershed management. Ongoing research and adaptive management should be emphasised to preserve the efficacy of these strategies (Kumar et al. 2021).

Integrating scientific research with this indigenous wisdom is crucial for the sustained effectiveness of EbA strategies. Therefore, it is vital to encourage cooperation between scientists, policymakers, and indigenous communities to guarantee the effective execution of EbA projects in the Himalayan region (Berkes 2012).

13.5.2 Case Study II: The Alps

13.5.2.1 Description of the Ecosystem

The Alps span roughly 1200 kilometres through eight European countries: Monaco, France, Switzerland, Italy, Liechtenstein, Germany, Austria, and Slovenia (Alpine Convention 2022). Unique in its ecological location, the Alps give rise to various environmental conditions, altitudinal gradients, and biological niches. Ecosystems from temperate deciduous woods at lower heights to alpine tundra at higher elevations are evident in this mountainous region (Körner 2003b). The floral and faunal components of the Alps are equipped with special adaptive features that are essential to cope with challenging climatic conditions such as high altitudes and fluctuating temperatures. For instance, the coniferous forests, abundant in the montane zone, play a vital role in capturing carbon and stabilising the soil. Norway spruce (*Picea abies*) and silver fir (*Abies alba*) are two of the most important species that can be found in these forests (Schirpke et al. 2023).

Above the tree line elevation, the subalpine and alpine zones harbour species adapted to withstand freezing temperatures and high levels of UV radiation. These adaptations include dwarf shrubs and perennial plants (Körner 2003b). The existence of endemic species is remarkable since they have developed in isolation due to the challenging topographical and environmental limitations. The Alpine ibex (*Capra ibex*) and the Apollo butterfly (*Parnassius apollo*) are a few examples of faunal diversity specific to this region (Grabherr 2009).

The hydrology of the Alps is of utmost importance owing to its vital water storage facility for a large part of Europe. The glaciers and snowfields, in addition to the copious precipitation, contribute to the major rivers such as the Rhine, Rhone, and Po, thus impacting water supply much beyond the confines of the mountains (Viviroli et al. 2007). The melting of glaciers provides water that is used for both generating hydropower and irrigating agricultural fields. This highlights the significance of these ecosystems in supporting regional economic activity (Beniston et al. 2011).

Nevertheless, the Alps are experiencing substantial environmental challenges due to climate change, which has altered species ranges, phenology, and ecosystem dynamics. The dwindling Alpine glaciers threaten biodiversity and water supplies (Haeberli 2004). In addition, tourism, urbanisation, and agriculture cause habitat fragmentation and pollution, resulting in increased ecosystem vulnerability (Tasser et al. 2007).

13.5.2.2 Specific Ecosystem-Based Adaptation Interventions

EbA is a crucial approach in the Alps' mountainous area to address the effects of climate change. The efficacy of EbA interventions has been demonstrated in several case studies conducted in this particular environment. Alpine meadows and pastures help provide ecological services to local agricultural communities by stabilising soil, controlling water cycles, and supplying food for cattle. Clearly, the need for protecting these ecosystems gains further significance in achieving EbA goals (Grêt-Regamey et al. 2012).

Furthermore, governance and rehabilitation of forest ecosystems have been crucial EbA strategies. Alpine forests have been known to safeguard against avalanches, landslides, and rockfalls. Research has indicated that mixed forests are more capable of withstanding climatic disturbances and offer improved protective capabilities compared to forests with only one type of tree (Brang et al. 2006). To retain forests' protective functions, sustainable forest management strategies such as selective logging, controlled burning, and replanting with native species (Lindner et al. 2010) are advised.

Wetland restoration is a successful EbA intervention. Alpine wetlands, such as peat bogs and floodplains, function as natural reservoirs for water, reducing the danger of floods and ensuring a consistent water supply during dry seasons (Tockner et al. 2010). Restoring the hydrological connection and rewetting drained wetlands can significantly improve their ability to mitigate extreme weather events and sustain biodiversity (Zedler and Kercher 2005).

Moreover, promoting sustainable agriculture practices must gain thrust to ensure better implementation of EbA strategies in the Alps. This includes approaches like agroforestry, rotational grazing, and organic farming, which enhance soil health and decrease agricultural carbon emissions (Altieri 1999). Incorporating additional strategies like traditional knowledge into contemporary conservation efforts may be a pivotal approach for the momentous preservation of the cultural landscape and ecological equilibrium of the region (Plieninger et al. 2013).

Establishing ecological corridors will ensure improved connectivity across habitats, which is vital for species movement and the exchange of genetic material. Given the context of climate change, species must relocate to higher elevations or latitudes to locate adequate habitats (Heller and Zavaleta 2009). In this regard, the Alpine Network of Protected Areas (ALPARC) strives to enhance connectivity and encourage collaborative management among countries (Ros and Mazzola 2012).

13.5.2.3 Outcomes and Conclusions

Multiple case studies have examined the effects of EbA interventions in the mountainous Alps area, showing substantial ecological, social, and economic advantages. EbA interventions have demonstrated notable efficacy in this area owing to its varied habitats and the crucial ecosystem services they offer. The collective data from these case studies highlight the many benefits of EbA initiatives in the Alps. These actions not only deal with current climate concerns but also enhance human and environmental systems' long-term sustainability and resilience (Scarano 2017). Incorporating EbA into regional and national adaptation programmes is essential for the future welfare of the Alps and its residents, given the ongoing difficulties posed by climate change (Seddon et al. 2020).

In summary, targeted EbA interventions in the Alps are vital for preserving ecological services, promoting biodiversity, and sustaining the lives of local communities. These initiatives, backed by scientific research and local expertise, provide a strong framework for tackling the issues presented by climate change in mountainous areas. By incorporating EbA into regional planning and policymaking, we can guarantee a future for the Alpine environment that is both sustainable and resilient (Köhler et al. 2022).

13.5.3 Case Study III: The Andes

13.5.3.1 Description of the Ecosystem

The Andes, spanning 7000 kilometres along the western coast of South America, is the most extensive mountain range on Earth and supports a vibrant and intricate ecology. The mountainous region in question extends over seven countries: Venezuela, Colombia, Ecuador, Peru, Bolivia, Chile, and Argentina. Andean ecosystem is characterised by variation in temperature and diversity in ecosystems, which ranges from tropical rainforests to arid deserts and glacial peaks. Striking altitudinal zonation is evident in this region, resulting in vertical species arrangement. The Andean slopes at lower levels are adorned with abundant montane forests, which gradually transform into cloud forests as the height rises. The cloud forests possess a high abundance of endemic species due to their isolated and distinctive ecosystems (Young et al. 2002).

The paramo habitats, exclusive to the northern Andes, are situated at altitudes ranging from 3000 to 4000 metres above the cloud forests. Paramos are distinguished by their humid, frigid temperature and unique flora, which consists of tussock grasses, shrubs, and cushion plants. High-altitude ecosystems are essential for regulating and supplying water to the lower-lying regions (Buytaert et al. 2006). The puna, located in the middle Andes at comparable elevations, has lower precipitation levels than the paramo and sustains a distinct collection of plant species, including several varieties of grasses and shrubs that have adapted to the more severe environmental circumstances (Anthelme et al. 2014).

These high Andes, also known as the puna grasslands, are situated at over 4000 metres. These areas are distinguished by their harsh environmental conditions, including low temperatures, powerful winds, and intense UV radiation. The vegetation in these locations is limited and primarily composed of resilient grasses, lichens, and mosses. The Andes also support a diverse array of fauna, most notable of which are camelids like llamas, alpacas, and vicuñas. These native domesticated animals are integral to Andean culture (Squeo et al. 2006).

The Andes also support regional hydrology with its prominent glacier systems, which function as reserve water sources during dry spells. However, rapidly melting glaciers due to climate change have threatened the region's water security (Vuille et al. 2008). Human activities like urbanisation, mining, and deforestation have seriously impacted Andean ecosystems, posing serious risks to their sustainability (Sarmiento 2000).

13.5.3.2 Specific Ecosystem-Based Adaptation Interventions

EbA measures in the Andean mountainous area have shown considerable promise in bolstering local populations' resilience to climate change's impacts. An effective measure is the reinstatement of indigenous flora, such as *Polylepis* woods, which are vital for water management and soil stabilisation. Research has demonstrated that replanting with indigenous plant species may boost the process of water seeping into the soil, decrease soil loss due to erosion, and promote the variety of living organisms in an ecosystem (Farley et al. 2013). In addition, agroforestry systems have improved agricultural production and promoted carbon sequestration, and they have been successful in diversifying livelihoods (Murgueitio et al. 2011).

Implementing infiltration ditches and terracing in high-altitude locations for water management and conservation has garnered significant positive results. Among other benefits of these techniques, collection and retention of precipitation have been significant. These are important during droughts and bolstering agricultural endeavours during arid periods (Browder 2019). Furthermore, it is essential to restore high Andean wetlands, also known as paramos, to preserve water regulation functions. Paramos function as natural water storage systems, and their rehabilitation can alleviate the effects of climatic variability on water provision (Buytaert et al. 2011).

The successful implementation of EbA solutions has also relied on community-based management systems. Involving local populations in decision-making guarantees that treatments are tailored to the unique setting and aligned with the local culture (Reid et al. 2010). Participatory mapping and planning have been done by merging indigenous knowledge with empirical data to identify areas appropriate for conservation and sustainable use (Bunce et al. 2010).

The effectiveness of EbA procedures has been further enhanced by integrating TEK with scientific study. The revival of indigenous approaches, such as the exploitation of traditional water management systems known as 'amounts', aims to improve water accessibility and reduce vulnerability to climate extremes (Hoefsloot et al. 2024). This process infiltrates water into the soil during the rainy season, recharges groundwater, and provides a reliable water source during dry seasons (Ochoa-Tocachi et al. 2019).

Furthermore, implementing sustainable land management practices, such as rotational grazing and utilising organic fertilisers, has enhanced soil fertility and productivity and reduced greenhouse gas emissions (Vallejo-Rojas et al. 2016). These methods make agricultural ecosystems more resilient, increasing their endurance and helping them recover from disruptions. This ensures the availability of food and the capacity of local communities to continue their livelihoods sustainably (Altieri et al. 2015).

In addition, EbA programmes have been effective in adapting measures and incorporating feedback into management plans. This method ensures the maintenance of long-term sustainability and resilience (Andersson et al. 2022). Moreover, this adaptable approach can address the identification of emerging environmental challenges and the adjustment of strategies (Sietz et al. 2019).

13.5.3.3 Outcomes and Conclusions

The Andes house diverse ecosystems specifically suited to the region's varied climatic and altitudinal gradients. The Andes' great biodiversity and intricate biological processes underscore the need for conservation endeavours to safeguard these distinctive habitats amidst increasing anthropogenic pressures and climate change (Cavieres et al. 2000).

To summarise, targeted EbA interventions in the Andes, such as the use of native species for reforestation, agroforestry practices, water conservation techniques, restoration of paramo ecosystems, community-based management approaches, incorporation of traditional knowledge, sustainable land management practices, and adaptive monitoring, have shown considerable promise in improving climate resilience. These actions, backed by empirical data and local participation, provide a complete framework for achieving sustainable adaptation in response to climate change. Table 13.1 provides a brief outline of the ecosystem pattern, specific EbA interventions, and their outcomes in the Himalayas, the Alps, and the Andes mountainous regions.

Table 13.1 Summary of the ecosystem, specific EbA interventions, and their outcomes in the Himalayas, the Alps, and the Andes mountainous regions

Region	Ecosystem description	Specific EbA interventions	Outcomes	References
The Himalayas	Altitudinal variation Diversified ecosystems Endemic species like Himalayan blue poppy and snow leopard Major rivers like the Ganges, Brahmaputra, Indus	Forest restoration and management Agroforestry, grassland and rangeland management Community-driven water management Wetland protection Sustainable tourism Watershed management	Resilience to climate change Stabilisation of slopes Improved water quality Reduced floods, soil fertility Agricultural productivity Water security Biodiversity conservation Sustainable tourism	Korner (2003a, b), Xu et al. (2009), Nautiyal et al. (2015)
The Alps	Ecosystems from temperate deciduous woods to alpine tundra Significant hydrological role Endemic species like Alpine ibex and Apollo butterfly	Protection of alpine meadows and pastures Governance and rehabilitation of forest ecosystems Wetland restoration Sustainable agriculture practices Establishment of ecological corridors	Substantial ecological, social, and economic benefits Enhanced soil stability, water cycles, biodiversity, agricultural productivity Long-term sustainability and resilience	Zedler and Kercher (2005), Brang et al. (2006), Ros and Mazzola (2012)
The Andes	Ecosystems from tropical rainforests to arid deserts and glacial peaks Endemic species like llamas, alpacas, and vicuñas Significant hydrological role in glacier systems	Reinstatement of indigenous flora Agroforestry systems Implementation of infiltration ditches and terracing Restoration of high Andean wetlands Community-based management systems Adaptive monitoring	Improved climate resilience Enhanced water management, soil fertility, agricultural productivity, biodiversity, and long-term sustainability	Squeo et al. (2006), Bunce et al. (2010), Reid et al. (2010)

13.6 Challenges and Opportunities in the Implementation of Ecosystem-Based Adaptations in Mountainous and Hilly Ecosystems

13.6.1 Technical and Financial Challenges

The effectiveness and longevity of EbA interventions face significant challenges due to the technical and financial constraints present in mountainous and hilly areas. The complex and varied nature of the terrains exacerbates these challenges. For example, due to differences in elevations, gradients, and local climates, the regions necessitate distinct strategies for each location, making it difficult to standardise practices and expand effective interventions (Hannah et al. 2013). In addition, the restricted accessibility and isolation of several mountainous regions impede the allocation of essential resources and knowledge, making project implementation more challenging (Naylor et al. 2020).

The lack of sufficient baseline data about these regions' ecological and socio-economic conditions poses a significant technical challenge. The formulation of efficient adaptation strategies is hindered by the absence of advanced climate forecasting systems like high-resolution climate models and comprehensive environmental monitoring (Beniston et al. 2011). This situation is further worsened by the inherent unpredictability of climate forecasts, which necessitates flexible administration and ongoing surveillance, which require significant resources (Schoe 2019).

The considerable upfront costs associated with EbA initiatives in mountainous regions can pose a fiscal challenge. Initiating soil stabilisation, reforestation, and water management projects requires significant capital. Nonetheless, the economic advantages of such investments tend to accrue over an extended duration and are not immediately quantifiable. This presents difficulties in obtaining funding from both public and private sources (Emerton 2006). In addition, the economic advantages of EbA are frequently underestimated in conventional cost-benefit assessments, which results in insufficient investment by stakeholders (Scarano 2017).

Furthermore, financial limitations worsen because of the absence of EbA-specific funding channels. Traditional funding sources such as government grants and international aid are usually inadequate and cannot be sustained over time. Naturally, alternative methods of financing arise, which include sources like public-private partnerships, payment for ecosystem services (PES), and climate bonds (Kabisch et al. 2017). For the effective execution of these strategies, it is crucial to prioritise governance frameworks and foster collaboration among all stakeholders (Frantzeskaki 2019).

13.6.2 Policy and Governance Issues

Many policy and governance-related problems hinder the implementation of EbA in mountainous and hilly environments and impede the effectiveness of adaptation initiatives. Fragmented governance systems in mountainous regions challenge coordinated action across many administrative levels and jurisdictions. Fragmentation frequently leads to inconsistent policies and methods, making it challenging to execute EbA treatments in a unified manner (Klenk et al. 2017). In addition, several stakeholders, such as local communities, regional authorities, and national governments, make decision-making processes intricate and complicated (Reed et al. 2014).

To implement EbA initiatives in mountainous regions properly, strategies should be more integrated and cross-sectoral. Policy frameworks often fail to integrate the advantages of ecosystem services into adaptation plans, instead prioritising sector-specific measures that might result in maladaptation (Munroe et al. 2012). The fragmented approach within sectors hinders the thorough evaluation and control of ecosystem services, which are vital for the durability of mountainous areas (Vignola et al. 2009).

Moreover, the scarcity of financial resources and the reliance on external finance provide substantial barriers to the long-term implementation of EbA policies. Financial limitations sometimes result in brief project durations, which are insufficient for addressing long-term climate changes and the sustainability of EbA projects (Jones et al. 2012). The lack of predictability in financing also impacts the capability to develop local capabilities and establish EbA practices within governance frameworks (Nalau et al. 2015).

In addition, there are difficulties associated with incorporating scientific information and local, indigenous knowledge systems into policymaking processes. The lack of connection between scientific research and local conditions might lead to adaptation techniques that do not entirely match mountainous regions' social and ecological settings (Berkes 2012). The involvement of local people in participatory governance processes holds immense significance for the successful implementation of EbA, which is often hindered due to a lack of effective methods (Fabricius et al. 2007).

In addition, certain mountainous areas experience political instability and inadequate institutional frameworks, rendering difficulties in the implementation of environmental legislation (Gomez-Echeverri 2013). These governance-related difficulties emphasise the necessity for enhanced institutional capabilities, consistent policy structures, and resilient funding mechanisms. This will ensure the consequent facilitation of sustainable execution of EbA in mountainous and hilly environments (Kumar et al. 2022). To tackle these difficulties and promote mountainous ecosystems' resilience and adaptability, stakeholders must formulate a comprehensive strategy by including solutions from ecological, socio-economic, and governance aspects (Folke et al. 2005).

13.6.3 Opportunities for Setting Up Ecosystem-Based Adaptation Practices

EbA is recognised as a valuable approach for increasing resilience and decreasing susceptibility to the impacts of climate change within mountainous and hilly ecosystems. A unique assortment of creatures and complex physical characteristics in these regions offer significant opportunities for implementing EbA techniques. Leveraging microclimatic and altitudinal variations can serve EbA objectives. Restoring degraded forest ecosystems and implementing sustainable management practices in alpine grasslands can yield many advantages, such as improved carbon sequestration, slope stabilisation, and enhanced water regulation (Locatelli et al. 2015). Furthermore, preserving and restoring wetlands and riparian buffers in these areas can decrease the probability of floods and improve water quality, thus positively impacting both species variety and human welfare (de Santiago-Martín et al. 2016).

Additionally, indigenous and local people living in mountainous and hilly regions have a vast amount of TEK that might be highly valuable in planning and implementing effective EbA strategies. These communities possess extensive knowledge of local ecosystems and have developed sustainable land management practices across several generations (Reid et al. 2017). Culturally appropriate adaptation strategies tailored to the specific local context can be created by integrating TEK with scientific understanding, which may potentially thrust the implementation of EbA approaches (Berkes et al. 2000).

The economic advantages of EbA in steep and hilly ecosystems are equally noteworthy. EbA can improve local agricultural practices by bolstering ecological services such as water purification, soil fertility, and pollination. As highlighted by Altieri and colleagues, ensuring food security is essential (Altieri et al. 2015). Furthermore, the endorsement of ecotourism, which prioritises the preservation of pristine natural habitats, might provide alternative revenue streams and foster initiatives to save the environment (Kiss 2004).

Nevertheless, challenges need to be surmounted in order to fully harness the potential of EbA in areas characterised by steep and mountainous terrains. To implement EbA strategies, robust institutional frameworks, adequate financing channels, and initiatives are required to improve competencies (Noble et al. 2015). The topmost priority should be given to promoting inclusiveness and fairness in EbA activities. It must be ensured that all stakeholders, especially those who are disadvantaged and vulnerable, enjoy the benefits of EbA (News 2010).

13.7 Conclusions and Future Perspectives

To summarise, EbA is a progressively encouraging approach to mitigate the impacts of climate change in mountainous and hilly ecosystems worldwide. The subject of EbA is currently seeing a notable convergence of state-of-the-art technology and community-led efforts. Remote sensing (RS) and global information system (GIS) technologies enhance the monitoring and management of ecosystems, hence improving the capacity to respond effectively to changes in climate (Munang et al. 2013a). Moreover, the sustainability and efficacy of EbA initiatives can be ensured by including local communities (Reid 2016). Climate-smart agriculture and land restoration approaches are gaining attention due to their vast socio-ecological advantages. These include the conservation of ecosystems, improvement of the benefits provided by natural systems, and enhancement of the well-being of local populations (Vignola et al. 2015).

Despite notable progressions, there are persistent lacunae in our understanding regarding the long-term effects and scalability of EbA projects. The inquiry should focus on developing a reliable approach to evaluate the effectiveness of EbA treatments under various climatic conditions and for extended periods (Doswald et al. 2014). Furthermore, it is essential to undertake comprehensive studies on the socio-economic advantages of EbA to guarantee that these strategies effectively tackle environmental issues while aiding the reduction of poverty and promoting social fairness (Scarano 2017). Additionally, it is crucial to incorporate ecological science with social sciences and policymaking in multidisciplinary research to develop techniques that are scientifically rigorous and socially feasible (Seddon et al. 2016).

Global case studies have demonstrated that practices like oil stabilisation, water management, and biodiversity protection are successful treatments for implementing EbA strategies to decrease vulnerability and improve adaptability (Elmqvist et al. 2003). Furthermore, incorporating indigenous knowledge into scientific research has improved the effectiveness and acceptance of EbA methodologies (Berkes 2012). EbA efforts' success generally relies on favourable legislative frameworks and sufficient finance, both of which are usually lacking in many mountainous and hilly landscapes (Jones et al. 2012). Therefore, governments, researchers, and local communities must adopt a collaborative approach to achieve EbA's benefits successfully.

In summary, filling up knowledge gaps and fostering interdisciplinary research supported by robust policies are the key factors for the success of EbA in mitigating climate change impacts in mountainous and hilly regions. Advancing and sustaining EbA initiatives require focusing on long-term impact studies, socio-economic analyses, and integrating local knowledge into new projects. Such measures can improve the performance of nature-based solutions, strengthening the resilience of ecosystems and communities against climate change.

References

Adger WN, Hughes TP, Folke C et al (2005) Social-ecological resilience to coastal disasters. Science 309:1036–1039. https://doi.org/10.1126/science.1112122

Alpine Convention (2022) Who we are. In: Alpconv. https://www.alpconv.org/en/home/. Accessed 16 Jul 2024d

Altieri MA (1999) The ecological role of biodiversity in agroecosystems. Agric Ecosyst Environ 74:19–31. https://doi.org/10.1016/S0167-8809(99)00028-6

Altieri MA, Nicholls CI, Henao A, Lana MA (2015) Agroecology and the design of climate change-resilient farming systems. Agron Sustain Dev 35:869–890. https://doi.org/10.1007/s13593-015-0285-2

Andersson E, Grimm NB, Lewis JA et al (2022) Urban climate resilience through hybrid infrastructure. Curr Opin Environ Sustain 55:101158. https://doi.org/10.1016/j.cosust.2022.101158

Anthelme F, Jacobsen D, Macek P et al (2014) Biodiversity patterns and continental insularity in the tropical high Andes. Arct Antarct Alp Res 46:811–828. https://doi.org/10.1657/1938-4246-46.4.811

Armitage SJ, Jasim SA, Marks AE et al (2011) The southern route "out of Africa": evidence for an early expansion of modern humans into Arabia. Science 331:453–456. https://doi.org/10.1126/science.1199113

Barry RG (2008) Mountain weather and climate. Cambridge University Press, Cambridge

Bawa K, Koh L, Lee T et al (2010) China, India, and the environment. Science (New York, NY) 327:1457–1459. https://doi.org/10.1126/science.1185164

Beniston M, Stoffel M, Hill M (2011) Impacts of climatic change on water and natural hazards in the Alps: can current water governance cope with future challenges? Examples from the European "ACQWA" project. Environ Sci Pol 14:734–743. https://doi.org/10.1016/j.envsci.2010.12.009

Berkes F (2012) Sacred ecology, 3rd edn. Routledge, New York

Berkes F, Colding J, Folke C (2000) Rediscovery of traditional ecological knowledge as adaptive management. Ecol Appl 10:1251–1262. https://doi.org/10.1890/1051-0761(2000)010[1251:ROTEKA]2.0.CO;2

Bertuzzo E, Carrara F, Mari L et al (2016) Geomorphic controls on elevational gradients of species richness. Proc Natl Acad Sci 113:1737–1742. https://doi.org/10.1073/pnas.1518922113

Bhattarai KR, Vetaas OR, Grytnes JA (2004) Fern species richness along a central Himalayan elevational gradient, Nepal. J Biogeogr 31:389–400. https://doi.org/10.1046/j.0305-0270.2003.01013.x

Brang P, Schönenberger W, Frehner M et al (2006) Management of protection forests in the European Alps: an overview. For Snow Landsc Res 80

Browder JO (2019) Fragile lands of Latin America: strategies for sustainable development. Routledge

Bunce M, Brown K, Rosendo S (2010) Policy misfits, climate change and cross-scale vulnerability in coastal Africa: how development projects undermine resilience. Environ Sci Pol 13:485–497. https://doi.org/10.1016/j.envsci.2010.06.003

Buytaert W, Célleri R, De Bièvre B et al (2006) Human impact on the hydrology of the Andean Páramos. Earth Sci Rev 79:53–72. https://doi.org/10.1016/j.earscirev.2006.06.002

Buytaert W, Cuesta-Camacho F, Tobón C (2011) Potential impacts of climate change on the environmental services of humid tropical alpine regions. Glob Ecol Biogeogr 20:19–33. https://doi.org/10.1111/j.1466-8238.2010.00585.x

Cardinale BJ, Duffy JE, Gonzalez A et al (2012) Biodiversity loss and its impact on humanity. Nature 486:59–67. https://doi.org/10.1038/nature11148

CARE Climate Change (2014) In: CARE climate change. https://careclimatechange.org/. Accessed 29 Dec 2024

Cavieres L, Peñaloza A, Kalin M (2000) Altitudinal vegetation belts in the high-Andes of central Chile (33°S). Revista Chilena De Historia Nat Rev Chil Hist Nat 73. https://doi.org/10.4067/S0716-078X2000000200008

Chen J, Yang H, Jin T, Wu K (2024) Assessment of terrestrial ecosystem sensitivity to climate change in arid, semi-arid, sub-humid, and humid regions using EVI, LAI, and SIF products. Ecol Indic 158:111511. https://doi.org/10.1016/j.ecolind.2023.111511

Chong J (2014) Ecosystem-based approaches to climate change adaptation: progress and challenges. Int Environ Agreem: Politics Law Econ 14:391–405

Choudhary A, Kumar S, Sharma C (2011) Constructed wetlands: an approach for wastewater treatment. Elixir Pollut 37:3666–3672

Convention on Biological Diversity (2007) Global biodiversity outlook 1. https://www.cbd.int/gbo1. Accessed 12 Jul 2024

Cousiño A, Penha-Lopes G (2021) Ecosystem based adaptation: concept and terminology in strategic adaptation planning (municipal and inter-municipal) in Portugal. Sustainability 13:6145. https://doi.org/10.3390/su13116145

de Santiago-Martín A, Guesdon G, Galvez R (2016) Plants for constructed wetlands as an ecological engineering alternative to road runoff desalination. In: Ansari AA, Gill SS, Gill R et al (eds) Phytoremediation: management of environmental contaminants, vol 4. Springer, Cham, pp 233–266

Dhyani S, Thummarukuddy M (2016) Ecological engineering for disaster risk reduction and climate change adaptation. Environ Sci Pollut Res 23:20049–20052. https://doi.org/10.1007/s11356-016-7517-0

Doswald N, Munroe R, Roe D et al (2014) Effectiveness of ecosystem-based approaches for adaptation: review of the evidence-base. Clim Dev 6:185–201. https://doi.org/10.1080/17565529.2013.867247

Duarte CM, Losada IJ, Hendriks IE et al (2013) The role of coastal plant communities for climate change mitigation and adaptation. Nat Clim Change 3:961–968. https://doi.org/10.1038/nclimate1970

Elmqvist T, Folke C, Nyström M et al (2003) Response diversity, ecosystem change, and resilience. Front Ecol Environ 1:488–494. https://doi.org/10.1890/1540-9295(2003)001[0488:RDECAR]2.0.CO;2

Emerton L (2006) Sustainable financing of protected areas : a global review of challenges and options. IUCN

Ensor TM, Surprenant AM, Neath I (2019) Increasing word distinctiveness eliminates the picture superiority effect in recognition: evidence for the physical-distinctiveness account. Mem Cogn 47:182–193. https://doi.org/10.3758/s13421-018-0858-9

Environment UN (2021) Ecosystem-based adaptation. UNEP – UN Environment Programme. http://www.unep.org/topics/climate-action/adaptation/ecosystem-based-adaptation. Accessed 13 Jul 2024

European Commission (2023) Good administration practices of the European Commission. https://commission.europa.eu/about-european-commission/service-standards-and-principles/ethics-and-good-administration/good-administration_en. Accessed 13 Jul 2024

Fabricius C, Folke C, Cundill G, Schultz L (2007) Powerless spectators, coping actors, and adaptive co-managers: a synthesis of the role of communities in ecosystem management. Ecol Soc 12

FAO (2022) Mountain agriculture. Opportunities for harnessing Zero Hunger in Asia |Policy Support and Governance| Food and Agriculture Organization of the United Nations. https://www.fao.org/policy-support/tools-and-publications/resources-details/en/c/1300326/. Accessed 15 Jul 2024c

Farley KA, Bremer LL, Harden CP, Hartsig J (2013) Changes in carbon storage under alternative land uses in biodiverse Andean grasslands: implications for payment for ecosystem services. Conserv Lett 6:21–27. https://doi.org/10.1111/j.1755-263X.2012.00267.x

Foley JA, DeFries R, Asner GP et al (2005) Global consequences of land use. Science 309:570–574. https://doi.org/10.1126/science.1111772

Folke C, Hahn T, Olsson P, Norberg J (2005) Adaptive governance of social-ecological systems. Annu Rev Environ Resour 30:441–473. https://doi.org/10.1146/annurev.energy.30.050504.144511

Folke C, Carpenter S, Walker B et al (2010) Resilience thinking: integrating resilience, adaptability and transformability. Ecol Soc 15. https://doi.org/10.5751/ES-03610-150420
Fox JL, Sinha SP, Chundawat RS (1992) Activity patterns and habitat use of Ibex in the Himalaya Mountains of India. J Mammal 73:527–534. https://doi.org/10.2307/1382018
Frantzeskaki N (2019) Seven lessons for planning nature-based solutions in cities. Environ Sci Pol 93:101–111. https://doi.org/10.1016/j.envsci.2018.12.033
Gadgil M, Berkes F, Folke C (1993) Indigenous knowledge for biodiversity conservation. Ambio 22
Gairola S, Rawal R, Todaria N (2008) Forest vegetation patterns along an altitudinal gradient in Sub-alpine zone of West Himalaya India, p 2
Girot P, Ehrhart C, Oglethorpe J (2012) Integrating community and ecosystem-based approaches in climate change adaptation responses
Gomez-Echeverri L (2013) The changing geopolitics of climate change finance. Clim Pol 13:632–648. https://doi.org/10.1080/14693062.2013.822690
Gowdy J, Howarth R, Tisdell C (2010) The economics of ecosystems and biodiversity: ecological and economic foundations
Grabherr G (2009) Biodiversity in the high ranges of the Alps: ethnobotanical and climate change perspectives. Glob Environ Chang 19:167–172. https://doi.org/10.1016/j.gloenvcha.2009.01.007
Grêt-Regamey A, Brunner SH, Kienast F (2012) Mountain ecosystem services: who cares? Mt Res Dev 32. https://doi.org/10.1659/MRD-JOURNAL-D-10-00115.S1
Grytnes JA, Vetaas OR (2002) Species richness and altitude: a comparison between null models and interpolated plant species richness along the Himalayan altitudinal gradient, Nepal. Am Nat 159:294–304. https://doi.org/10.1086/338542
Haeberli W (2004) Glaciers and ice caps: historical background and strategies of world-wide monitoring. In: Payne AJ, Bamber JL (eds) Mass balance of the cryosphere: observations and modelling of contemporary and future changes. Cambridge University Press, Cambridge, pp 559–578
Hannah L, Roehrdanz PR, Ikegami M et al (2013) Climate change, wine, and conservation. Proc Natl Acad Sci 110:6907–6912. https://doi.org/10.1073/pnas.1210127110
Heller NE, Zavaleta ES (2009) Biodiversity management in the face of climate change: a review of 22 years of recommendations. Biol Conserv 142:14–32. https://doi.org/10.1016/j.biocon.2008.10.006
Hewitt GM (2004) Genetic consequences of climatic oscillations in the Quaternary. Philos Trans R Soc Lond Ser B Biol Sci 359:183–195.; discussion 195. https://doi.org/10.1098/rstb.2003.1388
Hoefsloot FI, Martínez J, Pfeffer K (2024) An emerging knowledge system for future water governance: sowing water for Lima. Territ, Politics, Gov 12:825–845. https://doi.org/10.1080/21622671.2021.2023365
Hoegh-Guldberg O, Mumby PJ, Hooten AJ et al (2007) Coral reefs under rapid climate change and ocean acidification. Science 318:1737–1742. https://doi.org/10.1126/science.1152509
Hooper DU, Adair EC, Cardinale BJ et al (2012) A global synthesis reveals biodiversity loss as a major driver of ecosystem change. Nature 486:105–108. https://doi.org/10.1038/nature11118
Immerzeel WW, van Beek LPH, Bierkens MFP (2010) Climate change will affect the Asian water towers. Science 328:1382–1385. https://doi.org/10.1126/science.1183188
Isbell F, Calcagno V, Hector A et al (2011) High plant diversity is needed to maintain ecosystem services. Nature 477:199–202. https://doi.org/10.1038/nature10282
Jones HP, Hole DG, Zavaleta ES (2012) Harnessing nature to help people adapt to climate change. Nat Clim Change 2:504–509. https://doi.org/10.1038/nclimate1463
Jones HP, Jones PC, Barbier EB et al (2018) Restoration and repair of Earth's damaged ecosystems. Proc R Soc B Biol Sci 285:20172577. https://doi.org/10.1098/rspb.2017.2577
Kabisch N, Stadler J, Korn H, Bonn A (2017) Nature-based solutions for societal goals under climate change in urban areas – synthesis and ways forward. In: Kabisch N, Korn H, Stadler J, Bonn A (eds) Nature-based solutions to climate change adaptation in urban areas: linkages between science, policy and practice. Springer, Cham, pp 323–336

Kiss A (2004) Is community-based ecotourism a good use of biodiversity conservation funds? Trends Ecol Evol 19:232–237. https://doi.org/10.1016/j.tree.2004.03.010

Klenk N, Fiume A, Meehan K, Gibbes C (2017) Local knowledge in climate adaptation research: moving knowledge frameworks from extraction to co-production. WIREs Clim Change 8:e475. https://doi.org/10.1002/wcc.475

Köhler J, Bauer A, Andreas D, Kuenzer C (2022) Towards forecasting future snow cover dynamics in the European Alps – the potential of long optical remote-sensing time series. Remote Sens 14:4461. https://doi.org/10.3390/rs14184461

Körner C (2003a) Carbon limitation in trees. J Ecol 91:4–17. https://doi.org/10.1046/j.1365-2745.2003.00742.x

Körner C (2003b) Alpine plant life. Springer, Berlin/Heidelberg

Körner C (2007) The use of 'altitude' in ecological research. Trends Ecol Evol 22:569–574. https://doi.org/10.1016/j.tree.2007.09.006

Körner C (2012) Alpine Treelines. Springer, Basel

Körner C, Ohsawa M (2005) Mountain systems. In: Ecosystems and human well-being: current state and trends, pp 683–716

Körner C, Asshoff R, Bignucolo O et al (2005) Carbon flux and growth in mature deciduous forest trees exposed to elevated CO2. Science 309:1360–1362. https://doi.org/10.1126/science.1113977

Kumar BM, Handa AK, Dhyani SK, Arunachalam A (2018) Agroforestry in the Indian Himalayan region: an overview. Temp Agrofor Syst:153–172. https://doi.org/10.1079/9781780644851.0153

Kumar P, Sharma P, Kumar P et al (2021) Agricultural sustainability in Indian Himalayan Region: constraints and potentials. Indian J Ecol 42:649–662

Kumar A, Ekka P, Patra S et al (2022) Geospatial perspectives of sustainable forest management to enhance ecosystem services and livelihood security. In: Advances in remote sensing for forest monitoring. Wiley, pp 10–42

Lindner M, Maroschek M, Netherer S et al (2010) Climate change impacts, adaptive capacity, and vulnerability of European forest ecosystems. For Ecol Manag 259:698–709. https://doi.org/10.1016/j.foreco.2009.09.023

Locatelli B, Catterall CP, Imbach P et al (2015) Tropical reforestation and climate change: beyond carbon. Restor Ecol 23:337–343. https://doi.org/10.1111/rec.12209

Meybeck M, Green P, Vörösmarty C (2001) A New typology for mountains and other relief classes: an application to global continental water resources and population distribution. Mt Res Dev 21:34–45

Millennium Ecosystem Assessment (Program) (2005) Ecosystems and human well-being: synthesis. Island Press, Washington, DC

Mills AJ, Tan D, Manji AK et al (2020) Ecosystem-based adaptation to climate change: Lessons learned from a pioneering project spanning Mauritania, Nepal, the Seychelles, and China. Plants People Planet 2:587–597. https://doi.org/10.1002/ppp3.10126

Mishra N, Bhatt J, Bhatt V (2018) Ecotourism as an activity for sustaining livelihood opportunities for the local communities in Uttarakhand Himalayas. South Asian J Tour Herit 11:100–114

Mitsch WJ, Bernal B, Nahlik AM et al (2013) Wetlands, carbon, and climate change. Landsc Ecol 28:583–597. https://doi.org/10.1007/s10980-012-9758-8

Moser SC, Ekstrom JA (2010) A framework to diagnose barriers to climate change adaptation. Proc Natl Acad Sci 107:22026–22031. https://doi.org/10.1073/pnas.1007887107

Munang R, Thiaw I, Alverson K et al (2013a) Climate change and ecosystem-based adaptation: a new pragmatic approach to buffering climate change impacts. Curr Opin Environ Sustain 5:67–71. https://doi.org/10.1016/j.cosust.2012.12.001

Munang R, Thiaw I, Alverson K et al (2013b) Using ecosystem-based adaptation actions to tackle food insecurity. Environ Sci Policy Sustain Dev 55:29–35. https://doi.org/10.1080/00139157.2013.748395

Munroe R, Roe D, Doswald N et al (2012) Review of the evidence base for ecosystem-based approaches for adaptation to climate change. Environ Evid 1:13. https://doi.org/10.1186/2047-2382-1-13

Murgueitio E, Calle Z, Uribe F et al (2011) Native trees and shrubs for the productive rehabilitation of tropical cattle ranching lands. For Ecol Manag 261:1654–1663. https://doi.org/10.1016/j.foreco.2010.09.027

Naidoo R, Balmford A, Costanza R et al (2008) Global mapping of ecosystem services and conservation priorities. Proc Natl Acad Sci USA 105:9495–9500. https://doi.org/10.1073/pnas.0707823105

Nalau J, Preston BL, Maloney MC (2015) Is adaptation a local responsibility? Environ Sci Pol 48:89–98. https://doi.org/10.1016/j.envsci.2014.12.011

Nautiyal S, Rajan KS, Shibasaki R (2015) Interaction of biodiversity and economic welfare –a case study from the Himalayas of India. J Environ Inf 6:111–119

Naylor D, Sadler N, Bhattacharjee A et al (2020) Soil microbiomes under climate change and implications for carbon cycling. Annu Rev Environ Resour 45:29–59. https://doi.org/10.1146/annurev-environ-012320-082720

News LW (2010) Stringer et al – 2009 – adapt to climate change, drought and desertification – local insights to enhance policy in S Africa. Future Agricultures. https://www.future-agricultures.org/news/stringer-et-al-2009-adapt-to-climate-change-drought-and-desertification-local-insights-to-enhance-policy-in-s-africa/. Accessed 18 Jul 2024

Noble I, Huq S, Anokhin YA, et al (2015) Adaptation needs and options, pp 833–868

Ochoa-Tocachi BF, Bardales JD, Antiporta J et al (2019) Potential contributions of pre-Inca infiltration infrastructure to Andean water security. Nat Sustain 2:584–593. https://doi.org/10.1038/s41893-019-0307-1

Pan Y, Birdsey RA, Fang J et al (2011) A large and persistent carbon sink in the world's forests. Science 333:988–993. https://doi.org/10.1126/science.1201609

Pandit MK, Sodhi NS, Koh LP et al (2007) Unreported yet massive deforestation driving loss of endemic biodiversity in Indian Himalaya. Biodivers Conserv 16:153–163. https://doi.org/10.1007/s10531-006-9038-5

Pant LP, Adhikari B, Bhattarai KK (2015) Adaptive transition for transformations to sustainability in developing countries. Curr Opin Environ Sustain 14:206–212. https://doi.org/10.1016/j.cosust.2015.07.006

Paracchini ML, Bulgheroni C, Borreani G et al (2015) A diagnostic system to assess sustainability at a farm level: the SOSTARE model. Agric Syst 133:35–53. https://doi.org/10.1016/j.agsy.2014.10.004

Pepin N, Bradley RS, Diaz HF et al (2015) Elevation-dependent warming in mountain regions of the world. Nat Clim Change 5:424–430. https://doi.org/10.1038/nclimate2563

Plieninger T, Dijks S, Oteros-Rozas E, Bieling C (2013) Assessing, mapping, and quantifying cultural ecosystem services at community level. Land Use Policy 33:118–129. https://doi.org/10.1016/j.landusepol.2012.12.013

Rahbek C (1995) The elevational gradient of species richness: a uniform pattern? Ecography 18:200–205. https://doi.org/10.1111/j.1600-0587.1995.tb00341.x

Reed MS, Graves A, Dandy N et al (2009) Who's in and why? A typology of stakeholder analysis methods for natural resource management. J Environ Manag 90:1933–1949. https://doi.org/10.1016/j.jenvman.2009.01.001

Reed M, Stringer L, Fazey I et al (2014) Five principles for the practice of knowledge exchange in environmental management. J Environ Manag 146:337–345. https://doi.org/10.1016/j.jenvman.2014.07.021

Reid H (2016) Ecosystem- and community-based adaptation: learning from community-based natural resource management. Clim Dev 8:4–9. https://doi.org/10.1080/17565529.2015.1034233

Reid W, Mooney H, Cropper A et al (2005) Millennium ecosystem assessment. Ecosystems and human well-being: synthesis

Reid W, Chen D, Goldfarb L et al (2010) Earth system science for global sustainability: grand challenges. Science 330:916–917. https://doi.org/10.2307/40931852

Reid AJ, Brooks JL, Dolgova L et al (2017) Post-2015 sustainable development goals still neglecting their environmental roots in the Anthropocene. Environ Sci Pol 77:179–184. https://doi.org/10.1016/j.envsci.2017.07.006

Ros G, Mazzola A (2012) Networking with landscape: local initiatives in an Italian Alpine Valley. Mt Res Dev 32:400–410. https://doi.org/10.1659/MRD-JOURNAL-D-12-00060.1

Sarmiento F (2000) Breaking Mountain paradigms: ecological effects on human impacts in managed Tropandean landscapes. Ambio 29. https://doi.org/10.1639/0044-7447(2000)029[0423:BMPEEO]2.0.CO;2

Scarano FR (2017) Ecosystem-based adaptation to climate change: concept, scalability and a role for conservation science. Perspect Ecol Conserv 15:65–73. https://doi.org/10.1016/j.pecon.2017.05.003

Schirpke U, Tasser E, Borsky S et al (2023) Past and future impacts of land-use changes on ecosystem services in Austria. J Environ Manag 345:118728. https://doi.org/10.1016/j.jenvman.2023.118728

Schoe A (2019) Taking the long view and acting now – prioritizing management of high elevation five-needle pines

Seddon AWR, Macias-Fauria M, Long PR et al (2016) Sensitivity of global terrestrial ecosystems to climate variability. Nature 531:229–232. https://doi.org/10.1038/nature16986

Seddon N, Chausson A, Berry P et al (2020) Understanding the value and limits of nature-based solutions to climate change and other global challenges. Philos Trans R Soc B 375:20190120. https://doi.org/10.1098/rstb.2019.0120

Sharma S, Hamal K, Khadka N et al (2021) Drought characteristics over Nepal Himalaya and their relationship with climatic indices. Meteorol Appl 28:e1988. https://doi.org/10.1002/met.1988

Shrestha UB, Gautam S, Bawa KS (2012) Widespread climate change in the Himalayas and associated changes in local ecosystems. PLoS One 7:e36741. https://doi.org/10.1371/journal.pone.0036741

Sietz D, Frey U, Roggero M et al (2019) Archetype analysis in sustainability research: Methodological portfolio and analytical frontiers. Ecol Soc 24:34. https://doi.org/10.5751/ES-11103-240334

Singh SP. Himalayan forest ecosystem services

Squeo F, Warner B, Aravena R, Espinoza D (2006) Bofedales: high altitude peatlands of the central Andes. Rev Chil Hist Nat 79:245–255. https://doi.org/10.4067/S0716-078X2006000200010

Stringer LC, Dougill AJ, Fraser E et al (2006) Unpacking "participation" in the adaptive management of social–ecological systems: a critical review. Ecol Soc 11

Tasser E, Walde J, Tappeiner U et al (2007) Land-use changes and natural reforestation in the Eastern Central Alps. Agric Ecosyst Environ 118:115–129. https://doi.org/10.1016/j.agee.2006.05.004

Temmerman S, Meire P, Bouma TJ et al (2013) Ecosystem-based coastal defence in the face of global change. Nature 504:79–83. https://doi.org/10.1038/nature12859

Thakur S, Negi VS, Dhyani R et al (2022) Influence of environmental factors on tree species diversity and composition in the Indian western Himalaya. For Ecol Manag 503:119746. https://doi.org/10.1016/j.foreco.2021.119746

Tilman D, Reich PB, Isbell F (2012) Biodiversity impacts ecosystem productivity as much as resources, disturbance, or herbivory. Proc Natl Acad Sci 109:10394–10397. https://doi.org/10.1073/pnas.1208240109

Tiwari P, Joshi B (2011) Environmental changes and sustainable development of water resources in the Himalayan headwaters of India. Water Resour Manag 26:883–907. https://doi.org/10.1007/s11269-011-9825-y

Tockner K, Pusch M, Borchardt D, Lorang MS (2010) Multiple stressors in coupled river–floodplain ecosystems. Freshw Biol 55:135–151. https://doi.org/10.1111/j.1365-2427.2009.02371.x

Vallejo-Rojas V, Ravera F, Rivera-Ferre MG (2016) Developing an integrated framework to assess agri-food systems and its application in the Ecuadorian Andes. Reg Environ Chang 16:2171–2185. https://doi.org/10.1007/s10113-015-0887-x

Vignola R, Locatelli B, Martínez-Alonso C, Pablo I (2009) Ecosystem-based adaptation to climate change: what role for policy-makers, society and scientists? Mitig Adapt Strateg Glob Chang 14:691–696. https://doi.org/10.1007/s11027-009-9193-6

Vignola R, Harvey CA, Bautista-Solis P et al (2015) Ecosystem-based adaptation for smallholder farmers: definitions, opportunities and constraints. Agric Ecosyst Environ 211:126–132. https://doi.org/10.1016/j.agee.2015.05.013

Viviroli D, Weingartner R, Messerli B (2003) Assessing the hydrological significance of the world's mountains. Mt Res Dev 23:32–40

Viviroli D, Dürr HH, Messerli B et al (2007) Mountains of the world, water towers for humanity: Typology, mapping, and global significance. Water Resour Res 43. https://doi.org/10.1029/2006WR005653

Vuille M, Francou B, Wagnon P et al (2008) Climate change and tropical Andean glaciers: past, present and future. Earth Sci Rev 89:79–96. https://doi.org/10.1016/j.earscirev.2008.04.002

Xu J, Grumbine R, Shrestha A et al (2009) The melting Himalayas: cascading effects of climate change on water, biodiversity, and livelihoods. Conserv Biol 23:520–530. https://doi.org/10.1111/j.1523-1739.2009.01237.x

Young KR, Ulloa CU, Luteyn JL, Knapp S (2002) Plant evolution and endemism in Andean South America: an introduction. Bot Rev 68:4–21

Zedler J, Kercher S (2005) Wetland resources: status, trends, ecosystem services, and restorability. Annu Rev Environ Resour 15:39–74. https://doi.org/10.1146/annurev.energy.30.050504.144248

Chapter 14
Translational Dynamics of Social-Ecological Management System of Community Forestry: Prospects for Ecosystem-Based Adaptation in Community-Managed Forests of Uttarakhand

Mehul Pant and Harshit Pant

Abstract Community forestry renders multiple pathways for action-based and preventive adaptation, often reinforcing community resilience towards the changes. However, the key decision factors, such as the centralisation of power in resource management, are often neglected. The study focuses on the prerogative of transfer of decision-making power to the lower-level central government authorities or other local authorities who are upwardly accountable to the central government or the upper level decision making systems. In an aspect of the state of Uttarakhand (India), which harbours the century-old institution of Van Panchayats (VPs), this is a prime issue. The translation of power dynamics through the revised Van Panchayat Act of 1976 that replaced the 1927 Act affected many aspects of local management, making forest councils increasingly dependent on the government; after 1997, approximately 8000 Van Panchayat were established, and the role and authority of Van Panchayat over village forest has substantially declined after the amendments in the years 1976, 2001, 2007 and 2012. The present study examines the diverse forest-people interactions in light of ecosystem-based adaptation by addressing changes in the rules and regulations of Van Panchayats (community-based forest management institutions) in Uttarakhand and their translational impact on the community.

M. Pant
National Law Institute University, Bhopal, Madhya Pradesh, India

H. Pant (✉)
Centre for Socio-economic Development, GBPNIHE, G.B. Pant National Institute of Himalayan Environment, Almora, Uttarakhand, India
e-mail: harshit.pantj@gbpihed.nic.in

Y. D. I. Khan et al. (eds.), *Ecosystem-based Approaches for Resilience Building in Himalayan Landscapes*, Disaster Resilience and Green Growth,
https://doi.org/10.1007/978-981-95-2007-7_14

Keywords Community-based forest management · Forest resources · Natural resource management · Forest Department · Ecosystem-based adaptation

Abbreviations

CAMPA	Compensatory Afforestation Fund Management and Planning Authority
CBFM	Community-based forest management
CLR	Comprehensive literature review
DFO	Divisional Forest Officer
EBA	Ecosystem-based adaptation
FD	Forest Department
FSI	Forest Survey of India
MODES	Media, Observation(s), Documents, Expert(s) and Secondary Data
PES	Payment for Ecosystem Services
REDD	Reducing emissions from deforestation and forest degradation
REDD+	Reducing emissions from deforestation and forest degradation and conservation of forest carbon stocks, sustainable management of forests and enhancement of forest carbon stocks
SES	Social-ecological systems
VP/VPs	Van Panchayat/Van Panchayats

14.1 Introduction

Community-based forests cater to the livelihoods of more than 2.4 billion people directly and indirectly (Gilmore 2016). They depend on the forest and its resources for income generation and a comprehensive assortment of social, economic, cultural, spiritual and environmental benefits for local communities. Community forest management is a system of management of local forests by the group or community of those living in that vicinity. The forest is their primary livelihood source, whether it is furnished with items that may bring them monetary benefits or simply provided with commodities to be consumed directly in the household (Landell-Mills and Porras 2002). In an aspect of community-based forest management in Uttarakhand, it is generally believed that, till the advent of British colonial rule in the early 1800s, local communities exercised direct control over cultivated and uncultivated lands with little interference from the rulers (Sarin 2005). Between 1823 and 1971, the British introduced a series of changes in the control over land and forest resources that had dramatic consequences for managing these natural resources. The British government introduced the permanent land settlement, and then the state started the reservation and heavy exploitation of forest lands for timbre, softwood and resin. This led to intense and even violent protests (Guha 1985). Eventually, common lands got split into three categories: Class II reserves (commercially valuable forest lands) that the Forest Department directly controlled, Class I reserves (commercially less valuable forest lands) that were open to all residents of Kumaun and Garhwal division while the Civil-Soyam lands which are legally under the control of Revenue Department, having open access and villagers have unlimited rights on these forests with no categorical demarcation (Bisht et al. 2020). In 1931, as a direct consequence of the protests of the preceding decades, the British passed a law (Kumaun Panchayat Forest Rules) enabling local communities in Kumaun to apply for and create Van Panchayats (hereby VPs; literally, forest councils) to autonomously manage specific patches of lands from the Class II reserves and Civil-Soyam lands. These VPs regulated grazing and harvesting of fuel wood, fodder and small timbre. They had the authority to grant a limited amount of timbre for domestic use and to fine those who violated its rules. The VPs were initially set up across the Kumaun division, including the Pauri district of the Garhwal division. By the time of Independence, about 400-odd VPs had been set up in the region. Although not without lacunae, the institution of VPs is considered to be one of the rare examples of state-supported yet substantially autonomous community-level forest management institutions in South Asia till recent times (Nagendra 2007; Ghate and Nagendra 2019; Gibson et al. 2005, 2007; Mukherjee 2003; Nagahama et al. 2016). The VP arrangement ameliorated some of the adverse effects of earlier British forest policy in pockets, but it did not modify the overall approach of 'modern' forest policy—a policy that continued after Independence. This approach was not only geared towards generating revenues for the state (through commercial extraction), as against the meeting of local needs, but also the manner of its pursuit deprived local communities of a significant share in the employment and profits generated from such commercial use. The emergence of the Chipko movement in the 1970s was correspondingly driven by a multiplicity of concerns—over environmental problems such as floods being triggered by unbridled forest exploitation and devastation due to mining, over inadequate access and availability of forest products such as fuel wood and fodder for subsistence use and over the exclusion of local communities from sharing in the benefits of the commercial use of forests (Guha 1994; Guha and Gadgil 1989; Mawdsley 1998). The time has come for effective forest conservation, reforestation and recovery; we must strengthen our local forest governance mechanisms. The terms of reference for state REDD (reducing emissions from deforestation and forest degradation) plan clearly says that their preparation and implementation must be overseen by joint forest management committees, community forestry groups, Van Panchayats and village forest protection committees. National plans have a strong component of capacity building, including training youth from local rural communities to build a large cadre of community foresters. A potentially cheaper alternative monitoring approach at the local implementation scale of REDD+ (reducing emissions from deforestation and forest degradation and conservation of forest carbon stocks, sustainable management of forests and enhancement of forest carbon stocks) is to engage local people in monitoring a local-based forest monitoring approach therefore hold promise for REDD+. In the present scenario, the role of community forests cannot be neglected, as they have vast reserves of carbon stored in their woody biomass and in the soil carbon pool where carbon remains sequestered for long duration in the deeper layers.

14.1.1 Existing Role of Community-Managed Forests in Ecosystem-Based Adaptation

Many socio-economic changes are happening in the Himalayan region, which directly affect community forestry management practices and the way gender-based power relations are produced and reproduced in everyday practices. For example, with increasing male outmigration, forest management and agricultural tasks (mainly for subsistence) are increasingly feminised in the context of Himalayan region which are posing further challenges to rural women to cope with the changing situation (Bhattarai 2020). The restoration of forest landscapes, particularly in terms of afforestation (the introduction of new trees and tree seedlings), reforestation (natural or intentional restocking of existing forest) and restoration of degraded forest land, often helps in regaining ecological functionality and enhancing human well-being across deforested or degraded forest landscapes.

Forest restoration plays a key role in addressing local communities' livelihood needs while contributing to global climate mitigation strategies (Urzedo et al. 2020). Forest restoration projects provide many benefits at the community level, including reductions in greenhouse gas emissions (Sheng et al. 2017); the provision of a range of socio-economic benefits, including income and employment; and the minimisation of disruptions to land management flexibility. However, achieving the benefits of forest conservation through afforestation, reforestation and forest protection requires effective community-level participation in management strategies, as such strategies can provide effective methods for increasing afforestation or reducing deforestation (Miyamoto 2020). Since stakeholders' participation is inherent to the success of any afforestation or forest protection program/project, engaging rural households in such projects increases social acceptance, as they are viewed as key beneficiaries and facilitators of the implementation (Elagib and Al-Saidi 2020). Thus, stakeholders' social acceptance encourages biodiversity, motivates them to implement conservation measures and establishes their trust, which produces the desired outcomes in the form of changes in the environment and livelihoods through a positive behavioural attitude (Burivalova et al. 2017).

To improve the effectiveness of community forest management interventions, it is essential to know the role of various factors and their impacts on community participation in forest landscape restoration (Ullah et al. 2021) as the role of community forest management can be accrued through the amalgamation of the ecosystem-based adaptation (EbA) benefits which are multidimensional as they improve the resilience of ecosystems and communities by enhancing people's well-being and maintaining ecosystem's health along with building community capacity. The EbA is a strategical framework of actions for adapting towards climate change that harnesses nature-based solutions and ecosystem services, which can also be utilised for the overall improvement and simultaneous management of forest management systems. The major elements of EbA which are, namely, use of biodiversity and ecosystem services, helping people and adaptation to climate change are an integrally inherent part of community-based forest management towards being a part of overall adaptation strategy. In respect of VPs which are the community-managed forests of the state and the sole example of this setup, their role towards EbA is pertinent towards designing effective adaptation interventions at local level that can help in managing forest's resilience in the long term. It also depends on a thorough understanding between both of the dependencies of community and interlinkages within the community's social-ecological system and eventually on the interactions of climatic and non-climatic degradation processes in the forests. The climate resiliency of the villages in Himalayan region plays a significant role in addressing the preparedness for EbA at local level (Rawal and Jugran 2024), which can be enhanced through amalgamation of EbA strategies and forest management practices.

The present chapter deals with the diverse forest-people interactions in light of ecosystem-based adaptation through addressing changes in rules and regulations of VPs in the Uttarakhand state and its translational impact on the community towards forest resource conservation.

14.2 Methodology

The present study is based on the assessment of temporal changes in rules and regulations of the management of VPs and its impact on the management practices. The assessment was conducted based on literature review. The methodology undertaken comprises of documentation, primary and secondary collection and assessment following the seven-step model (Onwuegbuzie and Frels 2015) which has been used in community-based forest management (CBFM) surveys (Dasgupta and Vira 2005; Gruber 2011; Nhem and Lee 2019). The seven-step model comprises the steps as follows:

1. Exploring beliefs and topics.
2. Initiating the search.
3. Storing and organising information.
4. Selecting/deselecting information.
5. Expanding the search to include one or more MODES (Media, Observation(s), Documents, Expert(s) and Secondary Data).
6. Analyse and synthesise information.
7. Present the comprehensive literature review (CLR) report.

These seven steps are multidimensional, interactive, emergent, iterative, dynamic, holistic and synergistic and help in data curation. Under the present study, this approach has been widely implemented.

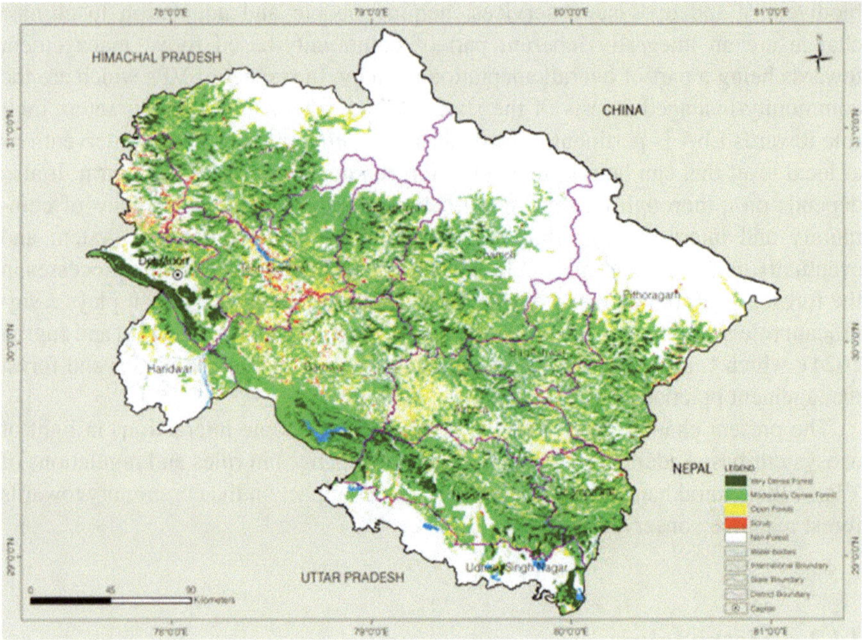

Fig. 14.1 Location map of the study area. (Source: Indian State of Forest Report 2015)

14.2.1 Description of Study Design

The study is conducted on a review basis for the state of Uttarakhand, and the focus was on rules and regulations of the management of VPs and its impact on the management practices. The data was collected from research papers and reports. The maps and graphical data were procured from Forest Survey of India (FSI) (2015). The study sites were the VPs located in the state of Uttarakhand (Fig. 14.1).

14.2.2 Objectives

The objectives of the present study are as follows:

- To conduct perception-based assessment of temporal changes in rules and regulations of Van Panchayats.
- To identify the potential drivers of changes in decentralised management of VPFs.

14.3 Results

The results showed that in the state of present-day Uttarakhand, there are a total of 12,089 VPs that are registered, which cover an area of 5.25 lakh hectares that calculates to 13% of the total land area in the state of Uttarakhand. The state is identifiable divided into two regions, *Kumaun* and *Garhwal*, both have a significant number of VPs in them with the latter taking a slight authority when compared. Kumaun division has 48% of the state's total VPs summing up to 5811, and Garhwal division has 52% of the state's total VPs' count adding up to 6278 (Fig. 14.2).

In reference to the area-wise distribution, the categorical distribution of VPs ranges between VPs lesser than 5 ha and VPs greater than 51 ha. Analysis showed that there are 28% VPs under 0–5 ha area, followed with 31% VPs under 6–15 ha area, 26% VPs under 16–50 ha area and 15% VPs under more than 51 ha area (Fig. 14.3). Moreover, the altitudinal distribution of these VPs shows that there are 27% VPs located between 0 and 1000 m elevation, majority (53%) of the VPs located between 1001 and 1500 m elevation, 19% of the VPs are located between 1501 and 2500 m elevation and 1% of the VPs are located above 2500 m elevation (Fig. 14.4).

The descriptive example of this can be given as per the VPs of the District of Almora, as there are 2199 registered VPs operating. When classified according to altitude of availability, it was found that there are 151 VPs operating in the altitude band of 0–1000 m, 1848 VPs operating in the altitude band of 1000–1500 m, 197 VPs operating in the altitude band of 1501–2500 m and 3 VPs operating in altitudes 2500 m and above. If one classifies the VPs according to the area under their custody, then we find that 545 VPs have an area of 0–5 ha, 517 VPs have an area of 6–15 ha, 772 VPs have an area of 16–50 ha and 365 VPs have an area of more than 51 ha (Fig. 14.5).

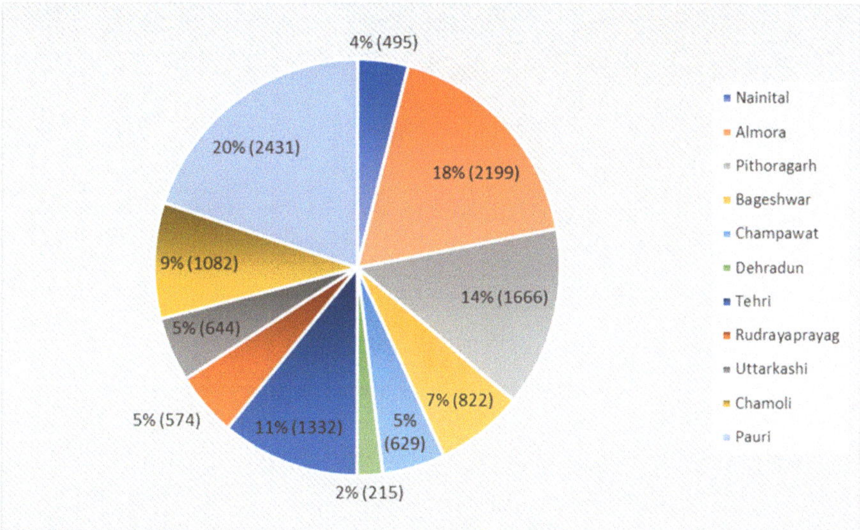

Fig. 14.2 District-wise distribution of VPFs in Uttarakhand

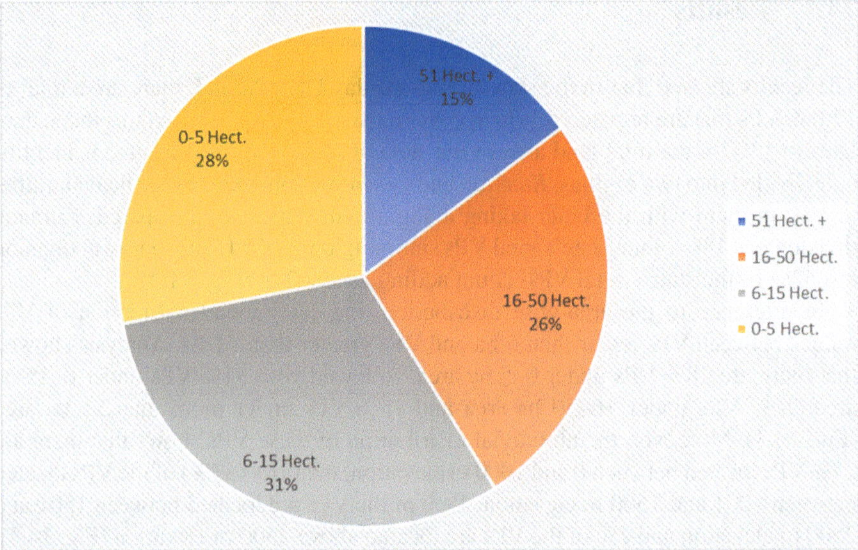

Fig. 14.3 Area (ha)-wise distribution of VPFs in Uttarakhand

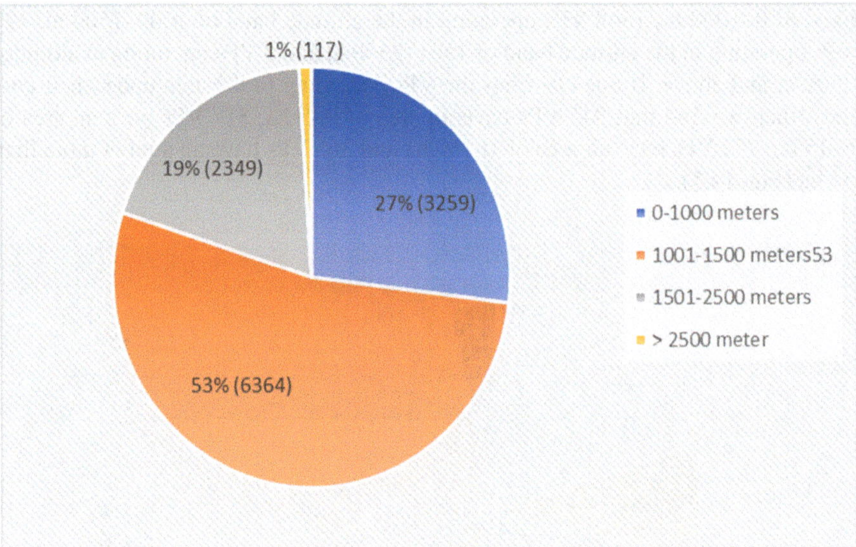

Fig. 14.4 Altitude-wise distribution of VPFs in Uttarakhand

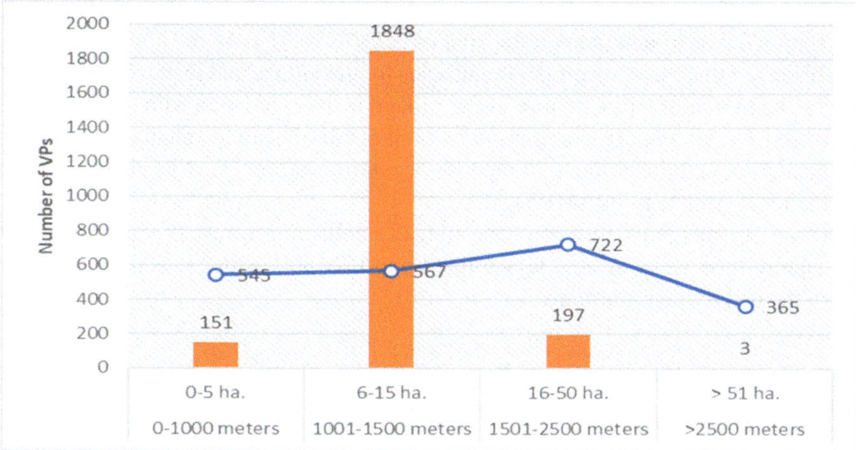

Fig. 14.5 Exemplar description of VPs in Almora District towards area (ha) and altitudinal gradient (m)

14.4 Discussion

It is observed that the VPs were allotted the area outside the protected and reserved forests and anywhere a VP could be formed. For this, a Gram Sabha should place an application before the divisional commissioner showing the need for the same, which should have been signed by not less than two-thirds (which is now made one-third) of the adult population who have been living in the villages under the Gram Sabha for not less than a period of 10 years. The managing committee of the same (VP) should consist of 5–13 members elected for 5 years whom the local people should elect by the simple method of raising hands, and nothing of the secret ballot is required. Currently, each VP has nine members; it is headed by a sarpanch and consists of a minimum of two women and two scheduled caste members (belonging to the officially designated most disadvantaged and marginalised socio-economic groups of people). Every family that resides in the village is a member of the VP. It is mandatory for the managing committee to meet a minimum of four times in the year with an equal time difference in the entire year. The member of the VP officer is the sub-divisional magistrate who is assisted in his work related to the Van Panchayat by the member of the VP inspectors from the Revenue Department.

14.4.1 Functions of the VPs

- To check against the arbitrary felling of trees.
- To distribute the forest and produce among the members equitably.

- To earmark trees fit for silvicultural operations.
- To prevent encroachment on the land of the VP.
- To ensure that there are proper boundaries for the village and they are maintained.
- To ensure that the functioning of the village is in accordance with the directions issued by the deputy commissioner or the sub-divisional magistrate, such as plantation activities under Compensatory Afforestation Fund Management and Planning Authority (CAMPA), etc.
- If the VP members find someone illegally collecting forest produce, they can capture their weapons and levy a fine up to INR 500, which needs the approval of the deputy commissioner.

14.4.2 Evolution, Challenges, and Prospects of VPs

Three key observations emerged from the detailed review of the literature, categorised as follows:

- Decline in institutional management over the years.
- Involvement of the Forest Department in resource provision and decision-making.
- Scoping for ecosystem-based adaptation in community-managed forests.

14.4.2.1 Decline in Institutional Management over the Years

One observed thing is that there has been a systematic decline in the authority of VPs since their commencement in 1931. The 1931 United Province Forest Act classified the forest into three categories—Reserve, Civil-Soyam and Panchayat. Reserve forests had their exclusive right by the Forest Department, whereas the other two categories, Civil-Soyam and Panchayat, came under the jurisdiction of the revenue department. As discussed earlier, certain procedures are to be followed if a VP is to be formed, and after approval, the Civil-Soyam Forest of that area is converted into a VP (Agrawal 1999). The main concept of VP, at the time of their establishment, was to bring up a people-centric institution. Local people have a better and deeper understanding of the ecological and environmental issues associated with the forests (Negi et al. 2012). Initially, two types of VPs were formed, one which was formed out of the reserve forest and governed by the Indian Forest Act, 1927, and the second one was sliced out of the Civil-Soyam land and governed by the guidelines in the District Schedule Act, 1931. However, this was changed with the VP regulations 2005, where both were consolidated and merged into 'Gram-Van'. This resulted in a consolidation of the authority of the Forest Department over all types of forests in the state of Uttarakhand.

14.4.2.2 Involvement of Forest Department in Resource Provision and Decision-Making

Over time, the forest officials have been assigned to look after the functioning of the VPs. The Divisional Forest Officer (DFO), who sits far away from the actual place of concern, has been vested with the authority to formulate the plan for the management of the forest under VPs. Several forest department officials have been responsible for looking after the VPs' affairs after the VP regulations of 2005 and 2012. As per the VP regulation 2005, the DFO must approve every rule made by the VP committee; going by the regulation of 2012, even the SDM has been given the power to give an extension of 6 months to the village management committee.

During the last 80 years, the institution of VP has seen a plethora of changes, and there have been four changes in the rules of VP in the years 1976, 2001, 2005 and 2012. These rules have changed the original concept of the institution of VP and have made it more centralised, and the control of power has shifted to the hands of the state's bureaucracy from the village's residents. An attempt is made in this study to understand the various rules that have been implemented since their inception and the effects that VPs are meted out. The major outcomes are described as follows.

(a) *Temporal Change in Rules and Regulations of Van Panchayats*

According to the rules of 1931, there was no government officer involved in the functioning of the VPs; by the rules of 1976, one special officer was made in charge, and this was raised to five by the rules of 2001, and innumerable officials have been assigned the duties from the rules of 2005. The role of government has changed from being a facilitator to a scrutiniser. Today, officials make rules and regulations for the VPs, and they are informed of what changes are being made. The participatory form of functioning has evolved into something where the stakeholders are treated as beneficiaries. The officials and bureaucrats should have a supportive and passive role in the formation of rules and regulations relating to the VPs and let those decide on the issues who are attached to the forests and have a sense of belongingness.

In the rules of VPs 1931, it was not mentioned who was responsible for formulating working plans for the Panchayat forests, which was given to the VPs in the rules of 1976. This orientation, however, changed completely with the rules of 2001, 2001 and 2012, and now the responsibility is vested in the Forest Department.

Today, the VP is ironically answerable to the government and not to the local people who formed it. It is disheartening to observe that the local residents have no say in the functioning of the VP, and the members of the Panchayat are accountable to those who sit far away from the actual site of concern.

(b) *Financial Concerns of the VP*

Initially, by the rules of 1931, the VPs were granted complete freedom regarding the distribution and utilisation of funds and resources. This was changed, and in successive amendments, official approval was made mandatory, which is often a cumbersome process.

The VPs have the following powers in relation to financial matters:

- Sell grass and twigs for firewood, stones and slates for utilisation by people.
- Tapping of resin and felling of trees and their auction with approval from the Forest Department.
- Auction of trees up to INR 5000, with the approval of DFO.
- The Forest Department conducts actions of amounts above INR 5000 with approval from the conservator of forests.

The money earned from the sales proceeds by the VPs is not fully available at their disposal and needs the government's permission to be utilised. Some VPs have reported that their royalty payment based on Chir-Pine Forest resin has not been provided to their VP accounts. Roughly around 40% of the total sales proceeds are given to the VPs, which is also not readily available for their disposal for development activities that may need to be undertaken in the VPs. Moreover, now the VPs are supposed to contribute 20% to the micro-plan they devised.

(c) *Fines and Forest Guard Vigilance in Van Panchayats*

The sarpanch of the VP has the authority to impose fines against the offenders. Various activities can attract imposition of fines like illicit wood cutting, grazing of cattle in restricted areas, encroachment and so on. But there is one big catch here, and that happens to be that not all offences are reported. Inspection in the forest is usually done to calculate the damage done by the perpetrators, and there is seldom any effort made to prevent it. Moreover, the offences conducted by local villagers are sometimes not reported due to obligations towards social cordial relations.

(d) *Disenchantment of the Local Community*

The forest cover in most of the region in Uttarakhand is depleting at a startling rate due to many factors, including the frequent occurrence of forest fires. Frequent forest fires cause a humongous loss to various species of animals, insects, birds and other creatures by either smouldering them down or snatching away their places of habitation. Also, forest fires translate to massive loss of precious timbre, fuel wood and natural vegetation. The forests of the Himalayan region are highly prone to forest fires, and coniferous species like chir-pine (*Pinus roxburghii*), fir (*Abies pindrow*), spruce (*Picea smithiana*), deodar or cider (*Cedrus deodara*), etc. are all exceedingly predisposed to fire. Forests in the other parts of the country are dominated by deciduous species, which are also prone to fire. Still, there are comparatively more coniferous species found in the Himalayan region. Fires, apart from monetary loss, cause a severe defeat of forests, leading to setbacks in soil fertility and resulting in the soil eroding easily.

14.4.2.3 Scoping for Ecosystem-Based Adaptation in Community-Managed Forests

The scoping of the EbA can be done by addressing the social-ecological aspect of community resilience in light of EbA, and the concept of social-ecological resilience is most relevant in the context of NbS. This is because NbS address societal goals by forming or augmenting place-based partnerships between people and nature, and successful NbS are embedded and built on a particular local social-ecological system (SES), and in an SES, ecosystems and people are interconnected, interdependent and fundamentally indivisible (Turner et al. 2022). Although multiple conceptualisations of resilience have come from different disciplines, a narrative strategy for the ecosystem-based adaptation of the community-managed forests of the Himalayan region can be described as a cumulative amalgamation of forest resource management and inculcation of proactive forest management (Table 14.1). A possible framework for the EbA in VPs is also presented in Fig. 14.6.

The EbA, through community involvement, can be applicable through a community-based approach coupled with enhancing social capital and institutional building, which is crucial for the success and sustainability of EbA. Moreover, the awareness about EbA among government officials looking after the VP management and the policymakers is equally important to accelerate the integration of the

Table 14.1 Ecosystem-based adaptation strategy for the community-managed forests of the region

Strategy 1	Maintaining or increasing the extent of forest ecosystems	Avoiding forest conversion to non-forest land uses such as illicit construction Reforestation of forest lands that have been deforested and afforest suitable land patches inside the forest Increase or implement agroforestry practices near the forest fringe regions
Strategy 2	Maintaining restoration in forest ecosystem	Addressing the impacts of soil and nutrient cycling in forest Maintaining or restoring forest hydrology through spring rejuvenation Prevent the introduction and establishment of invasive plant species and remove existing invasives Reduce competition for moisture, nutrients and light for better forest yield
Strategy 3	Addressing the forest disturbance	Restore or maintain forest fire in fire-adapted ecosystems such as pine-dominated forests Establish natural or artificial fuel breaks to slow the spread of catastrophic fire Altering the forest structure or composition to reduce the risk, severity or extent of wildfire Reducing the risk of tree mortality from biological or climatic stressors in fire-prone forests
Strategy 5	Prioritising the forest management	Prioritise management of locations that provide high aesthetical value across the landscape Prioritise low-vulnerability sites for maintaining or enhancing forest productivity through the plantation of MPTs
Strategy 6	Upgrading the existing forest ecosystem	Introducing the species or genotypes that are expected to be adapted under the climatic conditions Promotion of community-based plantation and 'care for tree' programmes for upgrading the existing forest ecosystem

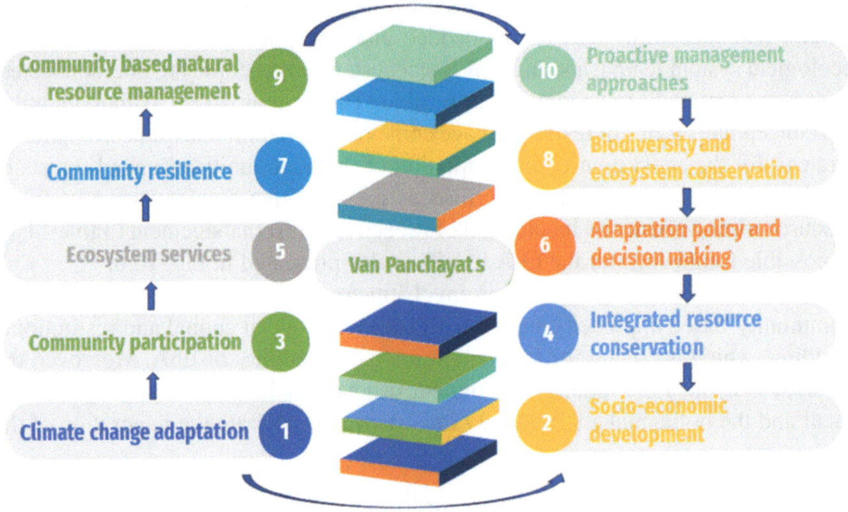

Fig. 14.6 A potential framework of ecosystem-based adaptations in VPs

approach into policies and strategies. The future of EbA planning, implementation and evaluation will require a combination of 'top-down' and 'bottom-up' approaches to ensure sustainable outcomes across various levels of implementation. Moreover, an integrated approach to social protection, climate change adaptation and widespread community awareness are necessary to implement EbA-based forest management in the region.

14.4.3 Major Identified Drawbacks in Management Scenario in Nexus with EbA

The forests of the Himalayan region have multiple roles to play, from fostering biodiversity to providing significant carbon sinks, preventing erosion of soils holding up mountains and supplying forest produce utilised in the day-to-day activities of people dependent on them directly. A study conducted by Pant (2017) and Sharma and Pant (2017) centred on the VPs of the Himalayan region lying in the state of Uttarakhand. It was based upon the impact of extreme climatic events, i.e. forest fires, increases in temperature and rainfall and the availability of resources from the community forests.

The responses collected from the members of the VPs indicated that they knew there was an increase in the temperature and the occurrence of climate change. They could also feel affected by this because it caused a delay in the growth of crops and numerous incidents of crops drying up and postponement in the ripening of fruits. All this was caused by a lack of rainfall and the prevalence of prolonged dry periods, which makes the occurrence of forest fires more possible. The same happens, causing damage to seedlings and saplings along with a dent in the availability of materials and employment, which could be generated by collecting firewood and other forest resources.

The results collected after discussion from the local people of VPs and the aftermaths of the fire showed that there was enough knowledge and awareness among the community about the change in climate, but there was a lack of resources to combat any incident of large-scale fire, and also, there is hardly any precautionary measure taken. Though the members of the forest VP committees take up measures in their capacity to combat forest fires, they are not accurate and nowhere near proper. The repeated fire incidents in regions of Pauri and nearby areas showed that distant areas can be challenging to manage in case of an outbreak of fire, which eventually results in large-scale devastation. One thing that can be easily concluded is that there exists a disconnection between the Forest Department, which has the means of controlling forest fires, and the common people, who get to know when and where there is a fire outbreak. There is an urgent need to combat this gap to prevent repeated fire rashes in forests, as Sharma and Pant (2017) suggested. Though the idea of community participation and decentralisation appears very interesting and captivating and is a measure that helps development at the grassroots level, it is full of its demerits. Some of these are highlighted below:

(a) *Lack of equitable support from the government authorities of both centre and state:* Over time and again, it has been seen that one of the most dominant reasons for the failure of institutions is dependent on the state or central government, which at the local level fails to provide timely support. The supervising authorities fail to provide technical assistance, which happens gradually, leaving the local institutions in their fate. In the long run, these institutions will be unable to manage themselves.
(b) *Lack of appropriate funds:* Lack of adequate financial resources is a problem that attacks nearly every institution in society, and the institution of one Panchayat is no exception. Various government schemes are not implemented at the village level simply because there are not sufficient funds and financial resources available for the disposal of authorities at the local level. The officials taking care of villages and setups placed at lower stands of the hierarchy of institutions are barely given proper monetary support. They are often treated as secondary or unimportant institutions.
(c) *Lack of actual decentralisation and capturing of power by those having authority:* There may be instances where the actual power in local-level institutions is captured by an individual or a group of individuals having a position to dominate the population of a village due to their position in the caste system; their social status in the village may be due to their ability to bring coercive means into action. This deprives people of the ability to decide freely on various matters related to the village. There may also be conditions where people are not even a part of the prices of decision-making and are just informed of the decisions taken by a group of authoritative individuals, which completely

defeats the purpose of decentralisation and the objective of community participation.

(d) *Inability of the government to respond to local needs in a timely fashion:* It is often seen that for various reasons, the government and the government institutions fail to provide support and remedies to needy people in a timely fashion. Often, this is due to a lack of resources, a delay in receiving information or simply because they lack personal incentive towards the situation brought before them. Thus, eventually, people lose interest in and hope for the institution of government.

(e) *Lack of actual accountability:* People working at village-level institutions or local bodies are often not really held accountable for their actions. This results in them being reluctant to do the job and ultimately causing terrible harm to the beneficiaries.

(f) *Bottom-up approach-based leadership:* It is simply not enough to carve out an institution for the participation of people at the local level. It should be ensured that these institutions work to their fullest capacity and achieve the objectives that were kept in mind while they were formed. A review of the basic literature and practical know-how about the functioning of these institutions is testimony to the fact that it is just not sufficient to mould our workplace; they need to be administered and supported in order to reach their objective.

(g) *Involvement of marginalised groups in decision-making:* It is evident that the two sections of society that are most susceptible to being left out of community participation are women and people belonging to the so-called lower caste. It is observed that women often lack the incentive or inclination to participate in the local-level institutions. They may be afraid of speaking out in public or unable to extract time from the household activities that keep them clenched. It, therefore, becomes the obligation of the local-level institution or even the government to ensure that women are helped to come out and take part in the decision-making of the village so that they give their valuable inputs.

14.5 Conclusion

On the basis of review-based studies, it can be suggested that for the successful and intended functioning of the VPs, it is important for community participation to be applicable in the reality of that particular VP rather than superficial representation. As mentioned by Agrawal (1998), successful collective action is not just about forming groups but about achieving the objectives for which the group was formed. Most studies of collective action have assumed that success in organizing a group and success in achieving the objective for which the group is being organised are the same. Under many conditions, the distinction might not be necessary, but in the case of VPs, successfully forming a group to protect village forest resources is different from succeeding in protecting these resources. Once formed, not only must the group create institutions to overcome first-order collective action dilemmas such as free riding and shirking, but it also faces the second-order dilemma of providing monitoring and sanctioning actions to enforce those institutions (Gupta 2014).

Agrawal and Ostrom (2001) and Gupta (2002) have reported that forest councils (VPs) in Uttarakhand represent one of the oldest examples of decentralisation initiatives around the world and present a helpful case study to assess the outcome of decentralisation initiatives in the management of natural resources; the present study showcases the participation of women and people from other communities; notably weaker sections should ensure that each representative section is given a chance to put forth their interests. Also, it is recorded that a community's devotion to a particular forest to a local deity is an attempt to inculcate the feeling of belonging and obligation, whichever works out, to ensure that the people look after the forest and take care of its resources and not just leave it for corrosion. Aesthetical and religious views of the community also pave the way for the proper management of the forest and can be helpful in aspects of VPs where the community's devotion plays a significant role.

It is also necessary to mention that women are a critical part of VPs as they are the primary gatherers of fuel, wood and fodder. They also carry out many protection duties of the forest area since they visit the forests regularly to collect the aforementioned products (Agrawal 2001; Agrawal and Chhatre 2006). Although the perceptions of women towards forest management issues are important, female villagers indicate they play a passive role in decision-making (Germaine et al. 2018). One of the objectives of the National Forest Policy of 1988 was to expand the scope of social forestry through a people's movement (with a particular focus on women) to carry out afforestation and other activities to meet fuel wood, fodder, grass and other needs of the people dependent on forests (Bahuguna and Upadhyay 2002).

Germaine et al. (2018) also support what others have reported that despite attempts to formalise their role and involvement in community and social forestry, women's participation in local community forestry institutions is mostly passive and limited in developing countries like India and Nepal (Agrawal and Chhatre 2006; Chhetri et al. 2013; Thapliyal 2013). In a study conducted in Uttarakhand, Pant and Pandey (2007) echoed about the same findings that although women conduct most of the community forestry work, their participation in forestry council committees was grossly inadequate.

As suggested by Germaine et al. (2018), the present study also concludes that as the villagers are ready and willing to improve the situation, unfortunately, they do not have the resources to make change unilaterally—they need professional and financial support from the Forest Department. The study views the viewpoint of Germaine et al. (2018) that decentralisation in natural resources management sounds good at the theoretical level. Still, it cannot be effectively implemented without professional guidance and resources. Regardless of the region or country, community forests require the expertise of foresters and natural resource managers to ensure that the users' long-term goals and objectives are met sustainably. In this aspect, the EbA approaches of harbouring the community needs and involvement of the community for decision-making can be incorporated by ensuring people's participation. As these VPs play a crucial role in conserving biodiversity and ensuring sustainable use of forest resources, the following are some ecosystem-based approaches that VPs can employ for sustenance:

- *Community Participation*: VPs involve local communities in managing and protecting forest resources. This participatory approach ensures that the people who depend on the forests for their livelihoods are directly involved in conservation efforts. Enhancing community participation can enhance the role of accruing benefits from ES and forest resource management.
- *Sustainable Resource Use*: Management practices that focus on the sustainable use of forest resources, such as controlled grazing, regulated fodder collection and sustainable harvesting of timbre and non-timbre forest products, will enhance the VPs' ecological and socio-economic strengths and can pave the way for the Payment for Ecosystem Services (PES).
- *Biodiversity Conservation*: VPs work towards conserving the region's rich biodiversity by protecting habitats and promoting the growth of native species. This includes measures to prevent overexploitation and restore degraded forest areas, eventually leading to restored habitats and native niches for the various floral and faunal species. This can be achieved through targeted plantation of native plant species, which can eventually enhance the region's micro- and macroclimate and combat the impacts of climate change.
- *Traditional Knowledge*: The EbA system integrates traditional ecological knowledge with modern scientific practices. This helps maintain the ecological balance and preserve the unique biodiversity of the region by sustaining and maintaining the TEK of the region. Moreover, to cope with changing climate, community needs to be more resilient, and for this purpose, communities can align their activities based on climatic events or by inventing unique practices to cope with continued varying climatic conditions, for the same purpose of forming an eco-sensitive environment and constructing more climate-resistant infrastructure (Rawal and Jugran 2024).
- *Adaptive Management*: VPs adapt their management practices based on the changing ecological conditions and the community's needs. This flexibility allows them to respond effectively to environmental challenges and ensure the long-term sustainability of forest resources. For instance, aligning governance and decision-making with the needs of marginalised groups can increase the resilience and adaptive capacity of these social-ecological systems and pave the way for robust EbAs.

14.6 Way Forwards

To ensure that VPs contribute to the emerging and increasing needs of the people, they must be managed based on scientific principles to enhance their ecological contributions and increase productivity. It is also noted that, unfortunately, the VP communities do not have the technical knowledge or the financial resources to develop a management plan for their community forest. On the other hand, the Forest Department does not have the resources or inclination to help them. Sometimes, the department does not even participate in the decision-making process meetings of the community group, and consequently, the communities are left to fend for themselves, resulting in degrading forest conditions and negligible technical resources. Moreover, several policies and plans are being implemented that have manipulated the structure of the institutions in the past, thereby contributing to the institutional weakening. Therefore, it is suggested that there is a need to thoroughly assess such plans and policies for the possible long-term impacts that they might have on institutions that will cater for the needs of the community towards conservation, adaptation and long-term sustainable management of the forest resources. Moreover, for the full functioning of VPs, as the EbA example in the Himalayan region, it is highly necessary to address the prospects of VPs by focusing on a few attributes of resilience: diversity, modularity and flexibility. As the VPs include multiple pathways for reactive and anticipatory adaptation, the management issues can be addressed by reinforcing community resilience. It can also be concluded that while the community manages ecological processes in VPs to enhance the overall resilience of the SES, the underlying social and political dynamics of VPs tend to be neglected in adaptation policy and planning. This prevents VPs from harnessing the benefits of ecological resilience to enhance their adaptive capacity. The contribution of ecological resilience to societal adaptation has been constrained by large-scale social and political drivers, especially bureaucratic structures underpinning the governance of forests in the region. Therefore, it is recommended that EbA is fully conversant with the local community and takes into account local power dynamics for sustenance in the Himalayan region context as harnessing the adaptation benefits of forest ecosystems and their related community institutions needs to consider complex and multi-layered interactions within social-ecological systems. It can be concluded that the translation of ecological resilience into societal adaptation in VPs has been constrained by bureaucratic structures and sociopolitical forces that overlook the local power dynamics at the VP level; therefore integrating resilient thinking with proactive approaches can pave the way for the adaptation policy and adaptive management of these VP resources.

References

Agarwal B (2001) Participatory exclusions, community forestry, and gender: an analysis for South Asia and a conceptual framework. World Dev 29(10):1623–1648

Agrawal A (1998) Group size and successful collective action: a case study of forests management institutions in the Indian Himalayas. In: Gibson C, McKean MA, Ostrom E (eds) Forests, trees and people programme, Forestry Department working paper no. 3 in Forest resource and institutions. FAO Corporate Document Repository, pp 23–30

Agrawal R (1999) Van Panchayats in Uttarakhand: a case study. Econ Polit Wkly 34:2779–2781

Agrawal A, Chhatre A (2006) Explaining success on the commons: community forest governance in the Indian Himalaya. World Dev 34(1):149–166
Agrawal A, Ostrom E (2001) Collective action, property rights, and decentralization in resource use in India and Nepal. Polit Soc 29(4):485–514
Bahuguna V, Upadhyay A (2002) Forest fires in India: policy initiatives for community participation. Int For Rev 4(2):122–127
Bhattarai B (2020) How do gender relations shape a community's ability to adapt to climate change? Insights from Nepal's community forestry. Clim Dev 12(10):876–887
Bisht D, Joshi VC, Yadava AK, Sundriyal RC, Jugran HP (2020) Community forest management in Uttarakhand: a comparative case study of Western Himalayan Van Panchayats. ENVIS Bull Himalayan Ecol 28:55
Burivalova Z, Hua F, Koh LP, Garcia C, Putz F (2017) A critical comparison of conventional, certified, and community management of tropical forests for timber in terms of environmental, economic, and social variables. Conserv Lett 10(1):4–14
Chhetri B, Johnsen F, Konoshima M, Yoshimoto A (2013) Community forestry in the hills of Nepal: determinants of user participation in forest management. Forest Policy Econ 30:6–13
Dasgupta P, Vira B (2005) Q methodology for mapping stakeholder perceptions in participatory forest management. Institute of Economic Growth Deli, Delhi
Elagib NA, Al-Saidi M (2020) Balancing the benefits from the water–energy–land–food nexus through agroforestry in the Sahel. Sci Total Environ 742:140509
Forest Survey of India (2015) Forest Survey of India, Ministry of Environment, Forest and Climate Change, Government of India (State of Forest Report 2011, 2013, 2015). http://F.nic.in/. Accessed on 15 July 2017
Germain R, Ghosh C, Jayasuriya M (2018) Community forestry in the state of Uttarakhand, India: not meeting the needs of the villagers. Small-scale Forestry 17(2):225–242
Ghate R, Nagendra H (2019) Building institutional foundations for community forest management, Centre for Science and Environment, New Delhi. Ajim Premji University
Gibson C, Williams JT, Ostrom E (2005) Local enforcement and better forests. World Dev 33:273–284
Gibson CC, Dodds D, Tuner P (2007) Explaining community-level forest outcomes: salience, scarcity and rules in Eastern Guatemala. Conserv Soc 5(3):361–381
Gilmore D (2016) Forty years of community-based forestry: a review of its extent and effectiveness. FAO Forestry Paper, 176. ISBN 978-92-5-109095-4
Gruber JS (2011) Perspectives of effective and sustainable community-based natural resource management: an application of Q methodology to forest projects. Conserv Soc 9(2):159–171
Guha R (1985) Scientific forestry and social change in Uttarakhand. Econ Polit Wkly 20:1939–1952
Guha R (1994) Forestry debate and draft forest act: who wins, who loses? Econ Polit Wkly 29:2192–2196
Guha R, Gadgil M (1989) State forestry and social conflict in British India. Past & Present 123:141–177
Gupta D (2002) Explaining enforcement mechanisms in collaborative natural resource governance: a study of cases of Van Panchayats from Central Himalayan region, India. Dissertation, Ohio State University
Gupta D (2014) Explaining enforcement mechanisms in collaborative natural resource governance: a study of cases of Van Panchayats From Central Himalayan Region, India (Doctoral dissertation, The Ohio State University)
Indian State of Forest Report (2015) Forest and tree resources in states and union territories. Forest Survey of India (Ministry of Environment & Forests), Dehradun
Landell-Mills N, Porras IT (2002) Silver bullet or fools' gold?: a global review of markets for forest environmental services and their impact on the poor. A research report prepared by the International Institute for Environment and Development (IIED), London. A component of the international collaborative research project steered by IIED: Instruments for sustainable private sector forestr

Mawdsley E (1998) After Chipko: from environment to region in Uttaranchal. J Peasant Stud 25(4):36–54. ISSN 0306-6150
Miyamoto M (2020) Poverty reduction saves forests sustainably: lessons for deforestation policies. World Dev 127:104746
Mukherjee P (2003) Community forest management in India: the Van Panchayats of Uttranchal. XII world forestry congress, Quebec City
Nagahama K, Saito K, Masuda M, Ota M, Gairola H, Kala SK, Rakwal R (2016) Forest commons use in India: a case study of Van Panchayat in the Himalayas reveals people's perception and characteristics of management committee. Environ Ecol Res 4(3):128–139
Nagendra H (2007) Drivers of reforestation in human-dominated forests. Proc Natl Acad Sci U S A 104:15218–15223
Negi B, Chauhan D, Todaria N (2012) Administrative and policy bottlenecks in effective management of Van Panchayats in Uttarakhand, India. Law Environ Dev J 8(1):141–148
Nhem S, Lee YJ (2019) Using Q methodology to investigate the views of local experts on the sustainability of community-based forestry in Oddar Meanchey province, Cambodia. For Policy Econ 106:101961
Onwuegbuzie AJ, Frels R (2015) Seven steps to a comprehensive literature review. Sage, Los Angeles
Pant H (2017) Case study of Uttarakhand in perspective of extreme climatic events: fire, ecosystem and livelihoods. In: Climate change research at universities. Springer, Cham, pp 519–531
Pant N, Pandey N (2007) Introducing JFM to rural women of Kumaun Hills—a case study. Indian For 133(12):1690–1694
Rawal R, Jugran HP (2024) Impact of climate change and mitigation plans for building climate-resilient village ecosystem: challenges and adaptive strategies. In: Warming mountains: implications for livelihood and sustainability. Cham, Springer, pp 197–211
Sarin M (2005) Laws, lore and logjams: critical issues in Indian forest conservation. International Institute for Environment and Development, London
Sharma S, Pant H (2017) Vulnerability of Indian Central Himalayan forests to fire in a warming climate and a participatory preparedness approach based on modern tools. Curr Sci 112(10):2100–2105
Sheng W, Zhen L, Xie G, Xiao Y (2017) Determining eco-compensation standards based on the ecosystem services value of the mountain ecological forests in Beijing, China. Ecosyst Serv 26:422–430
Thapliyal J (2013) Van Panchayats script a success story. Jharkhand Panchayat Women Resource Centre. http://jpwrc4.blogspot.com/2013/01/
Turner B, Devisscher T, Chabaneix N, Woroniecki S, Messier C, Seddon N (2022) The role of nature-based solutions in supporting social-ecological resilience for climate change adaptation. Annu Rev Environ Resour 47(1):123–148
Ullah A, Sam AS, Sathyan AR, Mahmood N, Zeb A, Kächele H (2021) Role of local communities in forest landscape restoration: key lessons from the billion trees afforestation project, Pakistan. Sci Total Environ 772:145613
Urzedo DI, Neilson J, Fisher R, Junqueira RG (2020) A global production network for ecosystem services: the emergent governance of landscape restoration in the Brazilian Amazon. Glob Environ Chang 61:102059

Chapter 15
Computing Payment of Water as an Ecosystem Service: Valuation and Management Practices in Western Himalaya, Uttarakhand, India

Subhasis Mahato, Dharmveer Kaintura, Atif Hussain, Kaiser Iqbal, and Rahul Kumar Singh

Abstract The article deals with a study conducted in the Western Himalayas to evaluate the willingness to pay (WTP) and local community-based approach to water resource conservation and sustainable use. The open-ended elicitation method of contingent valuation method was employed in this empirical work. The households to be surveyed were randomly selected, and 50% of the sampled households (70 households) were surveyed in selected villages. It was found that family size, income, landholding, animal unit, and education status are important factors in willingness to pay for an environmental service. A regression model was obtained using dependent and independent variables explaining maximum WTP variation. The mean willingness to pay was calculated to be Rs. 1,99,206.00 for the whole study area. Seasonal streams and mountain sources were the primary sources of water. The household water management system demonstrates a collective commitment to fairness, ensuring that local elites do not receive preferential treatment in accessing water. Villagers have established a conservation forest area, prohibiting tree cutting, land clearing, and hunting to safeguard the vital water source. Implementing a sound strategy could potentially quantify these valuable ecosystem services, aiding

---

S. Mahato (✉) · R. K. Singh
Department of Forestry and Natural Resources, HNB Garhwal University, Srinagar, Uttarakhand, India

Department of Forestry, P.R.M.S. Mahavidyalaya, Bankura, West Bengal, India

D. Kaintura
Department of Forestry and Natural Resources, HNB Garhwal University, Srinagar, Uttarakhand, India

A. Hussain · K. Iqbal
Department of Forestry and Natural Resources, HNB Garhwal University, Srinagar, Uttarakhand, India

Department of Wildlife Protection, Wildlife Division, Shopian, J&K, India

Y. D. I. Khan et al. (eds.), *Ecosystem-based Approaches for Resilience Building in Himalayan Landscapes*, Disaster Resilience and Green Growth,
https://doi.org/10.1007/978-981-95-2007-7_15

in the sustainable management of resources and curbing their excessive exploitation in the future.

Keywords Willingness to pay · Contingent valuation method · Factor analysis · Principal component analysis · Ecosystem services

15.1 Introduction

Ecosystem services refer to the benefits that humans derive from natural ecosystems. Human survival and well-being rely heavily on a wide range of Ecosystem Goods and Services (EGSs), which include provisioning resources, regulatory functions, supporting processes, and cultural benefits, as outlined in the Millennium Ecosystem Assessment (MEA 2005). There is a growing recognition of the critical role ecosystems play in sustaining the well-being of both current and future generations, thereby contributing significantly to sustainable development (MEA 2005; Turner et al. 2010). Assigning economic value to these services is increasingly seen as vital for conservation efforts and ecosystem-based management (Plummer 2009; Stenger et al. 2009), and it is gaining relevance in shaping local, national, and global policy decisions (Turner et al. 2010).

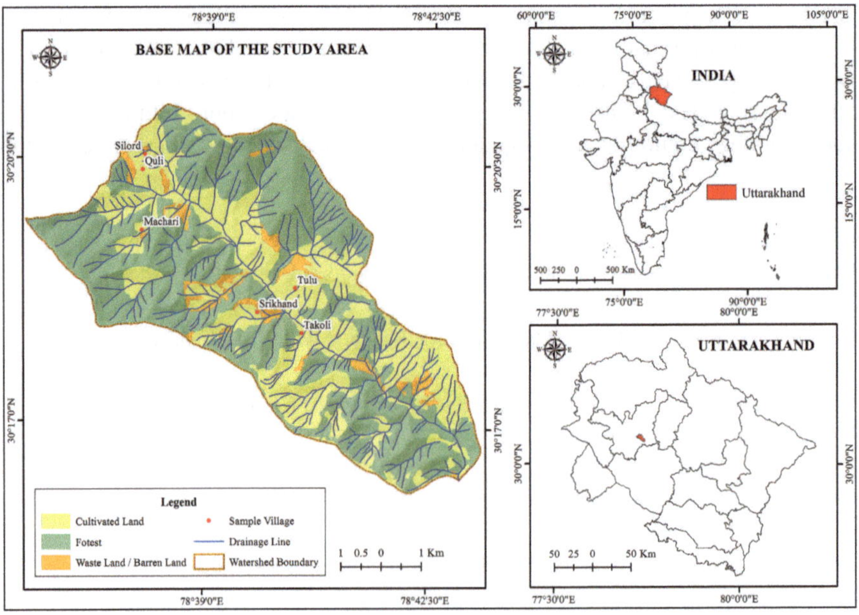

Fig. 15.1 Location map of the study area

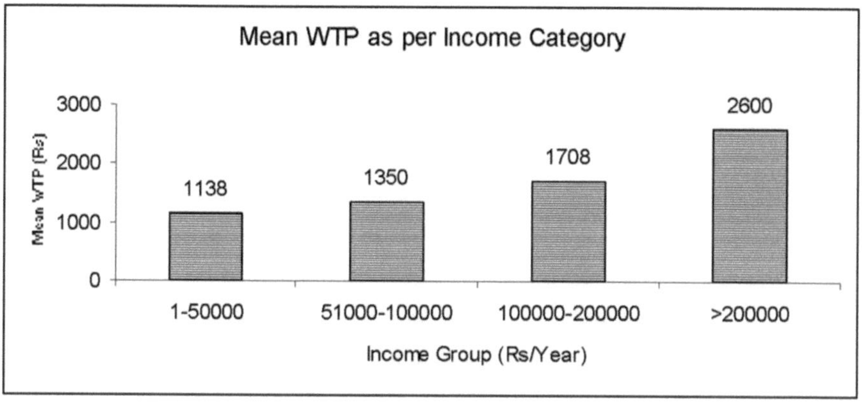

Fig. 15.2 Mean WTP as per income

However, most ecosystem services lack formal markets and observable prices, and their economic significance often goes unrecognized. This lack of awareness can result in the degradation and overuse of these vital resources (Ojea et al. 2012). As a result, the ecosystem service framework proposed by the MEA is gaining traction in both academic and policy circles. Despite this growing acceptance and the surge in research adopting the MEA approach (Fisher et al. 2009), accurately quantifying the extent and value of ecosystem services remains a complex challenge (Bateman et al. 2011; Turner et al. 2010).

A recent review by De Groot et al. (2010) highlights that one of the most pressing unresolved challenges is understanding and quantifying how ecosystems generate services. Moreover, considerable confusion persists among both policymakers and scholars across disciplines regarding the legitimacy and implications of valuing ecosystem services (Spangenberg and Settele 2010; Turner et al. 2010). Several studies argue that the Millennium Ecosystem Assessment (MEA) lacks clarity in differentiating between ecosystem functions—defined as the underlying ecological processes—and the services derived from them (Haines-Young and Potschin 2010). Therefore, further research is essential to examine the limitations of applying the MEA framework directly for economic valuation and to explore strategies to overcome these challenges.

Water is a fundamental element for ecosystem functioning and is vital to sustaining human well-being (Falkenmark 2003). A range of ecosystem services, collectively known as hydrological ecosystem services (HES), are derived from freshwater systems. These services are often regulated by terrestrial ecosystems and include provisioning services such as water for drinking, energy generation, industrial processes, and irrigation (Brauman et al. 2007), as well as regulating services like water purification and erosion control (De Groot et al. 2010). The availability and quality of freshwater, along with the delivery of HES, are closely linked to land and resource management practices. Key factors such as vegetation cover, land use, and climate play a significant role in determining how rainfall is partitioned into green and blue

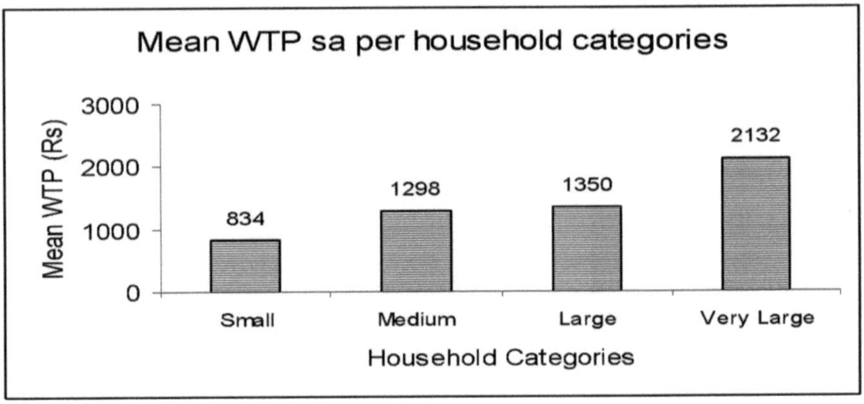

Fig. 15.3 Mean WTP as per household categories

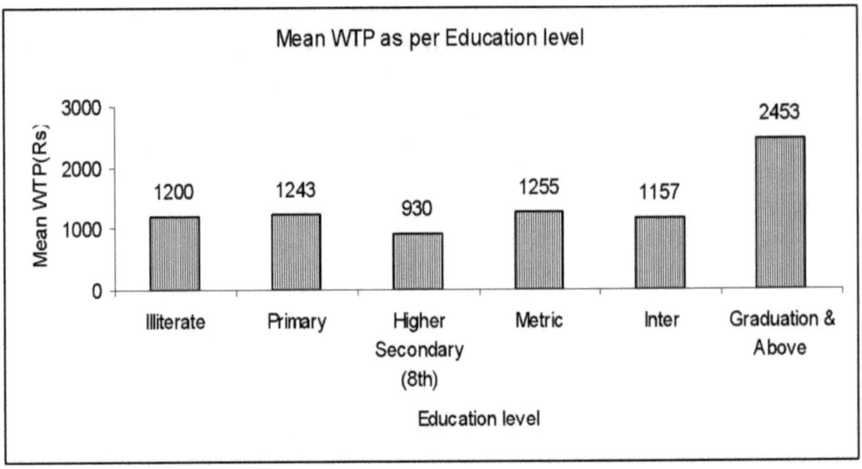

Fig. 15.4 Mean WTP as per education level

water flows (Zhang et al. 2001; Calder 2005). Forests, in particular, contribute to ecosystem services by maintaining the quantity, quality, and stability of water flows. Substantial evidence shows that land use changes can significantly disrupt the water cycle by altering both the volume and condition of freshwater flows within catchments (Farley et al. 2005; Gerten et al. 2005; Scanlon et al. 2007).

Payment for ecosystem services (PES) has emerged as a new approach to managing the valuable services derived from ecosystems in the last decades. Developing countries have increasingly adopted this method to assess environmental quality in the past decade. It has been widely used to measure willingness to pay (WTP) for public projects that provide essential services like safe drinking water and sanitation (e.g., Whittington et al. 1991; Briscoe et al. 1990; Johnson and Baltodano 2004;

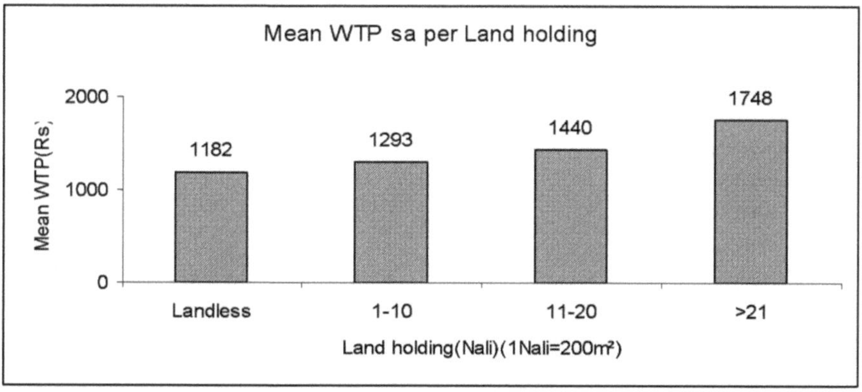

Fig. 15.5 Mean WTP as per landholding size

Choe et al. 1996; Lauria et al. 1999; de Oca and Bateman 2006; Biswas and Venkatachalam 2015; Chandrasekaran et al. 2009; Devi et al. 2009; Griffin et al. 1995; Chatterji 1997; Roy et al. 2004; Karthikeyan et al. 2009; Dutta et al. 2010).

Additionally, the method has been applied to estimate the nonmarket value of instream flows for recreational use (Daubert and Young 1981; Bishop et al. 1987; Brookshire and Smith 1987; Sanders et al. 1991; Duffield et al. 1992; Weber and Berrens 2006).

In developing countries, the contingent valuation method (CVM) is increasingly being used to estimate the value of environmental goods and services, as evidenced by several studies (Hadkar et al. 1997; Shultz et al. 1998; Whittington 1998; Raje et al. 2002). Notably, researchers such as Bernard et al. (2009), Biao et al. (2010), and Nahuelhual et al. (2007) have evaluated the value of freshwater forests in relation to drinking water supply. However, limited attention has been given in developing countries to valuing ecosystem services in complex natural systems like rivers particularly those involving indirect use and non-use values (Venkatachalam 2004). Some of the most advanced applications of ecosystem service valuation are found in Payment for Ecosystem Services (PES) programs, which have been implemented across a variety of global contexts, including in China, Costa Rica, Mexico, and France (Gong et al. 2010; Honey-Rosés 2009; Muñoz-Piña et al. 2008; Pagiola 2008; Perrot-Maître 2006). Despite this progress, there is a notable gap in the literature concerning the valuation of water resources in the Western Himalaya.

In response to this gap, a study was conducted to estimate the nonmarket value of water as an ecosystem service derived from the surrounding forest area, using the contingent valuation method (CVM). The study aimed to assess consumers' willingness to pay (WTP) for water-related environmental services and to explore the influence of various socioeconomic factors on respondents' WTP.

Table 15.1 Standardized loading of various socioeconomic variables for WTP

Parameters	Component I	II	III
Education level	**0.761**	−0.317	−0.077
Agriculture landholding	**0.649**	0.196	−0.109
Environmental awareness	0.570	0.078	0.415
Income source	−0.377	0–0.045	0.065
Animal unit	−0.144	**0.759**	−0.259
Household size	0.131	**0.722**	0.455
Annual income	0.467	0.585	0.213
Source of drinking water	−0.230	0.011	**0.838**
Eigenvalues	1.98	1.546	1.050
% variation	24.799	19.319	13.125
Cumulative %	24.799	44.118	57.242

Note: Highlighted variables with eigenvectors ≥±0.70 are considered significant

15.2 Materials and Methodology

15.2.1 Study Area

The study was conducted in the sub-Himalayan region in Tehri Garhwal, Uttarakhand's districts, in 2008. The region is home to a rich diversity of plant and animal life, supporting a variety of forest types that differ in species composition based on elevation and latitude. The study area lies between the parallels of 30°18′15.5″ to 30°20′40″ N latitude and 78°40′36.1″ to 78°37′40.4″ E longitude. Two Gram Sabha (Tokoli and Quli) were randomly selected in Tehri. Three villages each from Quli (Machari, Quli, and Silord) and Tokoli (Srikhanda, Tokoli, and Tulu) were selected (Fig. 15.1). The total land of these villages is 627 ha, which represents water bodies (4.78 ha), forest (285 ha), settlement (4.395 ha), agriculture land (176.7 ha), and degraded land (156.2 ha). The forest area in Tokoli Gram Sabha was dominated by *Pinus roxburghii* and its associate species, and Quli Gram Sabha was dominated by *Quercus leucotrichophora* species. Chandrabhaga stream is the primary source of water and irrigation in four villages (Tokoli, Tulu, Srikhanda, and Machari). At the same time, Pajani and Butas were the primary water sources of Silord and Quli villages.

15.2.2 Methodology

In the study area, three villages, each from two sites (Gram Sabha Tokoli and Quli) from the Takolighard watershed, were selected for data collection. The criterion for selection of Gram Sabha was due to their village community household water management system, which is maintained by the local community. These villages were rich in biodiversity, and people depended on forests (fuelwood, fodder, NTFP) and water for their livelihood, which was critical to their household economy and willingness to pay for water services. The households to be surveyed were randomly selected, and a 50% sampling percentage (70 households) was carried out due to the number of households (small village size) and low population in the area. These sites were selected to assess the "willingness to pay (WTP)" and local community-based approach to this resource's conservation and sustainable use. The contingent valuation method (CVM) was used to capture mainly the nonconsumptive direct values, such as the benefit of forests in providing drinking water. The WTP elicitation employed an open-ended format, asking respondents to state the maximum amount they were willing to pay. Following the recommendations of the National Oceanic and Atmospheric Administration (NOAA) panel (1993), the questionnaire underwent thorough pre-testing prior to the main survey. The final contingent valuation method (CVM) survey was conducted through face-to-face interviews.

To minimize potential sponsor bias, respondents were informed at the outset that the research was being conducted by university students. This helped ensure neutrality, as universities are generally perceived as impartial institutions, reducing any pressure on respondents to please the interviewer or seek social approval. This approach also aimed to limit protest zero responses and opposition to elements of the valuation scenario. Prior to eliciting responses, each participant was presented with a scenario highlighting the role of forests in conserving water sources, thereby reducing non-response rates and helping respondents understand the value of the service being assessed. To further minimize bias and obtain realistic valuations, respondents were briefed on the broader ecological connection between forests and water supply. To avoid scenario misspecification biases, all interviewers received training in conducting contingent valuation method surveys and were prepared to address likely questions from respondents. Responses were excluded from the analysis if a zero WTP was classified as a "protest zero," as defined by Carson et al. (2003). However, genuine zero responses—those reflecting a true lack of willingness to pay—were retained in the dataset.

The household interview consisted of two sections. The first was a detailed household questionnaire to elicit people's willingness to pay. This section also consisted of additional questions to know the socioeconomic profile of individual households. The socioeconomic questions comprised several family members, livestock, education status, environmental awareness, occupation, etc. The second section of the questionnaire dealt with the key people familiar with what was happening in the region. The questionnaire also aimed to get information on land use practices, water sources, seasonal availability, history of water services, payment for water use, and any management practices being followed for the conservation of water service.

To estimate the economic value of water services, the following regression model was developed. The model incorporates variables that represent both the perceived utility of water services and key indicators of a household's socioeconomic status. These variables are expected to explain individual household preferences for improved water quality and are included in the estimation framework, specified as follows:

$$\text{WTP} = f\left(\text{INC,EDULV,ENVAWR,HHM,AU,LH}\right)$$

where WTP = willingness to pay, INC = the annual income of the entire family (expressed in rupees), EDULV = the education level of the respondent, ENVAWR = environmental awareness, HHM = household member, LH = total landholding of entire family, and AU = adult cattle unit (adult cattle units (ACU) of cow, buffalo, and sheep were calculated as per the conversion formula used by Tyagi and Shankar 1995, while as ACU for horse, mule, ox, and hen were calculated as per the report of FAO 2005).

15.3 Results

15.3.1 Socioeconomic Profile of Respondent Households

Family size, income, landholding, animal unit, and education status are important factors in the willingness to pay for environmental services. Household size ranged from 1 to 22 persons, and the average household size was 6.8. Most respondents were found to be educated, and the overall literacy rate was 80%. While agriculture is the dominant source of livelihood across villages, individual households often rely on a diverse range of occupations, including agricultural work, private-sector employment, and government jobs. The average annual income per household varied among the households. The average cultivated land was 15.10 Nali (1 Nali is equal 200 m^2 in Uttarakhand). 80% of households have livestock. These livestock include buffaloes, cows, goats, etc. The total adult cattle unit (ACU) holding was 233.1, with an average ACU of 3.33 per household. The major water sources for most households are tap water and rivers.

15.3.2 Contingent Valuation Method Analysis

The contingent valuation method (CVM) was employed to elicit the monetary value that society is willing to pay for water services. The data collected were analyzed using regression techniques, with WTP as the dependent variable and various quantitative and qualitative factors serving as independent variables. The model was obtained using dependent and independent variables as follows:

$$\text{WTP} = 157.55\,\text{EDULV} + 0.00212\,\text{INC} - 35.55\,\text{AU}$$
$$-2.35\,\text{LH} + 437.39\,\text{ENVAWR} + 119.61\,\text{HHM}\left(R^2 = 0.33\right)$$

The variation in WTP (33%) was attributed to six variables, viz., the education level of the respondent, annual income, environmental awareness, animal unit, landholding, and household members. The mean value of the dependent variable (WTP) was calculated as Rs. 1355.14 (n = 70), and the maximum value for WTP (Range) was observed to be Rs. 6720. The estimated value the people are willing to pay collectively to derive the water service from the ecosystem annually in six villages (n = 147) was calculated to be approximately Rs. 199,206.00. The influence of various independent variables on the mean willingness to pay has been studied in the following directions.

15.3.3 WTP Based on Household Income

Income significantly correlates with the willingness to pay any society to derive any ecosystem service to fulfil their basic requirements. For estimating the willingness to pay, the households were categorized into various income groups as follows: Group I (1–50,000 Rs.), Group II (51,000–100,000 Rs.), Group III (100,000–200,000 Rs.), and Group IV (>200,000 Rs.). The highest number of households (54%) was found in the lower-income group (Group I), followed by income Group II (23%) and income Group III (18%). It has been found that only a few households (5%) were found in the higher-income group (Group IV)—the mean value of willingness to pay increases with the increasing income groups. The range in the mean value for willingness to pay varies in different income groups, and the value ranges between Rs. 2400 and 6720. It has been found that Group IV (>200,000 Rs.) has the highest WTP (Rs 2600) than that of others. The increase may be because higher-income respondents spend more money to fulfil their basic needs than the lower groups (Fig. 15.2).

15.3.4 WTP Based on Household Family Size

The household size also plays an important role in the willingness to pay to derive the ecosystem services in the rural areas of the Himalayas. For estimating the willingness to pay, the households were categorized into various household size groups as per the number of members in the households: small (1–4 individuals), medium (5–7 individuals), large (8–10 individuals), and enormous (>10 individuals). It was found that the maximum number of households was in medium-sized families (38%), and the minimum was in very large families (12%). The number of households in small and large families was 27% and 23%. The willingness to pay varied in different household family size classes. It has been found that their willingness to pay increases as the size of family members increases. The highest mean WTP of Rs. 2132.00 was found in huge families and the lowest in small families (Rs 834 per year). The reason could be that large households need more water than smaller ones. These families also face water scarcity problems more in summer than smaller families (Fig. 15.3).

15.3.5 WTP Based on Education Level and Environmental Awareness of Households

The education of any community creates environmental awareness and the link between biodiversity and sustainable livelihood. The education level of the respondents has been clubbed into the following categories: Category 1, Illiterate; Category 2, Primary; Category 3, Higher Secondary; Category 4, Metric; Category 5, Inter; and Category 6, Above Graduation. It has been found that the WTP increased from illiterate to above graduation category. The highest numbers of respondents were found in Categories 1 and 2, whereas the lowest were found in Categories 6 and 5. The maximum WTP was in Category 6 (Rs. 2453.00), and the lowest was in Category 3 (Rs. 930), followed by Category1 (Rs. 1200). The reason for the highest WTP in Category 6 may be due to the higher education level which generates some knowledge of importance related to forest in providing water resource and improving their conditions. Figure 15.4 shows the mean willingness to pay in context to the level of education. Most respondents (81%) were aware of environmental-related issues, whereas only 19% were unaware. The mean WTP was higher (Rs. 1494) in the respondents who were aware of environmental issues, whereas it was lower (Rs. 462) in respondents who were unaware of environmental issues. Education was the main reason for higher WTP in awareness of environmental issues.

15.3.6 WTP Based on Landholding of Households

The households were categorized as per their landholding (1 Nali = 200 square meters in Uttarakhand) to calculate the WTP for water services as follows: Category 1 (Landless), Category 2 (1–10 Nali), Category 3 (11–20 Nali), and Category 4 (>21 Nali). The majority of households (48%) were found in Category 2, and a lower number of households (10%) were found in Category 1 (Landless), whereas 21% of each number of households were found in Categories 3 and 4. The WTP varied in different households as per their landholdings. It has been found that WTP increases as the landholding of households increases. The highest WTP was found in Category 4, followed by 3 and 2, whereas it was lowest in Category 1 (Fig. 15.5). The water used to irrigate their fields may be the reason for the highest WTP as the landholding increases.

15.3.7 Factorial Analysis of the Variables for WTP

Principal component analysis (PCA) was employed to extract the factor loadings for various parameters. The correlation matrix revealed strong interrelationships among several measured variables. Given the high degree of correlation, PCA was utilized to identify the key variables influencing the willingness to pay. Eigenvalue analysis indicated that three principal components should be retained to represent WTP. The Kaiser-Meyer-Olkin (KMO) measure of sampling adequacy (0.553) and the significance of Bartlett's test of sphericity confirmed the suitability of the data for principal component analysis. Only principal components with eigenvalues greater than 1 were selected to determine the main components. The component matrix of various parameter variables showed that three principal components were retained, which accounted for 57.2% variation. In Principal Component I, two (2) parameters were loaded heavily on the component, and variables included education level (0.761) and agriculture landholding (0.649). This component explains educational awareness has a significant effect on willingness to pay. Principal component II shows that two (2) parameter variables were loaded heavily on the component, and variables included animal unit (0.759) and household size (0.722). This component explains human animal capital as a factor for willingness to pay. Principal component III shows that only one parameter loaded heavily on the component, and the variable is the source of drinking water (0.838), which explains the type of water source as a primary factor for the willingness to pay (Table 15.1).

15.4 Discussion and Conclusion

The Himalayan region has a rich legacy of water harvesting traditions that have developed over thousands of years. However, water stress is intensifying across India, with per capita water availability decreasing from 1986 m^3 in 1998 to 1731 m^3 in 2005 and projected to fall further to 1140 m^3 by 2050 (Gupta and Deshpande 2004). Climate change is expected to significantly impact water availability during the dry season (Shrestha and Aryal 2011). If left unaddressed, water scarcity could severely affect food production (Rasul 2012), while increasing population and land use changes may further drive-up demand for drinking water (Pouget et al. 2012). Additionally, watershed degradation reduces groundwater recharge, leading to the drying up of springs, streams, and other vital water sources (Tiwari and Joshi 2012). This situation has resulted in critical water shortages for drinking, irrigation, and other livelihood needs in the Himalayan region, posing a serious threat to the survival of marginalized rural mountain communities who already experience widespread poverty and have limited capacity to respond to such crises. In this context, the concept of payments for ecosystem services (PES) for water resource conservation is gaining relevance as a potential solution.

Community-managed household water systems in the region can be broadly classified into two types: systems with water fees and those without. In the first type, households pay a fee that covers the cost of water provision, which is managed through the *Gramkosh* (village fund) overseen by the village head. To safeguard the water source, villagers have designated areas as conservation forests, where activities such as tree cutting, land clearing, and hunting are strictly prohibited. In contrast, communal water systems without fees rely on gravity-fed supply from springs and creeks, which channel water to storage tanks and households. These systems depend heavily on collective responsibility, requiring strong social cohesion and cooperation among users to manage and maintain the infrastructure and prevent free-riding behavior. During the hot and dry season (March to June), water becomes scarce. In response, villagers are restricted to using the communal tap (a shared water supply point akin to municipal connections) solely for drinking and cooking. Activities such as bathing and washing clothes must be carried out at alternate water sources located outside the village.

Among the various types of payments for environmental services, schemes related to water services have received notable attention over the past few decades. Historically, people have accessed ecosystem services freely, making the shift toward accepting paid services a gradual process that requires changes in perception and behavior. In the Himalayan region, implementing PES presents unique challenges, particularly in the economic valuation of ecosystem services and the efficiency analysis of alternative ecosystem management strategies. These challenges call for the careful selection, design, and administration of appropriate valuation techniques. In this case study, the contingent valuation method (CVM) was employed to assess households' willingness to pay (WTP) for water-related ecosystem services. The findings revealed that respondents were willing to pay an additional Rs. 82.92 per month for a continuous and reliable water supply, indicating a potential avenue for revenue generation. A comparable study in Kathmandu found that people were also willing to pay more than their current expenditure on coping strategies—an amount estimated at 1% of their average monthly income (Pattanayak et al. 2005). The analysis further revealed that income and family size significantly influence household WTP. Higher-income levels corresponded with increased WTP, while larger families exhibited the highest mean WTP. This trend may reflect their stronger dependency on natural resources and closer connection to the local environment. Households with more educated heads, greater environmental awareness, and larger landholdings also demonstrated a higher WTP. Education likely enhances awareness of the critical role forests play in sustaining water resources. Similarly, larger landholdings may drive higher WTP due to increased water needs for irrigation and agricultural activities.

PES refer to various mechanisms through which beneficiaries of ecosystem services compensate those who manage or preserve these services, ensuring their continued availability and sustainability. If the value of ecosystem services in the

Himalayan region were properly accounted for and integrated into the global economic system, it would significantly alter current pricing structures. Commodities that rely directly or indirectly on ecosystem services would carry much higher prices, reflecting their true environmental cost. Consequently, the distribution of factor payments—such as wages, interest, and profits—would shift substantially. Moreover, the magnitude and composition of global Gross National Product (GNP) would look markedly different if it fully incorporated the economic value of ecosystem services. Hence, if a proper strategy (formulation and implementation of policy) is suggested, these valued ecosystem services can be quantified, which would help regulate the indiscriminate use of these resources.

References

Bateman IJ, Fezzi Grace GM, Atkinson C, Turner GK (2011) Economic analysis for ecosystem service assessments. Environmental & Resource Economics 48:177–218

Bernard F, de Groot RS, Campos JJ (2009) Valuation of tropical forest services and mechanisms to finance their conservation and sustainable use: a case study of Tapanti National Park, Costa Rica. Forest Policy Econ 11:174–183

Biao Z, Wenhua L, Gaodi X, Yu X (2010) Water conservation of forest ecosystem in Beijing and its value. Ecol Econ 69:1416–1426

Bishop RC, Boyle KJ, Welsh MP, Baumgartner RM, Rathbun PR (1987) Glen canyon dam releases and downstream recreation: an analysis of user preferences and economic values, U.S. Bureau of Reclamation, Glen Canyon Environmental Studies, Washington, DC

Biswas D, Venkatachalam L (2015) Farmers' willingness to pay for improved irrigation water – a case study of Malaprabha Irrigation Project in Karnataka, India. Water Econ Policy 1(1). https://doi.org/10.1142/S2382624X14500040

Brauman KA, Daily GC, Duarte TK, Mooney HA (2007) The nature and value of ecosystem services: an overview highlighting hydrologic services. Ann Rev Environ Res 32:67–98

Briscoe J, Furtado de Castro P, Griffen C, North J, Oslen O (1990) Towards equitable and sustainable rural water supplies: a contingent valuation study in Brazil. World Bank Econ Rev 4(2):115–134

Brookshire DS, Smith VK (1987) Measuring recreation benefits: conceptual and empirical issues. Water Resour Res 23(5):931–935

Calder IR (2005) The blue revolution, land use and integrated water resources management. Earthscan, London, p 208

Carson RT, Mitchell RC, Hanemann M, Kopp RJ, Presser S, Ruud PA (2003) Contingent valuation and lost passive use: damages from the Exxon Valdez oil spill. Environ Resour Econ 25:257–286

Chandrasekaran K, Devarajulu S, Kuppannan P (2009) Farmers' willingness to pay for irrigation water: a case of tank irrigation systems in South India. Water 1(1):5–18

Chatterji D (1997) Economic evaluation of urban water supply schemes. Economic and Political Weekly of India, June 28

Choe K, Whittington D, Lauria DT (1996) The economic benefits of surface water quality improvements in developing countries: a case study of Davao. Philippines Land Economics 72(4):519–537

Daubert JT, Young RA (1981) Recreational demands for maintaining instream flows: a contingent valuation approach. Am J Agric Econ 63:666–676

De Groot RS, Alkemade R, Braat L, Hein L, Willemen L (2010) Challenges in integrating the concept of ecosystem services and values in landscape planning, management and decision making. Ecol Complex 7(3):260–272

de Oca GSM, Bateman IJ (2006) Scope sensitivity in households' willingness to pay for maintained and improved water supplies in a developing world urban area: investigating the influence of baseline supply quality and income distribution upon stated preferences in Mexico City. Water Resour Res 42(7):Art. No. W07421

Devi JS, Joseph B, Karunakaran K, Anurdha B, Devi KR (2009) People's attitudes towards paying for water: India review article. Curr Sci 97(9-10): 1296–1302

Duffield JW, Neher CJ, Brown TC (1992) Recreation benefits of instream flow: application to Montana's big hole and bitterroot rivers. Water Resour Res 28(9):2169–2181

Dutta KR, Thakur B, Konar TS, Chakrabarty SN (2010) Rapid evaluation of water supply project feasibility in Kolkata, India. Drink Water Eng Sci 3:29–42

Falkenmark M (2003) Freshwater as shared between society and ecosystems: from divided approaches to integrated challenges. Phil Trans R Soc B: Biol Sci 358:2037–2049

FAO (2005) FAOSTAT data. Food and Agricultural Organisation, Rome

Farley KA, Jobbagy EG, Jackson RB (2005) Effects of afforestation on water yield: a global synthesis with implications for policy. Glob Change Biol 11:1565–1576

Fisher B, Turner KR, Morling P (2009) Defining and classifying ecosystem services for decision making. Ecol Econ 68(3):643–653

Gerten D, Hoff H, Bondeau A, Lucht W, Smith P, Zaehle S (2005) Contemporary "green" water flows: simulations with a dynamic global vegetation and water balance model. Phys Chem Earth 30:334–338

Gong Y, Bull G, Baylis K (2010) Participation in the first clean development mechanism Forest project: the role of property rights, social capital and contractual rules. Ecol Econ 69(6):1292–1302

Griffin CC, Briscoe J, Singh B, Ramasubban R, Bhatia R (1995) Contingent valuation and actual behavior: predicting connections to new water systems in the state of Kerala India. World Bank Econ Rev 9(3):373–395

Gupta SK, Deshpande RD (2004) Water for India in 2050: first order assessment of available options. Curr Sci 86:1216–1224

Hadkar N, Sharma S, David A, Muraleedharan TR (1997) Willingness-to-pay for Borivali National Park: evidence from a contingent valuation. Ecol Econ 21(2):105–122

Haines-Young R, Potschin M (2010) The links between biodiversity, ecosystem services and human well-being. In: Raffaelli D, Frid C (eds) Ecosystem ecology: a new synthesis, BES ecological reviews series. CUP, Cambridge

Honey-Rosés J (2009) Reviewing the arguments for market based approaches to water distribution: a critical assessment for sustainable water management in Spain. Sustain Dev 17:357–364

Johnson NL, Baltodano ME (2004) The economics of community watershed management: some evidence from Nicaragua. Ecol Econ 49:57–71

Karthikeyan C, Devarajulu S, Kuppannan P (2009) Farmers' willingness to pay for irrigation water: a case of tank irrigation systems in South India. Water 1:5–18

Lauria DT, Whittington D, Choe K, Turingan C, Abiad V (1999) Household demand for improved sanitation services: a case study of Calamba, Philippines. In: Bateman IJ, Willis KG (eds) Valuing environmental preferences: theory and practice of the contingent valuation method in the US, EU, and developing countries. Oxford University Press, New York

MEA (2005) Millenium ecosystem assessment, ecosystems and human well-being: current state and trends. Island Press, Washington, DC, p 155

Muñoz-Piña C, Guevara A, Torres JM, Braña J (2008) Paying for the hydrological services of Mexico's forests: analysis, negotiations and results. Ecol Econ 65:726–737

Nahuelhual L, Donoso P, Lara A, Nunez D, Oyarzun C, Neira E (2007) Valuing ecosystem services of Chilean temperate rainforests. Environ Dev Sustain 9:481–499

National Oceanic and Atmospheric Administration (NOAA) (1993) Report of the NOAA panel on contingent valuation. Fed Regist 58(10):4601–4614
Ojea E, Martin-Ortega J, Chianai A (2012) Defining and classifying ecosystem services for economic valuation: the case of forest water services. Environ Sci Policy 19:1–15
Pagiola S (2008) Payments for environmental services in Costa Rica. Ecol Econ 65(4):712–724
Pattanayak SK, Yang J, Whittington D (2005) Coping with unreliable public water supplies: averting expenditures by households in Kathmandu, Nepal. Water Resour Res 41:1–11
Perrot-Maître D (2006) The Vittel payments for ecosystem services: a 'perfect' PES case? Project Paper No. 3. IIED, London
Plummer ML (2009) Assessing benefit transfer for the valuation of ecosystem services. Front Ecol Environ 7(1):38–45
Pouget L, Escaler I, Guiu R, Mc Ennis S, Versini P (2012) Global change adaptation in water resources management: the water change project. Sci Total Environ 440:186–193
Raje DV, Dhobe PS, Deshpande AW (2002) Consumer's willingness to pay more for municipal supplied water: a case study. Ecol Econ 42:391–400
Rasul G (2012) Contribution of Himalayan ecosystems to water, energy, and food security in South Asia: a nexus approach. ICIMOD, Kathmandu
Roy J, Chattopadhyay S, Mukherjee S, Kanjilal M, Samajpati S, Roy S (2004) An economic analysis of demand for water quality: case of Kolkata. Econ Polit Wkly 39(2):186–192
Sanders LD, Walsh RG, McKean JR (1991) Comparable estimates of the recreational values of rivers. Water Resour Res 27(7):1387–1394
Scanlon BR, Jolly I, Sophocleous M, Zhang L (2007) Global impacts of conversions from natural to agricultural ecosystems on water resources: quantity versus quality. Water Resour Res 4(3). https://doi.org/10.1029/2006WR005486
Shrestha A, Aryal R (2011) Climate change in Nepal and its impact on Himalayan glaciers. Reg Environ Change 11(1):S65–S77
Shultz S, Pinazzo J, Cifuentes M (1998) Opportunities and limitations of contingent valuation surveys to determine national park entrance fees: evidence from Costa Rica. Environ Dev Econ 3:131–149
Spangenberg JH, Settele J (2010) Precisely incorrect? Monetising the value of ecosystem services. Ecol Complex 7:327–337
Stenger A, Harou P, Navrud S (2009) Valuing environmental goods and services derived from the forests. J For Econ 1(1):1–14
Tiwari PC, Joshi B (2012) Environmental changes and sustainable development of water resources in the Himalayan headwaters of India. Water Resour Manag 26:883–907
Turner RK, Morse-Jones S, Fisher B (2010) Ecosystem valuation, a sequential decision support system and quality assessment issues. Ann N Y Acad Sci 1185:79–101
Tyagi RK, Shankar V (1995) Forage supply demand analysis for Saurashtra and Kachchh. Forage Res 16(2):45–57
Venkatachalam L (2004) The contingent valuation method: a review. Environ Impact Assess Rev 24(1):89–124
Weber MA, Berrens RP (2006) Value of instream recreation in the Sonoran Desert. J Water Resour Plan Manag 132(1):53–60
Whittington D (1998) Administering contingent valuation surveys in developing countries. World Dev 26(1):21–30
Whittington D, Lauria DT, Mu XM (1991) A study of water vending and willingness to pay for water in Onitsha. Niger World Dev 19(2/3):179–198
Zhang L, Dawes WR, Walter GR (2001) Response of mean annual evapotranspiration to vegetation changes at catchment scale. Water Resour Res 37:701–708

Chapter 16
Protected Areas (PAs) and Ecotourism in the Transboundary Landscape of Khangchendzonga: Needs, Opportunities, and Impacts

Jhony Lepcha, Kailash S. Gaira, Hemant K. Badola, Rajesh Joshi, and Nakul Chettri

Abstract Protected area (PA) connectivity in transboundary regions enhances ecotourism potential by linking biodiversity-rich areas, fostering conservation, and supporting local economies. In such settings, shared cultural and natural resources create opportunities for transboundary ecotourism, benefiting communities on both sides. Strengthening connectivity, PAs promotes sustainable tourism and encourages collaborative conservation efforts, driving socio-economic development while preserving ecological and cultural heritage in the Khangchendzonga Landscape (KL). Realising the importance, the current study explores the potential for ecotourism development in the surrounding villages of two important Indian protected areas (PAs)—Singalila National Park (West Bengal) and Barsey Rhododendron Sanctuary (Sikkim). Both PAs are part of a transboundary region shared between India and eastern Nepal, characterised by deep sociocultural ties. The villages in this transboundary zone are fragmented and largely dependent on the PAs, emphasising the need for enhanced connectivity to create an ecotourism corridor that can improve the socio-economic conditions of local communities and ensure biodiversity

J. Lepcha · K. S. Gaira (✉) · R. Joshi
G.B. Pant National Institute of Himalayan Environment, Sikkim Regional Centre, Gangtok, Sikkim, India

H. K. Badola
G.B. Pant National Institute of Himalayan Environment, Sikkim Regional Centre, Gangtok, Sikkim, India

D-408, Aditya Doonshire Apartments, Sailok Phase II, GMS Road, Dehradun, Uttarakhand, India

N. Chettri
International Centre for Integrated Mountain Development, Kathmandu, Nepal

Regional Office Eastern Himalaya -Northeast India, Ashoka Trust for Research in Ecology and the Environment (ATREE), NH 10, Tadong, Gangtok, Sikkim, India

Y. D. I. Khan et al. (eds.), *Ecosystem-based Approaches for Resilience Building in Himalayan Landscapes*, Disaster Resilience and Green Growth,
https://doi.org/10.1007/978-981-95-2007-7_16

conservation. Drawing on Participatory Rural Appraisal (PRA) techniques, household surveys conducted in seven villages, and secondary literature, this study reveals that ecotourism in and around these PAs is necessary and an opportunity to support rural livelihoods and promote biodiversity conservation. The research outcomes propose community-based strategies for promoting, managing, and developing ecotourism, ensuring long-term benefits for local communities and contributing to ecological conservation. Further, the study advocates for stronger interstate (West Bengal and Sikkim) and transboundary (India and Nepal) collaborations to enhance the socio-ecological value of the region. In addition, the result findings support assisting policymakers in developing appropriate policies based on local perspectives, ensuring long-term benefit-sharing, and promoting community-led activities within the targeted landscape while initiating transboundary ecotourism between India and Nepal.

Keywords Biodiversity · Ecotourism · Khangchendzonga landscape · Participatory rural appraisal · Protected area

16.1 Introduction

Parks are major tourist assets, and if managed well for tourism, can be self-financing and motivate initiatives for conservation (Davies 2024; Lindsey et al. 2021; Eagles and Hillel 2008). Considering its advantage for protected areas (PAs), especially targeting the economic benefits for the marginalised communities, nature conservation, and PA management, ecotourism has been an emerging concept worldwide to nature lovers and policymakers (Rahman et al. 2013; Kuniyal et al. 2004; Bookbinder et al. 1998). The Vth IUCN World Parks Congress suggested, '*protected areas should not exist as islands, divorced from the social, cultural and economic context in which they are located*' (Recommendation 5.29, Vth IUCN World Parks Congress convened in Durban, South Africa, from 8 to 17 September 2003). That means it is important if the PAs are considered for a much broader interconnected conservation planning where nature conservation and sociocultural and economic entities could be given equal space and time (Leung et al. 2015; Badola and Subba 2012). Earlier, while prioritising the conservation issues of PAs, the human and nature-based issues were considered separately (Ghimire and Pimbert 1997), which excluded the forest dwellers from PAs (McLean and Straede 2003). The gradual shift from this protectionist to a people-oriented approach to the conservation of PAs has opened up a new dimension in integrating fringe villages as the main pillar supporting the conservation of a targeted area. It is well understood that people's livelihood development and conservation cannot be handled independently, and integrating both with similar objectives has become necessary (Scherl et al. 2004). Many aspects of people's livelihood dependencies need exploration, including biological resource utilisation in sustaining life or alleviating poverty (Badola and Aitken 2010). Amongst them, ecotourism has emerged as one of the most beneficial tools for the economic upliftment of people and the conservation management of areas (Samal and Dash 2023; Lepcha et al. 2015; Kuniyal et al. 2004). At the same time, there are views that tourism can draw some funding to support system management of PAs (McCool and Spenceley 2014) comprising over 13% of Earth's terrestrial area (Bertzky et al. 2012).

Developing ecotourism within and around PAs can help better conserve biodiversity and sustain area importance for local people, substantially boosting the regional economy (Samal and Dash 2023; Admasu 2020; Spenceley 2015). Consequently, the PAs may serve as a revenue generator and should also provide ecosystem-based paybacks like health benefits, accelerated local development, and maintaining peace in the transboundary context (McCool and Spenceley 2014). By combining nature conservancy and sustainable development strategy, a practical approach can be realised through ecotourism, attracting tourists to nature without causing harm (Drumm and Moore 2005; Eagles et al. 2002). Such approaches are more inevitable in the Himalaya, where unsustainable resource use has been leading to natural degradation, increasing human-wildlife conflicts and rapid alteration in land use, etc. (ICIMOD et al. 2017a; Bhattacharjee and Parthasarathy 2013; Sharma and Chettri 2005; Rai and Sundriyal 1997); however, the climate change is yet another threat. Nature-based sustainable tourism has been viewed as an important livelihood option for the forest fringe village people at the same time, which will motivate them to conserve the resources (Bennett et al. 2012). Further, ecotourism inherently addresses environmental and cultural conservation and ensures the benefits and well-being of the local communities (TES 1991). If managed judicially, ecotourism can significantly help minimise ecological and cultural degradation in mountain regions and offer socio-economic benefits to the community (Kiss 2004; Nepal 2002). From both short-term and long-term perspectives, ecotourism has been vital for helping conserve and enrich indigenous cultures for generations (Trofimov and Soimu 2011).

To involve local people in conservation initiatives, they must benefit from the conservation interventions (Phuntsho et al. 2012). In such a process, there is a need to take priority responsibilities by conservation managers and tourism entrepreneurs for establishing an effective relationship between the community and nature sustainably, ensuring direct benefits to the local communities from ecotourism as a significant ecosystem services provider. Therefore, it is essential to assess the possibilities of potential ecotourism development in a critical area, especially natural settings with unique attractions, everyday use of resources in the surroundings, and the livelihood dynamics of community people. However, except for a few researches on tourism, that too from other parts of Sikkim (Joshi and Dhyani 2009; Rai and Maharana 2000; Maharana et al. 2000; Rai and Sundriyal 1997), there is hardly any publication available from our study areas in Khangchendzonga Landscape (KL)-India (ICIMOD et al. 2017a).

India, Nepal, and Bhutan share the Khangchendzonga Landscape (KL) in the Eastern Himalaya as a transboundary landscape (TL) setup, which is home to unique biodiversity and cultural diversity, with vast mountain ecosystems, indigenous cultures, and traditional lifestyles (Gurung et al. 2019). A *transboundary landscape* (*TL*) refers to a geographic area that spans political borders, often shared by two or more nations, characterised by ecological, cultural, and social continuity across borders despite political boundaries. TLs often face challenges related to differing national policies, governance structures, and resource uses, resulting in neglected parts for livelihood and environmental concerns. It offers opportunities for collaboration in biodiversity protection, sustainable tourism, and cultural exchange through collaborative management efforts. In addition, the interstate transboundary region also encloses a governance mismatch between the two states, resulting in deprived corridor connectivity for both conservation and socio-economic development. However, global climate change has evidenced multiple effects on biodiversity and the socio-economy of the Himalayan ecosystem, which is sensitive compared to other parts of the world.

Realising the importance of ecotourism in the context of sustainable livelihood improvement and conservation of nature and culture within a transboundary setup, the present study was focused on the Barsey-Singalila site of Khangchendzonga Landscape in India, having the transboundary setting with Nepal. The primary objectives are : (i) to assess and analyse the possible ways to strengthen the conservation of two protected areas, the Singalila National Park and Barsey Rhododendron Sanctuary, and (ii) to analyse the opportunities offered for the development of sustainable ecotourism to enhance and sustain livelihoods of the communities inhabiting around PAs.

16.2 Material and Methods

16.2.1 Study Area

The present paper deals with a case study from Khangchendzonga Landscape, located between 26°21′40.49″ E and 28°7′51.25″ E latitude and 87°30′30.67″ N and 90°24′31.18″ N longitude, which is a unique transboundary landscape covering a spatial area of 25,086 km^2 spatial area, which extends from the southern stretch from Mount Khangchendzonga (8586 m asl) spreading over to varied ecological zones in eastern Nepal, Sikkim, and northern West Bengal in India and western Bhutan (ICIMOD et al. 2017a, b; Badola et al. 2015, 2016a; Sharma and Chettri 2005). The region is recognised as one of the richest global biodiversity hotspots and one of the most critical centres of biodiversity (Mittermeier et al. 2004), characterised by ten terrestrial eco-regions (ICIMOD et al. 2017a). The landscape comprises 18 protected areas (PAs), of which 6 are transboundary, and 4 are in the Indian part (ICIMOD et al. 2017a).

The Indian part of the Khangchendzonga Landscape is located between 26°29′13.56″ N and 28°7′51.6″ N latitude, and 87°59′1.32″ E and 89°53′42.96″

Elongitude, ranging from 40 to 8586 m asl, covering an area of 14,126.36 km² of 14,126.36 km²area (Gaira et al. 2017). The Khangchendzonga Landscape-India comprises Sikkim and four districts of West Bengal (Darjeeling with recently bifurcated district Kalimpong, Jalpaiguri, and Alipurduar (excluding blocks along Bangladesh borders). KL-India has 16 notified PAs, including recently inscribed (17 July 2016) UNESCO's world heritage site, the Khangchendzonga National Park in Sikkim. KL-India has unique vegetation features, extending from tropical to the alpine region (Gaira et al. 2022; Sinha et al. 2018).

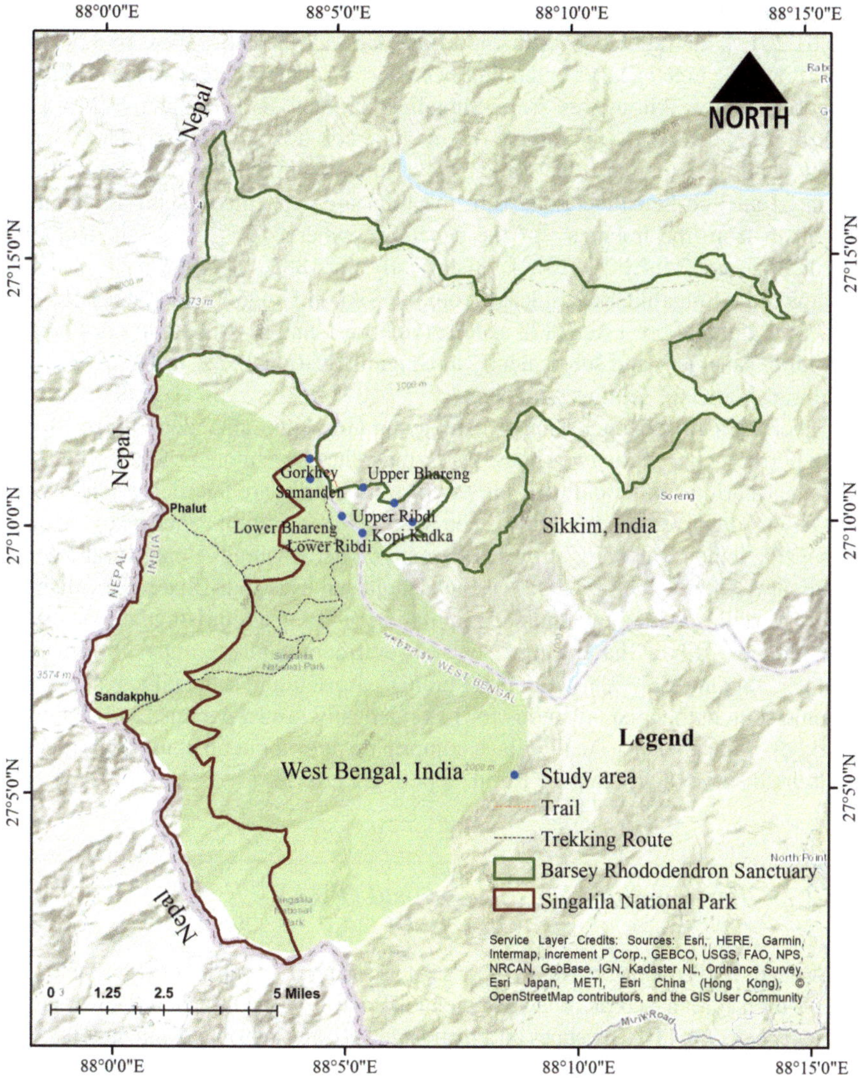

Fig. 16.1 The location map of the study area

Our study focused on two adjoining areas from KL-India, located in and around two PAs, i.e. Ribdi-Bhareng village area near Barsey Rhododendron Sanctuary (West Sikkim) and Gorkhey-Samanden village area within and around Singalila National Park (Darjeeling district, West Bengal) in Khangchendzonga Landscape, India (Fig. 16.1). The total area of Singalila National Park (IUCN category II) is 78.60 km^2, with reported over 383 flowering plants, 500 birds, and 26 mammal species, whereas the Barsey Rhododendron Sanctuary (IUCN category IV, established the year 1988) covers 104 km^2 area and reported to have 141 flowering plants, 550 birds and 22 mammals species, 200 species of amphibians, 70 species of snakes, and greatly diversified insects (MoEFCC 2014, 2017a, b; Shakya and Joshi 2008).

For Singalila National Park, through the draft notification No. 286.S.O.67 (E) [10.01.2017], the Indian Government declared an Eco-Sensitive Zone of 52.60 km^2 area, covering nine villages extending from zero (Nepal border at west and Sikkim at north) to two kilometres from Singalila National Park boundary (MoEFCC 2017b), targeted with its unique focus on ecotourism. Similarly, for Barsey Rhododendron Sanctuary, the Indian Government, through the Gazette of India, declared an Eco-Sensitive Zone (not the total area specified), with an extent of 25–50 m from the boundary of the sanctuary, where 42 villages fall, including Ribdi-Bhareng (MoEFCC 2014). Historically, in 1882, the British Government occupied the Singalila area on lease from the Sikkim Kingdom and later notified it under the Indian Forest Act 1878 as Reserve Forest, and in 1986, it was declared as a wildlife sanctuary and subsequently as Singalila National Park in 1992. Both PAs hold transboundary settings with Nepal.

River Ramam separates Ribdi-Bhareng and Gorkhey-Samanden villages, forming a border between the two states. Both study sites have more or less similar rural settings, sharing similar cultures and traditions, but they are interdependent in various day-to-day livelihood activities. The Ribdi-Bhareng Gram Panchayat Unit comprises five Panchayat wards: Lower Bhareng, Upper Bhareng, Upper Ribdi, Lower Ribdi, and Khopi Kharka. As per our household survey, Ribdi-Bhareng is inhabited by 318 families, with a total human population of 1319, comprising 646 males and 673 females. Ribdi-Bhareng represents several community groups (Sherpa, Rai, Manger, Tamang, Subba, Chettri, and Pradhan). Agriculture is the villagers' primary income source; the area is famous for its organically grown potatoes and cabbages. Gorkhey-Samanden is a small village comprising only about 65 households, per our household survey/Participatory Rural Appraisal exercise.

16.2.2 Participatory Rural Appraisal (PRA)

PRA stands for Participatory Rural Appraisal. It is an approach and method to enable local people to share, enhance, and analyse their knowledge of life and conditions, plan, and act (Chambers 1994). PRA generates information directly involving the community, a popular method in rural research and project planning. The information generated is more reliable and relevant to community interest than conventional social research methods (Mosse

1994; Chambers 1983, 1991). It is a friendly exploratory approach (Freudenberger 1999; Chambers 1994) comprising various tools and techniques. We have used different Participatory Rural Appraisal tools like a historical timeline, trend line, natural resource mapping, pairwise ranking, matrix ranking, seasonal calendar, and problem tree analysis (Badola et al. 2016b). Participatory Rural Appraisal was conducted in Ribdi-Bhareng and Gorkhey-Samanden with the help of a team (native language speakers). The local community of the study area was targeted as participants during PRA studies.

16.2.3 Household Survey

The household survey was also undertaken to collect quantitative information using a structured questionnaire form. About 60% of the total households in both study areas are involved in this study (Table 16.1). Community people involved in the ecotourism business, such as homestay owners, guides, porters, horse owners, and farmers, were mainly targeted in the survey as respondents. The questionnaire used in the survey was prepared in the local language (Nepali) to ensure the information's quality and avoid misunderstandings between the enumerator and the respondent.

Further, we used a Relative Importance Index method (Gündüz et al. 2012; Kometa et al. 1994; Sambasivan and Soon 2007) to determine the relative importance/priority of reasons behind tourist visits in both sites using quantitative data collected through household survey, as

$$\text{RII} = \frac{\sum_{i=1}^{k} W_i fi}{A+N} * 100$$

where

w_i = the weight assigned to the ith responds category(eg.1= low, 2 = Moderate, 3 = high)

Table 16.1 The sample size of the surveyed villages in Khangchendzonga landscape, India

Sl. no.	Name of the villages	Total households	Surveyed households (no.)	Surveyed households (%)
Gorkhey-Samanden, West Bengal				
1	Gorkhey	33	20	60.60%
2	Samanden	32	20	62.5%
Ribdi-Bhareng, Sikkim				
1	Khopi Kharka	75	45	60%
2	Lower Ribdi	85	51	60%
3	Upper Ribdi	52	31	59.61%
4	Lower Bhareng	55	33	60%
5	Upper Bhareng	51	31	60.78%

fi = the number of respondents selecting the ith category
A = the highest possible weight on the scale (3 for a three-point scale)
K = the total number of responds category
N = the total number of respondents.

16.2.4 Focused Group Discussions (FGDs)

Focused group discussions (FGDs) were conducted with the selected numbers of the community residing in the study area. It is considered an important process in participatory-based studies, where comprehensive qualitative information is collected from a group of people, and it is also helpful in validating previously collected information through other methods. A total of 45 (20 female and 25 male) participants in Gorkhey-Samanden and 40 (28 female and 12 male) in Ribdi-Bhareng participated in 4-h-long FGDs. The discussion was conducted in a friendly way to ensure the active participation of the community.

16.3 Results

Gorkhey-Samanden is poorly connected to the rest of West Bengal; consequently, the villagers rely more on the Sikkim part for their daily needs and livelihood activities. These villages lack basic facilities like electricity connections, road connectivity, education centres, health centres, and markets, whereas Ribdi-Bhareng enjoys all kinds of them. Thus, the villagers of Gorkhey-Samanden largely depend on Ribdi-Bhareng, and their only means of communication is a 3-km-long walking trail located between Ribdi-Bhareng and Gorkhey-Samanden village, which serves as a backbone for the development of the region. Although the inhabitants of Gorkhey-Samanden can visit other nearby places in West Bengal, namely, Rimbik, Ramam, and Darjeeling town, because of their closeness, they prefer Ribdi-Bhareng. Gorkhey-Samanden people visit the primary health sub-centre at Ribdi for essential health services. In major health cases, they use this existing trail to Sombarey and Namchi district hospitals of Sikkim, Darjeeling, and Siliguri. The villagers frequently visit Sombarey, Jorethang, and Namchi townships for marketing and banking. The people's mobility map suggests the importance of the above trail and is the only way of linking two states to meet their socio-economic needs.

16.3.1 Livelihood and Importance of Ecotourism

In Ribdi-Bhareng, Participatory Rural Appraisal (PRA) tools suggested that the local people mostly rely on agricultural practices and livestock rearing for their livelihood. They generate farmyard manure for agriculture, but the ban on free grazing (imposed in 1998) and also on the use of chemical fertiliser/pesticides imposed by the State Government under Sikkim Organic Mission (Notification No. 21/ Home/2010, dated 26.10.2010) has posed considerable challenges to their livelihood strategies, as meeting out stall-feed requirement of fodder and procurement of adequate organic manure. Besides, owing to insufficient storage facilities, frequent crop damage by wildlife, and dwindling livestock rearing, their agriculture (productivity and practices) has been a problem.

Local people firmly believe that ecotourism has the potential to offer an alternative livelihood option for them, and they also assume it will lessen extraction pressure on forests by diversifying forest-based livelihood to ecotourism-based livelihood. Here, homestays are coming up as income-generative initiatives; so far, Gorkhey-Samanden has 7 operational homestays and 20 operational homestays around Ribdi-Bhareng. Lately, residents have realised the importance of sacred natural entities, viz. springs, rivers, trees, and mountains (Devisthan, Deorali, Chorten, Gumpa forest, all managed a sacred space) to promote agro-tourism, as an ecotourism attraction. As per the timeline tool of Participatory Rural Appraisal, ecotourism in Ribdi-Bhareng and the adjoining area is not a recent activity but has been practised for over three decades, when the Yaks were the only modes of transportation. As per the trend line, tourist flow gradually increased from 2010 onwards. In the same year, the Sikkim Government celebrated the International Rhododendron Festival (Mainra et al. 2010) at Hilley in Barsey Rhododendron Sanctuary, 3 km away from the Ribdi-Bhareng, as one of the significant initiatives for ecotourism and rhododendrons' conservation. Thereafter, the event is celebrated as Lali Gurans (Rhododendron) Festival.

Gorkhey-Samanden village serves as a place of rest/stay for visitors and trekkers via Sandakphu (3636 m asl) and Phalut (3595 m asl). Significantly, these two points also offer open access to two countries, India and Nepal; however, no data is available for such unrecorded tourist movements. The seasonal calendar tool suggested the tourist flow in the study area is maximum during mid-March to mid-June, coinciding with Rhododendrons' flowerings, which spurt again from mid-September to December end, a local festive season. Cultural festivals celebrated in the study site are standard on the other side of the border in Nepal. *Dasai and Tihar* (Dusshera and Dipawali) are celebrated during autumn, the most fascinating season for tourists and native inhabitants. These festivals, linked with the victory of good prevailing over evil and eschewing the darkness, respectively, are celebrated with the union of all the close relatives and dear ones. The Buddhists of the region celebrate the Losar/ Losoong in both countries with great gaiety and a sense of high spirituality. Close relatives, family members, and friends meet and enjoy these occasions, giving them equal importance. Besides, despite their ethnic or regional differences, all community people practise many traditional rituals, to which relatives from Nepal are also invited. Likewise, these festivals unite the near and far ones, manifesting their strong cultural linkages.

Tourists' flow gradually declined and remains almost nil from mid-June to mid-September each year (breeding season of different fauna) because of restrictions imposed by the area manager for any entry inside the Singalila National Park. The mobility map of tourists also indicated the tourists' flow channelised towards Rimbik, West Bengal, despite its longer route across the main town compared to the congenial Ribdi-Bhareng route. This is due to the inconvenient and longer formal channel in getting inner line permits from government departments, experienced by the tourists entering the Sikkim part and foreign tourists at the remote site. The pairwise ranking tool suggested that the maximum economic benefits from ecotourism activities go to the lodge owners, homestay owners, guides, porters, horse owners, and other village farmers.

16.3.2 Mobility of Tourists from Ribdi-Bhareng (Sikkim) and Gorkhey-Samanden (West Bengal)

As per the mobility map of tourists, though the least frequently visited, the furthest destination tourists prefer to go from Ribdi-Bhareng by trekking is Uttarey, which takes 3–4 days to complete. Phalut and Sandakphu are RB's most frequently visited destinations, taking about 7 h on foot. Barsey Rhododendron Sanctuary is a less frequently visited destination than Phalut-Sandakphu. Tourists also frequently visit Gorkhey-Samanden but less than Barsey Rhododendron Sanctuary, a rhododendron-rich PA and an excellent habitat for red panda. Tourists trek for around 2 h to reach Gorkhey-Samanden from Ribdi-Bhareng. Other destinations visited by tourists from Ribdi-Bhareng are Daramdin, Lamsing, Hilley, Deonigale Dhap, Okhrey Monastery, and Ribdi Monastery. On the other side, in Gorkhey-Samanden, frequent movements of tourists were observed towards Siri River/Rimbik and beyond because they did not want to take the same route back to Phalut. Gorkhey-Samanden villages beset some naturally beautiful sites for tourist attention, viz. Gorkhey River, Ramam River, Gorkhey to Ribdi-Bhareng eco-trail, and Phalut, the transboundary ridge connecting to Nepal. The Phalut-Singalila trek is important not only for nature lovers, adventure tourists, and wildlife enthusiasts but also for people from India and Nepal, who are a common point of interest. People from both countries use this area to come to each other. Ecotourism, with expected logistics and support from park management, if developed well, may offer viable support to biodiversity conservation by the community itself besides ensuring monitoring.

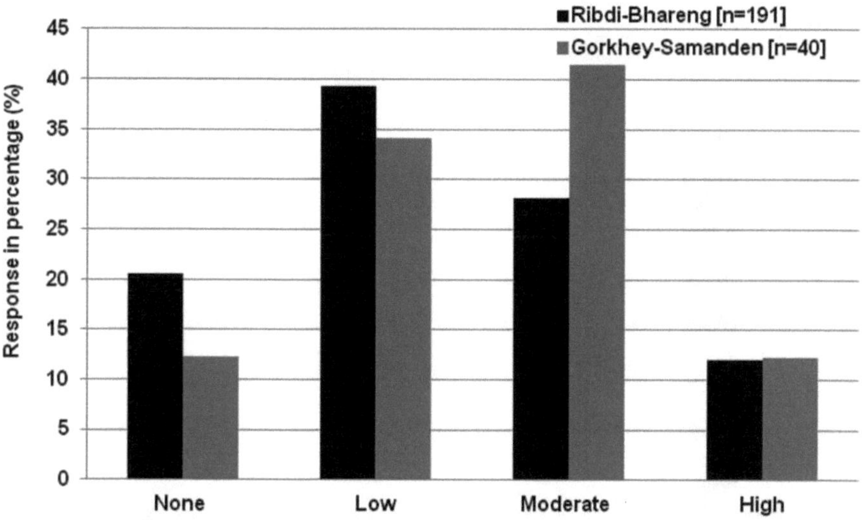

Fig. 16.2 Impact of ecotourism on the socio-economic upliftment of people, as per their perception

16.3.3 Community Perceptions Towards Ecotourism

Hitherto, ecotourism has not impacted the socio-economic upliftment of community people in either site (Fig. 16.2). Only 11.97% of households in Ribdi-Bhareng and 12.2% in Gorkhey-Samanden concurred that ecotourism plays a high role in their society's socio-economic upliftment, the lowest response amongst the other choices.

Conversely, 93.0% and 100% of households in Ribdi-Bhareng and Gorkhey-Samanden considered ecotourism activity beneficial for village development (Fig. 16.3). The results in Figs. 16.2 and 16.3 differed from each other. The local community of either study site supports ecotourism. But they are still unable to take adequate benefits from ecotourism owing to prevailing constraints such as the lack of proper publicity, the inconvenient process of getting an inner line permit, the lack of skilled workforce (guides and homestay operators), and limited numbers of accommodations (homestays, lodges) in Ribdi-Bhareng. In Gorkhey-Samanden, constraints like frequent strikes in the Darjeeling district for separate Gorkhaland state, lack of electricity, poor telecommunication, bleak tourism promotion, and lack of resourceful guides directly or indirectly hinder the ecotourism development in either area. These constraints appeared in pairwise ranking, revealing that bleak tourism promotion mainly hinders ecotourism in Gorkhey-Samanden. Meanwhile, in Ribdi-Bhareng, proper publicity mainly affects ecotourism development, followed by the inner line permit problem, lack of workforce (guides and homestay operators), and limited accommodations. Yet, 79.31% and 90% of households in Ribdi-Bhareng and Gorkhey-Samanden believe ecotourism can be a promising option for their livelihood and habitat conservation.

The Relative Importance Index (RII) value indicates that nature-based activities, including trekking and adventure, are the main reasons for tourists' visits, followed

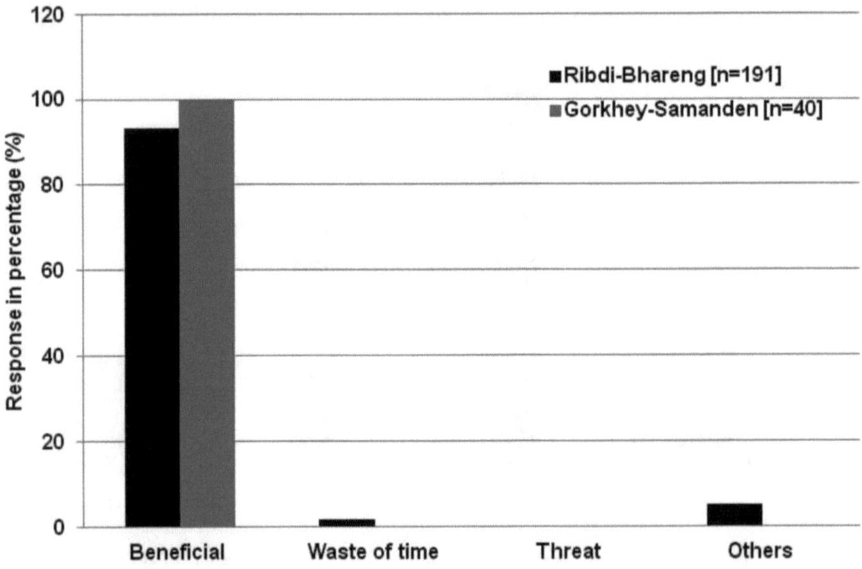

Fig. 16.3 Peoples' perception of the ecotourism activities benefited them in study villages

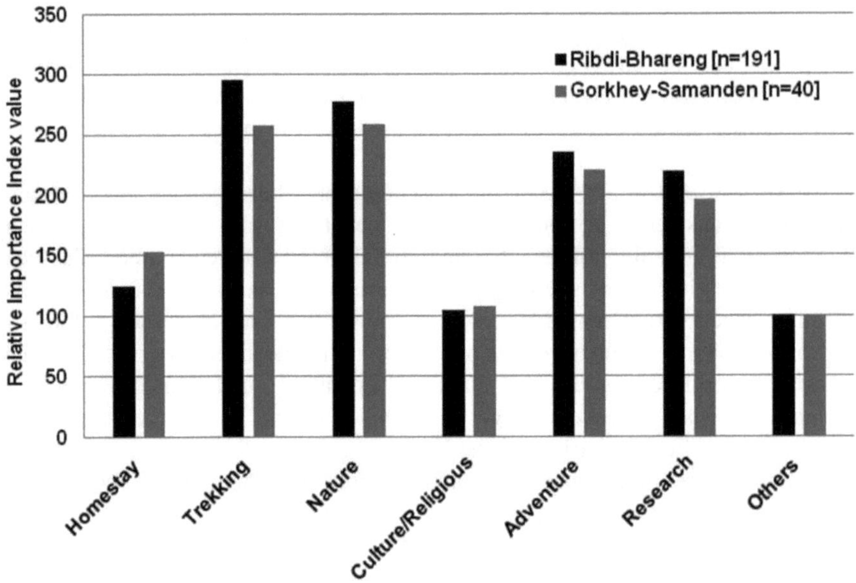

Fig. 16.4 Reasons for tourist visits based on the priority of the respondents (villagers and tourism service provider) in study villages

by other purposes (research, homestay/rural stay, culture/religious, and others in Fig. 16.4) which is further validated through matrix ranking exercise in both areas (Table 16.2). Some homestays, lodges, and guest houses offer accommodation facilities as well as a glimpse of local culture, ethnic foods, and drinks to visitors,

Table 16.2 Matrix ranking of potential attraction/area to be encouraged for ecotourism development based on local community perception and field observations

The potential attraction of the study area	Ribdi-Bhareng (RB), Sikkim		Gorkhey-Samanden (GS), West Bengal	
	Tourism	Matrix ranking	Tourism	Matrix ranking
1. Culture and tradition	1.1 Sherpa community-dominated area with other ethnic groups (Rai, Manger, Tamang, Subba, Chettri, and Pradhan) 1.2 Local or ethnic cuisines, e.g. wines (wheat, rice, millet, maize, tapioca, ginger, rhododendron, and guava) and food (Wachipa, Aril, Choburiki, Shyoti, Rigirildok, Chang ku, Phal ghi, Alum, and baber) 1.3 Monasteries	***	1.1 Tamang, Sherpa, Chettri, Rai, and Subba/Limboo communities	***
2. Protected areas (PAs)	2.1 Barsey rhododendron sanctuary (BRS)	****	Singalila National Park (SNP)	****
3. Flora	3.1 Rhododendron spp. (*flagship species*); *Aconitum ferox* (threatened beautiful flowered herb)	***	3.1 Rhododendron spp. (*flagship species*)	***
4. Fauna	4.1 Red panda (*Ailurus fulgens*) and other avian species	***	Red panda (*Ailurus fulgens*), blood pheasant, and satyr tragopan (*Tragopan satyra*)	***
5. Trek	5.1 RB to Uttarey, RB to Phalut, and RB to Sandakphu	****	5.1 Gorkhey to Phalut and Phalut to Sandakphu	****
6. Organic agriculture products	6.1 Famous for the production of local good-quality organic potatoes and cabbages	**	6.1 The agricultural products grown are insufficient to feed tourists. All local products are imported from RB and Sikkim's adjoining markets	*
7. Weather	7.1 Pleasant, March to November, and cold, December to February	***	7.1 Pleasant, March to November, and cold, December to February	***
8. Transboundary	8.1 Transboundary area with Nepal	***	8.1 Transboundary area with Nepal	***
9. Homestay	Operational homestay: 20	**	Operational homestay: 7	***

(continued)

Table 16.2 (continued)

The potential attraction of the study area	Ribdi-Bhareng (RB), Sikkim		Gorkhey-Samanden (GS), West Bengal	
	Tourism	Matrix ranking	Tourism	Matrix ranking
10. Existing trail/route	10.1 Between RB and GS (about 80% of local people of RB believed the promotion of existing trail had an immense potential to boost ecotourism)	***	Between RB and GS (about 55% people of GS believed the promotion of existing trails could be a big ecotourism booster	***
11. Festivals and other events	11.1 Rhododendron festival (nature-based, state government supported)	***	11.1 Nil	*
	11.2 Dasai and Tihar festivals celebrated during October to November		11.2 Dasai and Tihar festivals celebrated during October to November	***
	11.3 Losar/Losoong (Buddhists' festival) in winter	***	11.3 Losar/Losoong (Buddhists' festival) in winter	***

supporting ecotourism in the study area. Blessed with vibrant biodiversity and natural beauty, this part of Khangchendzonga Landscape offers tourists the diverse potential of recreational choices; the most potent aspects are focused on in Table 16.2. Their proximity to PAs and the opportunity to trek into woods are given greater preference to be responsible for the tourist flow by the community in both study areas. Other factors, such as the decadent and unique floral and faunal elements in this part of the landscape, transboundary location, and pleasant weather conditions, are some other important factors that attract greater tourist flow.

16.3.4 Opportunities, Weakness, and Solutions

Many unique features make the study area a potential ecotourism destination. The natural settings of study villages are mesmerising; countless tourist attractions can be added to the list. The cultures and traditions varied in their uniqueness and importance, drawing the attention of visitors aspiring for such experiences. The study sites also offer to learn ground for organic farming practices, a potential attraction. However, a weakness needs to be focused and desired initiatives would be required for the same, parallel to the steps taken for the development activities. For this, forming a local ecotourism committee of the villagers, including local self-help groups, non-governmental organisations, etc., and providing basic knowledge and proper training to all the service providers on various aspects of ecotourism is a prerequisite (Fig. 16.5).

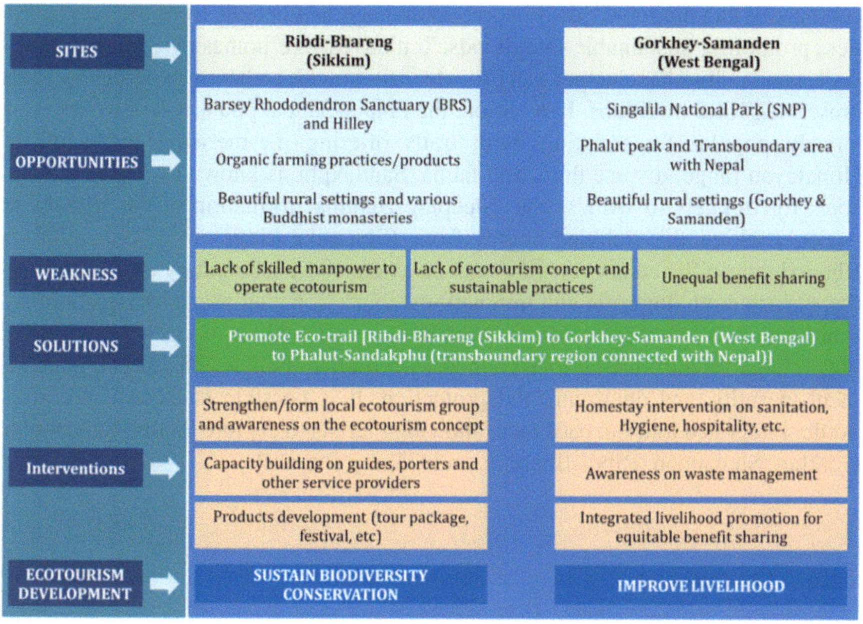

Fig. 16.5 Community-based ecotourism development plan for the study area in KL, India

16.3.5 Transboundary Tourism Potential

The potential for transboundary ecotourism between Phalut and Sandakphu (Darjeeling) and Ribdi-Bhareng and Okhrey (Sikkim) is significant, given the rich biodiversity, cultural heritage, and unique landscapes of the region. The Phalut and Sandakphu (Darjeeling) are known for stunning panoramic views of Himalayan peaks like Khangchendzonga, Everest, and Makalu. The region is rich in biodiversity, forming part of the Singalila National Park, a hotspot for flora and fauna like the endangered red panda, leopards, and over 120 bird species, including a variety of rhododendrons, magnolias, and oak forests (Sinha et al. 2018). In contrast, Ribdi-Bhareng and Okhrey (Sikkim) are gateways to the Barsey Rhododendron Sanctuary, famous for its variety of rhododendron species and rare wildlife such as the Himalayan black bear, red panda, and leopards. The sanctuary offers eco-friendly trekking routes and bird-watching opportunities, making it a promising destination for nature enthusiasts. Linking these regions as part of a larger ecotourism circuit can create a seamless, enriched experience that blends the best of nature, wildlife, and cultural heritage.

The areas surrounding Phalut, Sandakphu, Ribdi-Bhareng, and Okhrey are inhabited by indigenous communities, including the Lepchas, Sherpas, and Bhutias, who have a deep connection to the land and its ecosystems. Community-based ecotourism can empower local people to manage and benefit from tourism activities, promoting sustainable livelihoods. Initiatives like homestays, guided nature walks, and cultural exchanges can provide visitors with authentic experiences while preserving local traditions. In addition, the Phalut-Sandakphu and Barsey areas are already popular for trekking, with trails offering mesmerising views of the Himalayan range, diverse flora, and fauna. Sandakphu is known as one of the best spots for trekkers to witness the 'Sleeping Buddha' formation of the Himalayan peaks. Linking the trekking routes from (Singalila Ridge) to Sikkim (Barsey Rhododendron Sanctuary) could create an extended transboundary trekking circuit that attracts both domestic and international tourists, which needs strong coordination between the governments of West Bengal and Sikkim along with the Nepal government by introducing transboundary ecotourism policies on tourism management, permits, and environmental protection. It is possible to experience cross-border integrated tourism packages, allowing tourists to explore Phalut, Sandakphu, Gorkhey-Samanden, Ribdi-Bhareng, and Okhrey eco-trail.

16.4 Discussion

Globally, the mountains are second in their popularity as tourist destinations after coastal regions (Mieczkowski 1995); in recent years, peoples' inclination towards mountains has rapidly accelerated significantly for exploring their high biological and natural diversity and solitude over coastal areas. To offer tourists wider choices, the region's available resources and prevailing organisations could be revamped in some new form of attractions that will help build small-scale entrepreneurship amongst the village womenfolk and thereby enhance the village economy. Perhaps a less expensive and feasible programme could be initiated for the purpose. In our study area, eco-friendly approaches would be beneficial. For example, we discovered by developing an eco-trail between Ribdi-Bhareng and Gorkhey-Samanden or between Barsey Rhododendron Sanctuary and Singalila National Park. The short, adventurous eco-trail would not only link two places for ecotourism development, but it can emerge as an ecotourism product. The visitors of the Barsey Rhododendron Sanctuary during the Rhododendron flowering season can be channelised to Singalila National Park, and the travellers from Phalut, halting in Gorkhey-Samanden, may have other options to visit Ribdi-Bhareng as well as Barsey Rhododendron Sanctuary. Moreover, eco-trails can be integrated with popular trekking routes such as Maneybhanjang to Meghma, Toughu, Gairibas, Sandakphu, and Phalut in the Darjeeling district of West Bengal (Bhutia 2014), as well as with transboundary trekking routes connecting to Nepal. Its significance is highly appreciated by the visitors for encountering locations like Sandakphu, Phalut, and Tonglu (3070 m asl), which offer mesmerising scenic views of high-altitude mountains, viz. Mt. Khangchendzonga, Mt. Lhotse (8516 m asl), Yalung Kang (8505 m asl), and Mt. Everest (8848 m asl).

As most of these routes pass through potential habitats of the flagship and endangered mammal red panda (*Ailurus fulgens*), and by promoting ecotourism, these habitats will come under direct conservation monitoring of communities, besides getting more attention from park management. Very recently, the Government of India recommended the development of a Red Panda Conservation Scheme (MoEFCC 2017a), which strengthens the conservation significance of Singalila National Park by promoting ecotourism. The recent notification of an Eco-Sensitive Zone covering 9 villages (MoEFCC 2017b) within 2 km extent and zero extends along the international border of Nepal (western side), and the Indian state Sikkim-northern side signifies the cross-border importance of Singalila National Park. Since the Gorkhey village is part of the fringe sanctuary and Samanden is within 1 km of the Eco-Sensitive Zone, the entire complex deserves special attention for promoting ecotourism. Also, three fascinating but locally threatened pheasant species, the blood pheasant (*Ithaginis cruentus*) and satyr tragopan (*Tragopan satyra*), and kalij (*Lophura leucomelanos*), and Asiatic black bear (*Ursus thibetanus*), Himalayan serow (*Capricornis sumatraensis thar*), yellow-throated marten (*Martes flavigula*), Himalayan Griffon vulture (*Gyps himalayensis*)· etc., which will be great attention to visitors and park managers for their conservation protection through ecotourism. Interestingly, Singalila National Park is an important bird and biodiversity area, with over 550 bird species and other rich faunal and floral diversity; it is an important resource for conservation exposure for students and a vital treasure for researchers (MoEFCC 2017b) under ecotourism initiative.

By notifying Gazette, the Government of India declared an Eco-Sensitive Zone for Barsey Rhododendron Sanctuary on 27.8.2014 (MoEFCC 2014), which placed a central governmental mechanism for its monitoring, called the State-Level Eco-Sensitive Zone Monitoring Committee (SESZMC) for the State of Sikkim. The Eco-Sensitive Zone notification effectively foresees PA conservation and encourages ecotourism activities. This is especially critical as the Ribdi-Bhareng village complex falls within the zone and will draw greater attention soon. Such activities include cottage industries, including village artisans, homestays, ropeways, funiculars, kiosks, rainwater harvesting, and many others (MoEFCC 2014). As recommended for Singalila National Park and Barsey Rhododendron Sanctuary, the Eco-Sensitive Zone does not mark any area adjoining the border with Nepal. The zone further extends to the outer bank of the Ramam River. Promoting the proposed eco-trail will not only boost tourist flow in either site but also strengthen the socio-cultural relationships between communities of the two states and between India and Nepal. The trekkers could enjoy the trail accompanied by local guides and explore the diverse landscape diversity in its sociocultural and natural biodiversity. They will get the opportunity to explore the wilderness. A key ecological challenge in the landscape is the overdominance of maling bamboo (*Yushania maling*), which alters forest structure and composition (Gaira et al. 2022). Managing this invasive bamboo by using it to produce handicrafts and other items could provide an economic link to ecotourism while helping to restore the natural forest balance.

Ribdi-Bhareng and Gorkhey-Samanden areas are located in India, yet they have a cultural resemblance to the adjoining regions of Nepal. The ethnic communities on both sides of the international border are the same, as are the linguistic, traditional, cultural, ritual, and other aspects of their everyday life. The border between these two countries is not a bar for intermarriages. In contrast, it is one of the region's desired and standard practices. The rituals and festivals are celebrated together, and the people often cross the borders to meet their relatives and in-laws. National and international boundaries could not delimit the ecological entities of the landscape. The political border does not interrupt the biological integrity of the region and the interdependency of communities from both sides over natural resources. The silhouette of the international border is now and then crossed by the people from both sides, dependent on the region's forest resources. The mutual understanding as a part of the cultural landscape and the flexibility of laws in either region (Visa not required for Indian and Nepalese citizens to travel each other's country) has always been important and given more area for good transboundary relationship and vis-à-vis bonding between the two, like the case in the Kailash Sacred Landscape (Badola et al. 2017). However, in Kailash Sacred Landscape-India, the Government's cross-cultural trade between India and Nepal is an excellent example to follow in the Khangchendzonga Landscape Indian part (Badola et al. 2017). Such geographical spaces have ceased to be an island of the yesteryears. Still, modern means of communication, transportation, and initiatives for transboundary tourism have become a cradle of newer cultures of human relations and existence.

Also, the homestays developing at either end of the eco-trail would allow visitors to know more about the places and the people and get a glimpse of the local culture and traditions. The goal would be to provide visitors with an enjoyable learning experience about nature, culture, and the region's uniqueness while emphasising nature's conservation value to visitors and the local people. Sikkim government pioneered in KL by declaring the Registration of Homestay Establishment Rules 2013, under which XIII Finance Commission funded the construction of 736 homestays in different villages of Sikkim under the rural tourism program (State Tourism Policy 2017). As per the data available in the Government of Sikkim records (as of June 2017), there are 343 registered and 367 non-registered homestays, of which 212 registered homestays are located in villages of West Sikkim. This is a considerable boost compared to 26 registered homestays during 2013–2014.

Linking people, places, and nature is one of the most effective and well-acknowledged tools for PA management (McLean and Straede 2003; McCool and Spenceley 2014). The international conference, '*Ecotourism in Protected Areas: Strengthening Livelihoods, Conservation, and Adaptation'*, considered "*Streamlining ecotourism in protected areas will go a long way in strengthening livelihoods, conservation, and adaptation*' (Outpost 2015). Sustainable nature-based tourism concerning conservation purposes in and around PAs can be turned into the main facet of the local economy (Brandt 2011). Through ecotourism, rural economic development, as well as ecosystem conservation, can be achieved simultaneously. The local people in our study area could also become aware of the importance and the need for the resources and behave as local guardians of the surrounding ecosystems (Bookbinder et al. 1998).

This study areas, Ribdi-Bhareng and Gorkhey-Samanden, offer excellent opportunities to promote ecotourism as a tool for PA conservation, as either side offers ample tourist value to the visitors. Although ecotourism activity in both sites has been in practice for decades, the area has yet to manage the desired tourist flow and be mapped to the international tourist circuit for its high ecotourism potential. The existing trail between both study areas tenders as an alternative route to famous tourist destinations like Phalut and Sandakphu, but the tourist movement from the route is not well known so far due to a lack of publicity and difficulties in getting inner line permit from the government for the overseas tourists. Our study suggests developing linkages between the two regions of Khangchendzonga Landscape under a generalised outlook for ecotourism development in cooperation with the Sikkim and West Bengal Governments. It is ideal to look at the socio-political possibilities of linking this area with neighbouring Nepal as transboundary cooperation. In such a case, the things of utmost importance rely on their eco-friendly concept of ecotourism and avoiding any damage to the rich ecosystem. Furthermore, socio-economic impact assessment and monitoring related changes are required from the initial planning, management, and tourism development stage in associated PAs (Liu et al. 2012). The system's sole effort cannot be taken as a holistic approach to conserving any PA; the result would be limited unless the system also includes the communities (Kiss 2004; Chettri and Shakya 2008). By including the agenda of community development through ecotourism, the conservation plans will also be strengthened in any community-dependent PA (Gaira et al. 2017). This will motivate community people in the conservation programme, especially for their greater awareness of the importance of the surrounding areas. The development of sustainable ecotourism will be a great need and an opportunity to gain long-term community support for conserving the two important PAs of Khangchendzonga Landscape in India.

In addition, Sikkim's transboundary ecotourism potential extends from Phalut and Sandakphu to Gorkhey-Samanden, Darjeeling, and up to Ribdi-Bhareng and Okhrey. This offers an integrated livelihood for the surrounding community and the conservation of biodiversity. For this reason, government-owned participatory planning is required to identify and develop joint strategies between two states (Sikkim and West Bengal) and, further, India and Nepal (Chettri et al. 2020).

16.5 Conclusion and Recommendations

Our paper examines the potential for ecotourism development in the Ribdi-Bhareng (Sikkim) and Gorkhey-Samanden (West Bengal) areas, extending to Phalut-Sandakphu within the Khangchendzonga Landscape (India). Through comprehensive analysis using Participatory Rural Appraisal (PRA), household surveys, and secondary data, we identify the pressing needs and unique opportunities that position this region as a promising ecotourism destination. The study highlights local communities' challenges, such as declining crop productivity, insufficient government support for infrastructure and services, and the critical need for diversified livelihood options. Despite these challenges, the landscape presents many opportunities, including its rich biodiversity, breathtaking scenery, proximity to protected areas (PAs), and vibrant cultural heritage—all essential for successful ecotourism. We also underscore the value of the trail linking Ribdi-Bhareng and Gorkhey-Samanden as a driver of ecotourism, sociocultural exchange, and economic connectivity. Developing this eco-corridor could deepen connections between major tourist sites, creating a vital social and tourism link that benefits travellers and residents. This corridor also opens the door to transboundary tourism with Nepal, fostering regional cooperation. The initiative promises to invigorate the local economy through community-driven revenue generation while enhancing conservation efforts. Establishing Eco-Sensitive Zones around the Barsey Rhododendron Sanctuary and Singalila National Park underscores the critical role of ecotourism in conservation, particularly in these PAs within the Khangchendzonga Landscape.

Our findings offer essential insights for policymakers aiming to create impactful regional strategies, with local perspectives ensuring fair benefit distribution and supporting community-led initiatives. Further research and capacity-building across stakeholder levels are crucial to advancing these ecotourism goals (McCool and Spenceley 2014). Success will hinge on collaborative efforts amongst the Sikkim and West Bengal state governments, scientific organisations, and local NGOs to achieve sustainable ecotourism development. Additionally, this initiative opens up the potential to promote a transboundary ecotourism circuit, linking Bhutan, India, and Nepal, as a pioneering model of regional ecotourism cooperation (Lama et al. 2019).

Acknowledgements We thank the Director of G.B. Pant National Institute of Himalayan Environment (GBPNIHE) for providing the necessary facilities and encouragement. We thank the International Centre for Integrated Mountain Development (ICIMOD) for providing financial support under the Khangchendzonga Landscape Conservation and Development Initiatives (KLCDI) India programme. We appreciate field cooperation from the state forest departments of the Governments of Sikkim and West Bengal and the community people of the study areas.

References

Admasu S (2020) Ecotourism as a tool to substantially funding conservation endeavors in Ethiopia. J Tourism Hosp 9:432

Badola HK, Aitken S (2010) Potential biological resources for poverty alleviation in Indian Himalaya. Biodiversity 11(3–4):8–18

Badola HK, Subba JB (2012) Khangchendzonga biosphere reserve (Sikkim). In: Palni LMS, Rawal RS, Rai RK et al (eds) Compendium on Indian biosphere reserves: progression during two decades of conservation. GBPIHED/MoEFCC (Govt. of India), Kosi-Almora/New Delhi, pp 133–142

Badola HK, Gaira KS, Dhyani PP (2015) Khangchendzonga landscape conservation and development initiative (KLCDI) in India. Preparatory phase. Project completion report (15.11.2013 to 31.3.2015). Submitted to ICIMOD by GBPIHED, Sikkim Unit, Gangtok, Sikkim, India, p 103

Badola HK, Gaira KS, Lepcha J et al (2016a) Feasibility study and bioresource management plan: Barsey-Singalila site (Khangchendzonga landscape, India). GBPNIHESD, India, p 56
Badola HK, Lepcha J, Gaira KS et al (2016b) Socio economic and bioresource assessment: participatory and household survey methods, tools and techniques (a training manual based on the experiences from the Khangchendzonga landscape, India). GBPNIHESD, Sikkim Unit/Highlanders Communications Private Limited, Gangtok/New Delhi, p 60
Badola E, Rawal RS, Dhyani PP (2017) Stories of success: narratives from a sacred land. G.B. Pant National Institute of Himalayan Environment & Sustainable Development, Kosi-Katarmal, Almora, p 95
Bennett N, Lemelin RH, Koster R et al (2012) A capital assets framework for appraising and building capacity for tourism development in aboriginal protected area gateway communities. Tour Manag 33(4):752–766
Bertzky B, Corrigan C, Kemsey J et al (2012) Protected planet report 2012: tracking progress towards global targets for protected areas. IUCN, Gland
Bhattacharjee A, Parthasarathy N (2013) Coexisting with large carnivores: a case study from Western Duars, India. Hum Dimens Wildl 18(1):20–31
Bhutia S (2014) The role of tourism for human resource development in Darjeeling District of West Bengal, India. J Tourism Hosp Manag 2(1):113–128
Bookbinder MP, Dinerstein E, Rijal A et al (1998) Ecotourism's support of biodiversity conservation. Conserv Biol 12(6):1399–1404
Brandt J (2011) Carrying capacity – how much tourism can protected areas cope with? In: Ostermann O (ed) Parks and benefits: guide to sustainable tourism in protected areas: ways to sustainable management. Socio-Economic Benefits of Conserving Nature, Rostock, pp 26–37
Chambers R (1983) Rural development: putting the last first. Longman Scientific and Technical, London
Chambers R (1991) Shortcut and participatory methods for gaining social information for projects. In: Cernea M (ed) Putting people first: sociological variables in rural development, 2nd edn. The World Bank; New York and Oxford: Oxford University Press, Washington, DC, pp 515–537
Chambers R (1994) The origins and practice of participatory rural appraisal. World Dev 22(7):953–969
Chettri N, Shakya B (2008) Species to landscape: a paradigm shift in biodiversity conservation through people's participation and policy reform. In: Rasul G, Karki M (eds) Policy priorities for sustainable mountain development: proceedings and selected papers from the regional policy workshop, September 18–20, 2006, Kathmandu. ICIMOD, Nepal
Chettri N, Lama AK, Gaira KS et al (2020) Transboundary cooperation for mountain tourism in the Kangchenjunga landscape. In: Wymann von Dach S, Ruiz Peyre F (eds) Vibrant mountain communities. Regional development mountains: realizing potentials, tackling disparities. Centre for Development and Environment (CDE), University of Bern, with Bern Open Publishing (BOP), Bern, Switzerland, pp 36–38
Davies G (2024) Rebalancing the nature tourism value chain. In: Mandić A, Spenceley A, Fennell DA (eds) Handbook on managing nature-based tourism destinations amid climate change. Elgar Online, pp 347–364. https://doi.org/10.4337/9781035311255.00030
Drumm A, Moore A (2005) Ecotourism development - A manual for conservation planners and managers. Vol 1. An introduction to ecotourism planning. The Nature Conservancy, Arlington
Eagles P, Hillel O (2008) Improving protected area finance through tourism. In Secretariat of the convention on biological diversity. Protected areas in today's world: their values and benefits for the welfare of the planet. Montreal, Technical Series no. 36, i–vii + 96 pages
Eagles PFJ, McCool SF, Haynes CDA (2002) Sustainable tourism in protected areas: guidelines for planning and management. IUCN, Gland/Cambridge. xv + 183pp. ISBN:2-8317-0648-3
Freudenberger KS (1999) Rapid rural appraisal (RRA) and participatory rural appraisal (PRA): a manual for CRS field workers and partners. Catholic Relief Services, Baltimore
Gaira KS, Lepcha J, Sinha S et al (2017) Khangchendzonga landscape conservation and development initiative (KLCDI)-India: implementation phase. GBPNIHESD, Sikkim, p 4

Gaira KS, Pandey A, Sinha S et al (2022) Maling bamboo (*Yushania maling*) overdominance alters forest structure and composition in Khangchendzonga landscape. East Himalaya Sci Rep 12(1):1–13

Ghimire KB, Pimbert MP (1997) Social change and conservation: an overview of issues and concepts. In: Ghimire KB, Pimbert MP (eds) Social change and conservation. Earthscan, London, pp 1–45

Gündüz M, Nielsen Y, Özdemir M (2012) Quantification of delay factors using the relative importance index method for construction projects in Turkey. J Manag Eng 29(2):133–139

Gurung J, Chettri N, Sharma E et al (2019) Evolution of a transboundary landscape approach in the Hindu Kush Himalaya: key learnings from the Kangchenjunga landscape. Glob Ecol Conserv 17:e00599

ICIMOD, WCD, GBPNIHESD, RECAST (2017a) Kangchenjunga landscape feasibility assessment report, ICIMOD working paper 2017/9. ICIMOD, Kathmandu

ICIMOD, WCD, GBPNIHESD, RECAST (2017b) Kangchenjunga landscape conservation and development strategy and regional cooperation framework, ICIMOD working paper 2017/2. ICIMOD, Kathmandu

Joshi R, Dhyani PP (2009) Environment sustainability and *tourism*: implications of trends synergies of *tourism* in Sikkim Himalayas. Curr Sci 97(1):33–41

Kiss A (2004) Is community-based ecotourism a good use of biodiversity conservation funds? Trends Ecol Evol 19(5):232–237

Kometa ST, Olomolaiye PO, Harris FC (1994) Attributes of UK construction clients influencing project consultants' performance. Constr Manage Econ 12(5):433–443

Kuniyal JC, Vishvakarma SCR, Badola HK et al (2004) Tourism in Kullu Valley: an environmental assessment, HIMAVIKAS Publ. vol 15. Bishan Singh-Mahendra Pal Singh, Dehradun, p 210

Lama AK, Kandel P, Chaudhary S et al (2019) Transboundary ecotourism in the Kangchenjunga landscape: opportunities for sustainabledevelopment through regional cooperation, ICIMOD working paper 2019/4. ICIMOD, Kathmandu

Lepcha J, Rai YK, Rai LK (2015) Fambonglho wildlife sanctuary (Sikkim, India) and promotion of ecotourism: perspectives from ground zero. Int J Multidiscipl Curr Res 3:1204–1209

Leung YF, Spenceley A, Hvenegaard G et al (2015) Tourism and visitor Management in Protected Areas: guidelines towards sustainability, Best practice protected area guidelines series no. XX. IUCN, Gland

Lindsey P, Baghai M, Bigurube G et al (2021) Attracting investment for Africa's protected areas by creating enabling environments for collaborative management partnerships. Biol Conserv 255:108979

Liu W, Vogt CA, Luo J et al (2012) Drivers and socioeconomic impacts of tourism participation in protected areas. PLoS One 7(4):e35420. https://doi.org/10.1371/journal.pone.0035420

Maharana I (2000) Economic benefits and conservation linkages from tourism development in the Sikkim Himalaya. PhD Thesis. North Bengal University

Maharana I, Rai SC, Sharma E (2000) Valuing ecotourism in a sacred lake of the Sikkim Himalaya, India. Environ Conserv 27(3):269–277

Mainra A, Badola HK, Mohanty B (eds) (2010) Proceeding, international conference, rhododendrons: conservation and sustainable use, FEWMD, Government of Sikkim, Gangtok-Sikkim, India. Printed at CONCEPT, Siliguri, p 100

McCool SF, Spenceley A (2014) Tourism and protected areas: a growing nexus of challenge and opportunity. Koedoe 56(2). https://doi.org/10.4102/koedoe.v56i2.1221

McLean J, Straede S (2003) Conservation, relocation, and the paradigms of parks and people management—a case study of Padampur villages and the Royal Chitwan National Park, Nepal. Soc Nat Resour 16:509–526

Mieczkowski Z (1995) Environmental issues of tourism and recreation. University Press of America, Lanham

Mittermeier RA, Gil PR, Hoffmann M et al (2004) Hotspots revisited: Earth's biologically richest and most endangered terrestrial ecoregions. CEMEX, Mexico

MoEFCC (2014) Notification. Available at http://www.moef.gov.in/sites/default/files/S.O.%202172%20%5B27.08.2014%5D%20Final%20ESZ%20Notification%20on%20Barsey%20Rhododendron%20Sanctuary%2C%20Sikkim.pdf. Accessed 29 Oct 2017

MoEFCC (2017a) Expert committee meeting of 26th for the declaration of eco-sensitive zone (ESZ) around wildlife sanctuaries/national parks. Indra Paryavaran Bhawan, New Delhi. Available at http://www.moef.gov.in/sites/default/files/Final_26th%20ESZ%20Meeting_Minutes1.pdf. Accessed 27 Oct 2017

MoEFCC (2017b) Notification. Available at http://www.moef.gov.in/sites/default/files/Singalila%20National%20Park%2C%20West%20Bengal.pdf. Accessed 29 October 2017

Mosse D (1994) Authority, gender and knowledge: theoretical reflections on the practice of participatory rural appraisal. Dev Chang 25(3):497–526

Nepal SK (2002) Mountain ecotourism and sustainable development: ecology, economics, and ethics. Mountain Res Dev 22(2):104–109

Outpost (2015) Streamline ecotourism in protected areas, say experts. Available at http://outpost.pk/international/416-streamline-ecotourism-in-protected-areas-say-experts.html. Accessed 7 Feb 2017

Phuntsho K, Chettri N, Oli KP (2012) Mainstreaming community-based conservation in a transboundary mountain landscape – lessons from Kangchenjunga. ICIMOD, Kathmandu, p 5

Rahman M, Roy B, Anik SI (2013) Ecotourism and protected area conservation in Bangladesh: a case study on understanding the visitors views on prospects and development. J For Environ Sci 29(1):15–28

Rai SC, Sundriyal RC (1997) Tourism and biodiversity conservation: the Sikkim Himalaya. Ambio 26:235–242

Samal R, Dash M (2023) Ecotourism, biodiversity conservation and livelihoods: understanding the convergence and divergence. Int J Geoheritage Parks 11:1–20

Sambasivan M, Soon YW (2007) Causes and effects of delays in Malaysian construction industry. Int J Proj Manag 25(5):517–526

Scherl LM, Wilson A, Wild R et al (2004) Can protected areas contribute to poverty reduction? Opportunities and limitations. World Conservation Union (IUCN), Gland and Cambridge, UK

Shakya B, Joshi RM (2008) Protected areas and biodiversity conservation in the Hindu Kush-Himalayan region with special reference to the Kangchenjunga landscape. In: Chettri N, Shakya B, Sharma E (eds) Biodiversity conservation in the Kangchenjunga landscape. ICIMOD, Kathmandu, pp 13–20

Sharma E, Chettri N (2005) ICIMOD's transboundary biodiversity management initiative in the Hindu Kush-Himalayas. Mountain Res Dev 25(3):280–283

Sinha S, Badola HK, Chhetri B et al (2018) Effect of altitude and climate in shaping the forest compositions of Singalila National Park in Khangchendzonga landscape, eastern Himalaya, India. J Asia-Pac Biodivers 11(2):267–275

Spenceley A (2015) Tourism and the IUCN World parks congress 2014. J Sustain Tour 23(7):1114–1116

State Tourism Policy (2017) Government of Sikkim. www.sikkimtourism.gov.in/Webforms/General/pdf/Policy.pdf. Accessed 26 Oct 2017

The Ecotourism Society (1991) Ecotourism guidelines for nature-based tour operators. The Ecotourism Society, Vermont

Trofimov V, Soimu O (2011) Ecotourism concept in the light of cultural diversity and regional development. Agric Econ Rural Dev 1:117–125

Chapter 17
Evaluating the Relationship of Ecosystem Provisioning Services with the Socio-economic Attributes Along an Altitudinal Gradient in the Western Himalayas, India

Surajit Dutta, Subhasis Mahato, and Ajay Maletha

Abstract Millions of rural people depend upon forests extensively for fuel, fodder, grazing, and numerous subsistence necessities, especially in the mountainous region. This study was conducted at three altitudinal zones in the Dudhatoli region (Garhwal Himalaya) to ascertain the amount of biomass extraction from forests in terms of the socio-economic attributes of local communities. In the study, it was found that the sampled village has an average of 1346.71 cattle that consume fodder quantity of 47.29 ± 31.69 kg (i.e. at higher altitude), 37.38 ± 28.82 kg (i.e. middle altitude), and 27.54 ± 24.71 kg (i.e. lower altitude) per day. Every household has an average of 6.91 ± 2.81 members, with a fuelwood consumption of an average of 2.65–2.80 kg/person/day (i.e. with increasing altitude) and 4.16 ± 3.22–2.4 ± 0.54 kg/person/day (i.e. with increasing family size). The study results provide an association of socio-economic status with different patterns of regulatory ecosystem services related to forest resources. The Dudhatoli region's ecosystem provisioning services require effective management using ecosystem-based approaches. These include local involvement, agroforestry, alternative energy sources, and government initiatives. Implementing these strategies can improve livelihoods and address climate change mitigation.

Keywords Garhwal Himalaya · Ecosystem services · Fuelwood · Fodder · Socio-economic attributes

---

S. Dutta (✉)
Department of Geography, Dr. Shyama Prasad Mukherjee University, Ranchi, Jharkhand, India

S. Mahato
Department of Forestry, PRMS Mahavidyalaya, Bankura, West Bengal, India

A. Maletha
Amity Institute of Forestry and Wildlife, Amity University, Noida, Uttar Pradesh, India

Y. D. I. Khan et al. (eds.), *Ecosystem-based Approaches for Resilience Building in Himalayan Landscapes*, Disaster Resilience and Green Growth,
https://doi.org/10.1007/978-981-95-2007-7_17

17.1 Introduction

The Himalayan region preserves a unique place in the mountain biomes of the world. The mountainous areas combine ecological, social, and cultural heterogeneity (Price and Thompson 1997). Forests and pasture lands, cultivable land, livestock, and human population constitute the significant structure of the hilly agroecosystem. All these constitute dynamic connections relating to energy production, consumption, and transfer (Tiwari 2000). The forest resource maintains the ecological system and supports the socio-economic development of any geographical region. Availability of fuel, fodder, and litter is crucial for rural communities, with about 90% of energy requirements fulfilled by biomass in the region (Sharma et al. 1999). In particular, for the needs of fuelwood and domestic animal fodder, the adjoining forests form the primary source of energy for the sustenance of these societies (Martins and Nautiyal 1993) and thus hold as the most crucial aspect for the sustainability of the mountain agricultural ecosystems (Tripathi and Sah 2001).

Forests play a significant part in the development process of human civilisation and are notable as primary sources of numerous vital provisional ecosystem services. The goods and services such as fuelwood, fodder, wild edibles, medicines, building materials, agricultural implements, religious sentiments, and other purposes are linked with local communities (Samant and Dhar 1997). Despite the richness of natural resources in the Himalayas, most rural inhabitants are economically poor, are marginalised, and have a subsistence way of life (Singh 2006). Forests are an essential community resource for fodder, fuelwood, grazing, and other subsistence requirements by villagers (Rahmani 2003). Forests have been an important livelihood source for most of the rural populations in India since the ages, and the same is true for the entire Himalayan range. For rural people, these are the most important sources of income because of the limited employment opportunities. The Indian Himalayan Region (IHR) farmers are comparatively economically poor, and they rely on biological resources for sustenance in one way or the other. In the rural Himalayan region, where people directly or indirectly depend considerably on these resources for their basic needs, the extraction of biomass in the form of fuelwood, non-timbre forest products (NTFPs), and grazing pressure may be the direct concern for the sustainability of the forest.

This study examines 14 villages across 3 distinct altitudinal zones within 5 catchments of the Dudhatoli region in the Garhwal Himalayas. The main object of the study was to analyse forest biomass extraction in its various forms, particularly fuelwood, fodder, and NTFPs, and their associations with socio-economic attributes. Understanding the associations between the socio-economy and forest resource use patterns would benefit better planning and management of valuable natural resources. The research also delineates an ecosystem-based strategy for preserving ecological sustainability.

17.2 Study Area

The study area of Dudhatoli is located in the lesser North East Garhwal Himalaya, a central part of the state of Uttarakhand, India. The geographical coordinates of the area are 29°56′13″ N to 30°10′53″ N latitude and 78°59′17″E to 79°19′14″E longitude. Dudhatoli High Mountain is a water tower of the Uttarakhand from which Western Nayar, Eastern Nayar, Bino (Tributary of Ram Ganga), Ram Ganga, and Ata Gad (Tributary of Pinder) and their tributaries originate. The total geographical area of this region is 493.11 km^2. The Dudhatoli region covers mountain terrain with an elevation of 1098 msl–3119 msl.

The annual mean temperature of the area is 14.6 °C, with the maximum and minimum temperatures ranging from 28.4 °C to 0 °C, respectively. The average annual precipitation of the region is about 1545.4 mm. The natural vegetation of the Dudhatoli is primarily represented by broadleaf and conifer forests. Due to high precipitation and effects of other physio-biological factors (viz. altitude, climate, geology, soil, and landforms), the whole Dudhatoli region comprises rich forest resources.

17.3 Materials and Methods

The early phases of the research included an exploratory survey conducted in the region. The complete survey was carried out from May 2015 to July 2017. The area of study was divided into three broad altitudinal zonations, viz. lower altitude (1098–1500 m), middle (1500–2000 m), and higher altitudinal zones (>2000 m). The data was collected from 14 sample villages from different altitudinal zones (Fig. 17.1). The socio-economic survey was conducted in 312 households in 14 villages in 5 catchments during the field study. A structured questionnaire was prepared for the households' forest resource usage pattern survey. The survey consisted of the following information: basic household data (viz. member particulars, educational qualification, landholdings, income, expenditures); livestock data (viz. units, fodder consumption, distance travelled); fuel statistics (viz. collection by quantity per day, species preference, source and place of collection); and NTFP data.

17.3.1 Data Analysis

The data was analysed to see the effect of altitudes, income, and household sizes on biomass extraction from Dudhatoli's region forests. The households were selected to represent all economic classes and household sizes equally. The household sizes were further categorised according to the number of individuals in the households, viz. small (1–3 members), medium (4–6 members), and large (>6 members). The

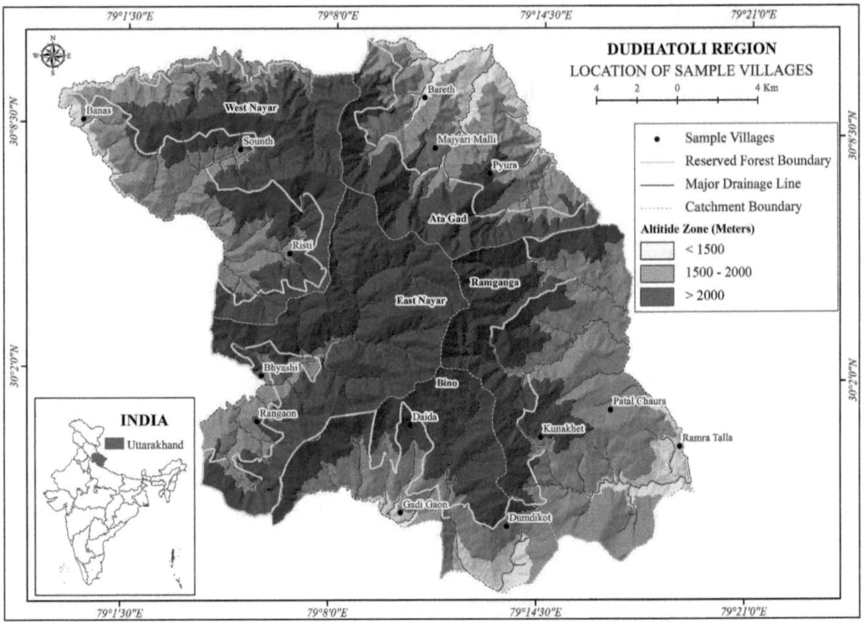

Fig. 17.1 The study area of the Dudhatoli region in the Garhwal Himalaya

impact of household sizes on the variations of fuelwood consumption was analysed for every three altitudinal zones. The income groups were categorised into four groups, viz. G1 < ₹30,000, G2 ₹30,000–60,000, G3 ₹60,000–90,000, and G4 > ₹90,000.

To analyse the quantity of fuelwood consumption, the 24-h weight survey method was conducted by household survey in different altitudinal zones following Bhatt et al. (1994). To quantify the fuel consumption pattern, the per capita fuelwood consumption (PCFC) was computed using the ratio PCFC = FC/p (Ramachandra et al. 2000), wherein FC is fuel consumed (kgs/day) and p is the number of adult equivalents. The livestock population was converted into adult cattle units per the formula used by the Animal Husbandry Statistical Report (Census of India 2011; Samal et al. 2003).

17.4 Result

17.4.1 Socio-economic Condition of Respondents

The age of the respondents was, on average, 44 ± 16 years. The total population of the surveyed households was 2162. The household size varied from 2 to 19 persons. The average household size differs from lower to higher altitudes. The household member size averaged 6.91 ± 2.81 members (Table 17.1). Of the total 312 respondents in the region, households comprised small (4.5%), medium (47.8%), and large families (47.8%). Out of the enumerated population, the maximum population was of the age group of 10–14 years, followed by the 15–19-year age group, while the minimum was found in the>75-year age group. The literacy status of the sample population was 75.93%; the average male and female literacy rates were 87.30% and 63.15%, respectively. The literacy percentage varied along the altitudinal gradient and was higher in lower altitudes (Table 17.1). This could be because of more accessibility and less distance from educational institutions. Most of the illiterate persons were older people. The majority of houses were of the traditional type. These were made of mud, wooden sleepers, stones, and straw, although modern houses are primarily seen in lower altitudes.

The primary sources of income in all three altitudinal zones were remittance, agriculture, and pasture. The maximum household income is generated from remittance (30.13%), followed by the agriculture sector (19.55%) and government employees (13.14%). The household's average income per annum differed among the villages. The average yearly income per household was found to be ₹90,816, which ranged from ₹21,600 to 540,000. The income data analysis indicated that families lay above the below poverty line (BPL) limit defined by the Government of India for rural areas (₹12,000). It was analysed that the households' average annual income decreased with the increasing altitudinal zone (Table 17.1).

17.4.2 Livestock and Fodder Consumption

Animal husbandry is one of the essential socio-economic components and a significant source of income in the study area. In the studied villages, most of the families owned livestock, which included ox (29.63%), cow (21.24%), buffalo (20.12%), cow-calf (9.06%), and goat (8.74%). These were domesticated for basic needs like milk production and ploughing the land in a terraced field with slopes and smaller areas. The total cattle unit in sample villages was 1346.71. Overall, the number of domestic animals per household on average was 5.18 (in the higher altitudinal zone), 4.12 (in the middle altitudinal zone), and 3.04 (in the lower altitudinal zone). The cattle unit shows an increasing trend with increasing altitude. The maximum cattle unit was found at a higher altitude due to open reserved forest land for pasturing on the top of the Dudhatoli range.

In the Dudhatoli region, cattle are generally stall-fed, although buffaloes, sheep, and goats are left for open grazing in surrounding forests. The collection of fodder is a daily routine, and it is mainly carried out by females (84.97%). The annual fodder consumption data has been updated to 44,55,993 kg, which was calculated by multiplying thenumber of village-wise sampled households, the average daily fodder consumption per household (in kg), and 365days (representing one year) (Table 17.1). The daily average household fodder consumption was estimated at 39.14 ± 30.06 kg. The fodder consumption per household per day (in kg) was, on

Table 17.1 Socio-economic profile of the sample villages

Catchments →	Ata Gad			Bino Nadi			Ramganga			E Nayar		W Nayar		
Villages →	Bareth	Majiyarimalli	Pyonra	Daira	Dumnikot	Godigaon	Kunkhet	Ramratalla	Patalchaura	Rangaon	Bhyashi	Risti	Saunth	Banas
Altitude (m.)	1400	1800	2060	2200	1850	1450	2030	1465	1860	2010	1840	2010	1980	1480
Sampled HH*	21	30	20	12	19	7	29	21	16	28	18	36	34	21
Total population	122	208	138	102	145	53	217	134	111	212	130	256	205	129
Average HH size	5.81 ± 1.84	6.93 ± 3.12	6.90 ± 3.06	8.50 ± 3.32	7.63 ± 3.20	7.57 ± 1.81	7.48 ± 7.37	6.38 ± 2.73	6.94 ± 3.59	7.57 ± 2.67	7.22 ± 2.96	7.11 ± 2.09	6.03 ± 2.33	6.14 ± 2.26
Sex ratio	1000	763	916	1170	959	963	991	1062	820	738	711	1098	767	1048
Average monthly income (Rs)	5428.57 ± 4569.70	6393.33 ± 6850.87	7075 ± 2344.95	5566.67 ± 2344.95	6542.57 ± 4826.47	4428.57 ± 1170.06	12,413.79 ± 11,804.26	11,904.76 ± 12,180.54	11,875 ± 10,229.86	5803.57 ± 4051.38	8444.44 ± 8390.98	5736.11 ± 4115.28	4914.71 ± 2742.39	9704 ± 11,367.74
Landholding (ha)	0.15	0.18	0.16	0.18	0.30	0.41	0.17	0.26	0.21	0.55	0.34	0.24	0.19	0.13
Literacy (%)	78.07	76.56	79.07	71.11	70.50	72.55	85.15	83.33	85.98	72.91	79.34	69.92	67.36	81.30
Male literacy (%)	84.75	90.83	89.86	77.50	79.71	88.46	95.10	95.24	92.98	78.69	91.67	86.09	80.36	95.00
Female literacy (%)	70.91	57.83	66.67	66.00	61.43	56.00	75.00	72.46	78.00	64.20	61.22	54.55	49.38	68.25
Cattle units	2.79 ± 1.18	4.82 ± 2.71	5.26 ± 2.39	11.45 ± 3.74	5.31 ± 3.96	8.66 ± 5.96	2.81 ± 1.87	2.11 ± 1.39	2.51 ± 2.24	5.32 ± 3.65	2.14 ± 1.23	4.83 ± 2.71	4.65 ± 3.71	2.34 ± 1.20
Average fodder consumption per HH/day(kg)	25.12 ± 10.65	43.35 ± 24.43	47.38 ± 21.47	97.31 ± 31.8	44.12 ± 34.77	73.61 ± 50.66	28.08 ± 18.69	21.11 ± 13.91	25.09 ± 22.39	50.58 ± 34.72	20.36 ± 11.64	43.38 ± 24.37	43.13 ± 33.39	21.02 ± 13.86
Distance travelled for fodder collection (km)	2.5–5	3–6	3.5–5	1–5	4–6	5–6	3–6	1.5–4	3–5	2–5	2–6	3–5	3.5–5	2.5–6

Villages →	Bareth	Majiyarimalli	Pyonra	Daira	Dumnikot	Godigaon	Kunkhet	Ramratalla	Patalchaura	Rangaon	Bhyashi	Risti	Saunth	Banas
Average fuelwood consumption per HH/day (kg)	16.74 ± 3.76	20.05 ± 4.52	20.6 ± 5.61	21.85 ± 4.76	20.75 ± 8.22	20.46 ± 6.1	22.13 ± 7.70	15.4 ± 4.22	15.99 ± 7.05	20.98 ± 6.64	17.65 ± 3.64	19.08 ± 3.63	17.33 ± 4.57	16.37 ± 4.40
Average fuelwood consumption/ capita/day (kg)	2.99 ± 0.54	3.20 ± 0.94	3.24 ± 0.74	2.77 ± 0.73	2.83 ± 0.81	2.74 ± 0.72	3.14 ± 0.72	2.74 ± 1.53	2.48 ± 0.73	2.91 ± 0.81	2.64 ± 0.56	2.81 ± 0.63	3.04 ± 0.69	2.86 ± 0.88
LPG using families (%)	19.05	0.00	0.00	0.00	9.52	4.76	57.14	90.48	28.57	23.81	14.29	28.57	9.52	52.38

* Sampled HH*- it mean Household

average, 47.29 ± 31.69 (higher altitude), 37.38 ± 28.82 (middle altitude), and 27.54 ± 24.71 (lower altitude). The differences were observed due to the varying number of domestic animals per household. The fodder collection shows an increasing trend with increasing altitude. The fodder collection depends on the number of cattle and forage availability in the nearby forest.

17.4.3 Pattern of Energy Consumption

In this study, fuelwood was the most common energy resource. Most people collected fuelwood from the village civil forest, but sometimes, people had to travel a long distance in the reserved forest. The average distance travelled by the villagers for fuelwood collection was 4.61 ± 1.04 km (Table 17.1). Generally, women were involved in fuelwood collection, although both genders had access to natural resources in a few instances.

In all three altitudinal zones, 98.5% of households used fuelwood as the primary energy source, mainly for cooking and heating. Local people usually collect fuelwood throughout the year except during snowfall in winter and rainy days in the monsoon season. Fuelwood consumption in surveyed households was 5310.78 kg/day, with the average consumption per household being 18.94 ± 5.76 kg/day (Table 17.1). Fuelwood consumption in the region was about 2.73 kg per capita per day (Table 17.1).

17.4.4 Altitudinal Variation in Fuelwood Consumption

The fuelwood consumption varied along the altitudinal gradient and differed by the village. At higher altitudes, the average fuelwood consumption was observed to be 20.72 ± 5.90 kg per household per day and about 2.80 kg per capita per day. In the middle altitude, the fuelwood consumption averaged18.45 ± 5.72 kg/household/day and 2.70 kg per capita per day (Table 17.1). In the lower altitude, average household fuelwood consumption was 16.60 ± 4.48 kg/day, and per capita per day was about 2.65 kg (Table 17.1). The results showed that fuelwood consumption rapidly increased with the increasing altitude of the region. It was observed that the temperature was reduced due to the high altitude compared to the other two lower altitudes, and hence, more fuelwood was required for various basic needs. Notably, high-altitude villages were located near the vicinity of the forests.

17.4.5 Family Size-Wise Variations in Fuelwood Consumption with Altitudinal Variation

In the lower altitude (<1500 m), families were comprised of small (5.71%), medium (57.14%), and large size (37.14%). In the middle altitudinal zone (1500–2000 m), families comprised small (7.69%), medium (47.86%), and large-size families (44.44%). In the higher altitudinal zone (>2000 m), families comprised small (0.8%), medium (42.4%), and large size (56.8%). The variations in fuelwood consumption relative to family size along the altitudinal categories are shown in Table 17.2.

In the lower altitude zone (<1500 m), per capita per day (kg) fuelwood consumption (PCFC) varies directly with family size classes (4.16 ± 3.22 in small, 2.99 ± 0.61 in medium, 2.41 ± 0.54 in large). The same trend of fuelwood consumption was estimated in the middle and higher altitude zones. Data analysis (Table 17.2) concluded that household fuel consumption is higher in large families in all the altitudinal zones.

Besides this, overall, the per capita fuel consumption in large families was not as much as that of small families.

17.4.6 Variations by Income in Fuelwood Consumption with Altitudinal Variation

The maximum number of households was in the income group G2 (55.45%), and the minimum was in the lower-income group (6.09%). The fuel consumption pattern was also correlated with income groups from different altitudinal zones. The maximum average day^{-1} $family^{-1}$ fuelwood consumption (19.10 ± 4.95 kg) was found in the income group of G3 followed by G1, i.e. 17.71 ± 5.25 kg. Besides this average lowest fuelwood consumption was found in G4 (16.78 ± 4.19 day^{-1} $family^{-1}$) and G2 income group (16.02 ± 4.43 kg day^{-1} $family^{-1}$) in lower altitudes (<1500 m). In middle altitude (1500–2000 m), higher average fuelwood consumption was estimated to be 18.98 ± 5.70, 18.59 ± 6.76, and 18.13 ± 4.04 kg day-1 family-1 G2, G4, and G3 income groups, respectively (Table 17.3). The higher altitudinal zone (>2000 m) shows a clear trend of increasing fuelwood consumption with increasing household income. The results showed no clear trend in fuelwood consumption, income groups, and altitudinal zone relations. However, it was observed that variations in average fuelwood consumption per household in specific income groups in different altitudinal zones were with a smaller amount. In the case of per capita fuelwood consumption, a negative trend of lowest per capita fuel consumption was observed in higher-income groups.

Table 17.2 Variations by family sizes in energy consumption in relation to altitudinal variation

Altitude	Family size	Number of household	Total annual fuelwood consumption (kg)	Average fuelwood consumption (kg)	Average daily fuelwood consumption per household (kg)	Per capita daily fuelwood consumption (kg)
<1500 m	Small	4	15,199.64	3799.91 ± 2080.10	10.41 ± 5.70	4.16 ± 3.22
	Medium	40	218,530.71	5463.27 ± 1109.39	14.97 ± 3.04	2.99 ± 0.61
	Large	26	190,425.71	7324.07 ± 1379.06	20.07 ± 3.78	2.41 ± 0.54
1500–2000 m	Small	9	31,207.50	3467.50 ± 944.35	9.50 ± 2.59	3.66 ± 1.08
	Medium	56	332,045.71	5929.39 ± 1270.27	16.24 ± 3.48	3.16 ± 0.61
	Large	52	424,573.21	8164.87 ± 1833.99	22.37 ± 5.02	2.47 ± 0.54
>2000 m	Small	1	2346.43	2346.43 ± 0.00	6.43 ± 0.00	2.14 ± 0.00
	Medium	53	322,842.03	6091.36 ± 1446.40	16.69 ± 3.96	3.36 ± 0.78
	Large	71	585,474.24	8246.12 ± 2274.64	22.59 ± 6.23	2.52 ± 0.63
Overall	Small	14	48,753.57	3482.40 ± 1295.27	9.54 ± 3.55	3.83 ± 1.88
	Medium	149	873,418.46	5861.87 ± 1312.46	16.06 ± 3.60	3.19 ± 0.69
	Large	149	1,200,473.16	8056.87 ± 2010.74	22.07 ± 5.51	2.49 ± 0.58

Table 17.3 Variations by income in energy consumption concerning altitudinal zones

Altitudinal zone (m)	Income group	Number of household	Annual total fuelwood consumption (kg)	Annual average fuelwood consumption	Average daily fuelwood consumption per household (kg)	Per capita daily fuelwood consumption (kg)
<1500	G1	6	38,794.29	6465.71 ±1916.99	17.71 ± 5.25	3.60 ± 0.97
	G2	42	245,566.79	5846.83 ±1615.46	16.02 ± 4.43	3.03 ± 1.13
	G3	6	41,818.57	6969.76 ±1806.18	19.10 ± 4.95	2.46 ± 0.42
	G4	16	97,976.43	6123.53 ±1530.43	16.78 ± 4.19	2.32 ± 0.50
1500–2000	G1	5	22,134.64	4426.93 ±994.89	12.13 ± 2.73	3.91 ± 1.09
	G2	65	450,305.71	6927.78 ±2081.17	18.98 ± 5.70	3.20 ± 0.58
	G3	21	138,960.71	6617.18 ±1473.04	18.13 ± 4.04	2.49 ± 0.54
	G4	26	176,425.36	6785.59 ±2467.41	18.59 ± 6.76	2.27 ± 0.52
>2000	G1	8	50,005.00	6250.63 ±1766.81	17.13 ± 4.84	2.97 ± 0.65
	G2	66	461,881.43	6998.20 ±1709.22	19.17 ± 4.68	3.23 ± 0.67
	G3	26	213,551.07	8213.50 ±1878.84	22.50 ± 5.15	2.69 ± 0.80
	G4	25	220,016.79	8800.67 ±2832.50	24.11 ± 7.76	2.60 ± 0.59
Total	G1	19	110,933.93	5838.63 ±1792.99	15.99 ± 4.91	3.42 ± 0.93
	G2	173	1,157,753.93	6692.22 ±1889.23	18.33 ± 5.18	3.17 ± 0.78
	G3	53	394,330.36	7440.2 ±1855.66	20.38 ± 5.08	2.58 ± 0.67
	G4	67	494,418.57	7379.38 ±2653.74	20.22 ± 7.27	2.4 ±0.56.

17.5 Discussion

Rural people of the Himalayan region have been entirely dependent on forests for livelihood generation (Malik et al. 2014). The extraction of biomass from forests relates to factors related to local people's socio-economic conditions. The mean household size was 6.91 persons per household in the study area. This was observed as a relatively larger household size compared to the entire state, i.e. 6.4 persons, and the entire country, i.e. 5.3 persons (Census of India 2011). In the study area, literacy status was 75.93%, less than Uttarakhand's literacy rate (78.82%) (Census of India 2011). The Garhwal region's remote location and non-approachable backward zone significantly cause this region's overall low literacy rate (Sharma et al. 2011; Chauniyal and Chauniyal 2017). The primary income sources in all three altitudes were remittance, agriculture, and pasture. A large portion of the income (30.13%) came from remittance income, which migrated inhabitants sent. This was also found in another study by Kumar and Sharma (2009). Annual income per household was, on average, ₹90,816, which indicated that families lie above the poverty line as per the definition by the Government of India.

According to socio-economic concerns, animal husbandry is important in the study region; Kumar (2010) support this statement. It is the primary occupation and livelihood source for about 70% of the population in the Western Himalayas (Dhyani and Maikhuri 2012). The overall number of cattle units per household was on an average of 5.18 (higher altitude), 4.12 (middle altitude), and 3.04 (lower altitude). However, it was observed that the local people left animal husbandry as their secondary occupation. The reason was attributed to the increasing trend of illiteracy. However, the new generation perceives a lack of enthusiasm for animal husbandry or any agricultural pursuits (Borda et al. 2023). This study analysed that females (84.97%) from each household frequently collected fodder as a household activity. Fodder requirements were observed to be uniform throughout all the seasons in a year, although green forage availability was scarce during wintertime and had always been a serious concern (Dhyani and Maikhuri 2012). The primary fodder sources during the winter season were observed to be the lopped oak tree leaves and dry grass, which were added to the agricultural crop residues.

In the summer, people migrate to the upper reaches and forested areas to graze their livestock, where villagers stay at specific locations called *Kharaks* (Chauniyal and Dutta 2018). This practice continued in some of the villages of the study area. It was perceived that livestock grazed in forested areas, and in higher altitudes, grazing continues even up to an elevation of the subalpine region (2700–2900 masl). The daily average household fodder consumption was estimated at 39.14 ± 30.06 kg, as found in other studies in Garhwal Himalaya by Pandeya and Yadava (2016) and Malik et al. (2014). The fodder species utilisation pattern varied from lower to higher altitudinal zones. It was analysed that fodder consumption per household per day (in kg) was, on average, 47.29 ± 31.69 (higher altitude), 37.38 ± 28.82 (middle altitude), and 27.54 ± 24.71 (lower altitude). The fodder collection depended upon the number of cattle units, which increased from lower to higher altitudes, and the availability of forage in the nearby forests. A positive relationship was obtained between the livestock number and altitude concerning fodder consumption (Malik et al. 2014).

Fuelwood is the primary energy source for many rural people. The major contributing factors of fuelwood consumption include household income, household size, site-specific ecological environments (i.e. forest type, climate, soil type), household land ownership, educational level, culture, economy, land tenure, and religious aspects (Rao and Reddy 2007). Around 98.5% of households in the study area acknowledged the use of fuelwood as a cooking fuel, which was observed to be relatively higher as compared to NSSO, i.e. 77% as reported (2007–2008) and recently for the whole country, i.e. 64% as reported (Das and Srinivasan 2012). The average distance travelled by the villagers for fuelwood collection was 4.61 ± 1.04 km.

The present study's fuel consumption per household was 18.94 ± 5.76 kg/day, similar to Pandeya and Yadava (2016). At the national level, in rural areas, the fuelwood consumption on average is 121.19 kg per month (i.e. 4.06 kg/household/day) as per the Centre for Development Finance (CDF), which on comparison shows that fuel consumption in the Dudhatoli region is relatively higher. However, the values were lower than those reported for Van Gujjar communities in Garhwal Himalaya (23.44 kg day^{-1} $family^{-1}$) reported by Hussain et al. (2012). This study's per capita fuel consumption was about 2.73 kg, similar to the values for the rural and tribal communities of Garhwal Himalaya (1.1–2.8 kg/capita/day) reported by Sharma et al. (2011). However, the values were relatively higher than as reported by Bhatt et al. (1994) (i.e. 1.49 kg/capita/day), Himalayan range of Nepal by Mahat et al. (1987) (i.e. 1.23 kg/capita/day), local communities of Garhwal Himalaya by Hussain et al. (2012) (i.e. 1.97 kg/capita/day), Himalayan villages by Bartwal (1987) (i.e. 1.53 kg/capita/day), Mishra et al. (1988) (i.e. 1.26–1.95 kg/capita/day) and by Saksena et al. (1995) (i.e. 0.76–1.21 kg/capita/day).

Elevation plays a critical role in fuelwood consumption. The results depicted that fuelwood consumption rapidly increased with the increasing altitude in the region. Our results are consistent with the studies of Negi et al. (1999) and Bhatt and Sachan (2004), which have found that fuelwood consumption increased with increasing altitudes. At higher altitudes in the Dudhatoli region, due to low temperatures, the winter season is more extended. Households on any given day require more hours of heating, and therefore, fuelwood consumption is higher. In the present study, the per capita fuelwood consumption varies from 2.65 to 2.80 kg per day with increasing altitude, which is relatively higher than as reported by Kumar and Sharma (2009) (i.e. 1.13–1.82 kg per day). The higher consumption in higher altitudes was observed due to location near the vicinity of the forest, low temperature, and remote location, among others.

A relationship was observed between fuel consumption and family size. The per capita per day (kg) fuelwood consumption (PCFC) varied from 4.16 to 2.41 for small to large families. The results clearly showed that in the Dudhatoli region, there was a steady decrease in PCFC with the increased family sizes. This study's result was similar to the study by Kumar and Sharma (2009) in the Garhwal Himalayas. The present study did not show any clear trend in fuelwood consumption in relation to income groups and altitudinal zones. The reason could be attributed to a study by Baland et al. (2007), which reports that wealth does not directly control the fuel collection rates without alternatives or fuel markets.

Over time, the Himalayan rural people have established an age-old practice of utilising varieties of selective forest resources for fuel based on their quality and accessibility (Rai et al. 2002). Trees were extensively cut down and lopped (Hussain et al. 2012). Due to these practices, many of these favoured and higher-quality species face extreme stress, which leads to alterations in species compositions and patterns of forest succession. In the present study, 20 species of trees were observed as the preferred fuelwood along different altitudinal zones (Table 17.4). These fuelwood species have also been reported as preferred species based on the characteristics and availability of other authors in the Himalayan region, such as Pandeya and Yadava (2016) and Kumar and Sharma (2009). The local people also extracted wild fruits, medicinal plants, lichens, etc. These were reported in various studies, such as Kumar (2010) and Bhat et al. (2013).

The clear trend of continuing and unsustainable use of resources in the region in the near prospect will be vindictive for the forest resources. The residents will make the most of the preferred species but will disperse future generations into dire conditions. Further, the socio-economic conditions of the local population are the foremost rationale for the total reliance on nearby forests for biomass extraction—the socio-economic and demographic characteristics influencing household reliance on forest resources. Agricultural revenue and off-farm income are inverse to forest dependency (Ali et al. 2020).

Education positively impacts individual education levels and household attainment, creating alternative work opportunities and reducing reliance on forest resources. Studies show that higher education improves employment prospects and livelihoods, redirecting resources from extraction. However, there is an inverse correlation between education and forest dependency (Godoy and Contreras 2001;

Table 17.4 Ranking of preferred fuelwood plant species

		Ranking (altitudinal zone)		
Plant species	Local name	<1500 m	1500–2000 m	>2000 m
Q. leucotrichophora	Banj	2	1	1
Rhododendron arboretum	Burans	4	2	2
Quercus semecarpifolia	Kharsu	–	6	5
Q. glauca	Faliyat	13	15	–
Alnus nepalensis	Utish	8	5	6
Q. floribunda	Tilanj	5	4	3
Toona ciliata	Tun	9	14	–
Celtis australis	Kharik	3	8	7
Lyonia ovalifolia	Ayar	7	3	4
Ficus auriculata	Timila	–	10	–
Berberis aristata	Kilmore	14	11	10
Pinus roxburghii	Pine	1	7	11
Pyrus pashia Buch.	Melu	12	13	14
Myrica esculenta	Kafal	11	10	8
Symplocos paniculata	Lodh	10	12	–
Fraxinus micrantha	Barara	15	–	15
Abies pindrow	Rasal	–	–	12
Cedrus deodara	Deodar	–	–	13
Carpinus viminea Lindl	Chamkharik	–	12	14
Persea duthiei	Kaul	–	10	14

Adhikari et al. 2004; Fonta and Ayuk 2013; Baiyegunhi et al. 2016; Soe and Yeo-Chang 2019).

Younger individuals often rely less on forest resources than elders due to career-oriented opportunities in urban areas (Jain and Sajjad 2015). Forest extraction practices are labour-intensive and require physical strength, reducing elderly reliance. The age of the household is inversely related to forest dependency (Fonta and Ayuk 2013; Thondhlana and Muchapondwa (2014).

Forest dependency is linked to household size, as larger families use more forest resources due to labour-intensive activities (Fonta and Ayuk 2013; Kabubo-Mariara 2013; Prado Cordova et al. 2013). Research conducted by Adhikari et al. (2004) shows that larger families have an increased demand for forest resources and greater availability of human resources to fulfil this demand.

As per Adam and El Tayeb (2014), agricultural lands are expected to depend less on forest resources, and households with higher agricultural income tend to reduce their dependence on forests. Off-farm income also negatively impacts forest dependency, as more off-farm options reduce households' reliance on forests (Angelsen and Kaimowitz 1999; Cavendish 2000; Escobal and Aldana 2003). Those engaged in lucrative off-farm activities also rely less on forest resources due to increased income from alternative sources.

Livestock, including goats, buffalo, cows, and oxen, rely on forest resources for grazing and feeding, causing ecological pressure. An increased livestock population leads to increased forest dependency (Jain and Sajjad 2015). Households with greater landholdings are anticipated to depend less on forests due to other livelihood opportunities provided by agricultural resources (Gunatilake 1998). An inverse correlation between landholding and forest reliance is assumed.

Thus, the information obtained from the present study could help understand the relational and behavioural practices in the mountain reaches. The study will help design suitable technologies and formulate management practices for conserving and using natural resources. Ecosystem-based approaches and ecosystem-based services use the benefits of sustainable management and restoration of natural resources.

17.6 Conclusion

The study of ecosystem provisioning services in the Dudhatoli region reveals high demand levels. Ecosystem-based approaches (EBA) can manage these services sustainably and effectively. These include engaging local villagers, mobilising human resources, investing in forested areas, and implementing coordinated grazing practices. Restrictions on heavy grazing during rainy seasons can help mitigate excessive grazing and regeneration of plant species. Public involvement in management practices like Joint Forest Management can also help. Agroforestry practices, such as cultivating multipurpose tree species, can help mitigate soil erosion and provide fuel and fodder. Alternative energy sources like solar, LPG, and biogas from cow dung can also help alleviate forest strain. The Indian government is advocating initiatives like Pradhan Mantri Ujjwala Yojana (PMUY), National Solar Mission (NSM), Unnat Chulha Abhiyan, Gram Ujala Yojana, Rural Electrification Programmes, National Biomass Cookstove Initiative (NBCI), etc. to reduce fuelwood consumption. Reducing fodder pressure on forests can be achieved through rotational and inter-cropping, multi-cropping, and mixed-cropping systems. Promoting higher education and expanding educational institutions in rural areas can alter the occupational landscape. Comprehending the oversight and administration of natural resources through ecosystem-based approaches is essential for improving livelihoods and addressing climate change mitigation and adaptation.

Acknowledgements The authors gratefully acknowledge the financial assistance received from the HNB Garhwal University in the form of a scholarship. The help provided by the villagers of the Dudhatoli region is also greatly acknowledged.

References

Adam YO, El Tayeb MA (2014) Forest dependency and its effect on conservation in Sudan: a case of Sarf-Saaid Reserved Forest in Gadarif State. Agric For 60(3):107–121

Adhikari M, Nagata S, Adhikari M (2004) Rural household and forest: an evaluation of household's dependency on community forest in Nepal. J For Res 9(1):33–44. https://doi.org/10.1007/s10310-003-0051-1

Ali N, Hu X, Hussain J (2020) The dependency of rural livelihood on forest resources in northern Pakistan's Chaprote Valley. Glob Ecol Conserv 22:e01001. https://doi.org/10.1016/j.gecco.2020.e01001

Angelsen A, Kaimowitz D (1999) Rethinking the causes of deforestation: lessons from economic models. World Bank Res Obs 14:73–98. https://doi.org/10.1093/wbro/14.1.73

Baiyegunhi LJS, Oppong BB, Senyolo MG (2016) Socio-economic factors influencing mopane worm (*Imbrasia belina*) harvesting in Limpopo Province, South Africa. J For Res 27:443–452. https://doi.org/10.1007/s11676-015-0168-z

Baland JM, Bardhan P, Das S, Mookherjee D, Sarkar R (2007) The environmental impact of poverty: evidence from firewood collection in rural Nepal. Revised version of paper presented at the Tenth conference of the International Association for the Study of Common Property, Oaxaca, Mexico 59(1), p 23–61. https://doi.org/10.1086/655455

Bartwal PS (1987) Fuelwood consumption pattern in rural areas-a case study. J Trop Forest 3:136–141

Bhat JA, Kumar M, Bussmann RW (2013) Ecological status and traditional knowledge of medicinal plants in Kedarnath Wildlife Sanctuary of Garhwal Himalaya. India J Ethnobiol Ethnomed 9(1):1–18. https://doi.org/10.1186/1746-4269-9-1

Bhatt BP, Sachan MS (2004) Firewood consumption along an altitudinal gradient in mountain villages of India. Biomass Bioenergy 27:69–75. https://doi.org/10.1016/j.biombioe.2003.10.004

Bhatt BP, Negi AK, Todaria NP (1994) Fuelwood consumption pattern at different altitudes in Garhwal Himalaya. Energy 19(4):465–468. https://doi.org/10.1016/0360-5442(94)90124-4

Borda AJ, Sárvári B, Balogh JM (2023) Generation change in agriculture: a systematic review of the literature. Economies 11:129. https://doi.org/10.3390/economies11050129

Cavendish W (2000) Empirical regularities in the poverty-environment relationship of rural households: evidence from Zimbabwe. World Dev 28:1979–2003. https://doi.org/10.1016/S0305-750X(00)00066-8

Census of India (2011). From http://www.censusindia.gov.in. Retrieved 11 July 2011
Chauniyal S, Chauniyal DD (2017) Level of socio-economic development in the mountain watershed of Uttarakhand state (India). In: Chand R, Nel E, Pelc S (eds) Societies, social inequalities and marginalization, Perspectives on geographical marginality. Springer, Cham. https://doi.org/10.1007/978-3-319-50998-3_6
Chauniyal DD, Dutta S (2018) Application of topographic position index for classification of landforms in Dudhatoli region of Garhwal Himalaya, Uttarakhand. J Indian Geomorphol 6:28–41
Das D, Srinivasan R (2012) Income levels and transition of cooking fuel among rural poor in India. Energy Sci Technol 4(2):85–91. https://doi.org/10.3968/J.EST.1923847920120402.226
Dhyani S, Maikhuri RK (2012) Fodder banks can reduce women drudgery and anthropogenic pressure from forests of Western Himalya. Curr Sci 103(7):763
Escobal J, Aldana U (2003) Are nontimber forest products the antidote to rainforest degradation? Brazil nut extraction in Madre De Dios, Peru. World Dev 31:1873–1887. https://doi.org/10.1016/j.worlddev.2003.08.001
Fonta WM, Ayuk ET (2013) Measuring the role of forest income in mitigating poverty and inequality: evidence from south-eastern Nigeria. For Trees Livelihoods 22(2):86–105. https://doi.org/10.1080/14728028.2013.785783
Godoy R, Contreras M (2001) A comparative study of education and tropical deforestation among lowland Bolivian Amerindians: Forest values. Environmental externality, and school subsidies. Econ Dev Cult Change 49:555–574. https://doi.org/10.1086/452515
Gunatilake HM (1998) The role of rural development in protecting tropical rainforests: evidence from Sri Lanka. J Environ Manag 53(3):273–292. https://doi.org/10.1006/jema.1998.0201
Hussain A, Bargali HS, Dasgupta S, Gore K (2012) Demography perception and impact of traditional pastoralists (Van Gujjars) on biotic resources of Corbett Tiger Reserve and the adjoining forest divisions. H.N.B. Garhwal University (A Central University) Srinagar Garhwal and the Corbett Foundation Ramnagar Uttarakhand. Report submitted to Uttarakhand State Forest Department, Srinagar
Jain P, Sajjad H (2015) Household dependency on forest resources in the Sariska Tiger Reserve (STR), India: implications for management. J Sustain For 35(1):60–74. https://doi.org/10.1080/10549811.2015.1099108
Kabubo-Mariara J (2013) Forest – poverty nexus: exploring the contribution of forests to household welfare in Kenya. Nat Res Forum 37:177–188. https://doi.org/10.1111/1477-8947.12003
Kumar B (2010) Ecological, social and commercial role of lichens in India with special reference to Garhwal Himalayas. Academia Arena 201, 1–118
Kumar M, Sharma CM (2009) Fuelwood consumption pattern at different altitudes in rural areas of Garhwal Himalaya. Biomass Bioenergy 33(14):1413–1418. https://doi.org/10.1016/j.biombioe.2009.06.003
Mahat TBS, Grigffin DM, Shepherd KP (1987) Human impacts on some forest of the middle hills of Nepal, part 4: a detailed study in southeast Sindhu Palanchock and northeast Kabhere Palanchock. Mt Res Dev 7:114–134
Malik ZA, Bhat JA, Bhatt AB (2014) Forest resource use pattern in Kedarnath wildlife sanctuary and its fringe areas (a case study from Western Himalaya, India). Energy Policy 67:138–145. https://doi.org/10.1016/j.enpol.2013.12.016
Martins PJ, Nautiyal JC (1993) Population supporting capacities: in the Central Himalayas. In: Khurana DK, Khosla PK (eds) Agroforestry for rural needs. ISTS, Dr. Y.S. Parmar University of Horticulture and Forestry, Solan, pp 439–462
Mishra NM, Mahendra AK, Ansari MY (1988) A pilot survey of fuel consumption in rural areas. Indian Forester 114(2):57–62
Negi AK, Bhatt BP, Todaria NP (1999) Local population impacts on the forests of Garhwal Himalaya, India. Environmentalist 19:293–303. https://doi.org/10.1023/A:1006641519415
NSSO (2007–2008) Energy used by Indian households, Report No 530, Department of Statistics, New Delhi

Pandeya H, Yadava AK (2016) Forest consumption pattern in relation to socio-economic arrangement of people in Western Ramganga watershed in central Himalaya, India. Uttarakhand Int J Recent Sci Res 7(1):8267–8275

Prado Cordova JP, Wunder S, Smith-Hall C et al (2013) Rural income and forest reliance in highland Guatemala. Environ Manag 51:1034–1043. https://doi.org/10.1007/s00267-013-0028-6

Price MF, Thompson M (1997) The complex life: human land uses in mountain ecosystems. Glob Ecol Biogeogr Lett 6:77–90. https://doi.org/10.2307/2997528

Rahmani A (2003) Conservation outside protected areas. In: Saberwal V, Rangarajan M (eds) Battles over nature: science and politics of conservation. Permanent Black, New Delhi, pp 117–138

Rai YK, Chettri N, Sharma E (2002) Fuelwood value index of woody tree species from forests of Mamlay Watershed South Sikkim India. For Trees Livelihoods 12:209–219. https://doi.org/10.1080/14728028.2002.9752425

Ramachandra TV, Subramanian DK, Joshi NV, Gunaga SV, Harikantra RB (2000) Domestic energy consumption patterns in Uttara Kannada district, Karnataka state, India. Energy Conver Manag 41:775–831. https://doi.org/10.1016/S0196-8904(99)00151-X

Rao MN, Reddy BS (2007) Variations in energy use by Indian households, an analysis of micro level data. Energy 32(2):143–153. https://doi.org/10.1016/j.energy.2006.03.012

Saksena S, Prasad R, Joshi V (1995) Time allocation and fuel usage in three villages of the Garhwal Himalaya India. Mt Res Dev 15:57–67. https://doi.org/10.2307/3673700

Samal P, Anubha SK, Tiwari S, Mehara B, Agarwal DP (2003) Indigenous animal health care practices in India central Himalaya: empirical evidence. Indian J Tradit Knowl 2(1):40–50. http://nopr.niscpr.res.in/handle/123456789/25919

Samant SS, Dhar U (1997) Diversity endemism and economic potential of wild edible plants of Indian Himalaya. Int J Sustain Dev World Ecol 4:179–191. https://doi.org/10.1080/13504509709469953

Sharma A, Prasad R, Saksena S, Joshi V (1999) Micro level sustainable biomass system development in central Himalaya: stress computation and biomass planning. Sustain Dev 7(3):132–139. https://doi.org/10.1002/(SICI)1099-1719(199908)7:3<132::AID-SD112>3.0.CO;2-2

Sharma CM, Butola DS, Gairola SK, Ghildiyal SK, Suyal S (2011) Forest utilization pattern in relation to socio economic status of people in Dudhatoli area of Garhwal Himalaya, forests. Trees Livelihoods 20:249–264. https://doi.org/10.1080/14728028.2011.9756712

Singh JS (2006) Sustainable development of the Indian Himalayan region: linking ecological and economic concerns. Curr Sci 90(6):784–788

Soe KT, Yeo-Chang Y (2019) Perceptions of forest-dependent communities toward participation in forest conservation: a case study in Bago Yoma, South-Central Myanmar. Forest Policy Econ 100:129–141. https://doi.org/10.1016/j.forpol.2018.11.009

Thondhlana G, Muchapondwa E (2014) Dependence on environmental resources and implications for household welfare: evidence from the Kalahari drylands, South Africa. Ecol Econ 108:59–67. https://doi.org/10.1016/j.ecolecon.2014.10.003

Tiwari PC (2000) Land use changes in Himalaya and their impact on the plains ecosystems, need for sustainable land use. Land Use Policy 17:101–111. https://doi.org/10.1016/S0264-8377(00)00002-8

Tripathi RS, Sah VK (2001) Material and energy flows in high hill, mid hill and valley farming systems of Garhwal Himalaya, agriculture. Ecosyst Environ 86:75–91. https://doi.org/10.1016/S0167-8809(00)00270-X

Chapter 18
Strengthening Springshed Management Through Nature-Based Solutions: The Potential of Spring Revival

Taufique Warsi and Khan Tahama

Abstract Springs are the lifeline for the people and a vital source of freshwater in mountainous and hilly regions, essential for maintaining ecosystems and empowering socioeconomic development. However, these vulnerable systems face various challenges, including overexploitation, climate change, encroachment, and land-use alterations, threatening their existence and sustainability. This chapter explores the concept and components of springsheds, highlighting their ecological and socioeconomic importance as lifelines for rural and urban communities in hilly regions. It also explores the problems that threaten springshed sustainability and emphasises how Nature-Based Solutions (NbS) can be essential in resolving them. Reflecting on a case study of Bhutan, where springshed management has been successfully implemented, the chapter demonstrates how ecological restoration and community-driven approaches can revive and sustain spring flows. Key techniques for spring revival, such as springshed mapping, capacity building and training, and monitoring, are detailed. The potential benefits of spring revival, ranging from biodiversity conservation to ecological restoration and enhanced water security to improved livelihoods, are examined. By integrating traditional wisdom with innovative practices, springshed management emerges as a holistic, sustainable water resource management strategy. This work advocates for the broader adoption of NBS in springshed management to ensure resilient ecosystems and equitable water availability for future generations.

Keywords Springshed · Hilly terrain · NbS · Ecosystem

T. Warsi (✉)
Raintree Foundation, Mumbai, Maharashtra, India

K. Tahama
KSK Geomagnetic Research laboratory, Indian Institute of Geomagnetism, Allahabad, Uttar Pradesh, India

Y. D. I. Khan et al. (eds.), *Ecosystem-based Approaches for Resilience Building in Himalayan Landscapes*, Disaster Resilience and Green Growth,
https://doi.org/10.1007/978-981-95-2007-7_18

18.1 Introduction

Springs are critical ecosystems and primary water sources for rural households in hilly regions, sustaining local communities by fulfilling domestic and agricultural needs (Jamaluddin et al. 2010; Nowreen et al. 2023). In the Hindu Kush Himalayas (HKH), springs serve as lifelines for rural and urban communities and hold significant cultural and religious importance (Prasad and Sharma 2019). However, these essential water sources are increasingly threatened by various environmental and anthropogenic pressures, including seasonal fluctuations in flow, reduced discharge, and the alarming trend of springs drying up (Buono et al. 2015; Rana and Gupta 2009). The primary contributors to the degradation of springs, including climate change, land use and cover changes, unplanned infrastructure development, and socioeconomic transformations, such as urbanisation and tourism, are pressing issues (Kumar et al. 2023). In the Himalayas, where potable water scarcity is becoming more prevalent, these pressures are exacerbated by unsustainable developments, which necessitate immediate and continuous monitoring of water sources and the implementation of sustainable groundwater management practices (Rathod et al. 2021; Thapa et al. 2023; Rawal et al. 2024). Springs function as groundwater discharge points where aquifers intersect the ground surface, allowing water to seep through rock pores, fissures, and fractures (Todd 1980; Soulios 2010; Nisa and Umar 2024). Their availability throughout the year is crucial for ensuring water security. However, forest conversion to agriculture, forest fires, declining rainfall trends, diminishing glacier coverage, and unplanned infrastructure growth reduce groundwater recharge and increase surface run-off (Panwar 2020). These changes severely impact spring recharge zones and contribute to water scarcity in mountainous regions.

Nature-based Solutions (NbS) provide a realistic solution to address these issues. NbS practices like reforestation, soil conservation, wetland restoration, and the construction of vegetative buffer zones all help promote groundwater recharge by improving soil infiltration and minimising surface run-off. Filtering contaminants and stabilising water flow to springs enhance the quantity and quality of the water. Additionally, NbS promote biodiversity and ecosystem health, which are critical for maintaining the natural hydrological cycles that support springs. Integrating these strategies can create a hopeful future where spring recharge zones are protected, and sustainable water access is ensured. For example, in Meghalaya, India, poor soil infiltration due to increased run-off caused by mining, deforestation, and urbanisation has further decreased spring discharge (Sharma and Laskar 2022). Securing dependable water availability is becoming more difficult due to changes in precipitation patterns and rising temperatures brought on by climate change, especially during times of scarcity. Springs are not just water sources but integral to the region's water needs and natural balance. Their depletion can disrupt river flows, underscoring their crucial role in maintaining the hydrological system and water supplies (Wheater and Gober 2015). Given their inherent connection to groundwater systems, practical groundwater management principles and Common-Pool Resources

(CPR) approaches are vital for restoration. A comprehensive hydrogeological strategy, bolstered by socio-ecological inputs, engineering surveys, decentralised governance, and Nature-based Solutions, can bolster the resilience of springs to climate variability and fortify community water management practices. Community participation is not just beneficial; it is indispensable for successful springshed management. By reforming policy and educating local populations about the science and governance of water resources, communities can be empowered to engage in the protection and sustainable management of springs actively (Stevens et al. 2021). This participatory approach ensures reliable access to water for daily needs and livelihoods. Regular hydrological monitoring and detailed hydrogeological mapping are key to understanding the spring aquifer's capacity, guiding decisions on water security plans, recharge enhancement, and resource governance. Apart from their practical utilisation, springs have significant cultural and historical significance, supporting ecological and social stability in the area (Prasad and Sharma 2019). Protecting these water sources is essential for the survival of the ecosystems and communities that rely on them and for preserving a healthy environment.

This chapter examines how Nature-based Solutions (NbS) can help enhance springshed management and revitalise springs. In hilly regions, springs are essential for ensuring water security, but unsustainable development, land degradation, and climate change pose an increasing threat to them. The chapter delves into the ecological and socioeconomic significance of springs, their difficulties, and how NbS—such as replanting, soil conservation, and wetland restoration—can support continued water flows and increase groundwater recharge. This chapter thoroughly explains how these solutions might protect springsheds by examining real-world NbS techniques, including a case study from Bhutan. Ultimately, the goal is to show the many benefits of spring revival for the environment and the communities that depend on these crucial water sources.

18.2 Understanding Springsheds

A springshed is the geographic area that contributes water to a spring, functioning similarly to a watershed but focused explicitly on groundwater flow. This area includes basins of surface and groundwater that provide water to one or more springs. Activities inside the springshed directly affect the quantity and quality of the water coming from a spring. Hence, managing it is essential to preserving the ecosystem and the water quality (Shrestha 2018; Lo Russo et al. 2021). Springsheds are critical for understanding spring hydrology because they connect several watersheds and aquifers into a cohesive system (Fig. 18.1a, b). Effective springshed management entails safeguarding these regions to maintain long-term water supply and quality, especially since many traditional spring sources are threatened by contamination and reduced flow owing to environmental changes. Understanding the key components of a springshed is essential for managing and conserving spring ecosystems. Here are the main components.

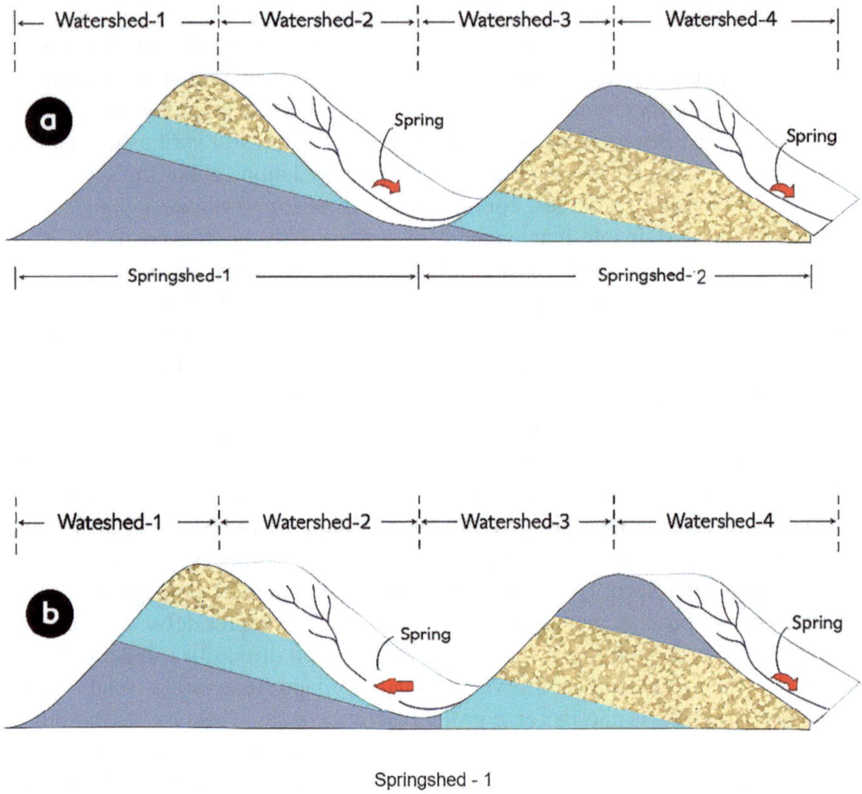

Fig. 18.1 Representation of springshed and watershed, (**a**) two springsheds each with two watersheds, (**b**) one springsheds with four watersheds. (Source: Shrestha et al. 2018)

(a) *Recharge Zones*

Recharge zones are essential areas where water penetrates the ground to replenish the aquifers that sustain springs, influencing flow and water quality. They encompass three main types: surface recharge areas, where rainwater or surface water directly infiltrates the soil; indirect recharge areas, where water accumulates in surface bodies like rivers or lakes before seeping into the groundwater; and focused recharge areas, where concentrated infiltration occurs, such as through sinkholes or fractures in karst landscapes (Fig. 18.2). Understanding and managing these zones is crucial for maintaining aquifer health and ensuring sustainable water resources.

(b) *Aquifers*

Aquifers are essential subsurface formations of water-bearing rock or sediment that store and transmit groundwater and are crucial to the hydrological cycle, especially in springsheds. Knowing the properties of aquifers is essential for proficient management of water resources. The term "permeability" describes a material of

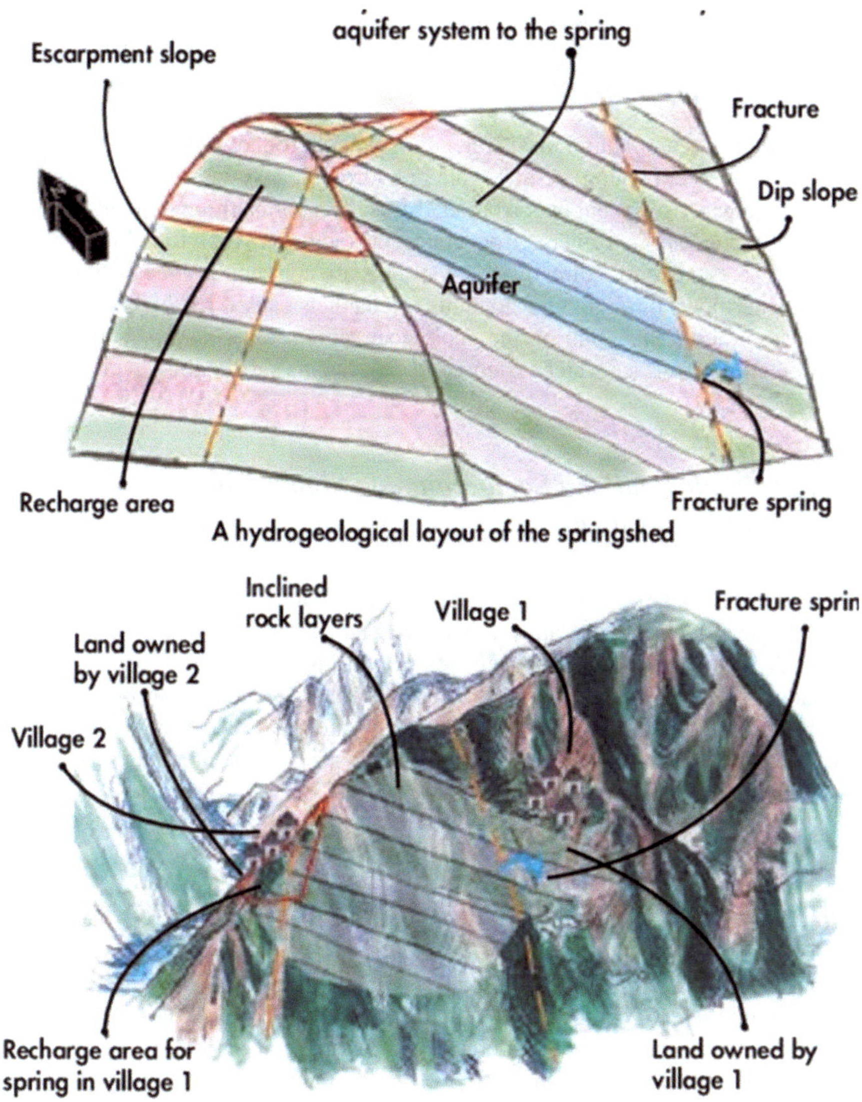

Fig. 18.2 A conceptual diagram showing the recharge zone, aquifer, and spring outlet. (Source: Shrestha et al. 2018)

the aquifer's capacity to transfer water, which is critical for maintaining springs and other surface water bodies. The smallest unit of an aquifer, porosity, is defined as voids or spaces in the rock or sediment. It impacts the amount of groundwater that can be held and released. Hydraulic conductivity determines the pace at which water flows through an aquifer, influencing water sources' availability and sustainability. Proper evaluation of these attributes is critical for aquifer sustainability, particularly in areas with high water demand or sensitivity to drought.

(c) *Spring Outlets*

Spring outlets are natural discharge points where groundwater flows from aquifers to the surface, playing a crucial role in hydrological systems. Seepage springs occur when water seeps out over a broad, diffuse area, often through permeable soil or rock. Fracture springs emerge from cracks or fractures in rock formations, allowing water to escape more directly. Karst springs are common in limestone terrains, where water flows through huge conduits or channels caused by the breakdown of carbonate rocks. Depending on the geological setting, these springs substantially contribute to the surface water systems and ecosystems surrounding them.

18.2.1 Ecological and Socioeconomic Significance of Springsheds

Springsheds are an interconnected system and are found to be of paramount importance in ecosystems and human communities. They play an essential role in sustaining biodiversity and regulating water cycles. A springshed includes recharge zones, groundwater flow paths, the springs themselves, and the entire hydrological catchment area that contributes water to a spring. This interconnected system is important for ecological health and human communities, especially in mountainous regions like the Himalayas.

18.2.2 Role of Springsheds in Biodiversity Management

Springsheds are lifelines for people and nature, critical in sustaining ecosystems and communities. In regions like the Indian Himalayas, around 200 million people depend on these natural water sources for drinking and agriculture. Many local communities hold springs in high regard, valuing them for their utility and natural purification properties. Springsheds not only quench the people's thirst but also help regulate stream flows, especially during the dry seasons, supporting river ecosystems and the diverse species that thrive there. When springs dry up, the impact is not confined to the water loss. Instead, it ripples through entire habitats, losing plant and animal species that rely on steady water availability. The loss of biodiversity disturbs the complete ecosystem, as many species tend to change as per the water supply. Protecting springsheds is essential for human survival, maintaining ecological balance, and preserving the natural world's resilience to climate change.

18.2.3 Role of Springsheds in Maintaining Water Cycles

Springsheds are vital in maintaining the local water cycle, acting as nature's groundwater and surface water flow regulators. It is the purest water supply by natural means and is filtered through the natural formation process. This natural filtration through soils removes the contaminants and ensures the water is safe for ecological and human needs. However, many springs are under threat due to deforestation, climate change, and unsustainable land-use practices. In regions like the Indian Himalayas, nearly half of the once-perennial springs have either dried up or turned seasonal, leaving tens of thousands of villages facing severe water shortages (Aayog 2017). This crisis not only threatens human populations but also disrupts the entire natural water cycle, leading to further environmental degradation. Effective springshed management and restoration efforts are essential to address these challenges. These efforts are crucial for reviving these springs and ensuring water security for the future.

18.3 Springshed Management

Managing springsheds requires a comprehensive approach to tackle the challenges they face. A crucial part of springshed management is protecting and restoring these vital areas. This includes preserving recharge zones and implementing ecological restoration measures like afforestation and soil conservation, which help improve water retention and recharge. These practices play a crucial role in their revival as springs dry up. By promoting efficient water conservation techniques, such as rainwater harvesting, and actively involving local communities in managing water resources, we can ensure that spring water sources remain sustainable for the long term. Community involvement is important for the success of springshed management because it creates a sense of responsibility and empowers residents to preserve their water sources actively. Finally, springsheds are critical for regulating the water cycle and maintaining biodiversity. Proper management promotes ecosystem health and the livelihoods of millions who rely on spring water. Addressing springshed concerns through sustainable practices and community participation is crucial to their resilience and long-term ecological contributions.

18.4 Challenges Facing by Springsheds

Numerous issues affect springsheds, threatening the communities that rely on them and their ecological health. One major problem is the over-extraction of groundwater resources, which diminishes spring discharge and disturbs the natural flow of springs. Land-use change, pollution from industry, inappropriate waste management, and agricultural run-off further deteriorate water quality, affecting human health and biodiversity. Changes in precipitation patterns, an increase in the frequency of droughts, a decrease in recharge rates, and the drying up of springs are all consequences of climate change. Encroachment and land-use changes also disrupt springsheds, leading to habitat loss and fragmentation (Shrestha et al. 2018; Buono 2019). The significant challenges are discussed in brief:

- *Declining Spring Discharge:* Several cases of declining spring discharge are caused by overexploitation, human development, and changes in the LULC in the catchment. Above all, climate change is adding insult to injury, and many perennial springs have either gone dry or become seasonal. A survey in Sikkim indicated that water production has declined in half of all springs in the state, signalling aquifer depletion in areas heavily reliant on springs for drinking water (Nowreen et al. 2023).
- *Water Quality Deterioration:* As humans approach the hilly region, sanitation and proper waste management have become critical concerns, leading to deterioration in water quality. Generally, in these kinds of terrain, people fulfil their domestic and drinking water needs from the springs, which leads to various waterborne diseases and compromised human health.
- *Climate Change:* Vagaries in rainfall and a constant rise in global temperature are highly responsible for altering the global hydrological scenario, and springs being one of the important water resources are also suffering on a big scale.
- *Lack of Specialized Knowledge:* All the primary and secondary stakeholders (inhabitants, government, and other working groups) lack specific springshed knowledge regarding groundwater management. This lack of understanding hampers efforts to find sustainable water sourcing and management solutions.
- *Institutional Challenges:* The institutional response to declining springs often focuses on improving infrastructure by making bigger tanks and removing the existing boxes rather than addressing the underlying causes of spring decline, such as catchment degradation. Therefore, the springshed should be thoroughly studied, and the recharge area must be protected.
- *Socio-Technical Governance Issues:* Springshed governance is often not up to the mark and poorly understood by a large margin, resulting in mismanagement and conflicts over water use. This includes gender and equity dimensions that are not adequately addressed.
- *Lack of Awareness:* Local communities often lack an adequate understanding of the importance of protecting recharge areas and the hydrogeological factors influencing spring health.
- *Data Gaps:* More systematic data and understanding of springshed are needed to link climate change, land use, land cover change, and spring discharge, complicating effective management and policy-making.

To address these challenges, sustainable and effective management approaches are essential. One promising strategy is the implementation of Nature-based Solutions (NbS), which leverage natural processes to enhance water security, restore ecological balance, and build resilience in springshed systems.

18.5 Nature-Based Solutions (NBS) for Springshed Management

The concept of Nature-based Solutions (NbS) was first mentioned in 2008 by the World Bank. The concept emerged from the search for innovative solutions to manage natural systems in a way that can balance the benefits for both nature and society (Sowińska-Świerkosz and García 2022). "*Solutions that are inspired and supported by nature, which are cost-effective, simultaneously provide environmental, social and economic benefits and help build resilience. Such solutions bring more, and more diverse, nature and natural features and processes into cities, landscapes and seascapes through locally adapted, resource-efficient and systemic interventions*". Nature-based Solutions are increasingly recognised as a practical approach to addressing various environmental and societal challenges, including water management. Springsheds, the areas that contribute to the recharge of springs, are crucial for ensuring water security, particularly in regions where springs are the primary water source. Globally, biodiversity, natural areas, and water security face significant threats due to climate change, unsustainable consumption, and pollution (Dekker et al. 2021). In this context, Nature-based Solutions offer a promising solution to address these challenges, aiming to reconcile economic development and ecosystem stewardship. Looking beyond Europe, North America, and Australia, there has been growing scholarly and practical attention to the benefits of adopting Nature-based Solutions within different socio-ecological and political contexts, such as in Africa and Asia, including countries like China, Singapore, Malaysia, and Indonesia (Kuller et al. 2022). Nature-based Solutions for water security are highly relevant for society and nature, as 4 billion people face severe water scarcity, and 1.32 trillion USD is needed to address water-related issues globally (Dekker et al. 2021). Nature-based Solutions in the Himalayas region, specifically for rejuvenating springs, hold great promise in enhancing water access and contributing to achieving sustainable development goals. An aquifer-based, springshed management approach that combines science, partnerships, and community participation has been identified as a promising strategy for the revival of springs.

18.5.1 Relevance to Springshed Management

Nature-based Solutions (NbS) are perfect for managing springsheds because they harness natural processes and ecosystems to improve water availability, quality, and sustainability. A springshed, the land area feeding water into a spring, relies on a delicate balance of water recharge, storage, and release, which requires careful management. We can enhance groundwater recharge by employing NbS techniques like reforestation, soil conservation, wetland restoration, and vegetative buffer zones through better soil infiltration and reduced surface run-off. These measures stabilise spring water flow and act as natural filters, improving water quality by removing pollutants. Besides managing water resources, NbS supports ecosystem health and biodiversity, which is essential for preserving the natural hydrological cycles that sustain springs. These technologies lessen the need for pricey infrastructure by providing economical, environmentally friendly alternatives to conventional designed systems. NbS contributes to the development of resilience to the effects of climate change, such as erratic rainfall and droughts, by cooperating with nature rather than against it. This ensures that springsheds remain reliable, clean water sources, benefiting communities and the surrounding ecosystems.

18.6 Case Study: Springshed Management—A Nature-Based Solution in Bhutan

18.6.1 Context and Challenges

Bhutan, the world's first carbon-negative nation, is facing a growing challenge as its mountain springs and lakes dry up, and its rainfall becomes increasingly unpredictable. These changes seriously threaten the country's water security and have made farming particularly difficult for residents of high-altitude areas like Dogar. Located at 2500 m above sea level, Dogar is a lively village that depends heavily on agriculture and dairy farming. However, with water sources dwindling, farming has become a struggle, and wildlife and livestock face water shortages. This situation brought up the urgent need for sustainable water management strategies. Therefore, a conscious effort was made to overcome the situation posed by humans and nature (NbS Bhutan).

18.6.2 The Springshed Management Solution

To address these pressing issues, Bhutan has turned to springshed management, a Nature-based Solution that combines traditional knowledge with scientific approaches to ensure sustainable water use and conservation. This strategy emphasises the interdependence of people and their environment and aims to revive drying water sources, promoting ecological and social resilience.

Various vital aspects are responsible for effective springshed management (Table 18.1). By ensuring the systematic assessment and targeted intervention, springshed thrives and sustains for extended periods. The collective efforts impart a better result in water resource rejuvenation, which eventually leads to a sustainable livelihood and builds the resilience of the community (Tables 18.2 and 18.3).

Springshed management in Bhutan's Dogar community exemplifies a multifaceted approach to water security that goes beyond economic gains to promote environmental sustainability, social well-being, cultural preservation, and resilient

Table 18.1 Key elements of springshed management solutions for sustainable water resource conservation

Key element	Description
Assessment of water sources	Hydrogeological surveys were conducted to identify the area's critical water sources and potential recharge zones
Natural rainwater harvesting	Rainwater harvesting structures, such as contour trenches and pits, were established in recharge potential zones to enhance groundwater recharge
Integration of traditional knowledge	Traditional water harvesting and filtration methods were integrated into the management strategy to improve water quality
Community-based governance	Local governance bodies, formed by community members, create rules, norms, and decision-making processes for equitable water use

Table 18.2 Outcomes and benefits of springshed management solutions

Results and benefits	Impact
Water resource rejuvenation	Springshed management has successfully revitalised drying springs, enhanced groundwater levels, and improved water availability
Ecosystem restoration	Preserving habitats and promoting biodiversity have helped restore ecological balance and maintain the health of surrounding ecosystems
Community empowerment	Engaging communities in water management helps build a sense of ownership, empowers them to participate in decision-making actively, and contributes to conservation efforts
Sustainable livelihoods	Improved water availability supports agriculture and livestock, providing sustainable livelihoods for the local population
Cost-effectiveness	Community engagement and natural processes reduce the need for expensive infrastructure, making it an affordable and sustainable solution
Cultural preservation	Integrating traditional knowledge preserves cultural heritage and strengthens the connection between communities and their environment
Conflict resolution	Springshed management brings people together and promotes a sense of shared responsibility among everyone involved

Table 18.3 Valuation of springshed management solutions

Valuation of the solution	Value created
Economic value	Improved agricultural yields, better food security, and job creation increase the local economy
Environmental benefit	Nature-based solutions rehabilitate ecosystems, conserve biodiversity, and control environmental degradation
Social value	Improved water security enhances living standards, health outcomes, and resilience to climate-related challenges
Cultural value	Traditional wisdom helps communities stay strong, builds a sense of identity, and nurtures pride in their heritage
Resilience building	Diversification of water sources and capacity building enhances community adaptability to droughts, floods, and other water-related risks

building. By leveraging Nature-based Solutions, traditional knowledge, and community empowerment, Bhutan is paving the way for comprehensive development and sustainable water management in the face of climate change.

18.7 Techniques for Spring Revival

18.7.1 Springshed Mapping

Spring management relies heavily on hydrogeologic surveys and mapping of the springshed to understand and sustain water resources, though a lack of expertise among stakeholders poses challenges. Fortunately, most individuals can learn and apply these skills effectively. The mapping process begins with on-site surveys, identifying rock types and geological structures like strike and dip, followed by the development of cross sections to assess subsurface features. Data on socioeconomic and environmental factors and hydrological measurements of discharge and water quality are collected. Tools such as Google Earth and topo maps delineate local watersheds and field data, which are integrated to create a conceptual model of the aquifer, catchment, and recharge areas. For example, in the North-Western Ghats on the Deccan Plateau, volcanic basalt layers vary in water storage capacity due to differences in cooling times and rock chemistry. Rainwater percolates through porous layers, encountering impermeable layers and moving laterally to emerge as springs on mountainsides. Lineaments, or fractures in the rock, can further facilitate water movement between layers. Springshed mapping requires a systematic and detailed approach, starting from the spring outlet and progressing uphill to map rock contacts and fractures, while satellite imagery reveals landscape-scale features. Springs are categorised based on location relative to landscape features, such as contour and valley positions, indicating different spring types, like contact or fracture springs. These observations are marked on satellite images and topo sheets to provide a comprehensive landscape-scale understanding of the hydrogeology. While tracer techniques like dye tracing offer precise mapping, the described approach is generally very effective, especially when executed by a well-trained field team familiar with local geology and hydrological processes, offering practical insights for sustainable water management.

18.7.2 Capacity Building of Resources

Hydrogeologic mapping is a specialised process requiring a solid grasp of geology and hydrology. Key tasks include identifying rock types based on their ability to store and transmit water and mapping these features relative to topography and spring locations. However, they can be translated and communicated to local communities through short training (Bamola et al. 2018). Engaging community members, such as Gram Panchayat, self-help groups, water user associations, and individuals, is required for the participatory process. Local knowledge and involvement in identifying traditional springs, mapping geology, and monitoring hydrology explain the science, providing a sense of ownership over water resources. Communities are trained to monitor discharge, rainfall, and water quality, empowering them to take responsibility for long-term management. These efforts have led to the emergence of para-professional hydrogeologists from within the communities. These individuals help their villages manage water resources, extend their knowledge to neighbouring communities, and train government and NGO staff. Farmers, women's groups, and last-mile workers are increasingly involved in spring mapping and management. Organisations such as ACWADAM in Pune and PSI in Uttarakhand now provide specialised training in these areas, strengthening community-based water resource management.

18.7.3 Community Action Plan

The community action plan is a collaborative approach integrating scientific methods with indigenous knowledge and practices (Rathod et al. 2021). Once the springshed is demarcated, the next step involves setting management goals and developing a plan. These goals may include restoration, protection, or enhancement of groundwater recharge. For example, if a community faces a drought, which leads to water shortages, the aim would be to look after the spring discharge. If the springshed is affected by human activities such as frequent burning, overgrazing, and deforestation, the focus would be on reducing these pressures. In restoration techniques, targeted replantation of native species in the recharge zone improves the soil cover and recharge rate, and structures like traditional rainwater-harvesting structures or staggered trenches are implemented to enhance infiltration. A detailed mapping of the springshed's components is needed to ensure effective management. A comprehensive degradation of the catchment area can decrease rainfall infiltration, reduce spring flow, and potentially turn a perennial spring into a seasonal one. Moreover, degradation of the spring outlet can adversely impact the local biodiversity, habitats, and water quality. Thus, preserving both the catchment and outlet areas is essential for sustaining the ecological and hydrological functions of the springshed.

18.7.4 Monitoring

Monitoring of spring is essential for data-driven decision-making and systematic spring management. Collecting spring discharge, rainfall, and water quality data is essential to understanding the hydrogeology and post-management impact. Monitoring also engages local communities and encourages participation through citizen science. Key data, such as discharge, is typically measured monthly via time-volumetric sampling and is recorded in litres per minute. This can be used for creating hydrographs and estimating annual discharge, which is critical for planning during low-flow periods. More frequent discharge measurements, especially during seasonal transitions, offer a deeper understanding of the aquifer characteristics, such as storativity and residence time, which affect water quantity and quality. Rainfall is another essential parameter and should be monitored near the spring catchment on an event basis, with data summed monthly and compared to discharge. Water quality parameters like pH, temperature, total dissolved solids (TDS), electrical conductivity (EC), and faecal coliform presence can be easily measured with field kits or portable tracer meters. Community involvement in collecting these parameters creates a sense of ownership and engagement. Overall, community and collective participatory monitoring promotes accurate data collection and empowers communities to contribute to the sustainable management of their water resources. Tools like SOURCE automate the analysis of spring data, allowing for the characterisation of aquifers based on parameters such as flow rate, temperature, and rainfall (Lo Russo et al. 2021).

18.8 Potential Benefits of Spring Revival

Reviving springs is an important method in water management and environmental restoration, and it provides numerous benefits that go far beyond protecting water resources. This approach creates healthier local ecosystems, ensures the community's resilience, and improves peoples' overall well-being. The possible benefits of spring revival are explored.

18.8.1 Water Security and Access

In the rainfed region, natural spring revival is a must to ensure water security. Initiatives to restore springs make more spring water available and give nearby populations a dependable and sustainable water supply. By utilising natural remedies like managing springsheds, these programmes aid in restoring and preserving aquifers, guaranteeing a consistent flow of water throughout the year. During the dry season, the basic needs of humans, agriculture, and livelihood suffer, so this approach assists the overall ecosystem health and promotes long-term resilience against water stress (Ranjan and Kumar Pandey 2020).

18.8.2 Improvement in Livelihoods

It is always beneficial to spare the resources in spring revival, which supports the local communities by improving their lives and ensuring access to reliable and sustainable water supplies. Revived springs give freedom for diversifying the farming strategy and improving the economic health of the inhabitants. Furthermore, better water availability reduces the time and labour required for water collection, allowing communities to focus on other income-generating activities and increasing socioeconomic stability (Buono 2019).

18.8.3 Biodiversity Conservation

Biodiversity conservation is essential for maintaining ecological integrity and reviving springs, particularly for aquatic and riverine ecosystems, and plays a pivotal role in this process. Springs are essential habitats for many unique plant and animal species that thrive in freshwater environments. Restoration of the spring ecosystem always strengthens the provisioning, regulating, and supporting services. Services and healthy connections are much needed to achieve ecosystem resilience and human well-being. However, the current level of spring understanding is not sufficient, and more of the management recommendations should come out as done by Jakub in Karst Landscape (Cíbik et al. 2022).

18.8.4 Community Empowerment and Social Cohesion

Community-based approaches to spring revival play a vital role in helping collective action among residents. The inclusion of all the groups (marginalised and wealthier) enhances social unity in water management decision-making processes. The participatory model promotes shared ownership, allowing community members to contribute to and profit from sustainable water resource management. Therefore, it addresses the issue of inequity among different groups. By promoting equitable resource allocation and collaborative decision-making, these initiatives contribute to long-term resilience and community empowerment (Khadka et al. 2023).

18.8.5 Resilience Against Climate Change

Restoring natural water systems, particularly springs, is essential for building resistance to climate change. Protecting this vital resource and continuous monitoring reverse the process from exploitation to sustainability and regenerative. These natural systems are critical for maintaining ecological balance and mitigating the unpredictable effects of climate change, such as changed rainfall patterns and extreme weather occurrences. Healthy springs assist in keeping water tables stable, support biodiversity, and sustain agriculture, allowing communities to better adapt to environmental changes, manage water resources, and limit the dangers associated with climate unpredictability (Kulkarni et al. 2021).

18.9 Conclusion

Spring restoration is critical for conserving springsheds and maintaining water security, especially in rural areas with mountainous terrain. Springsheds are significant in maintaining ecological balance and promoting socioeconomic well-being by providing communities with dependable and sustainable water sources. Despite their importance, springsheds face numerous challenges, including over-extraction, deforestation, land degradation, and climate change impacts. NbS provides a practical framework for effectively supporting the ecosystems without compromising the quality of human health and environmental health, and this approach is also very prominent for reviving natural springs and, hence, ensuring a sustainable springshed. NbS uses various integrated measures to address the core ecological issues that contribute to preventing spring depletion, including reforestation, soil conservation, and aquifer recharge. The case study demonstrates the success and usefulness of these measures by proving the improvement in the local ecosystem and resilience in the community. Combining indigenous knowledge with advanced scientific methods supports a comprehensive development and revival of springshed. These efforts have multiple benefits, including enhanced water availability, improved biodiversity, and strengthened livelihoods.

References

Aayog NITI (2017) Inventory and revival of springs in the Himalayas for water security. Dept of Sci Tech, Government of India, New Delhi

Bamola V, Ngullie A, Kumar R, Sharma A, Gautam A (2018) Community based springshed development in response to drying springs in Nagaland. In: Climate change: impact, adaptation & response in the eastern Himalayas, pp 40–51. ISBN: 978-93-88237-16-1

Buono J (2019) Spring protection and management: context, history and examples of spring management in India. In: Groundwater development and management: issues and challenges in South Asia. Springer, Cham, pp 227–241

Buono J, Thomas R, Kulkarni H, Mahamuni K, Karandikar M, Ghate K, Kulkarni K, Mahesh N, Ambrale D, More A (2015) Ecohydrologic description of springs in the North Western Ghats, Maharashtra. J Ecol Soc 28:8–24

Círbik J, Beracko P, Bulánková E, Zaťovičová ZČ, Gregušová K, Kodada J, Krno I (2022) Are springs hotspots of benthic invertebrate diversity? Biodiversity and conservation priority of rheocrene springs in the karst landscape. Aquat Conserv Mar Freshwat Ecosyst 32(5):843–858. https://doi.org/10.1002/aqc.3802

Dekker I, Sharifyazd S, Batung E, Dubrawski KL (2021) Maximizing benefits to nature and society in techno-ecological innovation for water. Sustainability 13(11):6400. https://doi.org/10.3390/su13116400

Jamaluddin M, Hassan MK, Miah MM (2010) Identifying livelihood patterns of ethnic minorities and their coping strategies different vulnerabilities situation in Chittagong hill tracts region, Bangladesh. Bangladesh Final Report CF, 7(08)

Khadka K, Bhuchar S, Dhakal M, Thomas S (2023) Reviving springs through community-based participatory action research in the Himalaya. AGU23

Kulkarni H, Desai J, Siddique MI (2021) Rejuvenation of springs in the Himalayan region. In: Water, climate change, and sustainability, pp 97–107. https://doi.org/10.1002/9781119564522.ch6

Kuller M, Farrelly M, Marthanty DR, Deletic A, Bach PM (2022) Planning support systems for strategic implementation of nature-based solutions in the global south: current role and future potential in Indonesia. Cities 126:103693. https://doi.org/10.1016/j.cities.2022.103693

Kumar M, Rathod R, Mukherji A (2023) Water security and spring conservation in the Himalaya. In: Climate risks to water security: framing effective response in Asia and the Pacific. Springer, Cham, pp 15–36

Lo Russo S, Suozzi E, Gizzi M, Taddia G (2021) SOURCE: a semi-automatic tool for spring-monitoring data analysis and aquifer characterisation. Environ Earth Sci 80(21):710. https://doi.org/10.1007/s12665-021-10027-8

Nisa FU, Umar R (2024) Spring water system classifications and their methods of study: an overview of the current status and future perspectives. J Earth Syst Sci 133(1):10. https://doi.org/10.1007/s12040-023-02218-7

Nowreen S, Misra AK, Zzaman RU, Sharma LP, Abdullah MS (2023) Sustainability challenges to springshed water management in India and Bangladesh: a bird's eye view. Sustainability 15(6):5065. https://doi.org/10.3390/su15065065

Panwar S (2020) Vulnerability of Himalayan springs to climate change and anthropogenic impact: a review. J Mt Sci 17(1):117–132. https://doi.org/10.1007/s11629-018-5308-4

Prasad C, Sharma RC (2019) Water resources and their traditional management in Kedarnath valley of Garhwal Himalaya, India. Int J Hydrol 3:194–203. https://doi.org/10.15406/ijh.2019.03.00180

Rana S, Gupta V (2009) Watershed management in the Indian Himalayan region: issues and challenges. In: World environmental and water resources congress 2009: great rivers, pp 1–12. https://doi.org/10.1061/41036(342)527

Ranjan P, Kumar Pandey P (2020) Reviving, development and protection of springs to increase water security in the Himalayan region. In: Proceedings of the 5th international conference on computers & management skills (ICCM 2019). North Eastern Regional Institute of Science & Technology (NERIST), Nirjuli, Arunachal Pradesh, India. January 9, 2020. https://doi.org/10.2139/ssrn.3516630

Rathod R, Kumar M, Mukherji A, Sikka A, Satapathy K K, Mishra A, Khan M (2021). Resource book on springshed management in the Indian Himalayan region: guidelines for policy makers and development practitioners. New Delhi: International Water Management Institute (IWMI)/NITI Aayog, Government of India/Swiss Agency for Development and Cooperation (SDC)

Rawal S, Singh D, Linda A (2024) Ensuring water security by reviving selected springs in Kangra region, Himachal Himalaya, India. Water Sci 38(1):239–255. https://doi.org/10.1080/23570008.2024.2333600

Sharma K, Laskar N (2022) Effect of climate change on spring discharge management system of the Himalayan region in India. In: Advanced modelling and innovations in water resources engineering. AMIWRE, pp 105–117. https://doi.org/10.1007/978-981-16-4629-4_9

Shrestha RB, Desai J, Mukherji A, Dhakal M, Kulkarni H, Mahamuni K, Bhuchar S, Bajracharya S (2018) Protocol for reviving springs in the Hindu Kush Himalayas: a practitioner's manual. International Centre for Integrated Mountain Development (ICIMOD), Lalitpur. ISBN 978-92-9115-606-1. https://doi.org/10.53055/icimod.735

Soulios G (2010) Springs (classification, function, capturing). Bull Geol Soc Greece 43(1):196–215. https://doi.org/10.12681/bgsg.11174

Sowińska-Świerkosz B, García J (2022) What are nature-based solutions (NBS)? Setting core ideas for concept clarification. Nat Based Solutions 2:100009. https://doi.org/10.1016/j.nbsj.2022.100009

Springshed management: nature-based solution in Bhutan https://www.nature basedsolutions.com/page/987/springshed-management-nature-based-solution-in-bhutan

Stevens LE, Aly AA, Arpin SM, Apostolova I, Ashley GM, Barba PQ, Barquín J, Beauger A, Benaabidate L, Bhat SU, Bouchaou L (2021) The ecological integrity of spring ecosystems: a global review. In: Earth systems and environmental sciences, reference module. Elsevier. https://doi.org/10.1016/B978-0-12-821139-7.00111-2

Thapa B, Bhattarai C, Dahal N, Tiwari S, Jacobsen D (2023) Drying of springs in the Himalayan region of Nepal: perspectives of local government leaders on causes, consequences, and conservation efforts. Mt Res Dev 43(4):R9–R15. https://doi.org/10.1659/mrd.2023.00007

Todd KD (1980) Groundwater hydrology, 2nd edn. Wiley, New York, p 535

Wheater HS, Gober P (2015) Water security and the science agenda. Water Resour Res 51(7):5406–5424. https://doi.org/10.1002/2015WR016892

Chapter 19
Disaster Risk Reduction in the Indian Himalayas: Strategies for Mitigating Impacts of Natural Disasters

Fatima Amin, Deeksha Katyal, and Anil K. Gupta

Abstract The Indian Himalayan Region (IHR) is a critical ecological and geopolitical zone, home to diverse ecosystems and communities. However, the region is highly prone to natural disasters such as earthquakes, landslides, floods, avalanches, and glacial lake outburst floods (GLOFs). Climate change, unregulated development, and deforestation exacerbate these risks. This chapter examines the unique challenges faced in the IHR, focusing on disaster risk reduction (DRR) strategies tailored to the Indian context. Using examples and case studies, it explores how India can enhance resilience through community-based approaches, infrastructure improvements, and policy interventions.

Keywords DRR · Himalayas · Mitigation · Resilience · Sustainability · Community

19.1 Introduction

The Indian Himalayan Region (IHR) is a vital geographical and ecological zone extending from the union territories of Jammu, Kashmir, and Ladakh in the west to Arunachal Pradesh in the east. This region, covering approximately 16.2% of India's total land area, is home to diverse ecosystems, rich biodiversity, and over 50 million people. It forms the cradle for several of India's major river systems, including the Ganga, Brahmaputra, and Yamuna, which sustain millions of lives in the plains.

F. Amin (✉) · A. K. Gupta
Integrated Centre for Adaptation, Disaster Risk Resilience & Sustainability, Indian Institute of Technology, Roorkee, Greater Noida Extension Centre, Greater Noida, India

D. Katyal
University School of Environment Management, Guru Gobind Singh Indraprastha University, New Delhi, India

Y. D. I. Khan et al. (eds.), *Ecosystem-based Approaches for Resilience Building in Himalayan Landscapes*, Disaster Resilience and Green Growth,
https://doi.org/10.1007/978-981-95-2007-7_19

However, the region's unique topographical and climatic conditions also make it one of the most disaster-prone areas in the country.

The IHR is characterised by a fragile environment shaped by its steep mountainous terrain, active tectonic plate movements, and extreme weather patterns. These factors contribute to its high susceptibility to natural disasters such as earthquakes, landslides, floods, glacial lake outburst floods (GLOFs), and avalanches. Human-induced pressures, including deforestation, unregulated construction, and the impacts of climate change, further exacerbate these risks.

Recent disasters have underscored the urgent need for robust disaster risk reduction (DRR) measures in the IHR. For instance, the Kedarnath disaster of 2013 was triggered by exceptionally heavy rainfall, which led to devastating floods and landslides, causing widespread destruction and claiming thousands of lives. Similarly, the 2021 Chamoli GLOF in Uttarakhand, caused by the sudden breach of a glacial lake, led to catastrophic flooding downstream, destroying infrastructure and taking a heavy toll on lives and livelihoods. These events are stark reminders of the region's vulnerabilities and the critical importance of implementing effective DRR strategies that are sensitive to the region's unique ecological and socio-economic contexts.

19.2 Natural Disaster Risks in the Indian Himalayas

Due to its complex geological, topographical, and climatic conditions, the Indian Himalayan Region (IHR) is inherently prone to various natural disasters. This section outlines the major natural hazards affecting the region, emphasising their causes, impacts, and lessons learned.

19.2.1 Earthquakes

The IHR lies predominantly in seismic zones IV and V, the most earthquake-prone regions in India as classified by the Bureau of Indian Standards (Fig. 19.1). These zones experience frequent tectonic activity because of the ongoing collision between the Indian and Eurasian tectonic plates forming the Himalayas.

A prominent example is the 2005 Kashmir earthquake (magnitude 7.6), which caused over 80,000 deaths, mainly in Pakistan-administered Kashmir, but also led to extensive loss of life and property in India's Jammu and Kashmir region. The earthquake damaged homes, schools, and hospitals, leaving hundreds of thousands homeless. One of the primary reasons for the catastrophic impact was the lack of earthquake-resistant infrastructure in the region, compounded by a delayed response and inadequate preparedness (Bilham et al. 2001). The event highlighted the urgent need for strengthening building codes, retrofitting existing structures, and improving emergency response mechanisms in the IHR.

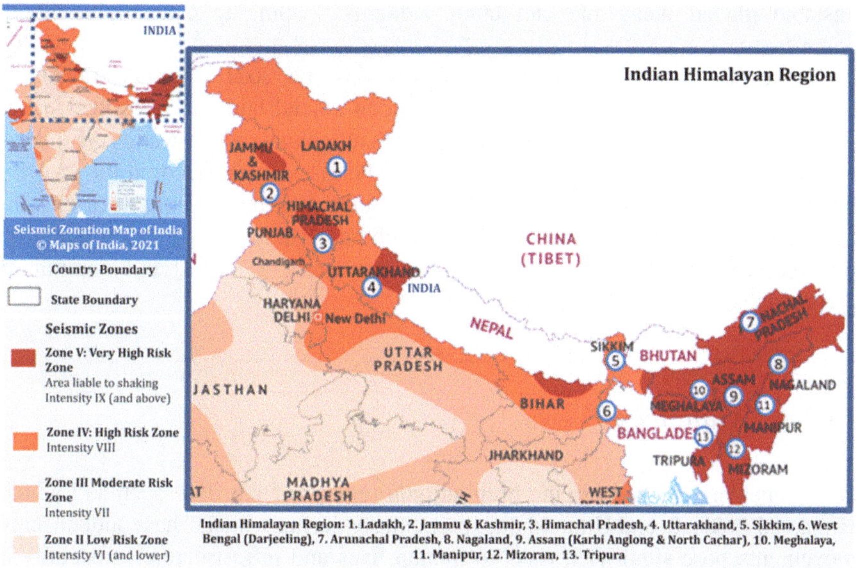

Fig. 19.1 Seismic zonation map of India. (Source: Adapted from © Maps of India (https://www.mapsofindia.com/maps/india/seismiczone.htm, last access: 1 April 2023))

19.2.2 Landslides

Landslides are one of the most frequent and devastating hazards in the IHR. They are primarily driven by the region's steep terrain, high rainfall, and widespread deforestation. Human activities such as road construction, mining, and unplanned urbanisation further destabilise slopes, increasing the frequency of landslides.

The 2016 Darjeeling landslides serve as a stark reminder of the risks posed by these hazards. Triggered by heavy monsoon rains, the landslides claimed 49 lives and caused significant damage to homes, roads, and agricultural land. They disrupted livelihoods and transport networks, isolating affected communities for days. The disaster also emphasised the role of deforestation and unscientific land-use practices in increasing slope instability. Preventive measures, such as afforestation, slope stabilisation, and proper drainage systems, are critical to reducing regional landslide risks.

19.2.3 Floods and Glacial Lake Outburst Floods (GLOFs)

The IHR is home to numerous rivers and glacial lakes, which are increasingly becoming sources of flooding due to intense rainfall, glacial melt, and glacial lake outbursts. Climate change is a significant driver, causing glaciers to retreat and form unstable glacial lakes that can burst suddenly, leading to catastrophic floods downstream.

One of the most notable recent disasters is the 2021 Chamoli tragedy in Uttarakhand. This event, attributed to a sudden glacial burst or landslide, caused devastating floods in the Rishiganga and Dhauliganga rivers, destroying hydro-power projects and claiming over 200 lives. The disaster underscored the growing threat of GLOFs in the context of climate change and the inadequacy of existing monitoring and early warning systems (Singh and Pandey 2024). Addressing this risk requires glacial lake monitoring, early warning systems, and community preparedness.

19.2.4 Avalanches

Avalanches are common in the high-altitude regions of the IHR, such as Ladakh, Himachal Pradesh, and Uttarakhand, particularly during winter. These sudden snow movements pose significant risks to human lives and infrastructure, especially in areas frequented by tourists, mountaineers, and military personnel.

For instance, in 2022, an avalanche in Uttarkashi, Uttarakhand, killed 28 mountaineers from a training expedition. Such incidents emphasise the unpredictability and severity of avalanches in the region. Despite establishing monitoring systems like the Snow and Avalanche Study Establishment (SASE), many vulnerable areas still lack adequate early warning mechanisms, evacuation plans, and practical training for communities to respond.

Investing in robust forecasting systems, increasing advanced technologies like remote sensing, and educating communities can significantly reduce avalanche-related risks.

19.3 Strategies for Disaster Risk Reduction in the Indian Himalayas

Effectively reducing disaster risks in the Indian Himalayan Region (IHR) requires a multidimensional approach combining advanced technology, community engagement, resilient infrastructure, climate adaptation, and strong governance. This section elaborates on key strategies to enhance disaster preparedness and resilience in the IHR, with real-world examples and case studies.

19.3.1 Strengthening Early Warning Systems (EWS)

Early warning systems mitigate disaster impacts by providing timely alerts to at-risk populations. In the IHR, where floods, landslides, and glacial lake outburst floods (GLOFs) are frequent, technological advancements have enhanced monitoring and prediction capabilities. Agencies like the Indian Meteorological Department (IMD) and the National Remote Sensing Centre (NRSC) utilise satellite imagery, hydrological models, and weather forecasting tools to predict and track disasters. However, while these systems have improved significantly, disseminating warnings to remote, vulnerable communities remains challenging.

In Uttarakhand, installing Doppler radar systems has enabled real-time weather monitoring, providing critical data for flood and landslide predictions. These systems, coupled with localised flood forecasting and communication networks, have helped improve preparedness among local authorities and communities. Expanding similar networks across the IHR and integrating local-level dissemination mechanisms, such as mobile alerts and community loudspeakers, can enhance the effectiveness of EWS.

19.3.2 Community-Based Disaster Risk Management (CBDRM)

Empowering local communities to participate in disaster risk management is vital for building resilience. Community-based disaster risk management (CBDRM) emphasises grassroots capacity-building, focusing on risk assessment, preparedness training, and response planning.

In Sikkim, village disaster management committees (VDMCs) have been established in high-risk areas to train residents in evacuation protocols, first aid, and disaster response techniques. This decentralised approach ensures that communities are better prepared to act during emergencies, thereby minimising loss of life and property.

Additionally, indigenous knowledge systems can play a crucial role in risk reduction. For example, the Apatani tribe in Arunachal Pradesh has developed traditional water management systems that mitigate the risks of flash floods and water shortages. By blending traditional practices with modern DRR frameworks, these strategies can serve as models for sustainable risk reduction in the IHR.

19.3.3 Infrastructure Resilience

The IHR's vulnerability to earthquakes, landslides, and other disasters necessitates resilient infrastructure. Buildings, roads, and public utilities must be designed to withstand the region's unique hazards. Retrofitting existing infrastructure, especially critical facilities like schools and hospitals, is equally important.

Post the 2005 Kashmir earthquake, retrofitting projects were initiated in Himachal Pradesh and Uttarakhand to strengthen the resilience of schools and healthcare centres. However, scaling these efforts across the IHR remains challenging due to financial and logistical constraints.

The Uttarakhand Disaster Recovery Project (UDRP), supported by the World Bank, provides a successful model. This initiative focused on rebuilding earthquake-resistant infrastructure in disaster-affected areas and training engineers and architects in resilient construction techniques. Expanding such initiatives across the IHR can significantly reduce future disaster risks.

19.3.4 Climate Change Adaptation

The IHR is highly sensitive to climate change, with rising temperatures accelerating glacial retreat and altering precipitation patterns. Integrating climate adaptation into disaster risk reduction (DRR) is essential for addressing long-term vulnerabilities.

Key strategies include reforestation, watershed management, and sustainable agricultural practices. Reforestation can reduce the risk of landslides and flash floods, while better water management helps mitigate droughts and waterlogging. Monitoring glacier dynamics and glacial lake stability is critical for preventing GLOFs.

The National Mission for Sustaining the Himalayan Ecosystem (NMSHE), a part of India's National Action Plan on Climate Change (NAPCC), promotes research on climate vulnerabilities in the region. Through pilot projects on sustainable agriculture and glacier monitoring, the mission aims to strengthen resilience against climate-induced hazards (Ministry of Environment, Forest and Climate Change 2019).

19.3.5 Policy and Governance

Effective disaster risk reduction in the IHR requires strong governance frameworks at both the national and state levels. India's Disaster Management Act (2005) provides a robust foundation by mandating disaster management plans and response mechanisms. However, there is a need for improved capacity building at the state level and enhanced collaboration between states in the IHR. Transboundary risks, such as GLOFs and shared river basins, also demand regional cooperation. Collaborative initiatives with neighbouring countries like Nepal, Bhutan, and China are crucial for sharing data, forecasting disasters, and managing shared ecosystems.

Joint river basin management programmes between India and Nepal have facilitated the exchange of hydrological data, improving flood management in the Ganga and Koshi river systems. Expanding such partnerships can strengthen disaster preparedness for shared risks.

19.4 Case Studies

Examining past disasters in the Indian Himalayan Region (IHR) provides valuable insights into the region's vulnerabilities and the effectiveness of response and mitigation strategies. This section discusses two significant disasters: the 2013 Kedarnath flood and the 2021 Chamoli glacial lake outburst flood (GLOF).

19.4.1 The 2013 Kedarnath Disaster

The 2013 Kedarnath disaster stands as one of the most devastating natural calamities in India's recent history (Fig. 19.2). Triggered by unprecedented heavy rainfall between June 14 and 17, the disaster caused the Chorabari glacier to breach, leading to massive flooding and landslides in the Mandakini River valley. The extreme rainfall during this period exceeded 375% of the average for the month, categorising it

Fig. 19.2 Kedarnath disaster 2013. (Source: Hindustan Times)

as an exceptional meteorological event. Over 5000 people lost their lives, and the Kedarnath temple, a significant religious site, was left surrounded by debris, though it miraculously withstood the deluge. Several key factors amplified the disaster, including the region's fragile ecology, where the geological instability of the area combined with unplanned urbanisation exacerbated the destruction. The proliferation of hotels, shops, and roads in floodplains and ecologically sensitive zones further hindered natural drainage, intensifying the damage.

In the aftermath of the tragedy, the state government of Uttarakhand introduced significant policy changes to mitigate future risks. Stricter land-use regulations were implemented, including restrictions on construction near riverbanks and in landslide-prone areas. Additionally, the Indian Meteorological Department (IMD) and other agencies supported installing improved hydrological and meteorological monitoring systems to provide better early warnings. The Char Dham Highway Project, aimed at enhancing connectivity, was re-evaluated to integrate ecological safeguards and disaster mitigation measures.

The Kedarnath disaster underscored the importance of aligning development with ecological constraints. Key lessons learned from the tragedy include integrating environmental impact assessments into development planning and strengthening disaster preparedness frameworks to better cope with such extreme events in the future.

19.4.2 The 2021 Chamoli GLOF

The Chamoli disaster, which occurred on February 7, 2021, highlighted the escalating threats posed by climate change in the Indian Himalayan Region (IHR). The sudden collapse of a section of the Nanda Devi glacier triggered a catastrophic flood in the Rishiganga and Dhauliganga rivers. This flood destroyed two hydroelectric power projects, caused extensive damage to infrastructure, and resulted in over 200 fatalities. Satellite imagery and analysis by the Indian Space Research Organisation (ISRO) revealed that a rock and ice avalanche from a hanging glacier initiated the event. Climate change was believed to have weakened the glacier's structure, making it more susceptible to collapse. The disaster also raised concerns about the placement of hydropower projects in ecologically sensitive areas, as evidenced by the destruction of the Rishi Ganga Hydroelectric Project, which was located along the flood path, underscoring the risks of infrastructure in high-hazard zones.

Technological interventions, particularly the use of satellite data and remote sensing, played a crucial role in understanding the causes of the disaster. ISRO's Cartosat-3 and Sentinel-2 satellites provided high-resolution imagery that tracked the event and its aftermath. In response to the disaster, discussions began on integrating real-time glacial lake monitoring systems and improving early warning mechanisms to prevent future glacial lake outburst floods (GLOFs). The Chamoli disaster prompted policy changes, including increased focus on strengthening glacial monitoring through advanced technologies like drones and ground-penetrating radar and re-evaluating the long-term viability of hydropower projects in the Himalayas.

The disaster emphasised the need for integrating climate adaptation strategies into disaster risk reduction frameworks. Key lessons learned include the importance of enhanced glacial monitoring, developing climate-resilient infrastructure, and imposing restrictions on construction in high-risk areas. These measures are crucial for reducing the likelihood of similar disasters in the future (Singh and Pandey 2024).

19.5 Challenges and Future Directions

Due to its unique geographical, ecological, and socio-economic conditions, the Indian Himalayan Region (IHR) faces complex challenges in implementing effective disaster risk reduction (DRR) strategies. Addressing these challenges requires a multi-pronged approach that balances development needs with environmental conservation, adapts to climate change, and fosters regional collaboration.

19.5.1 Resource Constraints

One of the significant barriers to effective DRR in the IHR is the lack of financial and logistical resources. Himalayan states, often with limited revenue streams, struggle to fund large-scale mitigation projects, such as retrofitting infrastructure, expanding early warning systems, and implementing robust land-use regulations. The steep terrain and remote locations further increase logistical costs for transporting materials and deploying emergency services.

Opportunities for Improvement

- *Increased Federal Funding:* Central government schemes like the National Disaster Response Fund (NDRF) and externally aided projects must prioritise the IHR. A dedicated Himalayan Development Fund could be established to cater specifically to the region's needs.
- *International Partnerships:* Collaboration with international organisations like the United Nations Development Programme (UNDP) and World Bank can bring technical expertise, funding, and innovative solutions. For example, the World Bank-supported **Uttarakhand Disaster Recovery Project (UDRP)** has shown the potential for transformative impact.

Nepal's earthquake-resistant housing programme, implemented with international support, offers a model for scaling resilient infrastructure in the IHR.

19.5.2 *Climate Change*

Climate change intensifies natural hazards in the IHR, with accelerated glacial retreat, erratic rainfall patterns, and increasing frequency of extreme weather events. Glacial lake outburst floods (GLOFs), previously rare, are becoming more common due to rising temperatures. Additionally, shifting precipitation patterns exacerbate droughts in some areas and floods in others.

Adaptive Strategies

- *Scenario-Based Planning:* Incorporate climate projections into DRR frameworks. Monitoring glacier dynamics and mapping potential GLOF sites can help identify high-risk areas (Singh and Pandey 2024).
- *Community Resilience Programmes:* Programmes promoting climate-resilient agriculture, water harvesting techniques, and afforestation can empower local communities to adapt.
- *Sustainable Infrastructure:* It is critical to build infrastructure that withstands climate extremes, such as flood-resistant housing and drought-proof water supply systems.
- The **National Mission for Sustaining the Himalayan Ecosystem (NMSHE)** under India's National Action Plan on Climate Change (NAPCC) focuses on understanding and mitigating climate vulnerabilities in the region.

19.5.3 *Balancing Development and Conservation*

Rapid urbanisation, construction of hydropower projects, deforestation, and unregulated tourism development are altering the fragile Himalayan ecosystem. These activities increase the risk of landslides, floods, and ecological degradation.

Solutions for Sustainable Development

- *Environmental Impact Assessments (EIAs):* Make EIAs mandatory for all projects in the IHR. These assessments should incorporate disaster risk evaluations and prioritise ecological integrity.
- *Eco-Tourism:* Promote low-impact tourism that benefits local economies while conserving biodiversity.
- *Hydropower Optimisation:* Hydropower is a renewable resource, but its development must consider cumulative environmental impacts. For example, strategically placing projects away from high-risk zones and adopting sediment management techniques can reduce risks.

Though controversial, the Tehri Dam project offers lessons on balancing energy needs with environmental and safety considerations through sedimentation studies and seismic-resistant design.

19.5.4 Transboundary Collaboration

The IHR shares its river basins, ecosystems, and disaster risks with neighbouring countries such as Nepal, Bhutan, and China. Effective DRR requires regional cooperation in data sharing, early warning systems, and coordinated response mechanisms.

Current Efforts and Opportunities

- *SAARC Disaster Management Centre (SDMC):* This platform facilitates regional collaboration on disaster management. Strengthening SDMC's role in data sharing and joint training programs is vital.
- *River Basin Agreements:* Transboundary water management, such as the India-Nepal Koshi River agreement, can reduce flood risks. Expanding similar frameworks for shared rivers like the Brahmaputra is critical.
- *Joint Monitoring Systems:* Establish collaborative networks for glacial and seismic monitoring, involving neighbouring countries for real-time data sharing.

The **Indus Waters Treaty** between India and Pakistan has demonstrated the potential of international cooperation for managing shared water resources, though its adaptation to modern challenges is necessary.

The IHR presents unique challenges for disaster risk reduction, driven by its fragile ecology, rapid development, and vulnerability to climate change. Addressing these requires a holistic approach combining resource mobilisation, adaptive strategies, sustainable development practices, and regional cooperation. By integrating scientific advancements, community involvement, and policy reforms, India can significantly enhance resilience and safeguard the livelihoods of the millions who call the Himalayas home.

19.6 Conclusion

The Indian Himalayan Region (IHR), home to over 50 million people, and the source of critical river systems like the Ganga, Brahmaputra, and Yamuna, faces complex disaster risks shaped by its unique geology, ecology, and socio-economic conditions. The combined pressures of climate change and unsustainable human activities, including deforestation, unplanned urbanisation, and large-scale infrastructure construction in fragile environments, further amplify these risks. The region's vulnerability highlights the urgent need for comprehensive disaster risk reduction (DRR) strategies tailored to its specific challenges.

Addressing disaster risks in the IHR requires a multifaceted approach that balances development, environmental conservation, and community resilience. Scientific innovation plays a crucial role, as advancements in satellite technology, early warning systems, and hazard mapping have significantly improved disaster prediction and mitigation. For instance, satellite monitoring by ISRO has been instrumental in understanding events like the 2021 Chamoli disaster, while Doppler radar systems in Uttarakhand have enhanced weather forecasting. Community engagement is equally important, with initiatives like Village Disaster Management Committees (VDMCs) and the incorporation of indigenous knowledge empowering local populations to become active agents of resilience. The Apatani tribe's traditional water management practices in Arunachal Pradesh exemplify how local or traditional knowledge can complement modern DRR efforts. Policy interventions, such as the Disaster Management Act (2005) and state-level land-use regulations, have laid the foundation for systematic DRR. However, more substantial implementation at the state and local levels is necessary to ensure ecological constraints are respected, and vulnerable populations are protected. Additionally, integrating climate adaptation into DRR is vital, given the growing impact of climate change on weather patterns, glacial retreat, and hydrological shifts. Programmes like the National Mission for Sustaining the Himalayan Ecosystem (NMSHE) reflect India's commitment to addressing these long-term challenges.

Analysing past disasters like the 2013 Kedarnath floods and the 2021 Chamoli GLOF offers critical disaster preparedness and response lessons. The Kedarnath disaster highlighted the devastating impact of unregulated development and underscored the need for robust land-use policies and resilient infrastructure. On the other hand, the Chamoli disaster demonstrated how climate change accelerates glacial hazards and emphasised the importance of real-time glacial monitoring and early warning systems.

India's Himalayan communities possess a wealth of traditional knowledge in sustainable agriculture and water management, which can be harnessed to complement scientific interventions. By integrating this indigenous knowledge into broader DRR strategies, India can enhance local resilience and position itself as a global leader in disaster management for mountain ecosystems. The IHR represents a microcosm of disaster-prone mountain regions' global challenges and opportunities. Strengthening DRR in this region is a national priority and a global imperative. India can develop a replicable, sustainable, resilient development model in mountainous areas by strengthening collaboration among governments, scientific institutions, civil society organisations, and local communities.

In conclusion, a proactive and inclusive approach that respects the delicate balance of the Himalayan ecosystem while addressing the needs of its people is essential. By integrating lessons from past disasters, leveraging technology, and empowering communities, India can mitigate disaster risks in the IHR and contribute to a safer, more resilient future.

References

Bilham R, Gaur VK, & Molnar P (2001) Himalayan seismic hazard. Science 293, 1442e1444

De SK (2017) Landslides and human interference in Darjiling Himalayas, India. Revista de Geomorfologie 19(1):44–57. https://doi.org/10.21094/rg.2017.014

Dimri AP, Allen S, Huggel C, Mal S, Ballesteros-Cánovas JA, Rohrer M, Pandey A (2021) Climate change, cryosphere and impacts in the Indian Himalayan Region. Current Science, 120(5), 774–790

Magotra R, Shaw M, & Tyagi A Disaster Vulnerability Assessment Action Plan to Strengthen Disaster Resilience of Gangtok, Sikkim. Disaster Development, 1

Malepati V, & Gowri CM (2011) The Role of Science and Technology for Disaster Risk Reduction Management with a Special Reference to India. Available at SSRN 1737839

Rai A, & Fulekar MH (2023) Climate Change—National Action Plan. In Climate Change and Sustainable Development (pp. 223-237). CRC Press

Rai P, & Khawas V (2019) Traditional knowledge system in disaster risk reduction: Exploration, acknowledgement and proposition. Jàmbá: Journal of Disaster Risk Studies, 11(1), 1–7

Rawat AS, & Semwal MM (2022) Unveiling the Himalayan Nexus: Unraveling Anthropogenic Triggers and Pursuing Sustainable Development in Uttarakhand

Shrestha AB, Steiner J, Nepal S, Maharjan SB, Jackson M, Rasul G, & Bajracharya B (2021) Understanding the Chamoli Cause, process, impacts, context of rapid infrastructure development. Cryosphere, 1–15

Singh G, Pandey A (2024) Climate change induced disasters and highly vulnerable infrastructure in Uttarakhand, India: current status and way forward towards resilience and long-term sustainability. Sustainable and Resilient Infrastructure, 9(2):145–167

Srivastava P, Namdev P, Singh PK (2022) 7 February Chamoli (Uttarakhand, India) rock-ice avalanche disaster: model-simulated prevailing meteorological conditions. Atmosphere, 13(2):267

Takefuji, Y. (2022). Enhancing disaster resilience by sustainable technologies. Urban Governance, 2(1), 197–199

Van Wyk de Vries M, Bhushan S, Jacquemart M, Deschamps-Berger C, Berthier E, Gascoin S, Kääb A (2022) Pre-collapse motion of the February 2021 Chamoli rock–ice avalanche, Indian Himalaya. Natural Hazards and Earth System Sciences, 22(10), 3309–3327

Chapter 20
The Effectiveness of Disaster Management Plans and Strategies at the Local Level: Evidence from Sikkim-Darjeeling Himalayan Region

Shikha Subba, Bishal Chhetri, Kabita Lepcha, and Pavan Tikhile

Abstract The International Frameworks for Disaster Risk Reduction have specified the role of national governments at all levels, including the local government. It is important to note that international agreements such as the Yokohama Strategy for Disaster Risk Reduction, the Hyogo Framework and the Sendai Framework for Disaster Risk Reduction mention community involvement in disaster management plan implementation. India also endorses international conventions to reduce the risk of disaster. According to the 2005 Disaster Management Act, both the central government and the states in India are in charge of disaster management. A question is whether a disaster management unit exists at all levels, making plans for disaster risk reduction and whether this makes a difference. The study areas were selected based on the severity of landslides from the state of Sikkim and the Gorkhaland Territorial Administration, consisting of Darjeeling and Kalimpong districts. The effectiveness of disaster management strategies that are implemented at the local level of this region is also discussed in this study. The study also discusses the applicability of disaster management strategies, concepts, plans and actions. It makes an effort to highlight their usefulness locally using the secondary and empirical data gathered from the sample households from the two study areas.

Keywords Landslides · Disasters · Disaster risk reduction · Hazard risk assessment

S. Subba
Independent Researcher, Darjeeling, West Bengal, India

B. Chhetri (✉)
Southfield College, Darjeeling, West Bengal, India

K. Lepcha
University of Gour Banga, Malda, West Bengal, India

P. Tikhile
Nawegaon Nagzira Tiger Reserve, Gondia, Maharashtra, India

Y. D. I. Khan et al. (eds.), *Ecosystem-based Approaches for Resilience Building in Himalayan Landscapes*, Disaster Resilience and Green Growth,
https://doi.org/10.1007/978-981-95-2007-7_20

20.1 Introduction

Disaster is a generic term, while landslide is an instance of a disaster. Disaster can be termed as the occurrence of something undesirable with high magnitude that results in deaths, injuries and damages to human property and environment which is not in the control of the government, or something that is difficult to manage. Disaster, when occurs suddenly and unexpectedly (which happens most of the time), will require emergencies and effective services, proper coordination from various government organisations, private sector, NGOs, etc., for quick recovery and meeting the needs of victims quickly (Taori 2005). Disaster is a serious disruption of the functioning of a community or a society causing widespread human, material, economic or environmental losses, which exceed the ability of the affected community or society to cope, using its own resources (UNISDR 2004).

The main aim of the chapter is to comprehend the disaster management at the local level by analysing the effectiveness of the local government particularly in landslide risk reduction. A question is whether a disaster management unit exists at all levels, making plans for disaster risk reduction and whether this makes a difference. Studies such as Scott and Tarazona (2011) and Maes et al. (2018) look at the role of decentralisation in disaster management. These studies show that decentralisation is expected to produce greater participation and more engaged decision-makers, incorporate local knowledge and funding into the planning and promote accountability. A decentralised government can be effective in all the three phases (pre-emergency, emergency and post-emergency) of disaster management. Messer (2003) argued that since disaster is mostly a local phenomenon, which rarely hits the entire country, use of local information, knowledge and resources is critical for effective prevention measures. India has been a signatory of the International Frameworks for Disaster Risk Reduction and the country has its own Disaster Management Plan and Act which focuses in prevention and mitigation measure for disaster risk reduction. The International Frameworks for Disaster Risk Reduction have specified the role of national governments at all levels, including the local government. The International Frameworks and the National Plans and Act also recognise the importance of local government in disaster risk reduction. It is important to note that international agreements such as the Yokohama Strategy for Disaster Risk Reduction, the Hyogo Framework and the Sendai Framework for Disaster Risk Reduction mention community involvement in disaster management plan implementation. India also endorses international conventions to reduce the risk of disaster. According to the 2005 Disaster Management Act, both the central government and the states in India are in charge of disaster management. The Indian Disaster Management Plan proclaims that the emphasis of disaster management has changed from a relief and rescue-centric strategy to a prevention and mitigation-centric one, i.e. greater attention has been given to minimising the risk of disaster.

20.1.1 Disaster Risk Reductions and International Frameworks

Disasters regularly occur in all countries due to climate change and high land use pressure (Zimmermann and Keiler 2015), unlike in the past when they were more or less sporadic. This is especially so in mountainous regions such as the Himalayas. Hence, there has been discussion at the global level since the 1990s on how to reduce disaster risks. From 1994 to 2015, there was a rapid rise in global cooperation for disaster risk reduction (Efiong et. al 2021). The United Nations proposed the Yokohama Strategy for a Safer World in 1994. The current global cooperation, the Sendai Framework for Disaster Risk Reduction, became operational in 2015. So far, there are three international frameworks for disaster risk reduction. In 1994, the first world conference on natural disasters was held in Yokohama, Japan. The conference adopted the Yokohama Strategy for a safer world, outlining strategies and plan of action for prevention, preparedness and mitigation. Subsequently, the Hyogo Framework for Action was formulated for 2005–2015. This framework emphasises building the resilience of nations and communities in addressing disasters by setting three strategic goals and five priority areas for integrating disaster risk reduction into sustainable development policies, capacity building, preparedness and vulnerability reduction. The Sendai Framework for Disaster Risk Reduction was adopted for 2015–2030. The framework was adopted to make stakeholders at all levels disaster-resilient and reduce the loss of lives and assets. All the above frameworks guide all countries in reducing risks from natural hazards, including landslides (UNISDRR 2015). Several studies evaluated these frameworks and gave a critical perspective, especially on the design and the extent to which they have made a difference at the local level. For instance, Zimmermann and Keiler (2015) assessed the frameworks of Hyogo and Sendai in a specific context that focused on challenges faced in mountain areas such as the Himalayas to reduce the risk of disasters. They argue that while the global frameworks provide valuable guidance to the international community, the guidelines need to be further adapted to the local context, especially in mountain areas, where disaster risk patterns keep changing due to climate change and increased pressure from human beings.

This is particularly the case regarding the role of local government and community in disaster risk reduction (DRR). All three frameworks describe the importance of local participation in the DRR. However, as argued in this study, there are differences between these frameworks and a marked shift from including local communities and knowledge towards a more top-down approach to disaster risk reduction. The Yokohama Strategy, formulated in 1994, values understanding local context as a key component of planning and implementing disaster risk reduction initiatives. Throughout the Yokohama Strategy, there are calls to strengthen Disaster Risk Reduction efforts by "mobilising" traditional expertise and increasing communities' self-confidence through "recognition and propagation of traditional knowledge, practices and values". The desired empowerment of local communities comes through understanding local circumstances and harnessing local expertise. These values for and cooperation with local actors are presented as "essential preconditions for reducing vulnerability". Such a deep value attached to local knowledge and recognition of the importance of a nuanced understanding of local circumstances seems to have been lost in the Hyogo Framework for Action (UNISDR 2005) and SFDRR (UNISDR 2015). The studies on the Hyogo framework expressed concerns over the extent to which the vision was translated into practice. Mid-term review (UNISDR 2011) of the Hyogo Framework for Action 2005–2015 advocated consistently for the importance of community involvement in DRR, resilience building and local implementation of the Hyogo Framework for Action. Tozier and Baudoin (2015) have expressed repeated concerns that "inclusion of gender perspective and effective community participation are the areas where the least progress seems to have been made". They also note that grassroots women's organisations remain marginalised and disconnected from decision-making. This "lack" of Hyogo Framework for Action implementation at the local level has led to "a significant gap between national and local level action" and "minimal" progress at the local level (UNISDR 2011). Tozier and Baudoin (2015) further argue that despite the caution expressed on the Hyogo Framework for Action by the mid-term review, the Sendai Framework for Disaster Risk Reduction appears to make an even more pronounced shift towards top-down advocacy of a DRR agenda rather than a shift towards more meaningful partnerships with local actors. Like Hyogo Framework for Action, SFDRR mentions the need for "people-centred" DRR, as well as engagement and partnership with "all of society" and "special attention to people disproportionately affected by disasters, especially the poorest" (UNISDR 2015, p. 8). This support entails providing incentives, "complementing" scientific knowledge with local knowledge "where appropriate" and "disseminating disaster risk information" (UNISDR 2015, p. 11). The framework calls for "assigning clear roles and tasks to community representatives"; such a language indicates a top-down approach rather than collaborative processes (UNISDR 2015). The Yokohama Strategy, and to some degree, the Hyogo Framework for Action, had valued community empowerment and local expertise as important components of disaster risk reduction. In the SFDRR, however, most references to the community are only made in passing, giving vague references to integration of "all levels"—from international to local. Rather than valuing local understanding, the emphasis is on supporting the most vulnerable (most exposed, poorest communities) through information or other external, often technology-based, expertise. These references are reminders to integrate local people into top-down plans rather than calls for partnerships and cross-scale collaborations that are necessary for improved disaster risk reduction (Tozier and Baudoin 2015).

The Sendai Framework (2015–2030) suggests that local communities are helpless and in need of externally driven efforts to prepare for, cope with and recover from natural hazards; such a position neglects the widely acknowledged fact (in academia and official reports) that local communities have been interacting with their environments for centuries, thus endowing them with a significant collective experience in risk reduction that is valuable to any disaster risk reduction framework. India and Disaster Risk Reduction India has been a signatory to the frameworks of Hyogo and Sendai and formulated policies and legislations to translate the guidance provided into action at the ground level. A crucial legislation is the Disaster Management Act of 2005, discussed below.

20.1.2 Disaster Management Act 2005

The origin of the Disaster Management Act 2005 can be traced back to the ideas, thinking and events that brought a paradigm shift in the approach of disaster management globally from relief and rehabilitation to prevention and mitigation through a comprehensive and holistic approach. The Disaster Management Act 2005 is India's first legal disaster management act. The focus shifted to policy, legislation and institutional arrangements as pertinent content for a holistic approach to disaster management. Legislation on disaster management will provide a proper basis for preparedness, mitigation, response and rehabilitation activities. Legal provisions will further support managerial arrangements and plans and assign responsibilities vital for properly implementing disaster mitigation actions. A legal framework creates legal authorities for programmes and organisations related to hazards, mitigation and risk management (Task Force, GOI 2013).

20.1.3 Disaster Management Act 2005 and the Local Authority

Since this study aims to understand the roles and responsibilities vested with the local government, we discuss below the Disaster Management Act 2005 regarding the local government. It must be noted here that in the Disaster Management Act 2005, the term "local authority" is used. As per the 2 (h) of the Act, the term local authority includes "Panchayati Raj Institutions, municipalities, district board, cantonment board, Town Planning Authority or Zilla Parishad or any other body or authority, by whatever name they are called, for the time being, invested by law, for rendering essential services or, with the control and management of civic services, within a specified local area". The responsibility of local authorities for disaster management is laid down in Chapter VI of the Disaster Management Act, 2005.

As per the Act, under the direction of the district authority, a local authority shall perform the following four functions:

1. Ensure that its officers and employees are trained in disaster management.
2. Ensure that resources relating to disaster management are so maintained as to be readily available for use in the event of any threatening disaster situation or disaster.
3. Ensure all construction projects under it or within its jurisdiction conform to the standards and specifications laid down for the prevention of disasters and mitigation by the National Authority, State Authority and District Authority.

4. Carry out relief, rehabilitation and reconstruction activities in the affected area according to the State Plan and the District Plan. The Task Force, Government of India, Ministry of Home Affairs report noted that local authorities' role, scope and function need to be clearly articulated in the Disaster Management Act 2005. The other significant functions that were recommended to be assigned to the local authority are ensuring compliance with building regulations and safe construction of buildings (new and old) and other structures, undertaking programmes to enhance community-based disaster mitigation, preparedness and other preventive measures (Task Force, GOI 2013).

Nevertheless, Section 42 of the Act makes clear the roles assigned to local authorities in the Disaster Management Act 2005 are limited to training functionaries, maintenance of resources required to provide relief to disaster-affected persons, ensuring that construction within the jurisdiction of the local authority does not come in conflict with disaster risk reduction and post-disaster activities such as relief and rehabilitation. The Act does not, however, provide any roles to the local government relating to identifying those vulnerable to disasters, making plans for disaster risk reduction for them and incorporating the same in line with the decentralisation philosophy. Assigning such functions to the local government would have had more relevance in the context of the 73rd and 74th Constitutional Amendments. Preparing a local disaster management plan for the local area should be considered one of the primary functions of the local authority. The 11th Schedule of the Constitution of the 73rd Constitutional Amendment Act (1992) assigned 29 functions to Panchayat Raj Institutions. Some of these functions are directly related to disaster risk reduction activities. The most important are soil conservation and land improvement, social forestry, minor irrigation and rural housing (safe shelter).

Considering the country's existing three-tier Panchayat System, the functions of local authorities during the relief and rehabilitation phase need to be defined. However, in some states (e.g. Gujarat), some provisions for emergency services have a mandatory function for local authorities. Including such emergency services as mandatory under the Disaster Management Act will help strengthen, upgrade and supplement ongoing efforts towards emerging and new vulnerabilities. The Report has further suggested that the role of the Mayor, Commissioner of the Municipal Corporation, Fire Services, Civil Defence and the community should be clearly defined in the Act (Task Force, GOI 2013). The Act provides the functions of local authorities such as training officers and employees, maintaining and keeping resources ready for use in the event of a disaster, ensuring that all construction projects conform to prescribed standards and carrying out relief, rehabilitation and reconstruction activities (Task Force, GOI 2013). Second, although the 73rd Amendment was enacted more than a decade before the Disaster Management Act came into being, the provisions of the 73rd Amendment have not been reflected in the Disaster Management Act. Evidence shows that local self-government has played vital roles in disaster management, for example, tsunami-affected Tamil Nadu and Kerala, cyclone-ravaged Orissa, flood-affected Bihar and earthquake-hit Gujarat (Pal and Shaw 2018).

The eighth point of the Prime Minister's ten-point agenda focuses on building local capacity for disaster risk reduction (NDMA 2014). The point further suggests making a specific locally designed plan for local implementation. The point also considers the importance of local traditional practices and Indigenous knowledge for reducing disaster risk.

20.1.4 The National Disaster Management Plan (2019)

The Disaster Management Act of 2005 requires the development of a National Disaster Management Plan (NDMP) for the entire country. As per the design, the National Disaster Management Plan, which provides a framework for government agencies in all phases of disaster management, should be revised regularly (every 5 years). According to the National Disaster Management Plan 2019 (GOI 2019), there are four sections of thematic plans for landslide management. The Landslide Management plans are divided into five themes: (i) understanding landslide risk, (ii) inter-agency coordination, (iii) investing in DRR-structural measures, (iv) investing in DRR-non-structural measures, and (v) capacity development. In understanding the landslide risk, the central areas of focus are geological and geotechnical investigation, hazard zoning and mapping, research and development and assessment of hazard and risk. Specific responsibilities are given to the local bodies from the state as per the National Disaster Management Plan. In the case of landslide disaster management, the PRIs are entrusted with the responsibility of consulting experts and stakeholders to strengthen the mechanism of studies on vulnerabilities, covering social, economic, physical, gender, ecological and equity aspects. In the case of inter-agency coordination, the main focus areas are overall disaster governance and effective coordination among different levels of government to ensure the effective dissemination of warnings and information. The PRIs are also responsible for providing regular updates to and from the risky areas. As far as investing in Disaster Risk Reduction Structures is concerned, these include measures that focus on human settlement protection, multi-hazard shelter construction and heritage structure protection. Regarding the roles of the PRIs, they are responsible for assisting the District Disaster Management Authority (DDMA) with the maintenance of roads in risk-prone areas and the identification of safe temporary shelters in high-risk zones. Investing in disaster risk reduction through non-structural measures covers areas such as licensing and certification of professionals and public-private partnerships. In this regard, the responsibilities are vested with the District and State Disaster Management Authorities, and local bodies don't have significant roles to perform. Capacity development focuses on various areas, such as training professionals, generating disaster awareness, conducting mock drills and empowering marginalised groups, women and differently abled people. Local authorities, including PRIs, are also responsible for these activities, such as creating awareness and sensitisation and developing local adaptation strategies. Capacity development includes landslide management, hazard risk assessment and a climate change adaptation programme.

20.1.5 Compendium of Laws on Disaster Management of Landslides: Mandates for Local Authority

Chapter 20, part D, 7 of the National Disaster Management law provides guidelines for managing avalanches and landslides. National Disaster Management Authority provides several guidelines for landslide management to be followed by the state/district of landslide-prone areas. The National Disaster Management Plan has formulated guidelines that include various administrative, scientific and planning apparatuses to address the different aspects of hazards and effectively reduce the losses from landslides. These apparatuses on the guidelines are described as nine elements for guiding the landslide-prone regions for risk reductions, and these are mentioned below:(i) landslide hazard, risk and vulnerability assessment; (ii) multi-hazard concept; (iii) remediation practice for landslides; (iv) early warning, monitoring and land development; (v) management of network and knowledge (IDKN); (vi) training and capacity building; (vii) public awareness and education on landslides; (viii) preparedness and response for emergencies; and (ix) enforcement of regulations.

As per the guideline, it has been mentioned that, like any other disaster, landslide hazards can be predicted, mapped out and restrained in a given condition, which comprises an aware community, the collaboration of detailed plans and the establishment of scientific research. The elements mentioned above provide more broad directives for prevention and mitigation. Analysis of the above nine elements brings to the comprehension of specifically restraining landslide hazards from becoming disastrous. The guidelines have given significance to the prevention and mitigation-centric (DRR) approach. An essential aspect of the guidelines is the inclusion of scientists, engineers, planners, local authorities and communities for prevention, mitigation, relief and rehabilitation. The National Disaster Management Response and paramilitary forces are assigned to respond to and rescue efforts to combat landslides (NDMA 2015). The elements under the law of disaster management signify that landslide disasters should be integrated into the multi-hazard disaster management plan at different levels for risk reduction, mitigation and response. The elements of landslide hazard, risk and vulnerability include the specific study of landslide areas and the development of inventories. It further gives directives for the demarcation of suspected hazard areas to know the status of landslide hazards and assess the resources as per the needs of the communities. The third element of the guideline focuses on building confidence among affected communities through landslide remediation strategies and mitigation technologies involving developing an early warning system and monitoring landslides. Capacity building and training must be given to geoscientist engineers and planners, public awareness programmes, audiovisual aid and media campaigns to be carried out at the community level. The compendium of laws for landslide management gives importance to enforcing compliance for assessing and mitigating hazards by establishing techno-legal mechanisms (NDMA 2015).

The framework for DRR at all levels, international, national, state and local, and the importance of community-based initiatives have been emphasised. It has been noted that while discussing disaster management at the local level, community-based disaster management cannot be alienated. Community-based disaster preparedness is a process that brings people together in a community to address disaster risk through collective action (NDMA 2013–14).

People's participation in community-based disaster preparedness is pertinent, as it will systematically mobilise a group of people to achieve resilience and safety from disaster and provide an opportunity for field practitioners to learn essential skills. Participants (youth leaders, volunteers, self-help groups and so on) will learn how to design and implement plans and programmes for building community capacity and reducing disaster risk. Participants will be able to learn about globally recognised projects and programmes on community-based disaster preparedness by participating in risk management planning. The course in disaster preparedness also deals with the issues of disaster mitigation from a developer perspective, adaptation of disaster preparedness practice, risk management plans, etc., with both government and non-governmental plans (NDMA 2013–14). The primary aim of community-based disaster preparedness is to reduce the vulnerability of the concerned community and strengthen its existing capacity to cope with disasters. The community-based disaster preparedness plans consider people's participation a necessary condition for disaster management (Perwaiz et al. 2008).

The community's involvement in the preparedness phase will increase the coordinated actions and help the community address the disaster issue collectively. Collective actions have yielded positive results, and they effectively lessen the impact of a disaster to a great extent.

The government of India and UNDP have reviewed different modes of community-based disaster preparedness conducted in the country. In some places, community-based disaster preparedness maintains disaster structure and has developed evacuation plans. The processes differ across the communities—some communities develop the plan and explain the components to the communities, while others prefer to develop plans involving the communities. However, these processes remain outside the government system, and there is an inherent danger of forgetting roles and responsibilities, particularly when they do not have to use the plans over time (Perwaiz et al. 2008).

20.2 Study Approach

This study looks at the local disaster management strategy in the specific context of landslides. It also discusses the applicability of disaster management strategies, concepts, plans and actions. It makes an effort to highlight their usefulness at the local level using the secondary data and empirical data gathered from the sample households from the two study areas.

The study areas were selected based on the severity of landslides from the state of Sikkim and the Gorkhaland Territorial Administration (GTA), consisting of Darjeeling and Kalimpong districts. The effectiveness of disaster management strategies that are implemented at the local level of this region is also discussed in this study (Fig. 20.1).

20.2.1 Mapping the Landslide Vulnerabilities

Two landslide zonation maps have been prepared based on information from the Inventory and GIS Mapping, Sikkim State Disaster Management Authority, Government of Sikkim. With the help of Arc GIS (10.1) version, vulnerable area plotting and map digitisation have been done. The latitudinal and longitudinal scale across the grids is 0.03° (2 min 30 s) (Fig. 20.2).

After the mapping exercise, four Gram Panchayats were purposively selected based on the severity of exposure to landslide disasters. With the help of the above

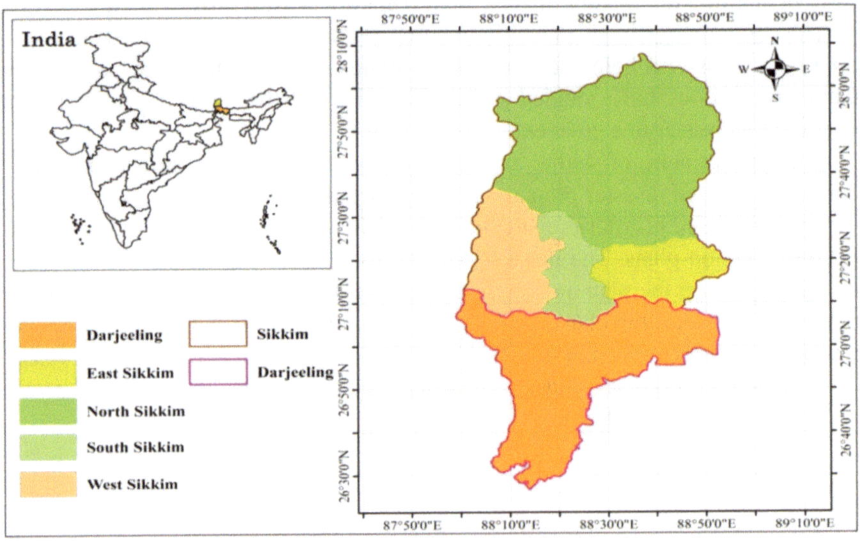

Fig. 20.1 Location of study areas of Sikkim-Darjeeling in the Himalayan region

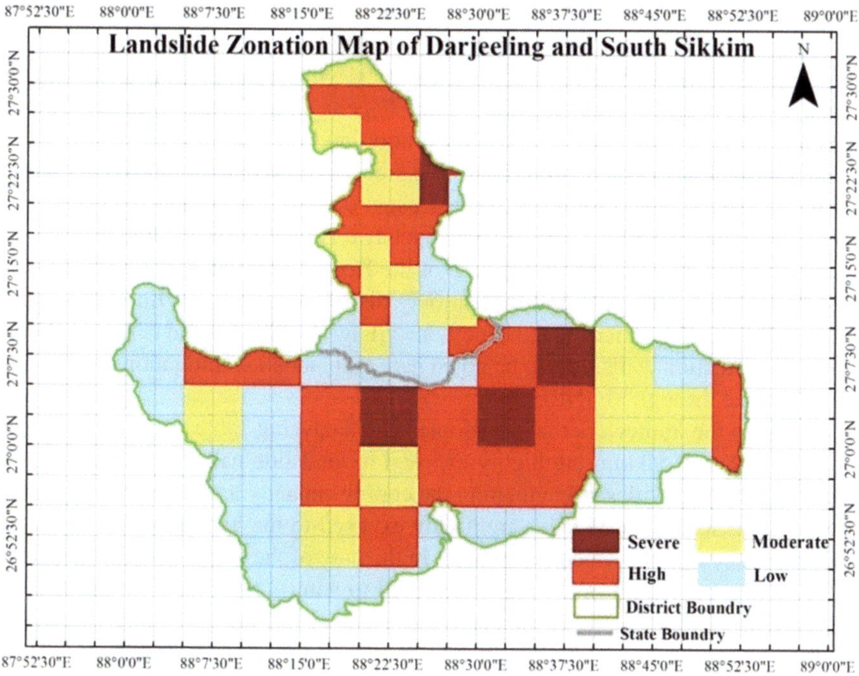

Fig. 20.2 Vulnerability to landslides in the two adjacent regions. (Source: Prepared by the author through Arc GIS)

map, four Gram Panchayats were selected—two from the high landslide zone and two lying in the low landslide zone. The two Gram Panchayats selected from South Sikkim are the following: (i) Lungchok-Kamarey, which falls under a low landslide zone, and (ii) Turung Mamring, which falls under a high landslide zone. The selected Gram Panchayats from Darjeeling are Rangli, which lies in the low landslide zone, and Sangse Gram Panchayat, lying in the high landslide zone. The forest and tea garden regions were excluded from the location, and we focused on the settled region. It is noted here that two Gram Panchayats lying in a high landslide zone, Turung Mamring (from South Sikkim) and Sangsay (Kalimpong), are called Red Zone. The Gram Panchayats under low landslide zone, Lungchok Kamaray (South Sikkim) and Rangli-Rongoliat (Darjeeling), are termed as Blue Zone. Three villages were randomly selected from each Gram Panchayat. The total number of households in these villages ranged from a low of 13 to a high of 24. Since the number of households in each village was lower than the initially decided number of 25, all households from the sample villages were surveyed. The total number of sample households is 254. In all the selected Gram Panchayats, an attempt was made to determine whether they had prepared any plans for disaster management in their jurisdictions. The representatives of all the Gram Panchayats informed that

Gram Panchayat Disaster Plans were not prepared and that they work under the direction of the District Disaster Management office.

Table 20.1 shows the activeness of the local government in different disaster management phases. In the Red Zone, Gram Panchayat of (Turung Mamring) South Sikkim, through MGNREGS, some prevention and mitigation activities are done mainly in the form of plantations and protection walls. In case of damage to houses during landslides, houses are being constructed through a rural housing scheme by the government of Sikkim. The village Panchayat is vital in forwarding the plea for grants to needy households. In the case of the Red Zone villages of Sangsay Gram Panchayat, GTA, the local government only distributes tarpaulins during the rainy season. The good practices in the Blue Zone villages of Sikkim and GTA are plantation and protection walls. It is to be noted here that landslide mitigation and prevention are done through MGNREGS.

Going by the framework that we adopted and analysing Table 20.1, the Red Zone villages of South Sikkim are highly exposed to landslide hazards and prevention, and mitigation activities strengthening the coping capacity are indeed being implemented. The Red Zone villages are highly exposed to the natural hazards of landslides, so they can be termed vulnerable. In the case of the Red Zone villages in the GTA region, the localities are highly exposed to natural hazards. Even then, these villages' prevention and mitigation activities are absent, leading to more vulnerability in these regions. As a result, in a similar setting, we have two categories of different vulnerable zones: highly vulnerable and vulnerable.

Similarly, in the case of Blue Zone villages, the exposure to natural hazards is less. In the Blue Zone villages of South Sikkim, the mitigation and prevention are well carried out, making the less hazardous region less vulnerable to landslides. Similarly, in less vulnerable villages of the GTA, mitigation activities are carried out, resulting in making the place less vulnerable. The primary factor behind Sikkim's performance compared to the GTA region, thus, can be attributed to the level of effectiveness of the local government. This finding also supports the study's

Table 20.1 Activities of the local government for DRR

Region	Prevention	Mitigation	Coping	Relief and rehabilitation
South Sikkim Red Zone	Plantations (MGNREGS), water harvesting tanks	Protection walls (MGNREGS)	–	House grants
GTA Red Zone	–		Tarpaulin	Tarpaulin
South Sikkim Blue Zone	Plantations (MGNREGS), rainwater harvesting	Protection walls (MGNREGS)		N.A
GTA Blue Zone		Protection wall, Jhora binding, small drainages (MGNREGS)		N.A

argument that in places where there is an effective PRI landslide, risk reduction will take place compared to those places where PRIs are dormant.

Table 20.2 shows the effectiveness of the local government's prevention and mitigation activities. In the Red Zone villages of Sangsay GTA, there are no prevention and mitigation activities through the local government. As a result, all the sample households believe that the local government is ineffective in reducing the risk of landslides. It is to be noted that there are no landslide risk reduction activities or practices at any level of government in the Sangsay, GP and GTA regions. Plantation work was done every Thursday by the Sri Satya Sai Organisation for a year in the villages of Sangsay through youth volunteers. The organisation has also constructed a Jeepable Road.

Meanwhile, in the Red Zone villages of Turung Mamring, South Sikkim, some plantation and protection wall activities are carried out through MGNREGS to protect the slopes. The participants don't consider these activities very effective, but 76% feel that the activities carried out through MGNREGS are less effective. The rest of the participants believe that the distributed species, such as Amliso (*Thysanolaena maxima*, broom plant), have weak roots and don't help prevent landslides. The participants believe protection walls are not provided to all households or are assigned biasedly. Therefore, 24% of the participants from the villages of Turung Mamring feel that the MGNREGS is ineffective in preventing and mitigating landslides.

In the Blue Zones, or the less vulnerable villages, the prevention and mitigation measures are well carried out compared to the Red Zone or highly vulnerable villages of both Sangsay GTA and Turung Mamring, South Sikkim. In Lunchuk-Kamerey, 91% of the sample households think plantation, rainwater harvesting and protection walls effectively prevent and mitigate landslides. In the case of less vulnerable villages of Rangli-Rongoliat, 31% of the participants said that the protective walls and drainage built-up are effective in mitigating landslides, whereas 21% of the participants believe that their sites are not being protected through MGNREGS, so they consider local government are not effective in mitigating or preventing landslides.

Our comprehension of the effectiveness of the DM plans is improved by the analysis above. Only MGNREGS's work can be utilised as a basis for preventative and mitigation strategies. As far as the governance barriers to successful DRR implementation are concerned, it is to be noted that the lack of a bottom-up approach leads to policy gaps and ineffective DRR situations. The exclusion of local or

Table 20.2 Percentages of participants on the effectiveness of the local government in reducing landslide problems

Region	Effective	Less effective	Not effective
South Sikkim Red Zone		76	24
GTA Red Zone			100
South Sikkim Blue Zone	91	7	2
GTA Blue Zone	31	58	21

community-level people from disaster management preparedness has created a chasm between the aspirations of the district administration and successful implementation. So, despite creating its disaster management plans and policies and being a founding signatory to the International Framework for Disaster Risk Reduction, what prevented those plans from being implemented locally?

20.3 Major Findings

This study identifies the following major categories of impediments to implementing DRR at the local level.

20.3.1 Non-formation of Disaster Management Committee

As per the suggestions in the National Disaster Management Plan through community-based disaster management, every village should have a Village Disaster Management Committee (VDMC) responsible for initiating disaster mitigation activities. This includes local elected representatives, NGOs, community-based organisations (CBOs), grassroots government, youth groups (e.g. NSS), etc. The initiatives mentioned above were missing in the villages surveyed, and committees were not formed. Even in the villages that are highly vulnerable to landslides of severe intensity, VDMCs are not set up. In Turung Mamring (South Sikkim), one to two people are taken for training, but the rest are unaware of it. According to sample households, NGOs are not actively involved in disaster mitigation or management. In the villages of the GTA region, the absence of local representatives, village-level volunteers or any training programmes has kept the community away from the purview of VDMCs.

20.3.2 Not Learning from the Past on Mitigation

There are no reviews or analyses of previous disasters by the local administration in the research area. According to the sample households from high-risk areas for landslides, they are well-informed about recent disasters that have affected their area and are knowledgeable about both large and small landslides in their communities. The comprehensive investigation of prior catastrophes on a common platform through community assemblies is still lacking. One of the key factors is the small population density; it takes 45 min to 1 h and 30 min to walk between each residence. It is necessary to review and analyse the location's previous disasters. This refers to prioritising disasters based on their frequency and predicted losses. The communities can evaluate the losses they suffered as a result of landslides and can discover the most effective techniques that have been used thus far (Bhagat and Devendra 2015).

20.3.3 Local Knowledge of Disaster Mitigation Is Not Used

In the case of the severe landslide region of our study area, the villagers can analyse the severity of landslides well. Still, the options left with them for good practices for mitigating landslides are minimal, which include making temporary drains through agricultural equipment, plantations, stone walls/protection walls, etc. These disaster management practices can be helpful and essential, but they cannot be relied upon wholly regarding life-saving strategies. From the surveyed households, we witnessed the early warning signs of landslides, viz. cracks in the walls and pillars, dislocation of windows and doors, etc. Participants are aware of these signs, but they consider their financial problems the major hindrance. They don't have enough funds to resettle in a safer location. Sangsay Gram Panchayat's (GTA) villages have no community shelter homes that can be used as safe shelter points during heavy rainy days. Therefore, despite being aware of the vulnerability of their houses, the villagers have no other option but to live with the risk of a disastrous event. The mapping of risk, vulnerabilities and coping capacity of the village is one of the most critical activities in the community-based disaster preparedness programme. It is straightforward to collect ground-level data. The mapping activity is mainly done through a Participatory Rural Appraisal (PRA) exercise. The villages were mapped under its jurisdiction in the Turung Mamring Gram Panchayat Unit, and the traces were seen on the wall of the Gram Panchayat office. In the case of GTA villages, no mapping activities were present. However, the map with the Gram Panchayat of Turung Mamring Gram Panchayat (South Sikkim) is fundamental, not necessarily the vulnerability or evacuation maps. The community is well aware of landslide disasters and their consequences due to previous landslides. However, certain state activities are not feasible for the fragile ecological zones, as per the respondents. The most common activities are the Hydel power dam across the River Teesta, underground tunnel construction for national security, etc. The other hindering factor in the region, particularly in Turung Mamring, is the road construction, which cuts the hills without sufficient protection walls. Road construction has led to the vulnerability of many houses. As per the research participants, it's the carelessness of the company named "Kamlesh". These activities have acted as some impediments to meeting the goals of disaster management plans and frameworks, which call for limiting intervention in fragile ecosystems. These factors made the work for the local government not feasible to implement inclusively.

20.4 Conclusion

The study provides an understanding of the disaster management plans and highlights the significance given by the global framework, the National Disaster Management Act, and plans for local-level implementation. The study examines the applicability of disaster management plans at the local level, specifically in the context of landslide disasters. While reviewing the Global Frameworks on DRR, the study finds that, while the first framework, the Yokohama Strategy on Disaster Risk Reduction, focused on the bottom-up approach, the later frameworks, the Hyogo and Sendai frameworks for Disaster Management, focused on the top approach of Disaster Management. The Sendai Framework advocates that the local community needs external support to understand and mitigate hazards. The study argues against this statement based on the facts acknowledged or widely accepted in academia, which state that local communities better understand disaster mitigation and have been cooperating with their environment. The study comprehends that local people have a good understanding of hazards that can turn into disastrous events. Still, the local community remains vulnerable due to a lack of finances to implement the mitigation measures.

Therefore, the intervention of the local government in landslide risk reduction can be an effective measure. The Disaster Management Act 2005 gives India a National Disaster Management Authority (NDMA), and through NDMA, National Disaster Management Plans are executed. The Disaster Management Act of 2005 lacks clarity on which specific functions are to be followed by local government for disaster risk reduction. As per the Act, the roles and functions entitled to local authorities are more post-disaster oriented. The local authorities must keep the resources readily available in case of any disaster and carry out relief and rehabilitation measures, etc. The core focus on post-disaster activities impedes achieving the aspirations of disaster risk reduction strategies. Under the Disaster Management Act 2005, the local authorities are given roles and responsibilities that are post-disaster oriented. Still, if the ethos of disaster management is to be achieved on a larger scale, the local authority should have a clear mitigation role for disaster risk reduction. In some cases, within a time frame, the Indian Disaster Management plans and policies don't seem to be made under the guidelines of the international framework, viz. the DM Act 2005 doesn't include the importance of local knowledge. In contrast, the Yokohama framework was advocating for the same. The lacuna is that the guidelines talk about the involvement of the community. Still, the plans don't give specific duties or roles to the local government, which is the base-level administrative unit for the community. The Government of India 2015 task force also discovered a lack of clarity in the Disaster Management Act 2005 regarding the local government's role, scope and function. The DM Act assigns four functions to the local authority: training the officers concerned, maintaining resources required for disaster relief, ensuring that construction is in line with the rules in the Act and undertaking rescue and relief activities. Among the four functions, three are based on post-disaster functions, and only one function (the third) mentions the local authority's prevention and mitigation role. The functions for prevention and mitigation are limited to ensuring that the construction activities of all projects adhere to the standards set by the district, state and national authorities. This ignores the disaster prevention and mitigation measures that can be vested in the local community if there is a disaster.

Risk reduction measures are to be viewed holistically and have not gained significance. The study further elaborates on the NDMP 2019 on local authorities and landslide management. The chapter emphasises that due to a lack of local community involvement, including their ideas and knowledge, awareness level and perspectives on mitigation in the planning process, as well as the limited role of the local government in prevention and mitigation measures, local communities are having difficulty emerging from highly vulnerable situations. The field evidence corroborates the chapter's argument or the arguments mentioned above.

In the survey, the good practices for LRR are insignificant in the case of the GTA region, where the PRIs are at a dormant stage. In the case of Sikkim, the LRR activities are slightly better but still demand more attention for prevention and mitigation measures. The participants from the Red Zone villages of the GTA region believe that local government or any other level of government is doing no good for landslide risk reduction. The information collected from the fieldwork also brings us to our understanding that the aspirations of an international framework for disaster risk reduction at the local level, the Indian Disaster Management Plans, including the Village Disaster Management Plans and Community-Based Disaster Preparedness, are yet to gain significance in the Sikkim-Darjeeling Himalayan region, and more specifically in the GTA region. In the GTA region, disaster management activities are more post-disaster relief oriented, whereas the decentralised government is almost dormant. The bottom-up approach is only when it comes to resources. The local authority can request resources if the local resources are not sufficient. The local government should deal with local disasters. In the event of a local disaster, the local government should be responsible for taking action to prevent and mitigate it. It should not be confined to the duty of distributing relief. One of the significant findings from this chapter is the role of MGNREGS in landslide risk reduction. The MGNREGS is doing well in the prevention and mitigation of landslide management. The disaster risk reduction at the village level has no significance even after two decades of the existence of the National Village Disaster Management Plan. Therefore, despite its shortcomings, disaster risk reduction activities are only done through MGNREGS.

References

Bhagat R, Devendra B (2015) Flood study of Wainganga River in Maharashtra using GIS and remote sensing techniques. Int J Sci Res 782–785

Efiong J, Eni DI, Obiefuna JN, Etu SJ (2021) Geospatial modelling of landslide susceptibility in Cross River State of Nigeria. Scientific African 14:e01032. https://doi.org/10.1016/j.sciaf.2021.e01032

Government of India (2019) Mainstreaming disaster risk reduction and climate change adaptation in the national flagship programmes

Maes J, Parra C, Mertens K, Bwambale B, Jacobs L, Poesen J, Dewitte O, Vranken L (2018) Questioning network governance for disaster risk management: lessons learnt from landslide risk management in Uganda. Environ Sci Policy 85:163–171. https://doi.org/10.1016/j.envsci.2018.04.002

Messer N (2003) The role of local institutions and their interaction in disaster risk mitigation. FAO. http://www.fao.org/3/a-ad710e.pdf

Ministry of Home Affairs Government of India Mainstream (2013) A review of the disaster management act, 2005. Task Force LII(42). Accessed 24 Nov 2017

National Disaster Management Authority (NDMA) (2013–14) Annual report 2013–14. Government of India

National Disaster Management Authority (NDMA) (2014) Annual report 2014–15. Government of India

National Disaster Management Authority (NDMA) (2015) Compendium of laws on disaster management. Government of India, January

National Disaster Management Authority (NDMA) (2016) National disaster management plan. Government of India, May

Pal I, Shaw R (Eds.) (2018) Disaster risk governance in India and cross-cutting issues (1st ed.). Springer Singapore. https://doi.org/10.1007/978-981-10-3310-0

Perwaiz A, Dey B, Trebucq D, Padmanabhan G, Stephan I, Mohapatra K, Sengupta R, Chaudhary S (2008) Local level risk management Indian experience. An initiative under GOI and UNDP. National Disaster Management Division. http://www.undp.org/content/dam/india/docs/llrm.pdf. Accessed 20 Nov 2017

Scott Z, Tarazona M (2011) Study on disaster risk reduction, decentralization and political economy. Global Assessment report on disaster risk reduction. United Nations

Taori K (2005) Disaster management through Panchayati Raj. Concept Publishing Company, New Delhi

Tozier de la Poterie A, Baudoin MA (2015) From Yokohama to Sendai: approaches to participation in international disaster risk reduction frameworks. Int J of Disaster Risk Sci 6(2):128–139. https://doi.org/10.1007/s13753-015-0053-6

UNISDR (2004) Living with Risk: A global review of disaster reduction initiatives (Vol 1). Geneva: UNISDR

UNISDR (United Nations International Strategy for Disaster Reduction) (2005) Hyogo framework for Action 2005–2015: building the resilience of nations and communities to disasters. Retrieved from. http://www.unisdr.org/files/1037_hyogoframeworkforactionenglish.pdf

UNISDR (United Nations International Strategy for Disaster Reduction) (2011) The Hyogo Framework for Action 2005–2015 mid-term review. Retrieved from http://www.preventionweb.net/files/18197_midterm.pdf

United Nations Office for Disaster Risk Reduction (UNISDRR) (2015) Global Assessment Report on Disaster Risk Reduction 2015: Making development sustainable – the future of disaster risk management. Geneva: UNISDRR. https://doi.org/10.18356/919076d9-en

Zimmermann M, Keiler M (2015) International frameworks for disaster risk reduction: Useful guidance for sustainable mountain development? Mt Res Dev 35(2):195–202. https://doi.org/10.1659/MRD-JOURNAL-D-15-00006.1

Chapter 21
Advancing Ecosystem Recovery: Cutting-Edge Research in Restoration Science

D. R. K. Saikanth, P. Prashanth, S. Kaviraju, M. Jaganmohan Reddy, P. P. Murugan, Shaik N. Meera, D. Uttej, C. Karpagam, P. Mooventhan, Vijay Kumar, M. R. Reddy, and V. Guhan

Abstract Ecosystem restoration has emerged as a crucial strategy to combat pressing global environmental challenges such as biodiversity loss, habitat degradation, and climate change. This exploration delves into the latest advancements in restoration science, highlighting key themes, including cutting-edge research, novel techniques, and integrating social and ecological perspectives. While conventional methods have shown success locally, the growing complexity and magnitude of ecosystem degradation call for transformative strategies and the adoption of technological innovations. Emerging tools like artificial intelligence, machine learning,

---

D. R. K. Saikanth (✉) · P. Prashanth
Agricultural Extension, Jayashankar Telangana Agricultural University (PJTAU), Hyderabad, Telangana, India

S. Kaviraju
Department of Agricultural Economics, Hyderabad, Telangana, India

M. J. Reddy
Extension Education Institute (EEI), Hyderabad, Telangana, India

P. P. Murugan
Tamil Nadu Agricultural University TNAU, Coimbatore, India

S. N. Meera · D. Uttej
ICAR-Agricultural Technology Application Research Institute Zone-X, Hyderabad, Telangana, India

C. Karpagam
ICAR-National Research Centre for Banana, Ariyavoor, Tamil Nadu, India

P. Mooventhan
ICAR-National Institute of Biotic Stress Management, Baronda, Chhattisgarh, India

V. Kumar · M. R. Reddy
ICAR-Directorate of Poultry Research, Hyderabad, Telangana, India

V. Guhan
Meteorological Centre, Hyderabad, Telangana, India

Y. D. I. Khan et al. (eds.), *Ecosystem-based Approaches for Resilience Building in Himalayan Landscapes*, Disaster Resilience and Green Growth,
https://doi.org/10.1007/978-981-95-2007-7_21

and bioengineering are revolutionising the planning, execution, and monitoring of restoration projects, enabling more precise and scalable solutions. Concurrently, incorporating Traditional Ecological Knowledge (TEK) and fostering community participation ensures that restoration efforts are culturally relevant and inclusive, enhancing their effectiveness and sustainability. However, significant challenges persist, such as existing knowledge gaps, scaling restoration efforts to meet global targets and finding a balance between restoration, conservation, and developmental objectives. Addressing these hurdles will require a collaborative, interdisciplinary approach, adaptive management strategies, and innovative policies that align restoration initiatives with broader sustainable development goals. Moving forward, the field of restoration science must expand its focus to include diverse ecosystems such as marine, coastal, and arid landscapes while refining long-term monitoring techniques to better track outcomes. The need for holistic frameworks that integrate socio-economic and ecological benefits to maximise restoration impact is paramount. Through real-world examples from forests, grasslands, coastal regions, and urban areas, it is evident that the success of large-scale restoration endeavours depends on fostering resilient ecosystems capable of adapting to changing environmental conditions. In conclusion, the future of ecosystem restoration lies in leveraging scientific advancements and inclusive methodologies to reverse ecological degradation and promote a harmonious coexistence between human societies and the natural world.

Keywords Restoration · Biodiversity · Ecosystems · Innovation · Sustainability · Community · Conservation

21.1 Introduction

21.1.1 Ecosystem Restoration

Ecosystem restoration is a scientific and practical approach to reviving and rehabilitating degraded, damaged, or destroyed ecosystems to their original structure, function, and biodiversity (Cheng and Li 2024). It encompasses various activities, from reforestation and wetland restoration to coastal recovery and urban greening. The practice of ecosystem restoration is not just limited to replanting native species but also involves re-establishing ecological processes and enhancing the resilience of these systems to environmental changes. Effective restoration practices can reverse the impacts of deforestation, soil erosion, habitat fragmentation, and species loss, thereby supporting ecological integrity and human well-being. Restoration efforts are rooted in understanding the complex interactions within ecosystems, including the relationships between flora, fauna, soil, water, and climate (Alexander et al. 2016; Gann et al. 2019). By restoring these interactions, practitioners aim to rebuild ecosystems that are self-sustaining and capable of supporting diverse biological communities.

Ecosystem restoration has gained global prominence as a key strategy for addressing environmental degradation and climate change, recognised by initiatives such as the United Nations Decade on Ecosystem Restoration (2021–2030), which emphasises the critical need for restoring ecosystems to achieve sustainable development goals (CEM 2021). With growing recognition of its potential, ecosystem restoration has evolved beyond traditional approaches to incorporate advanced scientific research and innovative techniques. The field now encompasses various interdisciplinary methods, such as restoration genomics, climate-smart restoration strategies, and integrating indigenous knowledge systems (Perring et al. 2015; Jones 2017).

21.1.2 Importance of Cutting-Edge Research in Restoration Efforts

Cutting-edge research is essential to the success of ecosystem restoration initiatives. While effective in specific contexts, traditional restoration methods often fail to address the full complexity of ecological systems or adapt to rapidly changing environmental conditions (Palmer et al. 2016). This is where contemporary research plays a transformative role by providing new insights into ecosystem dynamics, enabling precise interventions, and expanding the toolkit available to restoration practitioners. One of the most significant contributions of modern research is the ability to use novel technologies and methodologies to better understand ecological processes at multiple scales. Advances in restoration genomics help scientists identify genetic bottlenecks in species populations and develop strategies to enhance genetic diversity, thereby boosting restored populations' resilience and long-term viability. Similarly, innovations in soil microbiome research reveal the critical role of microbial communities in nutrient cycling and plant health, guiding more effective soil restoration practices (Breed et al. 2019; Theissinger et al. 2023).

Cutting-edge research enables the development of climate-smart restoration strategies, crucial for ensuring that restored ecosystems can withstand future climatic changes. This includes selecting species and designing interventions based on predictive models of climate scenarios. Technological tools such as remote sensing, drone-based monitoring, and artificial intelligence (AI) have revolutionised how restoration projects are planned, implemented, and monitored, offering greater precision and scalability. Beyond ecological outcomes, cutting-edge research also addresses the socio-economic dimensions of restoration, exploring how restoration projects can deliver co-benefits for local communities, improve livelihoods, and promote social equity. This holistic perspective is crucial, ensuring restoration is ecologically effective and socially sustainable. Thus, integrating cutting-edge research into restoration practices is not merely an enhancement but a necessity for overcoming current and future environmental challenges (Munawar et al. 2021).

21.2 Foundations of Ecosystem Restoration Science

21.2.1 History and Evolution of Restoration Science

The ecosystem restoration field is rooted in early conservation movements that recognised the need to repair damaged natural landscapes (Macdonald and King 2018). Restoration as a scientific discipline began to take shape in the early twentieth century, with foundational work by ecologists such as Aldo Leopold and George Perkins Marsh, who emphasised the importance of land health and ecological balance. During this period, restoration efforts were primarily focused on reforestation and soil conservation, aiming to counteract the negative impacts of deforestation and agricultural expansion (Gardner 2013; Rodriguez Franco and Conje 2023). The formalisation of restoration ecology as a scientific discipline gained momentum in the late twentieth century, especially with the establishment of the Society for Ecological Restoration (SER) in 1987. SER provided a platform for collaboration between scientists, practitioners, and policymakers and set the stage for developing key principles and methodologies for ecological restoration. Over the following decades, the scope of restoration science broadened to include a variety of ecosystems, such as wetlands, grasslands, and coastal habitats. It began incorporating multidisciplinary perspectives from fields like hydrology, soil science, and landscape ecology. The evolution of restoration science has been marked by a shift from localised, site-specific projects to large-scale, landscape-level approaches. Early restoration efforts often focused on individual species or small habitat patches, with success measured by short-term metrics like plant survival and vegetation cover. Today, restoration science emphasises holistic strategies considering entire ecosystems, their complex interactions, and long-term sustainability. This evolution has been driven by an increased understanding of ecological processes, advancements in scientific research, and the growing urgency to address global environmental challenges such as climate change and biodiversity loss.

The emergence of new research tools and methodologies, such as genomics, remote sensing, and predictive modelling, has further transformed restoration science (Cavender-Bares et al. 2022). These innovations enable a more nuanced understanding of ecosystem dynamics and allow for restoration interventions that are more effective and more adaptive to future changes. As the field continues to mature, restoration science is poised to play a central role in global conservation efforts, particularly as part of initiatives like the United Nations Decade on Ecosystem Restoration, which aims to restore 350 million hectares of degraded landscapes by 2030.

21.2.2 Core Principles of Ecosystem-Based Restoration

Ecosystem-based restoration is guided by several core principles emphasising ecological integrity, biodiversity, and resilience. These principles are the foundation for designing, implementing, and evaluating restoration projects, ensuring that interventions repair visible damage and restore underlying ecological processes and functions.

21.2.2.1 Ecological Integrity

Ecological integrity refers to the wholeness and functioning of an ecosystem, encompassing its native species composition, structural complexity, and natural processes. In restoration practice, ecological integrity is achieved by re-establishing species assemblages and environmental conditions that closely resemble those of reference ecosystems, such as undisturbed ecosystems of the same type that serve as benchmarks for restoration goals. Restoration efforts prioritising ecological integrity aim to create self-sustaining systems that maintain their structure and function over time, even in the face of disturbances. Achieving ecological integrity requires considering multiple ecosystem components, including vegetation structure, soil health, hydrological regimes, and the interactions between biotic and abiotic factors. For example, forest restoration might involve not only planting native tree species but also restoring soil microbial communities, hydrological patterns, and the connectivity between habitats. By focusing on ecological integrity, restoration projects enhance the overall health and stability of ecosystems, making them more capable of providing essential ecosystem services, such as water filtration, carbon storage, and habitat for wildlife.

21.2.2.2 Biodiversity Preservation

Biodiversity preservation is a fundamental principle of ecosystem-based restoration, as biological diversity is crucial for the resilience and functionality of ecosystems. Restoration projects prioritising biodiversity seek to reintroduce various native species, plants and animals, thereby enhancing genetic diversity and ecosystem complexity. High levels of biodiversity are associated with greater productivity, stability, and resistance to invasive species and environmental stressors. In practice, biodiversity preservation involves strategies such as planting diverse species assemblages, re-establishing keystone species, and creating habitat features that support a wide range of organisms (Nautiyal et al. 2015). It also includes addressing threats to biodiversity, such as invasive species and habitat fragmentation, which can undermine restoration efforts. By fostering diverse biological communities, restoration projects contribute to ecosystems' long-term stability and ability to adapt to changing conditions.

21.2.2.3 Resilience and Adaptive Capacity

Resilience refers to an ecosystem's ability to absorb disturbances and recover its essential structure and function. At the same time, adaptive capacity is its potential to adjust to changing conditions, such as climate change and novel stressors (Buma and Schultz 2020). In restoration science, building resilience and adaptive capacity is crucial for ensuring the sustainability of restored ecosystems. This principle is increasingly essential in global environmental changes, where ecosystems face unprecedented shifts in temperature, precipitation patterns, and human pressures. Enhancing resilience includes restoring ecological processes (e.g. fire regimes, hydrological flows), promoting genetic diversity, and creating buffer zones that protect core habitats from external pressures. Adaptive management involves monitoring and adjusting restoration strategies in response to new information and changing conditions and is a key tool for building resilience. By incorporating resilience and adaptive capacity into restoration planning, practitioners can create more stable ecosystems that can evolve in response to future changes.

21.2.3 Role of Restoration Science in Addressing Environmental Challenges

Ecosystem restoration is increasingly recognised as a powerful tool for addressing global environmental challenges (Gann et al. 2019). Restoration science can contribute to climate change mitigation and adaptation by repairing degraded ecosystems, combat habitat loss, and improving soil and water health. These benefits are particularly significant in the context of the Anthropocene, where human activities have led to widespread ecological degradation and biodiversity loss.

21.2.3.1 Climate Change Adaptation

Restoration science plays a critical role in enhancing the capacity of ecosystems to adapt to climate change. By restoring native vegetation and ecological processes, restoration projects can increase ecosystem resilience to extreme weather events, such as floods, droughts, and storms. For example, restoring coastal wetlands and mangroves can provide natural buffers against storm surges and rising sea levels. Reforesting upland areas can reduce the risk of soil erosion and landslides during heavy rainfall. Climate-adaptive restoration strategies often involve selecting plant and animal species likely to thrive under future climate scenarios, promoting genetic diversity, and creating migration corridors that facilitate species movement in response to changing environmental conditions. Such strategies ensure that restored ecosystems are better equipped to cope with future changes, thereby protecting biodiversity and maintaining ecosystem services in a warming world.

21.2.3.2 Habitat Loss and Fragmentation

Habitat loss and fragmentation are major drivers of biodiversity decline and ecosystem degradation (Hald-Mortensen 2023). Restoration science addresses these issues by reconnecting fragmented landscapes, enhancing habitat quality, and re-establishing ecological corridors that support species movement and genetic exchange. Forest and grassland restoration projects can help create contiguous habitats supporting more significant, viable wildlife populations. In highly fragmented landscapes, restoration efforts may focus on creating stepping-stone habitats or enhancing the quality of matrix habitats—the areas between core natural habitats—to facilitate the movement and survival of species. This approach not only restores habitat but also mitigates the negative impacts of fragmentation, such as inbreeding and local extinctions, thereby contributing to broader conservation goals.

21.2.3.3 Water and Soil Degradation

Degradation of water and soil resources is a critical challenge for ecosystems worldwide, leading to reduced agricultural productivity, loss of biodiversity, and increased vulnerability to environmental stressors. Restoration science addresses these issues by implementing techniques that restore hydrological function and soil health. For example, wetland restoration can improve water quality by filtering pollutants and regulating water flow, while reforestation and agroforestry practices can enhance soil structure, prevent erosion, and increase water infiltration. Soil restoration often involves techniques such as the reintroduction of native vegetation, organic amendments, and the restoration of soil microbial communities. These interventions promote soil fertility and stability, making the land more productive and less prone to degradation. By focusing on restoring water and soil systems, restoration science supports sustainable land management and contributes to the overall health and resilience of ecosystems.

21.3 Cutting-Edge Research Themes in Ecosystem Restoration

21.3.1 Restoration Ecology and Its Innovations

As a scientific discipline, restoration ecology focuses on the theoretical and practical aspects of returning damaged ecosystems to their pre-degraded states. It is the backbone of ecosystem restoration, integrating ecological principles with on-the-ground management strategies (Stanturf and Mansourian 2020). The field has seen rapid advancements driven by new technologies and a deeper understanding of ecological dynamics. Innovations in restoration ecology now encompass novel approaches to species reintroduction and sophisticated techniques to stabilise and bolster populations of plants and animals (Hobbs and Harries 2001; Perring et al. 2015).

21.3.1.1 Novel Approaches to Species Reintroduction

Species reintroduction is a central component of restoration projects, aiming to re-establish populations of native species that have been lost or severely diminished due to human activities. Traditional reintroduction methods often relied on a "single-species" approach, focusing on one species at a time. Recent research has highlighted the limitations of this method, prompting the development of more holistic and dynamic strategies. One such approach is "multispecies reintroduction", where several interdependent species are reintroduced simultaneously to rebuild community structures and ecological interactions. The concept of "assisted migration" has emerged as a cutting-edge strategy for species reintroduction in the face of climate change. Assisted migration involves relocating species to areas outside their historical range where future climate conditions are expected to be more suitable. This proactive approach can prevent species from becoming stranded in no longer viable habitats, ensuring their survival as environmental conditions change. For example, assisted migration has been used to help certain tree species in North America move to higher elevations or latitudes as temperatures rise. Another innovation is the use of captive breeding and release programmes that incorporate advanced genetic management. This involves selecting individuals for reintroduction based on their genetic profiles to maximise genetic diversity and adaptability. Such efforts are being employed in species recovery programmes for endangered animals like the California condor and the black-footed ferret, where maintaining genetic health is crucial for population stability.

21.3.1.2 Techniques for Enhancing Plant and Animal Population Stability

Achieving stable, self-sustaining populations is a key goal in restoration ecology. For plant populations, techniques such as seed enhancement—coating seeds with nutrients, symbiotic fungi, or protective materials—have shown promise in increasing germination rates and early establishment. Another innovative method is using "nurse plants" or facilitation techniques, where hardy, early-successional plants are used to create favourable microenvironments that support establishing more sensitive target species. In animal restoration, population stability is enhanced through habitat enrichment and management of predator-prey dynamics. One example is using "soft-release" techniques, where animals are acclimated to the release site in a controlled environment before full release, which reduces stress and improves survival rates. New tracking technologies, such as GPS collars and automated camera traps, allow for close monitoring of reintroduced populations, enabling adaptive management and timely interventions to support population stability.

21.3.2 Restoration Genomics

Restoration genomics is an emerging subfield that leverages genetic and genomic tools to inform and enhance restoration practices. It addresses critical questions about the genetic health of populations, inbreeding, and the adaptation of species to changing environments. By incorporating genetic insights, restoration genomics helps ensure that restored populations are resilient and capable of long-term survival.

21.3.2.1 Genetic Tools for Biodiversity Conservation

Genetic tools like whole-genome sequencing and genetic barcoding transform how conservationists approach species reintroductions and habitat restoration (Neaves et al. 2022). These tools allow researchers to assess genetic diversity within and between populations, identify inbreeding risks, and select individuals that maximise genetic variability. For example, in restoring wild populations of endangered species like the Florida panther, genetic tools have been used to introduce individuals from related populations, thereby increasing genetic diversity and reducing the adverse effects of inbreeding depression. Genetic techniques detect cryptic species that may not be distinguishable based on morphology alone, ensuring that restoration efforts target the right species and genetic lineages accurately. This is particularly important in ecosystems where closely related species occupy distinct ecological niches and play different roles in maintaining ecosystem functions.

21.3.2.2 Leveraging Genetic Data for Population Health and Resilience

Genomic data can inform the selection of individuals for reintroduction or breeding programmes based on traits linked to climate resilience, disease resistance, or other adaptive characteristics. By incorporating these data, restoration practitioners can enhance the adaptive capacity of restored populations, making them more robust in the face of environmental change. This approach is known as "adaptive genetic management"; it is being used to guide restoration in species like the American chestnut, where breeding for disease resistance is a primary focus. Another application is using environmental DNA (eDNA) to monitor biodiversity and assess the success of restoration projects. eDNA involves collecting genetic material from environmental samples, such as soil and water, to detect the presence of multiple species. This non-invasive technique provides a powerful tool for tracking species recovery and evaluating the health of restored ecosystems.

21.3.3 *Advances in Soil Microbiome Research*

The soil microbiome of microorganisms living in the soil plays a crucial role in ecosystem functioning, influencing nutrient cycling, plant health, and soil structure (Hartmann and Six 2023). Recent advances in soil microbiome research reveal how microbial communities can be harnessed to support ecosystem restoration.

21.3.3.1 Role of Microbial Communities in Soil Health

Microbial communities are essential for maintaining soil fertility and supporting plant growth. Beneficial microbes, such as mycorrhizal fungi and nitrogen-fixing bacteria, form symbiotic relationships with plants, enhancing nutrient uptake and improving plant resilience to stress. Re-establishing these microbial communities in restoration projects is as essential as replanting native vegetation, as a healthy soil microbiome accelerates plant establishment and ecosystem recovery. Researchers are now using DNA sequencing to profile soil microbial communities and identify key microbial taxa associated with healthy, functioning ecosystems. By understanding these relationships, practitioners can develop targeted strategies to restore the soil microbiome, such as inoculating soils with beneficial microbes or managing soil conditions to favour the growth of native microbial communities.

21.3.3.2 Techniques for Restoring Soil Function and Structure

Restoration of soil function often involves physical, chemical, and biological interventions. Biochar, organic amendments, and green manure can improve soil structure, increase water retention, and promote microbial activity. Researchers are experimenting with "microbial transplants", where soil from healthy ecosystems is transferred to degraded areas to jump-start the recovery of microbial communities. These techniques are applied in projects ranging from agricultural land restoration to rehabilitating post-mining landscapes. By restoring soil health, these interventions create a foundation for long-term ecosystem recovery, supporting plant establishment and increasing resilience to environmental stressors (Goswami et al. 2023).

21.3.4 *Remote Sensing and Spatial Analysis*

Technological advancements in remote sensing and spatial analysis have revolutionised restoration projects' planning, monitoring, and evaluation (Camarretta et al. 2020). These tools provide critical insights into landscape dynamics, vegetation health, and habitat quality, enabling more precise and large-scale restoration efforts.

21.3.4.1 Using Drones and Satellite Technology for Monitoring

Drones and satellites equipped with high-resolution cameras and sensors are now widely used in restoration projects to collect data on vegetation cover, species distribution, and habitat conditions. Drones, in particular, are valuable for surveying difficult-to-access areas and creating detailed maps of small-scale restoration sites. On the other hand, satellite imagery provides a broader view, allowing for monitoring landscape-scale changes over time.

These technologies enable the detection of subtle changes in vegetation health, soil moisture, and other indicators critical for assessing the progress of restoration efforts. For example, spectral analysis from satellite imagery can detect stress in plant communities before visible signs appear, providing an early warning system for restoration practitioners.

21.3.4.2 GIS-Based Tools for Mapping and Planning Restoration Projects

Geographic Information Systems (GIS) are powerful tools for mapping and analysing spatial data, helping to identify priority areas for restoration and model potential outcomes. GIS-based tools can integrate a wide range of data, such as topography, land use, hydrology, and species distribution, allowing practitioners to develop site-specific restoration plans that optimise resource use and ecological benefits. In addition, GIS models can simulate different restoration scenarios, helping stakeholders visualise the potential impacts of various strategies and make informed decisions. These tools are increasingly used to plan large-scale restoration initiatives, such as watershed rehabilitation and reforestation projects, where spatial complexity and multiple land uses must be considered.

21.3.5 Climate-Smart Restoration Strategies

As climate change reshapes ecosystems worldwide, restoration strategies must be designed with future conditions in mind. Climate-smart restoration involves anticipating and planning for these changes, ensuring restored ecosystems are resilient to future climate scenarios.

21.3.5.1 Designing Restoration Projects for Climate Resilience

Climate-smart restoration begins with selecting species and communities expected to thrive under future climate conditions. This might involve planting drought-tolerant species in areas predicted to experience reduced rainfall or selecting tree species with wider climate tolerances for reforestation projects. Climate resilience can also be enhanced by promoting genetic diversity and using adaptive management techniques that allow for adjustments as conditions change.

21.3.5.2 Adaptive Management in a Changing Climate

Adaptive management is a key component of climate-smart restoration, involving continuous monitoring and the flexibility to change strategies based on new information. This approach acknowledges that restoration projects must be dynamic, responding to unexpected outcomes and changing environmental conditions. By building adaptability into restoration planning, climate-smart strategies ensure that restored ecosystems can remain functional and resilient in the face of climate change (Table 21.1).

21.4 Innovative Restoration Techniques

21.4.1 Assisted Migration and Translocation

Assisted migration and translocation are emerging strategies used in restoration to help species adapt to changing environmental conditions. These techniques involve moving species, such as seeds, young plants, or animals, from their current habitats to new locations where conditions are more favourable due to climate change or habitat degradation in their native range.

21.4.1.1 Concept and Applications

Assisted migration, also known as managed relocation, aims to proactively address the mismatches between species' natural ranges and shifting environmental conditions caused by climate change. This technique involves moving species to areas outside their historical range that are expected to be more suitable for their survival in the future. For example, certain tree species in North America are being planted at higher altitudes or latitudes to ensure their long-term viability as temperatures rise. Similarly, plants and small mammals are being relocated to cooler habitats in alpine ecosystems as their native ranges warm. Translocation is often used to restore species to areas locally extirpated due to habitat destruction, overexploitation, or other human impacts. This method has been successfully used to re-establish populations of threatened species, such as the red wolf in the United States and the western swamp tortoise in Australia. Translocation is typically applied in the context of endangered species recovery, where establishing new populations in suitable habitats can protect against further losses.

Table 21.1 Cutting-edge research themes in ecosystem restoration

Theme	Description	Research focus areas	Emerging trends
Climate resilience and adaptation	Strategies to improve ecosystem resilience against climate change impacts	1. Climate adaptation frameworks 2. Species migration patterns	Nature-based solutions Climate-smart restoration
Biodiversity recovery	Techniques for enhancing species diversity and ecosystem complexity	1. Genetic diversity restoration 2. Rewilding approaches	Assisted migration Habitat connectivity
Restoration ecology	Understanding ecological processes for effective restoration of degraded areas	1. Soil health and microbiome 2. Ecological succession	Microbiome manipulation Mycorrhizal network research
Socio-ecological integration	Incorporating human dimensions and local knowledge into restoration practices	1. Community-based restoration 2. Indigenous knowledge systems	Participatory governance Co-management models
Technological innovations	Application of emerging technologies like AI, drones, and remote sensing for ecosystem assessment	1. AI for biodiversity monitoring 2. Drone-assisted planting	Predictive modelling Precision restoration techniques
Urban ecosystem restoration	Rehabilitation of green spaces and biodiversity in urban areas to enhance livability and ecosystem services	1. Green infrastructure 2. Urban wildlife corridors	Urban rewilding Green roofs and vertical ecosystems
Ecosystem function and services	Restoration aimed at reviving ecosystem functions such as carbon storage, water purification, and pollination	1. Ecosystem service valuation 2. Functional trait analysis	Ecosystem service quantification Multifunctional landscapes
Restoration genomics	Use genetic tools to aid restoration by selecting species and traits resilient to future conditions	1. Genomic selection for resilience 2. Conservation genomics	Genetic rescue Adaptive genetic restoration
Novel ecosystems	Understanding and managing ecosystems that have been significantly altered may not return to historical states	1. Hybrid ecosystems 2. Novel management strategies	Ecological engineering Acceptance of non-native species

21.4.1.2 Risks and Ethical Considerations

Despite their potential benefits, assisted migration and translocation have significant risks and ethical challenges. Moving species outside their native range can disrupt local ecosystems, introduce new diseases, or even lead to invasive behaviours in the new environment. These unintended consequences can negatively impact the translocated and native species in the receiving ecosystem. Ethical considerations also arise around the interventionist nature of these techniques. Critics argue that assisted migration represents a human-centric approach to nature that manipulates species distributions in ways that may have unforeseen ecological ramifications. As a result, any decision to employ assisted migration or translocation must be grounded in rigorous ecological assessments, stakeholder engagement, and careful consideration of the trade-offs between action and inaction.

21.4.2 Rewilding and Trophic Reintroduction

Rewilding is a restoration strategy focused on reintroducing missing species, particularly apex predators and keystone species, to restore natural ecological processes and promote biodiversity. By reviving top-down trophic interactions, rewilding aims to re-establish the balance and functionality of ecosystems that human activities have disrupted.

21.4.2.1 Reviving Functional Ecosystems Through Top-Down Processes

Rewilding often involves reintroducing large carnivores or herbivores that play critical roles in shaping ecosystems through their interactions with other species. For example, the reintroduction of wolves to Yellowstone National Park is a well-known rewilding success story. By reintroducing this apex predator, ecologists could control overpopulated elk herds, reducing overgrazing and allowing riparian vegetation to recover. This cascade of effects, known as a trophic cascade, improved habitat quality for a range of other species, including birds and beavers, and helped restore the ecological health of the entire region. In Europe, the rewilding of large herbivores such as European bison and wild horses has been used to restore grassland ecosystems. These animals maintain open habitats by grazing on shrubs and young trees, preventing the spread of forests into traditionally open landscapes. Such trophic reintroduction efforts highlight the importance of reinstating key species to achieve ecological balance and support biodiversity.

21.4.2.2 Case Studies and Successful Examples

Successful rewilding projects include reintroducing sea otters along the Pacific coast of North America, which has led to the recovery of kelp forests by controlling sea urchin populations. In Africa, reintroducing rhinos and elephants to private reserves has restored traditional grazing patterns and promoted a more diverse and resilient savanna ecosystem. Rewilding is also being applied in marine ecosystems, where the reintroduction of oysters and other filter-feeding shellfish has improved water quality and enhanced the recovery of degraded estuaries. Each of these case studies demonstrates how the strategic reintroduction of key species can drive large-scale ecological restoration and support the recovery of entire ecosystems.

21.4.3 Nature-Based Solutions for Water and Soil Management

Nature-based solutions use natural processes and ecosystems to address environmental challenges, such as water management, soil erosion, and climate adaptation. These solutions often involve restoring or mimicking natural systems to provide multiple benefits, including improved biodiversity, enhanced ecosystem services, and increased resilience to climate change.

21.4.3.1 Techniques for Erosion Control and Water Retention

Techniques for controlling soil erosion and retaining water include using vegetative cover, terracing, and constructing check dams. In riparian zones, planting native vegetation such as willows and reeds stabilises stream banks and reduces the impact of floods. Similarly, constructing small-scale check dams in arid regions helps to slow water run-off, promote groundwater recharge, and reduce soil loss.

In agricultural landscapes, contour farming, agroforestry, and cover crop use can reduce erosion and improve soil moisture retention. These techniques create a more stable soil structure, reduce waterway sedimentation, and enhance landscape resilience.

21.4.3.2 Implementing Green Infrastructure for Ecosystem Recovery

Green infrastructure refers to using natural and semi-natural systems to manage water resources and support ecosystem restoration in urban and rural settings. Examples include the creation of wetlands to filter pollutants, restoring floodplains to provide natural flood control, and constructing rain gardens to capture and infiltrate storm water. Green roofs, permeable pavements, and bioswales are increasingly used in urban areas to manage storm water while providing habitat for urban wildlife. Implementing green infrastructure supports ecosystem recovery and enhances human well-being by providing recreational spaces, improving air and water quality, and reducing the urban heat island effect. As such, green infrastructure represents a powerful nature-based solution that integrates ecological restoration with broader environmental and social goals.

21.4.4 Habitat Creation and Enhancement

Habitat creation and enhancement involve constructing or modifying environments to support target species and promote biodiversity. These techniques are often used in areas where natural habitats have been completely lost or severely degraded.

21.4.4.1 Artificial Reef Construction

In marine and coastal environments, artificial reefs are constructed to provide habitat for fish, corals, and other marine life. These reefs are made from various materials, including concrete, metal, and even repurposed structures like old ships. The aim is to create complex structures that mimic natural reef habitats, providing shelter, feeding grounds, and breeding sites for marine organisms. Artificial reefs have been used successfully in places like the Gulf of Mexico and the Mediterranean Sea, where they have increased fish abundance and diversity. Careful planning is required to ensure these structures do not disrupt local ecosystems or become traps for marine debris.

21.4.4.2 Urban Greening and Ecological Corridors

In urban environments, habitat creation often takes the form of urban greening initiatives, such as developing parks, green roofs, and ecological corridors. Urban greening provides critical habitat for pollinators, birds, and small mammals while also enhancing the quality of life for residents. Ecological corridors connect fragmented habitats, allowing wildlife to move safely between green spaces and supporting genetic exchange and species persistence.

Projects like the High Line in New York City and the Bosco Verticale (Vertical Forest) in Milan demonstrate how integrating green spaces into urban design can provide habitat for biodiversity while offering recreational and aesthetic benefits for people.

21.4.5 Restoration Through Traditional Ecological Knowledge (TEK)

Traditional Ecological Knowledge (TEK) refers to Indigenous and local communities' cumulative knowledge, practices, and beliefs about their surrounding environments. TEK is increasingly recognised as a valuable resource for ecological restoration, offering insights into sustainable management and long-term ecosystem health.

21.4.5.1 Integrating Indigenous Practices and Knowledge

Integrating TEK into restoration involves collaborating with Indigenous communities to apply traditional land management practices, such as controlled burning, sustainable harvesting, and habitat creation. For example, Indigenous fire management techniques in Australia have been used for millennia to maintain healthy landscapes, prevent catastrophic wildfires, and promote biodiversity. These practices are now being reintroduced as part of contemporary restoration efforts to enhance ecosystem resilience.

21.4.5.2 TEK as a Model for Sustainable Restoration

TEK offers a model for sustainable restoration that emphasises the interconnectedness of all life forms and the importance of maintaining balance within ecosystems. This holistic perspective complements scientific approaches by providing a deeper understanding of ecosystem dynamics and long-term sustainability. By integrating TEK with modern science, restoration practitioners can develop strategies that are not only ecologically sound but also culturally inclusive and respectful of the traditional custodians of the land (Purohit et al. 2025).

21.5 Integration of Technology in Restoration Science

Technological advancements are revolutionising the field of restoration science, providing new tools to understand ecosystems better, predict responses to interventions, and plan restoration efforts at unprecedented scales by integrating technologies such as artificial intelligence (AI), machine learning, bioengineering, and decision support systems (DSS); restoration practitioners can make more informed decisions, increase the efficiency of their interventions, and monitor outcomes with greater precision.

21.5.1 Role of Artificial Intelligence and Machine Learning

Artificial intelligence (AI) and machine learning (ML) have become powerful tools in ecological research and restoration planning (Shivaprakash et al. 2022). These technologies can analyse large datasets, identify complex patterns, and predict ecosystem responses to restoration efforts. Their ability to process and learn from diverse data sources such as remote sensing imagery, genetic information, and climate models makes AI and ML invaluable for tackling the intricacies of ecosystem restoration.

21.5.1.1 Predictive Modelling for Ecosystem Responses

Predictive modelling involves using algorithms to forecast how ecosystems might respond to different restoration scenarios, allowing practitioners to select the most effective strategies before implementing them. AI-driven models can incorporate various variables, such as species interactions, habitat requirements, soil characteristics, and climate trends, to simulate ecosystem dynamics under various restoration interventions. This approach helps predict outcomes such as species recovery rates, vegetation structure changes, or invasive species spread. For example, machine learning models have been used to predict the success of plant reintroductions in degraded grasslands based on soil conditions, historical climate data, and species traits. Similarly, AI models are being applied in coral reef restoration to forecast coral growth and survival under different restoration methods and environmental stressors, such as temperature fluctuations and acidification. Predictive modelling is used in forest restoration to optimise tree species selection for future climate conditions. By simulating growth and survival rates under projected climate scenarios, AI models guide practitioners in choosing species that will thrive in a changing environment, enhancing restoration efforts' long-term success and resilience.

21.5.1.2 AI-Driven Restoration Planning and Management

AI-driven tools are increasingly used for restoration planning, site selection, and adaptive management. By analysing spatial data and integrating various ecological and socio-economic factors, AI can identify priority areas for restoration, predict potential restoration outcomes, and even suggest optimal strategies for resource allocation. For example, AI algorithms can evaluate landscape connectivity and recommend locations for planting to maximise habitat corridors and biodiversity benefits. In adaptive management, AI can process real-time monitoring data to detect early signs of success or failure in restoration projects. Automated systems equipped with AI capabilities can trigger alerts when indicators such as vegetation health or animal population trends deviate from expected trajectories, allowing for timely adjustments to management strategies. This dynamic approach enables practitioners to respond proactively, minimising risks and maximising restoration effectiveness. It is used to develop automated species identification systems that streamline biodiversity assessments. Using computer vision and deep learning, these systems can analyse images or audio recordings to identify plant and animal species, reducing the time and labour required for field surveys. Such innovations increase efficiency and improve the accuracy of biodiversity monitoring, providing more robust data for evaluating restoration outcomes.

21.5.2 Bioengineering and Ecological Design

Bioengineering and ecological design focus on using scientific principles and technological innovations to create or modify ecosystems to achieve specific restoration goals. This interdisciplinary field combines biology, engineering, and environmental science to develop novel solutions for restoring degraded environments and enhancing ecosystem services.

21.5.2.1 Using Biomimicry in Restoration Projects

Biomimicry, drawing inspiration from nature to solve human problems, is increasingly being applied in ecological restoration. In the context of restoration science, biomimicry involves designing interventions that mimic natural processes and structures to restore ecosystem function. For example, river restoration projects may use bioengineered structures replicating the form and function of natural river meanders to stabilise banks, reduce erosion, and improve habitat quality. Another application of biomimicry is in creating "bio-inspired" reefs using materials and shapes that mimic the complexity and functionality of natural coral reefs. These artificial reefs provide shelter and breeding grounds for marine life, supporting biodiversity and promoting ecosystem resilience. Similarly, biomimicry is used in designing wetlands, replicating natural systems' water filtration, and nutrient cycling functions, enhancing water quality and providing habitat for various species. Biomimicry also extends to soil restoration, where techniques that emulate natural nutrient cycling processes rehabilitate degraded soils. The design of plant communities that mimic native ecosystems helps restore soil structure and fertility, facilitating the re-establishment of self-sustaining plant and microbial communities.

21.5.2.2 Developing Engineered Ecosystems for Specific Environmental Goals

Bioengineering creates "engineered ecosystems" to address specific environmental challenges, such as pollution remediation, carbon sequestration, and habitat creation. One example is the development of bioengineered wetlands optimised to remove pollutants from agricultural run-off. These constructed wetlands are designed with specific plant and microbial communities that enhance the breakdown and uptake of contaminants, improving water quality before it enters rivers and lakes. In coastal areas, bioengineered solutions such as "living shorelines" stabilise eroding coastlines while providing habitat for marine organisms. These shorelines often incorporate native vegetation and engineered structures that dissipate wave energy, reduce erosion, and promote sediment deposition. Such projects restore ecological function and protect human communities from storm surges and sea-level rise. Bioengineering is also being applied in carbon sequestration projects, where ecosystems are designed to maximise carbon uptake and storage. For example, bioengineered forests are planted with fast-growing tree species with high carbon absorption capacities. At the same time, biochar is being added to soils to increase long-term carbon storage. These engineered ecosystems provide targeted solutions for mitigating climate change while enhancing biodiversity and ecosystem services.

21.5.3 Decision Support Systems (DSS) and Restoration Planning

Decision support systems (DSS) are integrated tools that assist restoration practitioners and policymakers make informed decisions based on complex data. By combining data analysis, modelling, and visualisation, DSS tools provide a comprehensive framework for planning, implementing, and evaluating restoration projects.

21.5.3.1 Data Integration for Informed Decision-Making

DSS tools integrate diverse datasets, including ecological, hydrological, socio-economic, and climatic information, to support holistic decision-making. For example, a DSS might incorporate data on land use, soil types, hydrological flows, and biodiversity in watershed restoration to identify areas where interventions will have the most significant impact. These tools can also model the interactions between different factors, such as how land use or vegetation cover changes affect water quality and quantity. Platforms often include interactive mapping and scenario analysis capabilities, allowing users to visualise the potential outcomes of different restoration strategies. This functionality enables stakeholders to explore trade-offs, prioritise actions, and allocate resources more effectively. DSS tools facilitate collaboration in complex projects involving multiple stakeholders by providing a shared platform for evaluating options and reaching consensus.

21.5.3.2 Case Studies in Effective Use of DSS Tools

One example of DSS in action is using the "InVEST" (Integrated Valuation of Ecosystem Services and Tradeoffs) tool, which helps restoration planners assess the benefits of different land-use scenarios for biodiversity, water quality, and carbon storage. InVEST has been used in projects worldwide, including in Belize's coastal restoration, where it guided decisions on mangrove restoration by modelling the trade-offs between fisheries productivity, tourism value, and coastal protection. Another example is the MARXAN tool, which is widely used for spatial conservation planning. MARXAN uses algorithms to identify optimal configurations of protected areas that maximise biodiversity benefits while minimising socio-economic costs. This tool has been employed in marine and terrestrial restoration projects to design reserve networks that protect critical habitats and species.

The "Resource Investment Optimisation System (RIOS)" tool, developed by *The Nature Conservancy*, is another DSS designed specifically for watershed restoration. Rios integrates data on hydrology, land use, and socio-economic factors to prioritise restoration activities that will provide the greatest benefits for water quality and quantity. In Brazil, Rios has been used to guide investments in reforestation and erosion control in the Atlantic Forest, resulting in improved water quality for downstream communities. By integrating cutting-edge technologies such as AI, bioengineering, and DSS, restoration science is entering a new era of precision and effectiveness. These tools not only enhance our ability to restore degraded ecosystems but also empower practitioners to plan and implement restoration efforts that are resilient, sustainable, and adaptive to future environmental changes.

21.6 Socio-ecological Aspects of Restoration Research

The success of ecosystem restoration extends beyond ecological recovery; it encompasses the social, cultural, and economic dimensions of human communities that interact with these ecosystems. Restoration science increasingly recognises that ecological and social systems are interconnected and that sustainable restoration requires integrating socio-ecological approaches.

21.6.1 Community Engagement and Participatory Research

Practical restoration efforts require the involvement and support of local communities, who often have a deep connection to the landscapes in which they live (Gann et al. 2019). Community engagement and participatory research allow local stakeholders to contribute knowledge, preferences, and labour to restoration projects. This collaborative approach fosters a sense of ownership and responsibility and enhances the social legitimacy and sustainability of restoration activities (Subramanyam et al. 2019).

21.6.1.1 Involving Local Communities in Restoration Projects

Involving local communities in restoration projects means actively engaging them in every process stage, from planning and decision-making to implementation and monitoring. Community members can contribute valuable insights about local environmental conditions, species behaviours, and traditional management practices that are often overlooked by scientific approaches. This local knowledge is particularly critical in areas where historical land use and cultural practices have uniquely shaped ecosystems. For example, farmers and Indigenous communities are frequently involved in selecting suitable plant species, identifying optimal sites, and implementing water management strategies in the restoration of wetlands and riparian zones. Their knowledge of seasonal water flows, soil types, and historical land-use patterns can significantly improve the effectiveness of restoration efforts. In urban restoration projects, involving residents in activities like tree planting, invasive species removal, and monitoring contributes to project success and strengthens community cohesion and environmental stewardship. Participatory research models, such as community-based participatory research (CBPR), empower local communities to generate and apply knowledge actively. CBPR involves community members as co-researchers, ensuring that restoration goals and methods align with their needs and values. By integrating scientific and local perspectives, participatory research can produce more context-specific and culturally appropriate restoration strategies (Nautiyal et al. 2023).

21.6.1.2 Benefits of Co-management and Collaborative Science

Co-management is a governance approach where local communities, governmental agencies, and other stakeholders share responsibility for managing natural resources. In restoration contexts, co-management can enhance project outcomes by blending local and scientific knowledge, improving decision-making, and promoting long-term stewardship. When local communities are involved in setting restoration goals, determining management practices, and monitoring outcomes, they are more likely to support and sustain restoration efforts over time. One of the primary benefits of co-management is the establishment of trust and mutual respect among stakeholders. Collaborative science involves scientists working directly with community members and helps bridge the gap between scientific research and local needs. By co-producing knowledge, collaborative science fosters shared understanding and commitment, reducing conflicts and enhancing the adaptive capacity of restoration projects. Successful examples of co-management include the restoration of fisheries and coastal habitats in the Philippines, where local fishing communities work with government agencies and NGOs to restore mangroves and coral reefs. These projects improve biodiversity and fish stocks and provide economic benefits through sustainable fisheries and ecotourism. Similarly, in forest restoration efforts in Latin America, community co-management has been key to preventing illegal logging and ensuring the success of reforestation initiatives.

21.6.2 Socio-economic Impacts of Restoration Activities

Restoration projects can have significant socio-economic impacts, influencing livelihoods, employment opportunities, and community well-being. Understanding and maximising these impacts is essential for designing restoration initiatives contributing to sustainable development and social equity. Restoration can serve as a tool for poverty alleviation, economic diversification, and community empowerment, making it a valuable strategy for achieving broader environmental and social goals.

21.6.2.1 Restoration as a Tool for Sustainable Development

Ecosystem restoration is increasingly seen as supporting sustainable development by creating jobs, enhancing ecosystem services, and improving the quality of life for local communities. For example, large-scale reforestation projects, such as the Great Green Wall initiative in Africa, aim to restore degraded landscapes, create employment opportunities, enhance food security, and combat desertification. These projects contribute to sustainable development by linking ecological recovery with economic and social benefits. Restoration activities such as tree planting, habitat rehabilitation, and sustainable agroforestry can generate income for local communities, particularly in rural areas with limited employment opportunities. Restoration projects that involve sustainable harvesting of non-timbre forest products (NTFPs), such as medicinal plants, honey, and wild fruits, provide additional income streams without compromising ecosystem health. In coastal areas, mangrove restoration can enhance fisheries productivity, supporting the livelihoods of small-scale fishers and providing long-term economic benefits.

21.6.2.2 Measuring Social and Economic Outcomes of Restoration Efforts

Assessing restoration projects' social and economic outcomes is crucial for demonstrating their value and guiding future investments. Metrics such as employment generation, income changes, food security, and community well-being can be used to evaluate the socio-economic impacts of restoration. For example, the number of jobs created, the duration and quality of employment, and the skills gained by local participants are essential indicators of economic benefits. Social outcomes, such as increased community cohesion, improved environmental awareness, and enhanced cultural identity, are more difficult to quantify but are equally important. Participatory monitoring and evaluation, where local communities help define and measure success, can provide a more nuanced understanding of these social impacts. Case studies from India's watershed restoration projects show that including women in decision-making and implementation improves project outcomes and promotes gender equity and community empowerment. Integrating economic valuation methods, such as cost-benefit analysis and ecosystem service valuation, can help quantify the broader benefits of restoration projects. These approaches capture the value of restored ecosystems in terms of services, such as water purification, carbon sequestration, and recreation, providing a more comprehensive picture of restoration's contributions to human well-being.

21.6.3 Policy and Governance Frameworks

Effective policy and governance frameworks are essential for supporting restoration activities locally, nationally, and globally. Policies can create enabling environments for restoration by providing funding, setting clear targets, and establishing legal and institutional frameworks. Governance frameworks promoting transparency, accountability, and stakeholder participation are key to ensuring restoration initiatives are inclusive, effective, and sustainable.

21.6.3.1 Policy Support for Restoration Science

Policy support for restoration can take various forms, including national restoration strategies, financial incentives, and regulatory measures that promote sustainable land use and conservation. For example, Costa Rica and Brazil have implemented payment for ecosystem services (PES) schemes that incentivise landowners to restore and maintain forest cover, contributing to biodiversity conservation and climate goals. National policies integrating restoration into broader environmental and development agendas can provide a strong foundation for scaling restoration efforts. The Green India Mission, part of the country's National Action Plan on Climate Change, aims to restore forest and tree cover to enhance ecosystem services and support livelihoods. Similarly, the European Union's Biodiversity Strategy 2030 includes ambitious targets for restoring degraded ecosystems and integrating restoration into agricultural and urban planning.

21.6.3.2 International Agreements and Restoration Targets

International agreements and frameworks, such as the Convention on Biological Diversity (CBD) and the United Nations Framework Convention on Climate Change (UNFCCC), provide global mandates for ecosystem restoration. The Aichi Biodiversity Targets, which include goals for restoring at least 15% of degraded ecosystems, and the Bonn Challenge, which aims to restore 350 million hectares of deforested and degraded land by 2030, are key drivers of restoration efforts worldwide. The UN Decade on Ecosystem Restoration (2021–2030) represents a global call to action, bringing together governments, NGOs, and communities to restore ecosystems at an unprecedented scale. This initiative highlights the need for coordinated action, scientific research, and policy support to achieve restoration goals. The UN Decade aims to catalyse a global restoration movement that benefits both people and the planet by aligning restoration efforts with international climate, biodiversity, and sustainable development targets.

21.7 Challenges and Future in Restoration Science

Restoration science has made significant progress over the past few decades, with an expanding body of knowledge and the development of innovative tools and techniques. The complexity and scale of environmental degradation present ongoing challenges that require further research, refinement of methodologies, and interdisciplinary collaboration.

21.7.1 Addressing Knowledge Gaps and Research Priorities

Despite significant advancements, restoration science still faces critical knowledge gaps that hinder the effectiveness and scalability of restoration projects. These gaps range from a limited understanding of complex ecological processes to challenges predicting long-term outcomes under changing environmental conditions. Addressing these gaps is essential for refining restoration techniques and ensuring that interventions are ecologically sound and resilient.

21.7.1.1 Unresolved Scientific Questions

Several unresolved scientific questions challenge restoration practitioners and researchers (Testa et al. 2017). One major question is how to define and measure the "success" of restoration projects. While traditional metrics, such as species richness and vegetation cover, provide some insights, they do not capture the full complexity of restored ecosystems, including functions such as nutrient cycling, hydrological processes, and resilience to climate change. Developing more holistic and standardised indicators for assessing ecosystem recovery is critical for future research. Another key question concerns the role of genetic and species diversity in promoting resilience and ecosystem stability. While high biodiversity is generally assumed to enhance resilience, the specific traits and interactions contributing to ecosystem function and adaptability remain poorly understood. Research is needed to identify which species combinations and genetic attributes are most beneficial for long-term restoration success, particularly in climate change and shifting environmental baselines. There is a lack of understanding of how restored ecosystems evolve and respond to novel disturbances, such as invasive species, extreme weather events, and human land-use changes. Long-term monitoring and experimental studies are required to explore these dynamics and inform adaptive management strategies that can respond to unexpected outcomes and changing conditions.

21.7.1.2 Expanding Restoration to New Ecosystems and Biomes

While restoration research has traditionally focused on terrestrial ecosystems, such as forests and grasslands, there is a growing recognition of the need to expand restoration efforts to underrepresented ecosystems and biomes, including marine, freshwater, and arid environments. These ecosystems present unique challenges due to their distinct ecological dynamics and human impacts. For example, restoring coral reefs and deep-sea habitats requires specialised techniques and a nuanced understanding of complex marine interactions. Similarly, restoration in arid and semi-arid regions faces water scarcity, soil erosion, and extreme temperature variability (Nautiyal et al. 2015). To address these challenges, restoration science must develop tailored approaches considering these diverse systems' specific ecological, hydrological, and climatic conditions. Expanding restoration to new biomes also necessitates incorporating a broader range of species, ecological processes, and socio-economic considerations. This calls for increased interdisciplinary collaboration and cross-sectoral partnerships to design and implement context-specific restoration strategies that are ecologically viable and socially acceptable.

21.7.2 Scaling Up Restoration Efforts

Given the scale of global environmental degradation, restoration efforts must move beyond small, localised projects to achieve landscape-level and even regional-scale restoration. Scaling up restoration presents technical, logistical, and financial challenges that require innovative solutions and coordinated action at multiple levels.

21.7.2.1 Techniques for Large-Scale Implementation

Scaling up restoration involves increasing the spatial extent of interventions and adapting techniques to work efficiently at larger scales. Mechanised planting, seed broadcasting using drones, and developing large-scale nurseries are emerging methods for scaling up reforestation and habitat rehabilitation. For example, "seed bombing" using drones has been successfully implemented in reforestation projects in Myanmar and Australia, allowing for rapid and cost-effective coverage of large areas. Another large-scale implementation technique is using "framework species" or "nurse plants" to accelerate natural succession and facilitate the recovery of complex ecosystems. In this approach, fast-growing, hardy species are planted first to stabilise soils, improve microclimatic conditions, and create suitable environments for establishing more diverse and sensitive species over time. Large-scale implementation also requires infrastructure and logistical planning, including transportation, water management, and labour coordination. Integrating local communities and stakeholders into planning and execution can help address these challenges, ensuring that large-scale restoration is feasible and sustainable.

21.7.2.2 Achieving Global Restoration Targets

The global restoration community has set ambitious targets, such as the Bonn Challenge, which aims to restore 350 million hectares of degraded land by 2030, and the UN Decade on Ecosystem Restoration, which calls for large-scale restoration to combat climate change, biodiversity loss, and land degradation. Achieving these targets requires scaling up individual projects and creating enabling environments through policy, funding, and international cooperation. Achieving these global targets will require unprecedented collaboration among governments, NGOs, the private sector, and local communities. Financial support and incentives, such as payment for ecosystem services (PES) schemes, carbon credits, and green bonds, are critical for mobilising resources. In addition, monitoring and reporting frameworks are needed to track progress, share lessons learned, and adapt strategies as needed.

21.7.3 *Balancing Restoration with Conservation and Development*

Restoration often occurs in landscapes with multiple, and sometimes conflicting, land uses. Balancing restoration goals with conservation priorities and development needs is a complex challenge that requires navigating trade-offs and finding synergies.

21.7.3.1 Navigating Trade-Offs and Synergies

Restoration can sometimes conflict with conservation objectives, mainly when modifying habitats or reintroducing species that could disrupt existing communities. For example, restoring historical conditions in a landscape adapted to new species assemblages may harm current biodiversity. Understanding these trade-offs and designing restoration strategies that minimise negative impacts while maximising overall benefits is essential. Restoration projects must consider socio-economic trade-offs, such as the displacement of agricultural activities or the impact on local livelihoods. Restoring large tracts of degraded land might compete with agricultural production or infrastructure development in areas where land is scarce. Identifying synergies, such as agroforestry systems that integrate tree planting with crop production, can help balance restoration with development needs, ensuring that restoration efforts contribute to ecological and human well-being.

21.7.3.2 Integrating Restoration in Broader Environmental Planning

For restoration to be successful at larger scales, it must be integrated into broader land-use and environmental planning frameworks. This involves aligning restoration goals with national and regional development plans, conservation strategies, and climate adaptation efforts. Landscape-scale restoration planning considers ecological recovery and social, economic, and cultural factors, ensuring that restoration contributes to sustainable development and resilience. Integrated planning approaches, such as landscape and seascape planning, watershed management, and regional biodiversity strategies, provide frameworks for coordinating restoration with other land uses. By embedding restoration into these broader contexts, practitioners can create multifunctional landscapes that deliver multiple benefits, from biodiversity conservation to food security and climate mitigation.

21.8 Case Studies in Cutting-Edge Restoration Science

This presents case studies from various ecosystems and biomes to illustrate the diverse approaches and innovations in restoration science. These examples highlight the application of cutting-edge research and techniques in Himalayan landscape, forest, coastal, grassland, and urban environments.

21.8.1 Himalayan Ecosystem Restoration

21.8.1.1 Forest Regeneration and Biodiversity Conservation

The Himalayan ecosystem restoration focuses on reforestation and biodiversity conservation to counteract deforestation and habitat loss. Efforts include afforestation programmes and the protection of native plant species. In Uttarakhand, India, initiatives such as the Green India Mission and community-led reforestation projects have contributed to soil stabilisation, carbon sequestration, improved water retention, supporting local wildlife and reducing climate vulnerability (Rao et al. 1999; Negi et al. 2022; Rao et al. 2024).

21.8.1.2 Glacier and Watershed Management Strategies

Glacier conservation is critical for sustaining freshwater resources in the Himalayas. Strategies include artificial glacier projects and watershed management to mitigate glacial retreat and water scarcity. In Ladakh, artificial glaciers, known as ice stupas, store winter water and provide irrigation during dry months, ensuring water security for local communities. These efforts help maintain ecological balance and support agriculture in high-altitude regions (Gladfelter 2018; Clouse 2024; Gupta 2024).

21.8.2 Innovative Approaches in Forest Restoration

21.8.2.1 Tropical Rainforest Recovery Using Assisted Natural Regeneration

Assisted natural regeneration (ANR) is an effective strategy for restoring tropical rainforests. This technique protects and nurtures natural seed sources and saplings to accelerate forest recovery. In the Philippines, ANR has been used to restore thousands of hectares of degraded forest, improving biodiversity and carbon storage while being cost-effective compared to traditional planting methods (Werden et al. 2024).

21.8.2.2 Boreal and Temperate Forest Restoration Techniques

Restoration techniques in boreal and temperate forests focus on re-establishing natural fire regimes, controlling invasive species, and planting climate-resilient species (Girona et al. 2023). For example, prescribed burns in Canada's boreal forests mimic natural fire cycles, promoting the regeneration of fire-dependent species and enhancing landscape resilience (Whitman et al. 2019; Fredriksson et al. 2023).

21.8.3 Coastal and Marine Ecosystem Restoration

21.8.3.1 Coral Reef Rehabilitation Projects

Coral reef restoration uses innovative methods such as coral gardening and micro fragmentation to re-establish coral populations (Precht 2006; Schmidt-Roach et al. 2020). Scientists grow coral fragments in underwater nurseries in Florida and the Caribbean and then outplant them onto degraded reefs, significantly speeding up coral recovery (Rinkevich 2006; Blanco-Pimentel et al. 2022).

21.8.3.2 Mangrove and Salt Marsh Recovery Strategies

Mangrove restoration involves hydrological modifications and planting mangrove seedlings to stabilise coastlines and provide critical habitat (Sunkur et al. 2023). In Bangladesh, large-scale mangrove planting has reduced coastal erosion and enhanced fishery productivity, benefiting biodiversity and local communities (Hossain et al. 2024.

21.8.4 Grassland and Savanna Restoration

21.8.4.1 Using Fire Ecology and Grazing Management

Fire ecology is essential for maintaining grassland and savanna ecosystems. In Africa, controlled burns prevent woody encroachment, maintain open habitats, and promote biodiversity. Grazing management, involving rotational grazing and wildlife corridors, supports the coexistence of livestock and native herbivores (Case and Staver 2017; Starns et al. 2020).

21.8.4.2 Plant Community Reconstruction in Degraded Grasslands

In North America, prairie restoration projects focus on reconstructing plant communities using native species that were historically present. Techniques include direct seeding, soil preparation, and cover crops to restore soil health and plant diversity (Smith 2014).

21.8.5 Urban Ecosystem Restoration

21.8.5.1 Restoring Green Spaces in Urban Environments

Urban ecosystem restoration involves creating parks, green roofs, and restored wetlands to enhance biodiversity and improve quality of life. In New York City, the High Line project transformed an abandoned railway into a vibrant green space, providing habitat and recreational opportunities.

21.8.5.2 Incorporating Biodiversity in City Planning

Incorporating biodiversity into urban planning involves designing green spaces that support native species and promote ecological connectivity. The city of Singapore's City in a Garden initiative integrates natural habitats throughout the urban landscape, creating a model for urban biodiversity conservation. By addressing the challenges supporting future directions in restoration science, and applying innovative techniques in diverse ecosystems, restoration science can continue to advance, offering hope for reversing environmental degradation and building a sustainable future for both people and nature.

21.9 Conclusion

Ecosystem restoration is critical for reversing environmental degradation, enhancing biodiversity, and supporting human well-being. At the same time, restoration science has made significant strides through technological advancements, community engagement, innovative techniques, and ongoing challenges such as scaling up efforts, addressing knowledge gaps, and balancing ecological and socio-economic priorities. The successful integration of restoration into broader conservation and development frameworks requires a multidisciplinary approach considering ecological, social, and policy dimensions. Future research must prioritise long-term monitoring, expand restoration into underrepresented ecosystems, and refine strategies to ensure resilience in climate change. By leveraging cutting-edge research, embracing collaborative partnerships, and aligning with global restoration targets, restoration science can help restore the health and functionality of ecosystems worldwide. Ultimately, achieving large-scale restoration success will depend on sustained commitment, interdisciplinary collaboration, and the active participation of local communities and stakeholders.

References

Alexander S, Aronson J, Whaley O, Lamb D (2016) The relationship between ecological restoration and the ecosystem services concept. Ecol Soc 21(1)

Blanco-Pimentel M, Evensen NR, Cortés-Useche C, Calle-Triviño J, Barshis DJ, Galván V et al (2022) All-inclusive coral reef restoration: How the tourism sector can boost restoration efforts in the Caribbean. Front Mar Sci 9:931302

Breed MF, Harrison PA, Blyth C, Byrne M, Gaget V, Gellie NJ et al (2019) The potential of genomics for restoring ecosystems and biodiversity. Nat Rev Genet 20(10):615–628

Buma B, Schultz C (2020) Disturbances as opportunities: learning from disturbance-response parallels in social and ecological systems to better adapt to climate change. J Appl Ecol 57(6):1113–1123

Camarretta N, Harrison PA, Bailey T, Potts B, Lucieer A, Davidson N, Hunt M (2020) Monitoring forest structure to guide adaptive management of forest restoration: a review of remote sensing approaches. New For 51(4):573–596

Case MF, Staver AC (2017) Fire prevents woody encroachment only at higher-than-historical frequencies in a South African savanna. J Appl Ecol 54(3):955–962

Cavender-Bares J, Schneider FD, Santos MJ, Armstrong A, Carnaval A, Dahlin KM et al (2022) Integrating remote sensing with ecology and evolution to advance biodiversity conservation. Nat Ecol Evol 6(5):506–519

CEM S (2021) Principles for ecosystem restoration to guide the United Nations Decade 2021–2030

Cheng C, Li F (2024) Ecosystem restoration and management based on nature-based solutions in China: research progress and representative practices. Nat Based Solut 100176

Clouse (2024) Building up water artificial glaciers in Ladakh. In: Cipriani L (ed) (Co) Designing hope. Routledge, pp 199–215

Fredriksson E, Wallgren M, Löfroth T (2023) Wildfire and prescribed burning impact moose forage availability and browsing levels in the northern boreal forest. Scand J For Res 38(1–2):58–69

Gann GD, McDonald T, Walder B, Aronson J, Nelson CR, Jonson J et al (2019) International principles and standards for the practice of ecological restoration. Restor Ecol 27(S1):S1–S46

Gardner RC (2013) Technological forests: engineering nature with tree planting on the Great Plains, 1870–1944. Montana State University

Girona MM, Aakala T, Aquilué N, Bélisle AC, Chaste E, Danneyrolles V et al (2023) Challenges for the sustainable management of the boreal forest under climate change. In: Boreal Forests in the face of climate change: sustainable management. Springer, Cham, pp 773–837

Gladfelter S (2018) Ladakh's artificial glaciers, ice stupas, and human-made ice reserves. Fulbright Nehru Student Research Grant program

Goswami M, Gupta AK, Kishan R, Baidya S, Khan YDI, Prakash S, Nautiyal S (2023) An evaluation of climate resilient agricultural practices in India: a narrative synthesis of literature. Environ Sustain 6(1):7–23

Gupta K (2024) The Ice man of Ladakh building artificial glaciers in the Himalayas. BBC. https://www.bbc.com/future/article/20240612-the-ice-man-of-ladakh-building-artificial-glaciers-in-the-himalayas. Accessed 20 Jan 2024

Hald-Mortensen C (2023) The main drivers of biodiversity loss: a brief overview. J Ecol Nat Resour 7(3):000346

Hartmann M, Six J (2023) Soil structure and microbiome functions in agroecosystems. Nat Rev Earth Environ 4(1):4–18

Hobbs RJ, Harris JA (2001) Restoration ecology: repairing the Earth's ecosystems in the new millennium. Restor Ecol 9(2):239–246

Hossain M, Saha C, Saha S (2024) Mangrove restoration in Bangladesh. Open J For 15(1):35–52

Jones TA (2017) Ecosystem restoration: recent advances in theory and practice. Rangeland J 39(6):417–430

Macdonald E, King EG (2018) Novel ecosystems: a bridging concept for the consilience of cultural landscape conservation and ecological restoration. Landsc Urban Plan 177:148–159

Munawar HS, Hammad AW, Waller ST, Thaheem MJ, Shrestha A (2021) An integrated approach for post-disaster flood management via the use of cutting-edge technologies and UAVs: a review. Sustainability 13(14):7925

Nautiyal S, Bhaskar K, Khan YDI (2015) Biodiversity of semiarid landscape. Switzerland: Springer International Publishing. https://doi.org/10.1007/978-3-319-15464-0

Nautiyal S, Gupta AK, Goswami M, Khan YDI (2023) The Palgrave Handbook of socio-ecological resilience in the face of climate change. Palgrave Macmillan Singapore. https://doi.org/10.1007/978-981-99-2206-2

Neaves LE, Ogden R, Hollingsworth PM (2022) Genomics and conservation translocations. Conserv Translocations 9:271. https://doi.org/10.1017/9781108638142.013

Negi VS, Pathak R, Dhyani V, Durgapal M, Joshi RK, Bhatt ID (2022) Land restoration in the Himalayan region: steps towards biosphere integrity. Land Use Policy 121:106317

Palmer MA, Zedler JB, Falk DA (eds) (2016) Foundations of restoration ecology. Island Press, Washington, DC, pp 364–364

Perring MP, Standish RJ, Price JN, Craig MD, Erickson TE, Ruthrof KX et al (2015) Advances in restoration ecology: rising to the challenges of the coming decades. Ecosphere 6(8):1–25

Precht WF (2006) Coral reef restoration handbook. CRC Press

Purohit S, Hande OM, Khan YDI, Krishanmurti (2025) Role of traditional ecological knowledge in the current climate crisis: case study of Mahadev Kolis in Maharashtra, India. In Mitigation and adaptation strategies against climate change in natural systems. Cham: Springer Nature Switzerland, pp 445–467. https://doi.org/10.1007/978-3-031-75968-0_23

Rao KS, Maikhuri RK, Saxena KG (1999) Participatory approach to rehabilitation of degraded forest lands: a case study in a high altitude village of Indian Himalaya. Int Tree Crops J 10(1):1–17

Rao KS, Semwal RL, Maletha A, Nautiyal S, Maikhuri RK, Saxena K (2024) Participatory forest restoration for sustainable livelihood: a case study from Himalaya

Rinkevich B (2006) The coral gardening concept and the use of underwater nurseries: lessons learned from silvics and silviculture. Coral Reef Restor Handb 1:291–301

Rodriguez Franco C, Conje J (2023) The evolution of the dialogue and perspectives on sustainable forest management with special emphasis on the United States of America. J Sustain For 42(8):747–791

Schmidt-Roach S, Duarte CM, Hauser CA, Aranda M (2020) Beyond reef restoration: next-generation techniques for coral gardening, landscaping, and outreach. Front Mar Sci 7:672

Shivaprakash KN, Swami N, Mysorekar S, Arora R, Gangadharan A, Vohra K et al (2022) Potential for artificial intelligence (AI) and machine learning (ML) applications in biodiversity conservation, managing forests, and related services in India. Sustainability 14(12):7154

Smith DD (2014) Prairie restoration: bridging the past and the future. Prairie Nat 22. https://digitalcommons.unl.edu/tpn/22

Stanturf JA, Mansourian S (2020) Forest landscape restoration: state of play. R Soc Open Sci 7(12):201218

Starns HD, Taylor CA, Garza NE, Tolleson DR (2020) Effects of fire exclusion on previously fire-managed semiarid savanna ecosystem. Rangel Ecol Manag 73(1):93–96

Subramanyam VB, Khan YDI, Bai MM, Mukate SV (2019) Towards community conservation and it's challenges: a study from breeding colony of painted storks in Veerapuram Village,

Ananthapuramu District, Andhra Pradesh, India. Ecology, Environment and Conservation. 25:295–303pp

Sunkur R, Kantamaneni K, Bokhoree C, Ravan S (2023) Mangroves' role in supporting ecosystem-based techniques to reduce disaster risk and adapt to climate change: a review. J Sea Res:102449

Testa JM, Kemp WM, Harris LA, Woodland RJ, Boynton WR (2017) Challenges and directions for the advancement of estuarine ecosystem science. Ecosystems 20:14–22

Theissinger K, Fernandes C, Formenti G, Bista I, Berg PR, Bleidorn C et al (2023) How genomics can help biodiversity conservation. Trends Genet 39(7):545–559

Werden LK, Cole RJ, Schönhofer K, Holl KD, Zahawi RA, Averill C et al (2024) Assessing innovations for upscaling forest landscape restoration. One Earth 7(9):1515–1528

Whitman E, Parisien MA, Thompson DK, Flannigan MD (2019) Short-interval wildfire and drought overwhelm boreal forest resilience. Sci Rep 9(1):18796

Part IV
Epilogue

Chapter 22
Epilogue: Lessons Learned and the Path Forward for Ecosystem-Based Resilience in the Himalayan Landscapes

Y. D. Imran Khan, Mrinalini Goswami, and Sunil Nautiyal

Abstract This epilogue synthesizes the key insights from this volume and highlights the integration of social, ecological, technological, and policy perspectives to enhance resilience in the fragile Himalayan region. Overall, the chapters of this volume together emphasise the importance of ecosystem-based adaptation (EbA) Approach, traditional knowledge and community-driven initiatives in addressing the growing impacts of climate change. Key themes include climate-induced ecosystem shifts, indigenous adaptation practices, and governance challenges in disaster risk reduction. The discussions highlight strengthening scientific–traditional knowledge linkages, scaling up EbA strategies, promoting sustainable livelihoods, and promoting trans-boundary co-operation. As we advance, holistic, inclusive, and innovation-driven approaches are key to protecting the Himalayan landscapes – ensuring ecological stability, cultural continuity, and long-term socio-economic resilience for future generations.

As we conclude this exploration of ecosystem-based approaches for building resilience in the Himalayan landscapes, it is crucial to reflect on the key insights gained and consider the way forward. In this regard, the integration of ecological and

Y. D. I. Khan
WOTR Centre for Resilience Studies (W-CReS), Watershed Organisation Trust (WOTR), Pune, India
e-mail: ikeducationkhan08@gmail.com

M. Goswami (✉)
Researcher in Sustainability Science, Bengaluru, Karnataka, India

S. Nautiyal
Centre for Ecological Economics and Natural Resources (CEENR), Institute for Social and Economic Change (ISEC), Bengaluru, Karnataka, India

G.B. Pant National Institute of Himalayan Environment, Kosi-Katarmal, Almora, Uttarakhand, India

Y. D. I. Khan et al. (eds.), *Ecosystem-based Approaches for Resilience Building in Himalayan Landscapes*, Disaster Resilience and Green Growth,
https://doi.org/10.1007/978-981-95-2007-7_22

sociological perspectives has revealed a promising path toward enhancing the resilience of these fragile mountain communities. With the background of ecological sensitivity (Nautiyal et al. 2016; Das and Meher 2019), increase in vulnerability to disasters, and need of proper assessments (IPCC 2007; Singh et al. 2011; Rasul et al. 2019; Gupta 2024), the chapters of this book, Ecosystem-Based Approaches for Resilience Building in the Himalayan Landscapes: Insights for Climate Change and Disaster Risk Reduction, have collectively explored the intricate relationships between climate change, ecosystem dynamics, traditional knowledge, and resilience-building strategies. This epilogue emphasizes on the core characteristics of resilience that have emerged from the research articles included in this book, highlighting the importance of effective governance, community involvement, transformative adaptation, and trans-boundary and regional cooperation. Furthermore, it examines the critical role of indigenous knowledge and the need to promote social-ecological systems for a more sustainable, resilient, and adaptive future for the Himalayas.

The discussions throughout this book underscore the urgent need for integrated, ecosystem-based approaches to address the challenges posed by climate change in the Himalayas. By weaving together scientific research, policy frameworks, and community-driven initiatives, the chapters illustrate how EbA approaches can enhance ecological stability while simultaneously supporting livelihoods and cultural heritage of this region. From restoring degraded landscapes to strengthening local governance and encouraging trans-boundary cooperation, the diverse strategies presented confirm that resilience building must be both holistic and inclusive (Nautiyal et al. 2016, 2023). As we look ahead, it is imperative to translate these insights into action, promoting collaboration among governments, researchers, and local communities. The Himalayas, with their unique biodiversity and cultural richness, demand context-specific solutions that recognize the interconnectedness of ecosystems and human well-being. Strengthening adaptive capacities, leveraging indigenous knowledge, and mainstreaming ecosystem-based approaches into development planning will be crucial in navigating an uncertain future. This epilogue not only reflects on the key takeaways from the book but also serves as a call for sustained commitment to protecting the ecological and socioeconomic fabric of this Himalayan landscape.

22.1 Climate Change and Ecosystem Dynamics

The first section of the book meticulously examined how climate change is reshaping the fragile ecosystems of the Himalayas. The cryosphere response to climate change, as discussed in Chap. 2, underscores the alarming retreat of glaciers and the associated risks of glacial lake outburst floods (GLOFs), which pose severe threats to communities and infrastructure in Ladakh. Similarly, Chap. 3 shed light on the impacts of climate variability on Himalayan agriculture, emphasizing how erratic precipitation patterns and rising temperatures are altering crop yields, thereby threatening food security. Chapter 4 highlights on Changpa community's pastoralism practices and their connection with the ecology and their economic activities.

The expansion of the Himalayan birch (*Betula utilis*), as examined in Chap. 5, provides a fascinating ecological case study on the shifting timberline due to warming temperatures. The study's findings suggest that while certain species may benefit from climate change, the overall biodiversity balance is precariously at risk. The research presented in Chaps. 6 and 7 on landslide susceptibility and vegetation dynamics further reinforces the urgent need for better predictive modeling and disaster mitigation strategies. Additionally, the assessment of the endangered species *Rauvolfia serpentina* (Chap. 8) highlights how conservation priorities must integrate climate resilience into biodiversity management.

22.2 Traditional Knowledge and Adaptive Strategies

The second section of the book presented a compelling discourse on how local knowledge and traditional practices continue to shape resilience strategies in the Himalayas. The perceptions and indigenous knowledge systems in Meghalaya (Chap. 9) serve as a crucial reminder that climate adaptation is deeply embedded within cultural frameworks. Chap. 10 discusses the *Jhum* cultivation practices in Northeast India. Chapter 11 highlights on role of cultural wisdom in shaping the sustainable ecosystems and making communities stewards to their ecosystems. Chapter 12 emphasizes how pastoralism (age-old systems) maintains a delicate ecological balance, despite facing pressures from modernization and climate change.

Moreover, Chap. 13 provided an insightful look into traditional ecological restoration practices in the high-altitude Darma Valley. The study underscored how community-led biodiversity conservation practices have long been resilient against environmental stressors, offering valuable lessons for contemporary conservation strategies. This interplay of local knowledge, cultural stewardship, and scientific innovation could pave the way for future ecosystem-based adaptation approaches.

22.3 Ecosystem-Based Approaches, Resilience Building, and Policy Interventions

The final section of the book explored ecosystem-based adaptation (EbA) as a fundamental strategy for resilience building in the Himalayas. The theoretical underpinnings and global case studies presented in Chap. 14 demonstrate the universal applicability of EbA in mountainous and hilly landscapes. The community forestry initiatives in Uttarakhand (Chap. 15) highlight the importance of community involvement in sustaining forest resources while enhancing climate resilience.

Water security, a critical concern for the region, was extensively discussed in Chap. 16, which provided an assessment of the valuation of water as an ecosystem service. The pressing need for sustainable water management has been amplified by the depletion of springs, as detailed in Chap. 19, where nature-based solutions for springshed revival have been explored.

The nexus between conservation and sustainable development was a recurring theme, particularly in Chap. 17, which examined the potential of ecotourism in the trans-boundary Khangchendzonga landscape. Similarly, the evaluation of ecosystem provisioning services along altitudinal gradients (Chap. 18) reinforced the idea that socioeconomic resilience is inherently linked to ecosystem health.

The book also provided crucial insights into disaster risk reduction (DRR) strategies in the Indian Himalayas. The findings in Chaps. 20 and 21 highlighted both the challenges and opportunities in strengthening disaster management frameworks, with a focus on integrating ecosystem-based strategies into policy and governance structures. At the end, Chap. 22 brought attention to cutting-edge research in ecosystem restoration, emphasizing the need for innovation-driven recovery strategies in response to environmental degradation.

22.4 Recommendations to Strengthen Resilience in the Himalayas

As the impacts of climate change intensify, the Himalayas stand at a critical crossroads. The knowledge and research presented in this book point to several key takeaways for the future: which includes traditional, scientific, technological, and policy.

Integrating Scientific Research with Traditional Knowledge: There is a growing recognition that traditional ecological knowledge is not just complementary but essential to modern scientific approaches. Policymakers, researchers, and local communities must work collaboratively to harness both forms of knowledge for more effective climate adaptation and conservation strategies.

Scaling Up Ecosystem-Based Adaptation (EbA) Strategies: The book highlights multiple case studies where ecosystem-based solutions have been successful. However, for EbA to be a transformative tool, there needs to be a systematic effort to scale up these approaches through policy integration, funding mechanisms, and cross-border cooperation.

Strengthening Governance and Policy Implementation: Despite existing policies and frameworks, governance remains a major challenge in resilience-building efforts. Strengthening institutional capacities, ensuring community participation, and bridging gaps between scientific research and policy implementation will be crucial for sustainable progress.

Promoting Sustainable Livelihoods and Ecotourism: As evidenced in multiple chapters, sustainable livelihoods rooted in traditional practices and ecotourism have the potential to support both economic growth and environmental conservation. Future resilience strategies must ensure that these practices are enhanced rather than undermined by modernization.

Advancing Research and Technological Innovations: Emerging technologies, including remote sensing, GIS modeling, and artificial intelligence, offer promising tools for monitoring ecosystem changes and predicting disaster risks. Investing in research and innovation will be key to staying ahead of the rapid changes affecting the region.

Trans-boundary and Regional Cooperation: The Himalayas span multiple countries, making cross-border cooperation imperative. Strengthening regional partnerships on climate action, biodiversity conservation, and disaster risk reduction will be essential for a collective and coordinated response to shared challenges.

22.5 Way Forward

The insights from this book underscore the urgency of adopting integrated, community-centered, and ecosystem-based approaches for resilience building in the Himalayas. While climate change presents unprecedented challenges, it also offers an opportunity to rethink conservation and adaptation strategies in ways that are inclusive, innovative, and sustainable. The resilience of the Himalayas will ultimately depend on the collective efforts of scientists, policymakers, conservationists, and local communities. By promoting collaborative knowledge-sharing and implementing nature-based solutions, the region can navigate its path toward a more resilient and sustainable future. Overall, this book serves as both a reflection on the current state of Himalayan ecosystems and a guidepost for the future—one that embraces the insight of the past, the urgency of the present, and the possibilities of the future. Let this be a call to action for continued research, policy reform, and community-driven initiatives to ensure that the Himalayas remain a thriving, resilient landscape for generations to come.

References

Das L, Meher JK (2019) Drivers of climate over the Western Himalayan region of India: a review. Earth Sci Rev 198:102935

Gupta M (2024) Study of flood vulnerability in Himalayan region: a case of Uttarakhand. Int J Sci Res Eng Manag. https://doi.org/10.55041/ijsrem32614

IPCC (2007) Climate change 2007: impacts, adaptation and vulnerability. In: Parry ML, Canziani OF, Palutikof JP, van der Linden PJ, Hanson CE (eds) Contribution of Working Group II to the fourth assessment report of the Intergovernmental Panel on Climate Change. Cambridge University Press, Cambridge, 976pp

Nautiyal S, Schaldah R, Raju KV, Kaechele H, Pritchard B, Rao KS (2016) Climate change challenge (3C) and social-economic-ecological interface-building—exploring potential adaptation strategies for bio-resource conservation and livelihood development. Springer. https://doi.org/10.1007/978-3-319-31014-5

Nautiyal S, Gupta AK, Goswami M, Khan YDI (2023) The Palgrave handbook of socio-ecological resilience in the face of climate change. Palgrave Macmillan, Singapore. https://doi.org/10.1007/978-981-99-2206-2

Rasul G, Saboor A, Tiwari PC, Hussain A, Ghosh N, Chettri GB (2019) Food and nutrition security in the Hindu Kush Himalaya: unique challenges and niche opportunities. In: Hindu Kush-Himalaya assessment. Springer, Cham, pp 301–338. https://doi.org/10.1007/978-3-319-92288-1_9

Singh SP, Bassignana-Khadka I, Karky BS, Sharma E (2011) Climate change in the Hindu Kush-Himalayas: the state of current knowledge. ICIMOD, Kathmandu

MIX
Papier aus verantwortungsvollen Quellen
Paper from responsible sources
FSC® C105338

If you have any concerns about our products,
you can contact us on
ProductSafety@springernature.com

In case Publisher is established outside the EU,
the EU authorized representative is:
Springer Nature Customer Service Center GmbH
Europaplatz 3, 69115 Heidelberg, Germany

Printed by Libri Plureos GmbH
in Hamburg, Germany